ARRAYS OF CYLINDRICAL
DIPOLES

ARRAYS OF CYLINDRICAL DIPOLES

RONOLD W. P. KING

Gordon McKay Professor of Applied Physics
Harvard University

RICHARD B. MACK

Research Physicist
Air Force Cambridge Research Laboratories

SHELDON S. SANDLER

Associate Professor of Electrical Engineering
Northeastern University
Research Fellow, Harvard University

CAMBRIDGE
AT THE UNIVERSITY PRESS
1968

CAMBRIDGE UNIVERSITY PRESS
Cambridge, New York, Melbourne, Madrid, Cape Town, Singapore, São Paulo, Delhi

Cambridge University Press
The Edinburgh Building, Cambridge CB2 8RU, UK

Published in the United States of America by Cambridge University Press, New York

www.cambridge.org
Information on this title: www.cambridge.org/9780521114851

First published 1968
This digitally printed version 2009

A catalogue record for this publication is available from the British Library

Library of Congress Catalogue Card Number: 67-26069

ISBN 978-0-521-05887-2 hardback
ISBN 978-0-521-11485-1 paperback

To
Tai Tsun Wu

Good reasons must, of force, give place to better.

Shakespeare

CONTENTS

PREFACE

Studies of coupled antennas in arrays may be separated into two groups: those which postulate a single convenient distribution of current along all structurally identical elements regardless of their relative locations in the array and those which seek to determine the actual currents in the several elements. Virtually all of the early and most of the more recent analyses are in the first group in which both field patterns and impedances have been obtained for elements with assumed currents. Pioneer work in the determination of field patterns of arrays of elements with sinusoidally distributed currents was carried out for uniform arrays by Bontsch-Bruewitsch [1] in 1926, by Southworth [2] in 1930, by Sterba [3] and by Carter et al. [4] in 1931. Early studies of non-uniform arrays are by Schelkunoff [5] in 1943, by Dolph [6] in 1946, and by Taylor and Whinnery [7] in 1951. The self- and mutual impedances of arrays of elements with sinusoidally distributed currents were studied especially by Carter [8] in 1932, by Brown [9] in 1937, by Walkinshaw [10] in 1946, by Cox [11] in 1947, by Barzilai [12] in 1948, and by Starnecki and Fitch [13] in 1948. A thorough presentation of the basic theory of antennas with sinusoidal currents was given by Brückmann [14] in 1939. Actually, the current in any cylindrical antenna of length $2h$ and finite radius a is accurately sinusoidal only when it is driven by a continuous distribution of electromotive forces of proper amplitude and phase along its entire length. It is approximately sinusoidal in an isolated very thin antenna $(a \ll h)$ driven by a single lumped generator primarily when the antenna is near resonance. When antennas are coupled in an array with each driven by a single generator or excited parasitically, it is generally assumed that (1) the phase of the current along each element is the same as at the driving point and (2) the amplitude is distributed sinusoidally. Both of these assumptions are reasonably well satisfied only for very thin antennas $(a \ll \lambda)$ that are not too long $(h \leq \lambda/4)$. Nevertheless, a very extensive theory of arrays has been developed based implicitly on one or both of these assumptions. Evidently it is correspondingly restricted in its generality.

The analysis of coupled antennas from the point of view of determining the actual distributions of current was studied for two antennas by Tai [15] in 1948 and extended to the N-element circular array by King [16] in 1950. A general analysis of arrays of coupled antennas has been given by King [17]. Unfortunately, the rigorous solution of the simultaneous integral equations for the distributions of current in the elements of an array of parallel elements is very complicated and no simple and practically useful set of formulas was obtained. As a consequence, the extensive study of the electromagnetic fields of antennas and arrays in this earlier work (chapters 5 and 6 in King [17]) was limited to arrays with currents in the elements that satisfied the assumptions of constant phase angle and sinusoidal amplitude. Similar restrictions are implicit in the fields calculated, for example, by Aharoni [18], Stratton [19], Hansen [20] and many others.

A practical method for obtaining solutions of the simultaneous integral equation for the distributions of current in the elements of a parallel array in a form that combines simplicity with quantitative accuracy was proposed by King [21] in 1959. In this analysis an approximate procedure was developed which provided simple, two-term trigonometric formulas for the currents in all of the arbitrarily driven or parasitic elements in a circular array of N elements in a manner that took full account of the effects of mutual interactions on the distributions of current. These formulas applied to elements up to one and one-quarter wavelengths long. The application of this new procedure to actual arrays and the experimental verification of the results were carried out in an extensive series of investigations by Mack [22]. The generalization of the method to curtain arrays was developed by King and Sandler [23, 24] in 1962. The extension of the method to parasitic elements in arrays of the Yagi type was verified experimentally by Mailloux [25] in 1964. A modification of the theory and its application to the optimization of Yagi arrays by the use of a high-speed computer were devised by Morris [26] in 1964. In 1967 Cheong [27] extended the theory to unequal and unequally spaced elements. (The several researches were supported in part by Joint Services Contract Nonr 1866(32), Air Force Contract AF19(604)-4118 and National Science Foundation Grants NSF-GP-851 and GK-273.)

A further improvement in the simplified trigonometric representation of the current in an isolated antenna was introduced by King and Wu [28] in 1964 and extended to arrays in the present work.

This book begins with an introductory chapter that reviews the foundations and limitations of conventional antenna theory. It then proceeds to derive the new two- and three-term formulas for the isolated antenna in chapter 2 and for two coupled antennas in chapter 3. Chapter 4 provides the complete formulation of the new theory for the N-element circular array; chapter 5 for the N-element curtain array of identical elements. The more difficult problem of treating elements of different lengths—notably in the Yagi array and the log-periodic antenna—is treated in chapter 6. Chapter 7 is devoted to planar and three-dimensional arrays that include staggered and collinear elements. Chapter 8 is concerned with the broad problems of measurement—currents, impedances, field patterns and the correlation of theory with experiment. In the appendices summaries of programmes are given for the computational analysis of circular, curtain, and Yagi arrays.

In the preparation of the manuscript, S. S. Sandler was responsible for chapters 1 and 5, R. B. Mack for chapters 4 and 8, and R. W. P. King for chapters 2, 3, 6, and 7 and for the co-ordination of the several parts.

The authors are happy to acknowledge the important contributions of Drs Robert J. Mailloux, I. Larry Morris, and W.-M. Cheong whose researches form the basis of chapter 6; and of V. W. H. Chang whose work underlies chapter 7. They are grateful to Professor Tai Tsun Wu for many valuable suggestions and to Mrs Dilla G. Tingley for continuing painstaking assistance with the preparation of the manuscript, the graphical representation, the computations, and the programmes. Mrs Barbara Sandler, Mrs Evelyn Mack, and Mr Chang also assisted with the programmes, Mrs S. R. Seshadri with the preparation of the manuscript. Miss Margaret Owens contributed greatly to the accuracy of the presentation with her meticulous reading of the proofs. She also had a major share in the preparation of the index.

R.W.P.K.
R.B.M.
S.S.S.

Cambridge, Mass.
January 1967

CHAPTER 1

INTRODUCTION

1.1 Fundamentals—field vectors and potential functions

Radio communication depends upon the interaction of oscillating electric currents in specially designed, often widely separated configurations of conductors known as antennas. Those considered in this book consist of thin metal wires, rods or tubes arranged in parallel arrays of circular or planar form. Electric charges in the conductors of a transmitting array are maintained in systematic accelerated motion by suitable generators that are connected to one or more of the elements by transmission lines. These oscillating charges exert forces on other charges located in the distant conductors of a receiving array of elements of which at least one is connected by a transmission line to a receiver. Fundamental quantities upon which such an interaction depends are the electromagnetic field and the driving-point admittance. But these are completely determined by the distribution of current in the elements of an array. In this first chapter the basic electromagnetic equations are formulated and applied to simple antennas and arrays in the conventional manner which is based on assumed rather than actual currents. The limitations of this approach are pointed out as an introduction to the more accurate formulation of the theory of antennas and arrays that is presented in subsequent chapters.

Consider first the very simple, physically realizable transmitting antenna shown in Fig. 1.1. It consists of a thin conductor extending from $z = -h$ to $z = h$ that is centre driven by a generator which maintains a periodically varying potential difference across its terminals at $z = \pm \frac{1}{2}b$. The transmission line consists of two wires that are separated by a distance b that is small compared to the wavelength λ so that $b \ll \lambda$. Its output end is connected to the adjacent terminals of the antenna. Owing to the complications involved in a small region comparable in extent with the line spacing b, where antenna and line are coupled, it is convenient in an introductory and elementary analysis of the field properties of linear antennas to replace the actual generator-transmission-line with an idealized so-called delta-function generator. This maintains the electric field $E_z = -V_0\delta(z)$ on the surface of the antenna. The properties of the delta function are:

$$\delta(z) = \begin{cases} 0, & z \neq 0 \\ \infty, & z = 0 \end{cases} \tag{1.1a}$$

and
$$\int_{-h}^{h} V(z)\delta(z)\,dz = V(0). \tag{1.1b}$$

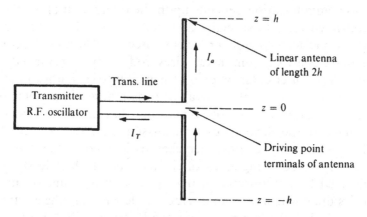

Fig. 1.1.   Practical antenna system.

The problem of relating the impedance obtained for an idealized delta-function source to that actually measured with a transmission line is discussed later. A simplified linear antenna is shown in Fig. 1.2. For this introductory study the conventional approach is followed and a sinusoidally distributed current is assumed along the antenna. Measurements of the current along very thin cylindrical antennas indicate that the current is distributed approximately sinusoidally especially when $h \leqslant \lambda/4$. Since the general shape of the field pattern of an isolated linear antenna does not depend critically on the distribution of current along the antenna, this approximation involves less error in the calculation of the major lobe of the far field pattern than in the evaluation of the minor lobe structure or the driving-point impedance. The assumption of a sinusoidal current implies that the distribution of current (but not its amplitude) is independent of the radius a of the antenna. Measurements show that the assumed sinusoidal current is a fairly good approximation near the first one or two resonant lengths ($h \sim n\lambda/4$, $n = 1, 3, ...$) of very thin antennas; it is not satisfactory near anti-resonant lengths ($h \sim n\lambda/4, n = 2, 4, ...$). The sinusoidal assumption is critically involved in the accuracy of the calculation of the driving-point impedance of an isolated antenna. When coupled antennas are considered, an assumed sinusoidal distribution of current proves to be a major source of inaccuracy in the calculations of both the driving-point impedance and the radiation pattern.

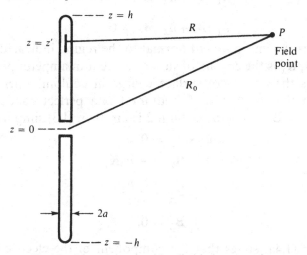

Fig. 1.2.   Linear antenna with cylindrical cross-section.

The interaction of charges and currents on conductors in space is governed by the well-known Maxwell–Lorentz equations which define the electromagnetic field. With an assumed time dependence $e^{j\omega t}$, they are

$$\nabla \times \mathbf{B} = \mu_0(\mathbf{J} + j\omega\varepsilon_0\mathbf{E}), \qquad \nabla \cdot \mathbf{B} = 0 \qquad (1.2a)$$

$$\nabla \times \mathbf{E} = -j\omega\mathbf{B}, \qquad \nabla \cdot \mathbf{E} = \rho/\varepsilon_0 = 0 \qquad (1.2b)$$

where the electric vector $\mathbf{E}$ is in volts per meter, the magnetic vector $\mathbf{B}$ in webers per square meter. Rationalized MKS units are used throughout this book. The volume density of current $\mathbf{J}$ in amperes per square meter is the charge crossing unit area per second. In a perfect conductor $\mathbf{J} = 0$. The volume density of charge ρ in coulombs per cubic meter is zero in the interior of all conductors. The universal electric and magnetic constants are, respectively, ε_0 and μ_0. They have the numerical values $\varepsilon_0 = 8 \cdot 854 \times 10^{-12}$ farads per meter and $\mu_0 = 4\pi \times 10^{-7}$ henrys per meter. The relevant boundary conditions at an interface between a conductor (subscript 1) and air (subscript 2) are expressed in terms of the tangential and normal components of the electric and magnetic fields. Thus

$$\hat{n}_1 \times \mathbf{E}_1 + \hat{n}_2 \times \mathbf{E}_2 = 0 \tag{1.3a}$$

$$\hat{n}_1 \times \mathbf{B}_1 + \hat{n}_2 \times \mathbf{B}_2 = -\mu_0 \mathbf{K}_1 \tag{1.3b}$$

and

$$\hat{n}_1 \cdot \mathbf{E}_1 + \hat{n}_2 \cdot \mathbf{E}_2 = \frac{-\eta_1}{\varepsilon_0} \tag{1.3c}$$

$$\hat{n}_1 \cdot \mathbf{B}_1 + \hat{n}_2 \cdot \mathbf{B}_2 = 0 \tag{1.3d}$$

where $\hat{n}$ is the unit outward normal to the region indicated by the subscript, $\mathbf{K}_1$ is the density of surface current in amperes per meter and η_1 is the density of surface charge in coulombs per square meter on the conductor. If medium 1 is a perfect conductor in which $\mathbf{E}_1 = \mathbf{B}_1 = 0$ and medium 2 is air, then (1.3) simplifies to

$$\hat{n}_2 \times \mathbf{E}_2 = 0 \tag{1.4a}$$

$$\hat{n}_2 \times \mathbf{B}_2 = -\mu_0 \mathbf{K}_1 \tag{1.4b}$$

$$\hat{n}_2 \cdot \mathbf{E}_2 = \frac{-\eta_1}{\varepsilon_0} \tag{1.4c}$$

$$\hat{n}_2 \cdot \mathbf{B}_2 = 0. \tag{1.4d}$$

Equation (1.4a) states that the component of the electric field in air tangent to the surface of a perfect conductor must be zero. Equation (1.4b) states that at the surface of a perfect conductor the tangential magnetic field in air is proportional to the surface density of current in the conductor.

A convenient method of solving the vector partial differential equations (1.2) is with the use of scalar and vector potentials, ϕ, $\mathbf{A}$. The defining relationships between the potentials and the electro-magnetic field vectors are obtained with the aid of Maxwell's equations. With the vector identity $\nabla \cdot (\nabla \times \mathbf{C}) = 0$ (where $\mathbf{C}$ is any vector) and the Maxwell equation $\nabla \cdot \mathbf{B} = 0$, the magnetic field may be expressed in the form

$$\mathbf{B} = \nabla \times \mathbf{A}. \tag{1.5a}$$

If (1.5a) is substituted in (1.2b) it follows that

$$\nabla \times (\mathbf{E} + j\omega \mathbf{A}) = 0. \tag{1.5b}$$

The identity $\nabla \times (\nabla \psi) = 0$, where ψ is a scalar function, then permits the definition of ϕ in the form

$$-\nabla \phi = \mathbf{E} + j\omega \mathbf{A}. \tag{1.5c}$$

The substitution of (1.5a) and (1.5c) into the remaining Maxwell equations leads to mixed vector equations for $\mathbf{A}$ and ϕ. The variables can be separated if the following condition relating $\mathbf{A}$ and ϕ is imposed:

$$\nabla \cdot \mathbf{A} = \frac{-j\beta_0^2}{\omega}\phi. \qquad (1.6)$$

This is known as the Lorentz condition. The resulting vector Helmholtz equations for $\mathbf{A}$ and ϕ in air are

$$(\nabla^2 + \beta_0^2)\mathbf{A} = -\mu_0 \mathbf{J} \qquad (1.7a)$$

and

$$(\nabla^2 + \beta_0^2)\phi = -\rho/\varepsilon_0 = 0. \qquad (1.7b)$$

The antenna theory developed in this book is concerned exclusively with thin cylindrical conductors all aligned in the z-direction in air so that it suffices to use only the axial component of the vector potential. With $\mathbf{A} = \hat{\mathbf{z}}A_z$ the vector Laplacian in (1.7a) is reduced to a scalar as in (1.7b). The simplified form of (1.7a) is

$$(\nabla^2 + \beta_0^2)A_z = -\mu_0 J_z. \qquad (1.7c)$$

Particular integrals of (1.7b) and (1.7c) are directly derivable with the use of the free-space Green's function. The integrals for A_z and ϕ for a thin cylindrical conductor of length $2h$ and radius a with its centre at $z = 0$ are

$$A_z = \frac{\mu_0}{4\pi} \int_{-h}^{h} I_z(z')\frac{e^{-j\beta_0 R}}{R}\, dz' \qquad (1.8a)$$

and

$$\phi = \frac{1}{4\pi\varepsilon_0} \int_{-h}^{h} q(z')\frac{e^{-j\beta_0 R}}{R}\, dz' \qquad (1.8b)$$

where, with $\Sigma = \pi a^2$ the area of cross-section, the total axial current is

$$I_z(z) = \int_{\Sigma} J_z\, d\Sigma + 2\pi a K_z \qquad (1.8c)$$

and the charge per unit length is

$$q(z) = 2\pi a \eta. \qquad (1.8d)$$

The wave number is $\beta_0 = 2\pi/\lambda_0$, where λ_0 is the free-space wavelength; $R = \sqrt{(z - z')^2 + a^2}$. For a perfect conductor $J_z = 0$. The one-dimensional Lorentz condition is

$$\frac{\partial A_z}{\partial z} = \frac{-j\beta_0^2}{\omega}\phi. \qquad (1.9)$$

The continuity equation expresses the condition of conservation of charge. For a thin cylindrical antenna it has the form

$$\frac{dI_z(z)}{dz} = -j\omega q(z). \tag{1.10}$$

The **E** and **B** fields for a finite cylindrical conductor are obtained from (1.5a) and (1.5c) with (1.8a) and (1.9). In the cylindrical co-ordinates ρ, Φ, z, they are $\mathbf{B} = \hat{\mathbf{\Phi}}B_\Phi$ and $\mathbf{E} = \hat{\rho}E_\rho + \hat{z}E_z$ where

$$B_\Phi = \frac{-\partial A_z}{\partial \rho} \tag{1.11a}$$

$$E_\rho = \frac{-j\omega}{\beta_0^2}\frac{\partial^2 A_z}{\partial \rho \, \partial z} \tag{1.11b}$$

$$E_z = \frac{-j\omega}{\beta_0^2}\left(\frac{\partial^2 A_z}{\partial z^2} + \beta_0^2 A_z\right). \tag{1.11c}$$

In the spherical coordinates r, Θ, Φ with origin at the centre of the antenna, the field is given by

$$E_r = E_z \cos\Theta + E_\rho \sin\Theta \tag{1.12a}$$

$$E_\Theta = -E_z \sin\Theta + E_\rho \cos\Theta. \tag{1.12b}$$

At sufficiently great distances from the antenna ($r^2 \gg h^2$ and $(\beta_0 r)^2 \gg 1$), the field reduces to a simple form known as the radiation or far field. It is given by

$$B_\Phi^r = E_\Theta^r/c \tag{1.13a}$$

where c is the velocity of light and

$$\mathbf{E}^r \doteq E_\Theta^r \hat{\mathbf{\Theta}}, \qquad E_\Theta^r = \frac{j\omega\mu_0}{4\pi}\sin\Theta\int_{-h}^{h} I_z(z')\frac{e^{-j\beta_0 R}}{R}\,dz'. \tag{1.13b}$$

The distance R from an arbitrary point on the antenna to the field point is given in terms of r and z' by the cosine law, viz., (Fig. 1.3)

$$R^2 = r^2 + (z')^2 - 2rz'\cos\Theta. \tag{1.14a}$$

In the radiation zone $r^2 \gg (z')^2$. If the binomial expansion is applied to (1.14a) and only the linear term in z' is retained, the following approximate form is obtained for R:

$$R \doteq r - z'\cos\Theta, \qquad (\beta_0 r)^2 \gg 1. \tag{1.14b}$$

The phase variation of $\exp(-j\beta_0 R)/R$ is replaced with the linear phase variation given by (1.14b), i.e. by $\exp(-j\beta_0 r + j\beta_0 z'\cos\Theta)$. The amplitude $1/R$ of $\exp(-j\beta_0 R)/R$ is a slowly varying function of z' and is replaced by $1/r$, where r is the distance to the centre of the antenna. Since r is independent of z', all functions of r may be removed from the integral in (1.13b) and the final form for E^r is

$$E^r_\Theta = \frac{j\zeta_0 I_z(0)}{2\pi} \frac{e^{-j\beta_0 r}}{r} F_0(\Theta, \beta_0 h) \qquad (1.15a)$$

where $\zeta_0 = \sqrt{\mu_0/\varepsilon_0} \doteq 120\pi$ ohms and

$$F_0(\Theta, \beta_0 h) = \frac{\beta_0 \sin \Theta}{2 I_z(0)} \int_{-h}^{h} I_z(z') \, e^{j\beta_0 z' \cos \Theta} \, dz'. \qquad (1.15b)$$

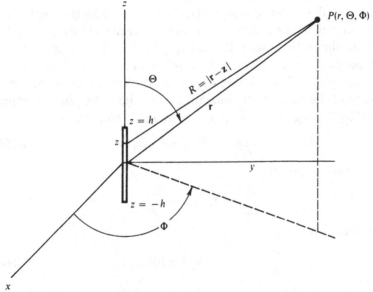

Fig. 1.3.   Coordinate system for calculations in the far zone.

The term $F_0(\Theta, \beta_0 h)$ contains all the directional properties of a linear radiator of length $2h$. It is called the field characteristic or field factor and will be computed for some commonly used current distributions. The magnetic field $\mathbf{B}^r$ in the far zone is at right angles to $\mathbf{E}^r$ and also perpendicular to the direction of propagation $\mathbf{r}$. It is given by (1.13a). Thus

$$\mathbf{B}^r = \hat{\mathbf{\Phi}} B^r_\Phi, \qquad B^r_\Phi = \frac{j\mu_0 I_z(0)}{2\pi} \frac{e^{-j\beta_0 r}}{r} F_0(\Theta, \beta_0 h). \qquad (1.15c)$$

Note that the field in the far zone depends on $F_0(\Theta, \beta_0 h)$ which is a function of the particular distribution of current in the antenna.

It is instructive to consider the instantaneous value of the field in (1.15a), which is obtained by multiplication with $e^{j\omega t}$ and selection of the real part. If the phase of the field is referred to that of the current

$$E_\Theta^r(\mathbf{r}, t) = \text{Re } E_\Theta(\mathbf{r})\, e^{j\omega t} \sim \frac{\sin(\omega t - \beta_0 r)}{r} = \frac{\sin \omega(t - r/c)}{r}.$$

(1.16a)

Note that the field at the point r at the instant t is computed from the current at $r = 0$ at the earlier time $(t - r/c)$. This is a consequence of the finite velocity of propagation c.

The equiphase and equipotential surfaces of $\mathbf{E}$ and $\mathbf{B}$ are spherical shells on which r is equal to a constant. There are an infinite number of such shells that have the same phase (differ by an integral multiple of 2π) but only one that has both the same amplitude and the same phase. The velocity of propagation is the outward radial velocity of the surfaces of constant phase where the phase is represented by the argument of the sine term in (1.16a), that is,

$$\text{phase} = \Psi = \omega t - \beta_0 r. \qquad (1.16b)$$

For a constant phase

$$\frac{d\Psi}{dt} = 0 = \omega - \frac{\beta_0 dr}{dt}. \qquad (1.16c)$$

It follows that

$$\frac{dr}{dt} = \frac{\omega}{\beta_0} = c = 3 \times 10^8 \text{ m/sec.} \qquad (1.16d)$$

Since the phase repeats itself every 2π radians, a wavelength is the distance between two adjacent equiphase surfaces. For example, if one surface is defined by $r = r_1$ and the other by $r = r_2$ then

$$\omega t - \beta_0 r_1 = 2\pi \quad \text{and} \quad \omega t - \beta_0 r_2 = 4\pi \qquad (1.17a)$$

or

$$r_2 - r_1 = \frac{2\pi}{\beta_0} = \lambda_0 \qquad (1.17b)$$

where λ_0 is the wavelength in air. The physical picture of the fields in the far zone is quite simple. The electric and magnetic vectors are mutually orthogonal and tangent to an outward travelling spherical shell. Thus, both components of the field are transverse to the radius vector $\mathbf{r}$; they have the same phase velocity $c = 3 \times 10^8$ m/sec, the velocity of light.

1.2 Power and the Poynting vector

In the conventional approach to antenna theory, the power radiated is usually determined by an application of the Poynting-vector theorem. An equation for the time-average power associated with a radiating antenna or array is readily derived from

$$\mathrm{Re}\tfrac{1}{2}\left\{\int_\tau \mathbf{J^*} . \mathbf{E}\, d\tau + \int_\Sigma \mathbf{K^*} . \mathbf{E}\, d\Sigma\right\} \qquad (1.18)$$

where Re indicates the real part. τ is the volume occupied by the currents $\mathbf{J}$ when imperfect conductors are considered; Σ is the surface of perfect conductors on which are the currents $\mathbf{K}$. The asterisk denotes the complex conjugate. It is, of course, clear that since $\mathbf{J}$ vanishes in air, the volume of integration may be enlarged to any desired size so long as the only contributions come from the antennas under study. When attention is directed to a single antenna isolated in space, τ may be extended to infinity.

The next step in the derivation of the desired power equation is the elimination of $\mathbf{J^*}$ and $\mathbf{K^*}$ from (1.18) by substitution from (1.2a) and (1.4b). Note first that with the vector identity

$$\nabla . (\mathbf{E} \times \mathbf{B^*}) = \mathbf{B^*} . (\nabla \times \mathbf{E}) - \mathbf{E} . (\nabla \times \mathbf{B^*}) \qquad (1.19a)$$

and the complex conjugate of (1.2b), the following equation can be obtained:

$$\mathbf{E} . (\nabla \times \mathbf{B^*}) = \nabla . (\mathbf{E} \times \mathbf{B^*}) + j\omega\mathbf{B} . \mathbf{B^*}. \qquad (1.19b)$$

With (1.19b) and (1.2a),

$$\mathbf{J^*} . \mathbf{E} = \mu_0^{-1}\mathbf{E} . (\nabla \times \mathbf{B^*}) + j\omega\varepsilon_0\mathbf{E} . \mathbf{E^*}$$

$$= \mu_0^{-1}\nabla . (\mathbf{E} \times \mathbf{B^*}) + j\omega(\varepsilon_0\mathbf{E} . \mathbf{E^*} + \mu_0^{-1}\mathbf{B} . \mathbf{B^*}). \qquad (1.20a)$$

From (1.4b),

$$\mathbf{K^*} . \mathbf{E} = -\mu_0^{-1}(\mathbf{\hat{n}} \times \mathbf{B^*}) . \mathbf{E} = \mu_0^{-1}\mathbf{\hat{n}} . (\mathbf{E} \times \mathbf{B^*}). \qquad (1.20b)$$

Since $\mathbf{E} . \mathbf{E^*} = E^2$ is real, it follows that

$$\mathrm{Re}\tfrac{1}{2}\left\{\int_\tau \mathbf{J^*} . \mathbf{E}\, d\tau + \int_\Sigma \mathbf{K^*} . \mathbf{E}\, d\Sigma\right\} = \mathrm{Re}\left\{\int_\tau \nabla . \mathbf{S}\, d\tau + \int_\Sigma \mathbf{\hat{n}} . \mathbf{S}\, d\Sigma\right\}$$

$$(1.21)$$

where
$$\mathbf{S} = (\mathbf{E} \times \mathbf{B^*})/2\mu_0$$

is the complex Poynting vector. In (1.21) τ is any region sufficiently large to contain the currents $\mathbf{J}$ and $\mathbf{K}$. Let the enclosing surface for τ be Σ_τ. Note that Σ represents surfaces across which $\mathbf{B}$ is discontinuous.

If the divergence theorem

$$\int_\tau \nabla . \mathbf{S}\, d\tau = \int_{\Sigma_\tau} \hat{\mathbf{n}} . \mathbf{S}\, d\Sigma \qquad (1.22)$$

where $\hat{\mathbf{n}}$ is the external normal to Σ_τ, is applied to the first integral in (1.21), all surfaces of discontinuity Σ in $\mathbf{S}$ must be excluded by an enclosing boundary since (1.22) is valid only for continuous functions. If this is done, the integrals over these surfaces of discontinuity Σ exactly cancel the last integral in (1.21). The result is

$$\mathrm{Re}\tfrac{1}{2}\left\{ \int_\tau \mathbf{J}^* . \mathbf{E}\, d\tau + \int_\Sigma \mathbf{K}^* . \mathbf{E}\, d\Sigma \right\} = \mathrm{Re} \int_{\Sigma_\tau} \hat{\mathbf{n}} . \mathbf{S}\, d\Sigma. \qquad (1.23)$$

The left side in (1.23) is now readily specialized to thin, current-carrying antennas. For simplicity, let these be perfect conductors in the interior of which $\mathbf{J}$ and $\mathbf{E}$ vanish. With rotational symmetry, the complex conjugate of the total axial current is

$$I_z^*(z) = 2\pi a K_z^*(z). \qquad (1.24)$$

In a tabular conductor there are no radial currents; if the ends are capped, the small radial currents on these may be neglected. It follows that (1.23) reduces to

$$\mathrm{Re}\tfrac{1}{2}\int_{-h}^{h} I_z^*(z)E_z(z)\, dz \doteq \mathrm{Re} \int_{\Sigma_\tau} \hat{\mathbf{n}} . \mathbf{S}\, d\Sigma. \qquad (1.25)$$

For an antenna driven by a delta-function generator at $z = 0$, the boundary condition at $\rho = a$ is $E_z(z) = -V_0\delta(z)$ so that the left side of (1.25) gives simply

$$P = \mathrm{Re}\tfrac{1}{2}I_z^*(0)V_0 = \mathrm{Re} \int_{\Sigma_\tau} \hat{\mathbf{n}} . \mathbf{S}\, d\Sigma \qquad (1.26)$$

where P is the time-average power supplied to the antenna at its terminals by the generator. Since there is no dissipation in the perfectly conducting antenna or the surrounding air, the integral on the right is the total radiated power. It is independent of the shape or size of the surface Σ_τ so long as this completely encloses the entire transmitting system consisting of antenna and generator and any connecting transmission line. It is important to note that the Poynting-vector theorem cannot be used logically to determine a path for the hypothetical flow of energy from a generator to a distant receiver. For example, let the surface Σ_τ in (1.26) be the surface of a pill box of infinitesimal thickness and radius a that encloses the entire delta-function generator at $z = 0$ but otherwise none of the centre-driven perfectly conducting antenna, that extends from

$z = -h$ to $z = h$. The same total radiated power is still transferred across Σ_t and, since $\mathbf{E} = 0$ in the interior of the conductor, the entire contribution to the integral comes from the ring bounding the delta function at $z = 0$ and $\rho = a$. This might be interpreted naively to mean that energy is transferred directly from the generator to the rest of the universe. However, since it is the currents in the antenna that maintain the electromagnetic field that exerts forces on charges in a distant receiving antenna and so do work, they cannot logically be excluded from the radiation process.

It is readily shown with reference to the simple transmitting system in Fig. 1.1 that, when Σ_t encloses only the transmitter oscillator, the total radiated power is obtained. On the other hand, if Σ_t is a closed surface around the antenna—of course crossed by the transmission line—the integral and, hence, the total power transferred outward is zero. Again, this might be interpreted naively as an indication that the generator radiates power and the antenna has nothing to do with it.

The real part of the integral of the normal component of the complex Poynting vector over any surface that completely encloses a complete transmitting system—antennas, transmission line (if there is one), and generator—correctly gives the total power transferred from the region inside the surface to the region outside. The conclusion that therefore the Poynting vector itself specifies the rate of flow of energy across each unit of area is without foundation. Nevertheless, it is a common assumption.

1.3   The field of the electrically short antenna; directivity

Consider first the radiation field of an electrically short linear antenna, where $(\beta_0 h)^2 \ll 1$, with a triangular current distribution which vanishes at $z = \pm h$. Actually, this is a special case of a sinusoidal distribution which is discussed later. A diagram of the triangular distribution is shown in Fig. 1.4, where the magnitude of the current is plotted along an axis perpendicular to the antenna. Since $(\beta_0 z')^2 \leqslant (\beta_0 h)^2 \ll 1$, the exponent in $F_0(\Theta, \beta_0 h)$ may be expanded through the linear term. Thus,

$$F_0(\Theta, \beta_0 h) \doteq \frac{\beta_0 \sin \Theta}{2} \int_{-h}^{h} \left(1 - \frac{z'}{h}\right)(1 + j\beta_0 z' \cos \Theta + ...) \, dz' \quad (1.27a)$$

$$F_0(\Theta, \beta_0 h) \doteq \frac{\beta_0 h \sin \Theta}{2}, \qquad (\beta_0 h)^2 \ll 1. \qquad (1.27b)$$

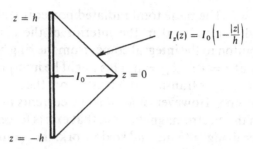

Fig. 1.4.   Linear antenna with triangular distribution of current.

Equation (1.27b) shows that the radiation field of a short linear antenna is proportional to sin Θ. Polar and rectangular graphs of the field are shown in Figs. 1.5a and 1.5b, normalized with respect to the maximum at $\Theta = 90°$.

The field quite near an electrically short antenna is readily evaluated from (1.8a) with $I(z) = I(0)(1 - |z|/h)$ and $R \doteq r$. This gives

$$A_z \doteq \frac{\mu_0 h I(0)}{4\pi} \frac{e^{-j\beta_0 r}}{r}. \tag{1.28}$$

The components of the field can be evaluated in the spherical coordinates r, Θ, Φ from (1.5a) and (1.5c) with (1.9). The results are

$$B_\Phi \doteq \frac{\mu_0 h I(0)}{4\pi} \left(\frac{j\beta_0}{r} + \frac{1}{r^2} \right) e^{-j\beta_0 r} \sin \Theta \tag{1.29a}$$

$$E_r \doteq \frac{\zeta_0 h I(0)}{4\pi} \left(\frac{2}{r^2} - \frac{j2}{\beta_0 r^3} \right) e^{-j\beta_0 r} \cos \Theta \tag{1.29b}$$

$$E_\Theta \doteq \frac{j\zeta_0 h I(0)}{4\pi} \left(\frac{\beta_0}{r} - \frac{j}{r^2} - \frac{1}{\beta_0 r^3} \right) e^{-j\beta_0 r} \sin \Theta. \tag{1.29c}$$

These expressions are valid subject to the conditions

$$(\beta_0 h)^4 \ll 1, \qquad (h/r)^3 \ll 1, \qquad (a/h)^2 \ll 1. \tag{1.29d}$$

They may be expressed in terms of the dipole moment $p_z = I(0)h/j\omega$ if desired. The electromagnetic power transferred across a closed surface is given by the integral of the normal component of the complex Poynting vector $\mathbf{S} = \frac{1}{2}\mu_0^{-1} \mathbf{E} \times \mathbf{B}^*$ over the surface. (The asterisk denotes the complex conjugate.) For an electrically small antenna $(\hat{\mathbf{n}} \cdot \mathbf{S}) \sim \sin^2 \Theta$. An angular graph of $\mathbf{S}$ is called a power pattern. Polar and rectangular graphs of the power pattern are shown in Figs. 1.5c and 1.5d. Note that because of symmetry both the field and power patterns are independent of the coordinate Φ.

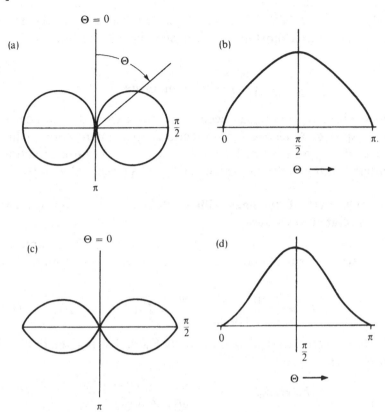

Fig. 1.5. (a) Field pattern, polar plot. (b) Field pattern, rectangular plot. (c) Power pattern, polar plot. (d) Power pattern, rectangular plot.

The half-power beamwidth Θ_{hp} is defined as the angular distance between half-power points on the radiation pattern referred to the principal lobe. The value of Θ_{hp} for the short linear antenna is 90° since the field is down by a factor of $\sqrt{2}/2$ at $\Theta = \pm 45°$. Another parameter useful in defining the directive properties of an antenna is the absolute directivity D. This parameter is a measure of the total time-average power transferred across a closed surface in the direction of the principal lobe. The time-average power transferred across a closed surface Σ is the integral of the normal component of $\mathbf{S}$. Thus

$$P = \int_{\Sigma} \hat{\mathbf{n}} \cdot \mathbf{S} \, d\Sigma \qquad (1.30)$$

where $\hat{\mathbf{n}}$ is the unit external normal to the surface. The directivity D

is the ratio of P with $\mathbf{S}$ set at its maximum value $\mathbf{S}_m$ to the actual value of P. For a short dipole with $|S| \sim \sin^2 \Theta$, the value of D is

$$D = \frac{4\pi}{\displaystyle\int_0^{2\pi} \int_0^{\pi} \sin^2 \Theta \sin \Theta \, d\Theta} = \tfrac{3}{2}, \qquad (1.31)$$

where the integration has been carried out over the surface of a great sphere. A nearly omnidirectional pattern requires a large value of Θ_{hp} and a nearly unity value of D. A more directional pattern requires a smaller value of Θ_{hp} and a larger value of D.

1.4   The field of antennas with sinusoidally distributed currents; radiation resistance

Conventional antenna theory applies specifically to antennas along which a sinusoidally distributed current is maintained. That is

$$I_z(z) = \frac{I_z(0) \sin \beta_0(h - |z|)}{\sin \beta_0 h} = I_m \sin \beta_0(h - |z|). \qquad (1.32)$$

For this current the field characteristic $F_0(\Theta, \beta_0 h)$ is given by (1.15b) with (1.32),

$$F_0(\Theta, \beta_0 h) = \frac{\cos (\beta_0 h \cos \Theta) - \cos \beta_0 h}{\sin \beta_0 h \sin \Theta}. \qquad (1.33a)$$

An alternative field characteristic $F_m(\Theta, \beta_0 h)$ is referred to the maximum value of the sinusoid, viz., $I_m = I_z(0)/\sin \beta_0 h$ which occurs at $h - \lambda_0/4$ when $\beta_0 h \geqslant \pi/2$.

$$F_m(\Theta, \beta_0 h) = \frac{\cos (\beta_0 h \cos \Theta) - \cos \beta_0 h}{\sin \Theta}. \qquad (1.33b)$$

The function $F_m(\Theta, \beta_0 h)$ is shown graphically in Fig. 1.6 for several values of h. It is seen that the pattern corresponding to $\beta_0 h = \pi/2$ ($h = \lambda_0/4$) is only slightly narrower than the pattern for $(\beta_0 h)^2 \ll 1$ which is shown in Fig. 1.5. Note that as $\beta_0 h$ is increased beyond π, minor lobes appear which successively become the major lobe and point in directions other than $\Theta = \pi/2$.

For an antenna with the simple sinusoidal distribution of current, the complete electromagnetic field can be evaluated in cylindrical coordinates for points near the antenna. This is accomplished with the substitution of the current (1.32) in the general integral for the vector potential (1.8a) and the subsequent use of this expression in

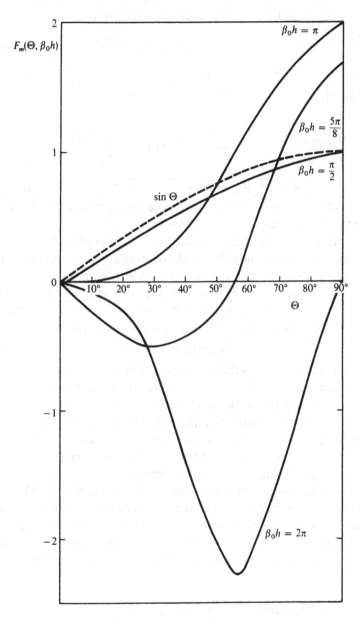

Fig. 1.6. Field factor of linear antenna.

(1.11a)–(1.11c). The indicated differentiations can be carried out directly without evaluating the integral. The results are:

$$B_\Phi = \frac{jI_m\mu_0}{4\pi\rho}[e^{-j\beta_0 R_{1h}} + e^{-j\beta_0 R_{2h}} - 2\cos\beta_0 h\, e^{-j\beta_0 R_0}] \tag{1.34a}$$

$$E_\rho = \frac{jI_m\zeta_0}{4\pi\rho}\left[\frac{z-h}{R_{1h}}e^{-j\beta_0 R_{1h}} + \frac{z+h}{R_{2h}}e^{-j\beta_0 R_{2h}} - \frac{2z}{R_0}\cos\beta_0 h\, e^{-j\beta_0 R_0}\right] \tag{1.34b}$$

$$E_z = \frac{-jI_m\zeta_0}{4\pi}\left[\frac{e^{-j\beta_0 R_{1h}}}{R_{1h}} + \frac{e^{-j\beta_0 R_{2h}}}{R_{2h}} - 2\cos\beta_0 h\,\frac{e^{-j\beta_0 R_0}}{R_0}\right] \tag{1.34c}$$

where $R_0 = \sqrt{z^2+\rho^2}$, $R_{1h} = \sqrt{(z-h)^2+\rho^2}$, $R_{2h} = \sqrt{(z+h)^2+\rho^2}$ are, respectively, the distances from the point where the field is evaluated to the centre and the two ends of the antenna. Their interpretation in terms of spheroidal waves is available elsewhere.†

It is often useful to relate the total power radiated by an antenna to the current at an arbitrary reference point. For a sinusoidally distributed current, the maximum value at $z = 0$ when $h < \lambda/4$ and at $z = h - \lambda/4$ when $h \geq \lambda/4$ is convenient. Since the total power radiated is given by (1.30), the desired relation is

$$\tfrac{1}{2}|I_m|^2 R_m^e = P \tag{1.35}$$

where the coefficient R_m^e is the so-called radiation resistance referred to I_m. When $\beta_0 h = \pi/2$, $R_m^e = 73 \cdot 1$ ohms. The value of R_m^e determined from (1.35) is not, in general, the driving-point resistance R_0 of a centre-driven antenna, although when $\beta_0 h = \pi/2$ so that $I_m = I(0)$, R_m^e does approximate R_0 when the antenna is sufficiently thin ($a/\lambda < 10^{-5}$ for an error of 5% or less). For very thin dipoles of resonant length the numerical values of R_0^e determined from (1.35) resemble the experimental results. As defined in (1.35) and with (1.30) evaluated over the surface of a great sphere, the radiation resistance strictly is a characteristic of the far field and only approximately and under special circumstances a circuit property of the antenna at its terminals. This is considered in greater detail in section 1.7.

1.5 The field of a two-element array

Variations in the directional pattern of a single antenna obtained by changes in its length are of very limited practical value. Much

† See [1] chapter V and [2] pp. 178–181.

more useful directional properties are made available when additional antennas are arranged physically displaced and parallel to the original element in a configuration called an array. The two-element array of Fig. 1.7a is an elementary example which illustrates important properties common to all arrays. The number of elements is denoted by $N = 2$, the distance between the elements is b, and the relative phases and magnitudes of the driving-point currents $I_1(0)$ and $I_2(0)$ can be adjusted. It is tacitly assumed in this elementary theory that the interaction of the currents in the coupled antennas does not affect the distributions of current along the elements. These are taken to be identical for all elements in the array. (A more correct theory must, of course, consider the actual distributions of current, which are not alike except in special cases. Note that in general it is not possible to specify completely the current along an element in an array by assigning the current at the driving point.)

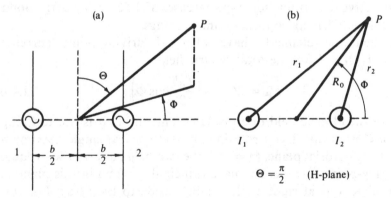

Fig. 1.7.   (a) Two-element linear array. (b) Two-element array in H-plane.

It is required to find the total electric field for the array of Fig. 1.7 by superposition of the individual contributions from the elements. The formula for the field is simplest in the plane $\Theta = \pi/2$, called the equatorial or H-plane. The field characteristic $F_0(\pi/2, \beta_0 h)$ is a constant in this plane and the total field is given by (1.15a). For two geometrically identical elements the result is

$$E_\Theta^r = \frac{j\zeta_0}{2\pi} F_0\left(\frac{\pi}{2}, \beta_0 h\right)\left[I_1(0)\frac{e^{-j\beta_0 R_1}}{R_1} + I_2(0)\frac{e^{-j\beta_0 R_2}}{R_2}\right] \quad (1.36)$$

where R_1 and R_2 are shown in Fig. 1.7b. The length b of the array is assumed small in comparison to the radial distance R_0 to the field point. The application of the law of cosines and the binomial expansion to the expressions for R_1 and R_2 yields the following simplified expressions:

$$R_{1,2}^2 = R_0^2 + \left(\frac{b}{2}\right)^2 \pm bR_0 \cos \Phi = R_0^2 \left[1 \pm \frac{b}{R_0} \cos \Phi + \left(\frac{b}{a}\right)^2 \frac{1}{R_0^2}\right]$$
$$\text{for} \quad b^2 \ll R_0^2 \tag{1.37a}$$

$$R_{1,2} \doteq R_0 \pm \frac{b}{2} \cos \Phi. \tag{1.37b}$$

In the exponents (phases) in (1.36) use is made of (1.37b); in the amplitudes let $R_1 \doteq R_2 \doteq R_0$. The result is

$$E_\Theta^r = C[I_1(0)\, e^{j(\beta_0 b/2)\cos \Phi} + I_2(0)\, e^{-j(\beta_0 b/2)\cos \Phi}] \tag{1.38}$$

where
$$C = \frac{j\zeta_0}{2\pi} F_0\left(\frac{\pi}{2}, \beta_0 h\right) \frac{e^{-j\beta_0 R_0}}{R_0}. \tag{1.39}$$

The term in brackets in (1.38) is the array factor $A(\Theta, \Phi)$. It is instructive to examine some special cases of (1.38) which correspond to different driving conditions and spacings.

When the elements have identical driving-point currents $I_0 = I_1(0) = I_2(0)$, the total electric field is

$$E_\Theta^r = 2C \cos\left(\frac{\beta_0 b}{2} \cos \Phi\right). \tag{1.40}$$

Since the electric field of an isolated antenna that is not too long $(\beta_0 h \leqslant \pi)$ always has a rotationally symmetrical single maximum in the equatorial plane, $\Theta = \pi/2$, the two-element array with equal driving-point currents will have principal maxima in this plane in the directions at right angles or broadside to the array. For this reason it is called a broadside array. Some representative patterns for such arrays are shown in Fig. 1.8. Due to symmetry in the H-plane the radiation pattern is always bilateral.

Now consider a two-element array in which the driving-point currents differ in phase by δ. That is, let

$$I_1(0) = I_{01}\, e^{-j\delta/2}, \qquad I_2(0) = I_{02}\, e^{j\delta/2}. \tag{1.41}$$

When the magnitudes of the driving-point currents are equal, $I_{01} = I_{02} = I_0$, the electric field is

$$E_\Theta^r = 2CI_0 \cos \tfrac{1}{2}(\beta_0 b \cos \Phi - \delta). \tag{1.42}$$

The main lobe may now be located at a desired angle Φ_n by an adjustment in the phase difference δ to make the argument of the cosine in (1.42) an integral multiple of π. That is, let

$$\tfrac{1}{2}(\beta_0 b \cos \Phi_n - \delta) = n\pi \qquad \text{or} \qquad \delta = 2n\pi - \beta_0 b \cos \Phi_n$$

$$n = 0, 1, 2, \ldots . \tag{1.43}$$

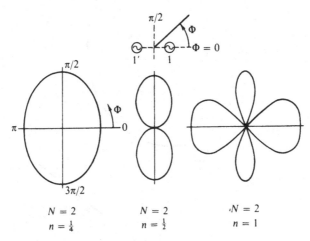

Fig. 1.8.   Representative broadside array factors $(N = 2)$.

The higher multiplicities of π indicate that an array with widely separated elements may have many maxima and, alternatively, an array with closely spaced elements can have only one maximum. An interesting example of an array with a specified phase difference between the driving-point currents is the endfire array, for which the major lobe in the field pattern is located at $\Phi = 0$. An important endfire array is the unilateral couplet for which $\beta_0 b = \pi/2$ and $\delta = \pi/2$. The H-plane field pattern shown in Fig. 1.9 is the familiar cardioid.

1.6   Fields of arrays of N elements; array factor

The directional properties of a two-element array are inadequate for many applications. Although the directivity of an array is roughly proportional to its overall length, it is not sufficient merely to increase the distance between elements in order to increase the directivity. A greater separation of the elements also assures more lobes in the field pattern. A greater directivity is best achieved by increasing the number of elements in the array. An array of $2N + 1$ elements is shown in Fig. 1.10. The electric field at a point R, Θ, Φ

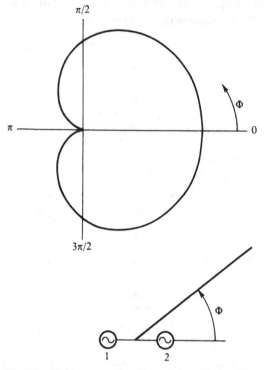

Fig. 1.9.   Field pattern of endfire couplet with $n = 1/4$.

in the far zone is a superposition of the contributions by all of the individual elements. Thus, if the total field is E_Θ^r and the contribution due to the ith element is $E_{\Theta i}^r$, and it is assumed that the distributions of current along all elements are identical, i.e. $I_{zi}(z) = I_{z0}(z)$,

$$E_\Theta^r = E_{\Theta 0}^r + \sum_{i=1}^{N} E_{\Theta i}^r + \sum_{i'=1}^{N} E_{\Theta i'}^r$$

$$= \left[\frac{j\omega\mu_0}{4\pi} \frac{e^{-j\beta_0 R}}{R} \int_{-h}^{h} I_{z0}(z') e^{j\beta_0 z' \cos \Theta} \sin \Theta \, d\Theta \right]$$

$$\left[1 + \sum_{i=1}^{N} \frac{I_{zi}(0)}{I_{z0}(0)} e^{j\beta_0 S_i} + \sum_{i'=1}^{N} \frac{I_{zi'}(0)}{I_{z0}(0)} e^{-j\beta_0 S_i} \right] \qquad (1.44)$$

where $S_i = ib \sin \Theta \cos \Phi$ and b is the distance between adjacent elements. This can be expressed in the simple form

$$E_\Theta^r = \frac{j\zeta_0 I_{z0}(0)}{2\pi} \frac{e^{-j\beta_0 R}}{R} F(\Theta, \beta_0 h) A(\Theta, \Phi) \qquad (1.45)$$

where $F_0(\Theta, \beta_0 h)$ is the vertical field function of an isolated element

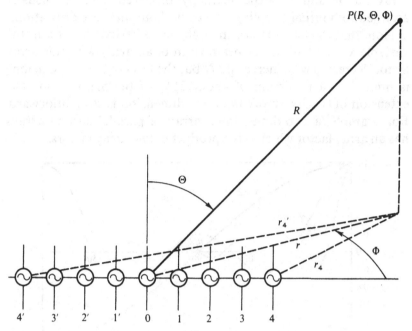

Fig. 1.10.   Geometry for curtain array.

and $A(\Theta, \Phi)$ is the array factor. The conventional study of linear arrays is concerned primarily with the nature of the array factor $A(\Theta, \Phi)$ since $F_0(\Theta, \beta_0 h)$ is a simple known function of Θ.

A so-called uniform array with equally spaced elements and with $|I_{zi}(0)| = |I_{z0}(0)|$ for all i's has the array factor

$$A(\Theta, \Phi) = \frac{\sin Nx}{\sin x} \qquad (1.46)$$

where $x = \pi(n \sin \Theta \cos \Phi - t)$, n is the distance between elements in fractions of a wavelength and t is the time delay from element to element in fractions of a cycle. The normalized array factor $A(x)/N$ is shown as a function of x for different values of N in Fig. 1.11. The curves for each value of N consist of major and minor maxima and minima. The major extreme values occur at $x = q\pi$, $q = 0, 1, 2, 3, \ldots$ and the minor ones at $x \doteq (p + 1/2)\pi/N$, $p = 1, 2, 3, \ldots$. Between each pair of extremes is a sharp null which indicates a perfect cancellation of the electric field in a definite direction. The mathematically simple result in (1.45) is seldom obtained in actual practice. The differences between the ideal array factor and an experimentally observed field are usually ascribed to 'mutual coupling effects' without further clarification.

Note that the array factor in (1.46) has a periodicity of π in the variable x. The half-power beamwidth of an array with this array factor decreases with increasing N but the level of the first side lobe is limited to a minimum of about 21% of the main beam. The extension of (1.45) to more than one dimension is straightforward. For example, a two-dimensional array of parallel curtain arrays has an array factor which is the product of two array factors.

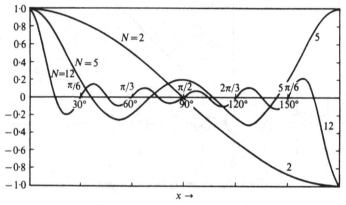

Fig. 1.11.   Normalized array factor ($N = 2, 5, 12$).

The conventional approach to the circuit properties of arrays follows by analogy with a low-frequency-circuit theory. Thus, an antenna is viewed as a circuit in which a driving voltage is impressed across a pair of terminals by a generator or transmission line and induced voltages are maintained at the same terminals due to coupling with other antennas. Self- and mutual impedances are required to relate the various driving and induced voltages and currents in the array. These impedances are usually computed from an energy-transfer formulation similar to (1.35).

1.7   Impedance of antenna; EMF method

The power radiated by an antenna is expressed in (1.30) in terms of the integral of the normal component of the complex Poynting vector over any surface Σ that completely encloses a transmitting system. In the evaluation of the radiation resistance R_m^e a great sphere was used since the formulas for the components of the radiation field are much simpler than those of the near field.

However, if the medium in which the antenna is immersed is non-dissipative, any other surface that encloses the transmitting system must yield the same result.

It is customary in determining the circuit properties of linear antennas to choose the cylindrical surface of the antenna as the surface of integration. Thus, if the small ends are neglected, (1.30) gives

$$P = 2\pi a \int_{-h}^{h} S_\rho \, dz = -\frac{\pi a}{\mu_0} \int_{-h}^{h} E_z B_\Phi^* \, dz \qquad (1.47)$$

where E_z and B_Φ^* are the values on the cylindrical surface of the antenna. Since

$$B_\Phi^* = \frac{I_z^*}{\mu_0 2\pi a} \qquad (1.48)$$

it follows that $$P = -\tfrac{1}{2} \int_{-h}^{h} E_z I_z^* \, dz. \qquad (1.49)$$

The power supplied to the antenna by a delta-function generator at its centre is

$$P = \tfrac{1}{2} V_0 I_0^* = -\tfrac{1}{2} \int_{-h}^{h} E_z I_z^* \, dz. \qquad (1.50)$$

In this form the Poynting vector theorem is known as the EMF method for determining the radiated power from the electromagnetic field and for defining the impedance of the antenna. Thus, since $V_0 = I_0 Z_0$, it follows that

$$Z_0 = -\frac{1}{|I_0|^2} \int_{-h}^{h} E_z I_z^* \, dz. \qquad (1.51)$$

When this formula is applied to a perfectly conducting cylinder which is centre-driven by a delta-function generator, the electric field is given by

$$E_z = -V_0 \delta(z) \qquad (1.52)$$

where $\delta(z) = 0$ except at $z = 0$. With (1.52) and the properties (1.1a, b) of the delta function, (1.51) becomes

$$Z_0 = \frac{1}{|I_0|^2} \int_{-h}^{h} V_0 I_z^* \delta(z) \, dz = \frac{1}{|I_0|^2} V_0 I_0^* = Z_0. \qquad (1.53)$$

Thus, (1.51) is simply an identity and not a means for determining Z_0.

In the usual application of (1.51) the boundary condition (1.52) for the tangential electric field is ignored and a sinusoidally distributed current is assumed. E_z due to the sinusoidally distributed current (1.32) is given by† (1.34c). It is

$$E_z = \frac{-jI_m\zeta_0}{4\pi}\left(\frac{e^{-j\beta_0 R_{1h}}}{R_{1h}} + \frac{e^{-j\beta_0 R_{2h}}}{R_{2h}} - 2\cos\beta_0 h\frac{e^{-j\beta_0 R_0}}{R_0}\right) \quad (1.54)$$

where, for a point on the surface of the antenna, $R_0 = \sqrt{z'^2 + a^2}$, $R_{1h} = \sqrt{(h-z')^2 + a^2}$, $R_{2h} = \sqrt{(h+z')^2 + a^2}$. If (1.54) and (1.32) are used in (1.51), with $I_0 = I_m \sin\beta_0 h$, the result is

$$Z_0 = \frac{j\zeta_0}{4\pi}\frac{1}{\sin^2\beta_0 h}\left\{\sin\beta_0 h\int_{-h}^{h}\cos\beta_0 z'\left[\frac{e^{-j\beta_0 R_{1h}}}{R_{1h}} + \frac{e^{-j\beta_0 R_{2h}}}{R_{2h}}\right.\right.$$
$$\left. - 2\cos\beta_0 h\frac{e^{-j\beta_0 R_0}}{R_0}\right]dz' - \cos\beta_0 h\int_{-h}^{h}\sin\beta_0|z'|\left[\frac{e^{-j\beta_0 R_{1h}}}{R_{1h}}\right.$$
$$\left.\left. + \frac{e^{-j\beta_0 R_{2h}}}{R_{2h}} - 2\cos\beta_0 h\frac{e^{-j\beta_0 R_0}}{R_0}\right]dz'\right\}. \quad (1.55)$$

The integrals

$$C_a(h, z) = \int_{-h}^{h}\cos\beta_0 z'\frac{e^{-j\beta_0 R_1}}{R_1}dz'$$
$$= \int_{0}^{h}\cos\beta_0 z'\left[\frac{e^{-j\beta_0 R_1}}{R_1} + \frac{e^{-j\beta_0 R_2}}{R_2}\right]dz' \quad (1.56a)$$

$$S_a(h, z) = \int_{-h}^{h}\sin\beta_0|z'|\frac{e^{-j\beta_0 R_1}}{R_1}dz'$$
$$= \int_{0}^{h}\sin\beta_0 z'\left[\frac{e^{-j\beta_0 R_1}}{R_1} + \frac{e^{-j\beta_0 R_2}}{R_2}\right]dz' \quad (1.56b)$$

have been tabulated with $z = h$ and 0. In (1.56a, b),

$$R_1 = \sqrt{(z-z')^2 + a^2}, R_2 = \sqrt{(z+z')^2 + a^2}.$$

With the tabulated integrals

$$Z_0 = \frac{j\zeta_0}{2\pi}\frac{1}{\sin^2\beta_0 h}\{\sin\beta_0 h[C_a(h, h) - \cos\beta_0 hC_a(h, 0)]$$
$$- \cos\beta_0 h[S_a(h, h) - \cos\beta_0 hS_a(h, 0)]\}. \quad (1.57)$$

In particular, when $\beta_0 h = \pi/2$,

$$Z_0 = \frac{j\zeta_0}{2\pi}C_a\left(\frac{\lambda}{4}, \frac{\lambda}{4}\right). \quad (1.58)$$

† See, for example, [1] p. 528.

The functions $C_a(h, h)$, $C_a(h, 0)$, $S_a(h, h)$ and $S_a(h, 0)$ may be expressed in terms of the generalized sine and cosine integral functions. When the radius a of the antenna is small ($a \to 0$),

$$Z_0 \doteq \frac{\zeta_0}{4\pi}[\text{Cin } 2\pi + jSi2\pi] = 73\cdot1 + j42\cdot5 \text{ ohms.} \qquad (1.59)$$

The real part, $R_0 = 73\cdot1$ ohms, is the value of R_m^e given in section 1.3. When $\beta_0 h = \pi$, $Z_0 = \infty$. Cin $x = \int_0^x u^{-1}(1 - \cos u)\, du$.

It is now necessary to inquire more deeply into the mechanism by which a sinusoidally distributed current can be maintained along an antenna. Since it is associated with a non-vanishing tangential electric field E_z along the surface of the antenna, it cannot be maintained solely by a single generator at $z = 0$ along a perfectly conducting cylinder. The boundary condition $E_z = 0$ applies to the total electric field and must be satisfied on the surface of a perfect conductor. Since E_z due to the currents in the conductor is not zero on the surface, it cannot be the total field. There must be an externally maintained field E_z^e of such magnitude and phase that

$$E_{z\,\text{total}} = E_z + E_z^e = 0, \qquad E_z^e = -E_z. \qquad (1.60)$$

In other words, the existence of a sinusoidally distributed current along a perfectly conducting cylinder implies the simultaneous existence of a continuous distribution of generators or their equivalent along the antenna. These maintain an impressed field that exactly cancels the field maintained by the currents and charges in the antenna.

If there is a continuous distribution of generators along the antenna, there can be no single pair of terminals at its centre through which all the power is supplied and across which a driving-point impedance can be defined. It follows that the impedance Z_0 given by (1.57) or (1.58) is not the impedance of a centre-driven antenna but simply the total complex power radiated by an antenna (in which a continuous distribution of generators maintains a sinusoidal current), divided by the square of the magnitude of the current at the centre of the antenna.

A centre-driven antenna with E_z on its surface given by (1.52) and an antenna with its distribution of current given by (1.32) and, therefore, with E_z on its surface given by (1.54), are two quite different models with different currents, different fields and different power-supplying devices. It is a common mistake to assume that *some* of the properties of each can be combined as though the two were, in fact, the same. Nevertheless, under special circumstances certain characteristics of the two models are comparable and no serious error is made if they are used interchangeably. But this is not true in general or of all significant quantities as is easily shown.

In Figs. 1.12 and 1.13 are shown the measured amplitude and phase of the current in a thin highly conducting base-driven monopole over a large metal ground screen together with the sinusoidally distributed current, respectively, for $\beta_0 h = \pi/2$ and π. When

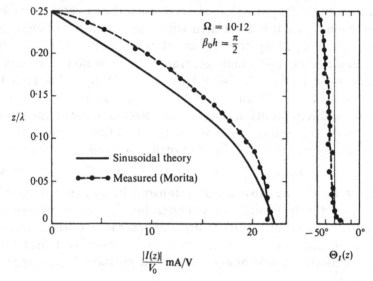

Fig. 1.12.   Distribution of amplitude and phase of current in half-wave dipole.

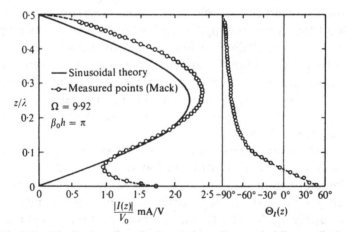

Fig. 1.13.   Distribution of amplitude and phase of current in full-wave dipole.

$\beta_0 h = \pi/2$ both the measured and sinusoidal values are reasonably alike so that quantities that depend directly on I_z, viz. I_0 (and hence Y_0 and Z_0) and the magnetic field B_Φ, must also be comparable. Since the electric field at great distances is linearly related to the magnetic field, it follows that the entire far field of the actual and the sinusoidal currents should be generally alike. On the other hand, when $\beta_0 h = \pi$ the measured distribution and the sinusoidal distribution differ greatly in both amplitude and phase near the centre of the antenna. It is clear that I_0 (and hence Z_0 and Y_0) have nothing in common—the one is quite large, the other is zero. Moreover, since the measured phase reverses at some distance from the centre, whereas all currents remain exactly in phase with the sinusoidal distribution, even the radiation field must differ significantly. The measured current in the centre-driven antenna must have a small minor lobe, whereas the sinusoidal current has none. Thus, when $\beta_0 h = \pi$, the only properties of the antenna with a sinusoidal current and the centre-driven antenna that are roughly comparable are the general nature of the major lobe in the far field.

Even though the currents, the near-zone magnetic fields, and the distant electric and magnetic fields are reasonably alike for the two differently driven antennas when $\beta_0 h = \pi/2$, this does not mean that all significant quantities are. That the associated distributions of charges differ significantly is shown in Fig. 1.14 where both

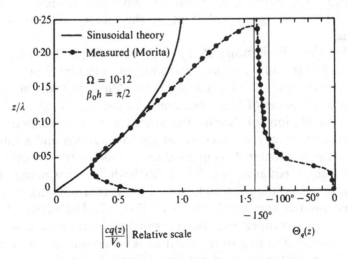

Fig. 1.14. Normalized distribution of charge in amplitude and phase for a half-wave dipole.

measured and cosinusoidally distributed charges per unit length are shown. It is seen that the charges near the centre and the ends have quite different magnitudes and that the phase reverses at some distance from $z = 0$ when the antenna is centre-driven but exactly at $z = 0$ when it has a sinusoidal current. Since the radial electric field E_ρ near the antenna is proportional to the charge per unit length, it must be quite different for the two models. Finally, it can be shown that E_z near the antenna depends not only on I_z but on $\partial q/\partial z$, i.e. on the slope of the charge curve. Since the values of $\partial q/\partial z$ are entirely different for the centre-driven antenna and the antenna with a distribution of generators that maintains a sinusoidal current, no correspondence in E_z for the two models can obtain. This explains how it is possible that for the centre-driven antenna $E_z = 0$ at $\rho = a$ except at $z = 0$, whereas for the antenna with sinusoidal currents E_z has a maximum at $\rho = a$ along its entire length.

These and other studies lead to the conclusion that the hypothetical antenna with a sinusoidally distributed current (and the implied continuous distribution of generators along its length) may be substituted for a centre-driven antenna in determining roughly corresponding values of I_z for all z including 0, the magnetic field at all points, and the radiation electric field provided the antenna is sufficiently thin ($a/h < 10^{-5}$) and its length is near resonance, i.e. $\beta_0 h \sim \pi/2$. However, no such correspondence exists for the charge distribution or the electric field near the antenna.

Within these limitations, Z_0 as given by (1.57), (1.58) and (1.59) may be used to obtain approximate values of the impedance of a centre-driven antenna. Specifically, when $h/a \sim 11,000$, the error in R_0 is of the order of 5%, when $h/a \sim 27$ the error is about 34%.

The correlation of linear antenna theory with experiment involves some important theoretical approximations and a knowledge of the particular driving conditions. Due to the difficulty in solving a single equation which includes both the transmission line and the antenna, the determination of the actual antenna current is based on the simplified model of Fig. 1.2. The effect of the inhomogeneous properties of the transmission line and the coupling of the transmission line to the antenna is taken into account by a lumped constant corrective network. Since the spacing between driving terminals of the antenna is finite, a correction must also be made for the missing section of conductor. This is discussed in chapter 7.

1.8 Coupled antennas; self- and mutual impedances

When an antenna is an element in an array, the Poynting-vector integration over the surface of the antenna in the form (1.47) includes contributions to E_z from other members in the array. By arranging the power terms in a certain sequence, the resulting expressions for the integrals resemble the standard equations for coupled circuits so that self- and mutual impedances can be defined in a manner analogous to that used in coupled-circuit equations.

Neglecting the ohmic loss in each antenna of Fig. 1.10, the total time-average power transferred across Σ_k, the surface of the kth antenna, is

$$P_k = \int_{\Sigma_k} \hat{n} . \mathbf{S}_k \, d\Sigma_k \qquad (1.61)$$

where

$$\mathbf{S}_k = \sum_i \mathbf{S}_{ki}. \qquad (1.62)$$

In (1.62) $\mathbf{S}_{ki}$ is the contribution of the ith element to the total Poynting vector at the surface of the kth element with the kth element as reference. The time-average Poynting vector for dipoles aligned in the z-direction is

$$\mathbf{S}_{ki} = \frac{1}{2\mu_0}(\mathbf{E}_{ki} \times \mathbf{B}_{ki}^*) = \frac{1}{2\mu_0}(\hat{\rho} E_{zki} B_{\Phi ki}^*) \qquad (1.63)$$

since on the cylindrical surface of antenna k

$$\mathbf{E}_{ki} = \hat{z} E_{zki} \qquad (1.64)$$

and

$$\mathbf{B}_{ki}^* = \hat{\Phi} B_{\Phi ki}. \qquad (1.65)$$

$\mathbf{S}_{ki}$ on the small ends is neglected.

At the surface of the kth element the only magnetic field encircling that element is due to the current in the element itself. Therefore,

$$B_{\Phi k}^* = I_{zk}^*/2\pi a \mu_0. \qquad (1.66)$$

The Poynting vector equations (1.61) may be expanded with (1.62)–(1.65):

$$P_k = \tfrac{1}{2}\int_{-h}^{h} E_{zk1} I_{zk}^* \, dz + \tfrac{1}{2}\int_{-h}^{h} E_{zk2} I_{zk}^* \, dz + \ldots + \tfrac{1}{2}\int_{-h}^{h} E_{zkk} I_{zk}^* \, dz + \ldots$$

$$+ \tfrac{1}{2}\int_{-h}^{h} E_{zkN} I_{zk}^* \, dz, \qquad k = 1, 2, 3, \ldots, N \qquad (1.67)$$

where $d\Sigma_k$ has been replaced by $(2\pi a)\,dz$ and N is the total number of elements. Equation (1.67) is used to define the complex driving-point impedance Z_{01} and the self- and mutual impedances Z_{ki}. Thus,

$$P_k = \tfrac{1}{2}I_0^2 Z_0 = \tfrac{1}{2}I_1 I_k^* Z_{1k} + \tfrac{1}{2}I_2 I_k^* Z_{2k} + \ldots + \tfrac{1}{2}I_k^2 Z_{kk} + \ldots + \tfrac{1}{2}I_N I_k^* Z_{Nk},$$
$$k = 1, 2, 3, \ldots, N \tag{1.68}$$

where
$$Z_{ik} = -\frac{1}{I_k^* I_i} \int_{-h}^{h} E_{zki} I_{zk}^* \, dz. \tag{1.69}$$

This is a generalization of (1.51) to include mutual impedances.

If the elements in an array are centre-driven (base-driven over a ground screen) the electric field along any element k is the superposition of the fields E_{zki} with $i = 1, 2, \ldots k, \ldots N$. The fields E_{zki} must be computed from the actual currents in the N elements and the resultant field must satisfy the boundary condition

$$\sum_{i=1}^{N} E_{zki} = -V_{0k}\delta(z). \tag{1.70}$$

Since these currents are unknown, this procedure for defining self- and mutual impedances is not directly useful.

On the other hand, if it is stipulated that the distribution of current along each element is sinusoidal irrespective of its location in the array, the impedances defined in (1.69) are readily evaluated. However, each antenna must then be driven by a distribution of generators with EMF's so disposed in amplitudes and phases that the postulated currents actually obtain. The impedances evaluated from (1.69) under these conditions are radiation impedances referred to a particular set of currents, those at $z = 0$. However, these are not driving-point currents, since the generators are not localized at $z = 0$.

As in the case of the isolated antenna, it is to be expected that there are certain circumstances under which *some* of the quantities that characterize an array are quantitatively similar for the two quite different sets of boundary and driving conditions: (1) Each element is centre-driven by a single generator, and (2) Each element is driven by a distribution of generators that maintains a sinusoidal current. Since it has been shown that the current maintained in a sufficiently thin antenna near resonance ($\beta_0 h \sim \pi/2$) by a single generator at $z = 0$ does not differ greatly in amplitude and phase from the cosinusoidal current, and since it is well known that the current induced in a parasitic antenna by an incident plane wave is approximately cosinusoidal when $\beta_0 h = \pi/2$,† it is reasonable to expect the resultant current in any very thin element in an array of parallel elements to be approximately cosinusoidal. The actual impedances for such arrays should then be approximated by those defined in (1.69) and the far fields should be comparable in an approximate quantitative sense. When the elements are electrically short, the current has a maximum at $z = 0$ and vanishes at $z = h$. It is to be expected that the far field and the impedances are not sensitive to the precise distributions of current in the short lengths of the elements.

As a simple example, the mutual impedance between two short Hertzian dipoles with a large separation will be computed. A triangular current is assumed. The current distribution and the relevant term of the electric field are

$$I_{zk}(z) = I_k\left(1 - \frac{|z|}{h}\right), \qquad |z| \leqslant h \ll \beta_0^{-1} \qquad (1.71)$$

$$E_{zki} = \frac{-jI_0 h \beta_0^2}{4\pi\omega\varepsilon_0 r^3}(r^2 - z^2)\,e^{-j\beta_0 r}, \qquad (\beta_0 b)^2 \gg 1. \qquad (1.72)$$

where
$$r = \sqrt{b^2 + z^2} \doteq b, \qquad b = b_{ki}. \qquad (1.73)$$

When (1.71) and (1.72) are substituted in (1.69) the result is

$$Z_{ik} \doteq \frac{j\zeta_0}{2\pi}\frac{\beta_0^2 h^2}{\beta_0 b}e^{-j\beta_0 b}, \qquad (\beta_0 b)^2 \gg 1. \qquad (1.74)$$

As another example of the computation of mutual impedance, consider the case of half-wavelength elements ($\beta_0 h = \pi/2$) with an assumed sinusoidal current given by

$$I_z(z) = I_0 \cos\beta_0 z, \qquad |\beta_0 z| \leqslant \pi/2. \qquad (1.75)$$

The electric field E_z due to this assumed current is given by (1.54) with $\beta_0 h = \pi/2$ and $I_m = I_0$. It is

$$E_z = \frac{-jI_0\zeta_0}{4\pi}\left(\frac{e^{-j\beta_0 R_{1h}}}{R_{1h}} + \frac{e^{-j\beta_0 R_{2h}}}{R_{2h}}\right) \qquad (1.76)$$

where
$$R_{1h} = \sqrt{(h-z)^2 + a^2}, \qquad R_{2h} = \sqrt{(h+z)^2 + a^2}. \qquad (1.77)$$

† [1] p. 475.

When (1.75) and (1.76) are substituted in (1.69), the result for the mutual impedance is

$$Z_{ik} = \frac{j\zeta_0}{2\pi} \int_0^{\lambda/4} \cos \beta_0 z' \left| \frac{e^{-j\beta_0 R_{1h}}}{R_{1h}} + \frac{e^{-j\beta_0 R_{2h}}}{R_{2h}} \right| dz'. \qquad (1.78)$$

The integral in (1.78) is one of the integral trigonometric functions tabulated by Mack and Mack†. It is $C_a(h, z)$ defined in (1.56a). When a is sufficiently small,

$$C_a\left(\frac{\lambda}{4}, \frac{\lambda}{4}\right) \doteq C_0\left(\frac{\lambda}{4}, \frac{\lambda}{4}\right) = \tfrac{1}{2}[\text{Si } 2\pi - j \text{ Cin } 2\pi],$$

so that $Z_{kk} = Z_0 = 73.1 + j42.5$ ohms as given in (1.59).

The mutual-impedance formula given by (1.78) with (1.56a) reduces to (1.58) for the self-impedance when the radius a is set equal to the radius of the cylindrical antenna. The final formula for Z_{ik} is

$$Z_{ik} = \frac{j\zeta_0}{2\pi} C_b\left(\frac{\lambda}{4}, \frac{\lambda}{4}\right) \qquad (1.79)$$

where

$b \equiv b_{ik}$ is the distance between the ith and k^{th} elements

$b_{ii} = a$ is the radius of each antenna. $\qquad (1.80)$

The mutual impedance between two half-wavelength elements in parallel is shown in Fig. 1.15. The calculations are based on (1.79). A comparison of the asymptotic formula (1.74) with assumed triangular current with the results of Fig. 1.15 shows that for large separation the mutual impedance is approximately independent of the current distribution.

1.9 Radiation-pattern synthesis

The far-field pattern represents the distribution of the electric field in space. Some applications, particularly point to point communication, require a large field directed within a small angular region. A rearrangement of the spatial distribution of the field is called shaping. The shaping of the radiation pattern of an antenna or array is accomplished by changes in the geometry and the relative currents in the antennas. A discussion of this problem lends insight into the effects of the actually unequal distributions of current in real arrays.

Consider first a single linear element aligned in the z-direction.

† [3] in chapter 5.

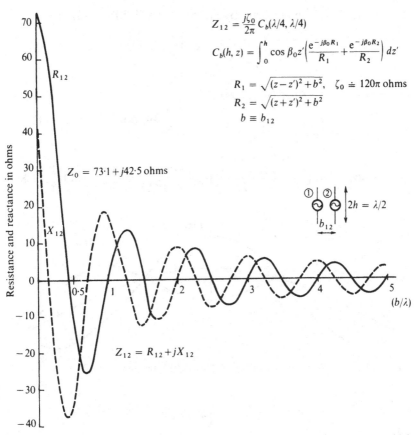

$$Z_{12} = \frac{j\zeta_0}{2\pi} C_b(\lambda/4, \lambda/4)$$

$$C_b(h, z) = \int_0^h \cos \beta_0 z' \left(\frac{e^{-j\beta_0 R_1}}{R_1} + \frac{e^{-j\beta_0 R_2}}{R_2} \right) dz'$$

$$R_1 = \sqrt{(z-z')^2 + b^2}, \quad \zeta_0 \doteq 120\pi \text{ ohms}$$

$$R_2 = \sqrt{(z+z')^2 + b^2}$$

$$b \equiv b_{12}$$

$$Z_0 = 73 \cdot 1 + j42 \cdot 5 \text{ ohms}$$

$$Z_{12} = R_{12} + jX_{12}$$

Fig. 1.15. Mutual impedance between two half-wavelength dipoles with sinusoidal current (EMF method).

Three different distributions of current will be used to calculate the far field. They are

$$(1) \quad I_{z1}(z) = \cos \beta_0 z \qquad\qquad\qquad (1.81)$$

$$(2) \quad I_{z2}(z) = \cos \beta_0 z + \tfrac{1}{2} \cos 3\beta_0 z \left.\vphantom{\begin{array}{c}1\\1\\1\end{array}}\right\} \beta_0 h = \frac{\pi}{2}. \qquad (1.82)$$

$$(3) \quad I_{z3}(z) = \cos \beta_0 z - \tfrac{1}{2} \cos 3\beta_0 z \qquad\qquad (1.83)$$

The distributions are shown in Fig. 1.16a. The conventional distribution is $I_z(z) = \cos \beta_0 z$ with a maximum at $z = 0$ and a monotonic decrease in magnitude to $z = \pm h$. The second and third distributions (1.82) and (1.83) have, respectively, a greater concentration of current near $z = 0$ and a maximum which is moved toward $z = h$. These distributions could be achieved with suitable generators placed along the element. The corresponding electric fields in the far zone by (1.15a) with (1.80)–(1.83) are

$$(1)\ E_\Theta^r = K_1 \left[\frac{\cos\left(\frac{\pi}{2}\cos\Theta\right)}{\sin\Theta} \right] \qquad (1.84)$$

$$(2)\ E_\Theta^r = K_1 \left[\frac{\cos\left(\frac{\pi}{2}\cos\Theta\right)}{\sin\Theta} - \frac{3\sin\Theta}{9-\cos^2\Theta}\cos\left(\frac{\pi}{2}\cos\Theta\right) \right] \qquad (1.85)$$

$$(3)\ E_\Theta^r = K_1 \left[\frac{\cos\left(\frac{\pi}{2}\cos\Theta\right)}{\sin\Theta} + \frac{3\sin\Theta}{9-\cos^2\Theta}\cos\left(\frac{\pi}{2}\cos\Theta\right) \right]. \qquad (1.86)$$

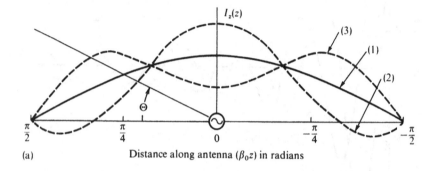

(a)   Distance along antenna $(\beta_0 z)$ in radians

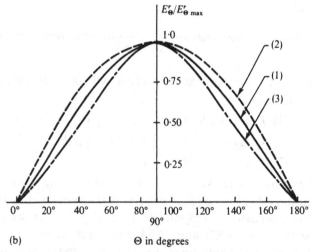

(b) Θ in degrees

Fig. 1.16. (a) Assumed current distributions on half-wave dipole. (1) $I_z(z) = \cos\beta_0 z$. (2) $I_z(z) = \cos\beta_0 z + \frac{1}{2}\cos 3\beta_0 z$. (3) $I_z(z) = \cos\beta_0 z + \frac{1}{2}\cos 3\beta_0 z$. (b) Radiation patterns for half-wave dipole with different current distributions.

The radiation patterns (1.84)–(1.86) are shown in Fig. 1.16b. The radiation pattern (1.85) has a half-power beamwidth that is increased when compared with the normal pattern for (1.81). Thus, when the current in an element is increased at the centre the beamwidth is increased. The opposite behaviour is shown by the radiation pattern for (1.86), which corresponds to (1.80). An increase in the magnitude of the current toward the ends of the element decreases the beamwidth of the radiation pattern. Although a single generator could not produce the currents shown in Fig. 1.16a, such distributions of current could be produced by coupling with other elements.

Consider again the half-wavelength element with a sinusoidally distributed current. This time let the current change linearly in phase along the element. That is, let

$$I_z(z) = \cos \beta_0 z \, e^{-j\delta z}. \tag{1.87}$$

The radiation field that corresponds to the current (1.87) is

$$E_\Theta^r = K \frac{\cos\left(\dfrac{\pi}{2}\cos\Theta - \delta/4\right)}{\sin\Theta}. \tag{1.88}$$

The maximum value of $E_\Theta^r(\Theta)$ now appears at a value of Θ given by

$$\Theta = \cos^{-1}\frac{\delta}{2\pi}. \tag{1.89}$$

It may be concluded that *both* the phase and the magnitude of the current in an element affect the radiation pattern.

A single half-wave element can be approximated by a number of collinear infinitesimal dipoles. Each infinitesimal dipole has a prescribed current equal to that in the original full-length dipole at this location. Fig. 1.17 shows such an arrangement for a half-wave

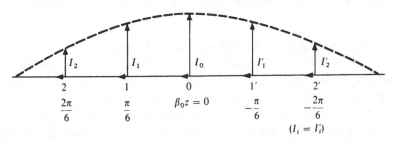

Fig. 1.17. Replacement of continuous current distribution with discrete sources.

dipole with a sinusoidally distributed current. The field factor $F_0(\Theta, \beta_0 h)$ is given by a sum instead of by an integral as in (1.15b):

$$F_0(\Theta, \beta_0 h) = \sin \Theta \left[I_0 + 2 \sum_{i=1}^{2} I_i \delta(z - d_i)\, e^{j\beta_0 d_i \cos \Theta} \right] / I_0 \quad (1.90)$$

$$= \sin \Theta [I_0 + 2I_1(e^{j\beta_0 d_1 \cos \Theta} + e^{-j\beta_0 d_1 \cos \Theta})$$
$$+ 2I_2(e^{j\beta_0 d_2 \cos \Theta} + e^{-j\beta_0 d_2 \cos \Theta})]/I_0 \quad (1.91)$$

$$F_0(\Theta, \beta_0 h) = \sin \Theta \sum_{i=1}^{2} 1 + I_i \cos (\beta_0 d_i \cos \Theta) \quad (1.92)$$

where for the currents of Fig. 1.17

$$I_1 = I_0 \sqrt{3}/2 \quad \text{and} \quad I_2 = I_0/2. \quad (1.93)$$

The field factor computed with (1.92) and (1.93) is compared to that for a continuous line source in Fig. 1.18. The agreement indicates that a sufficient number of discrete sources can approximate the radiation pattern of a continuous line source.

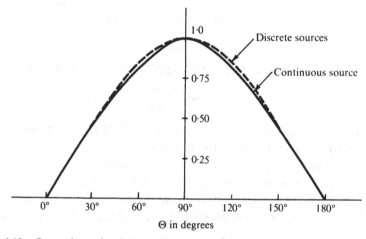

Fig. 1.18. Comparison of radiation patterns of half-wave dipole from continuous and discrete sources.

If the infinitesimal dipoles of Fig. 1.17 are each rotated 90° clockwise, the result is a crude picture of a curtain array similar to that shown in Fig. 1.10. The problem of synthesis is to determine how the currents in the array must be adjusted to generate a desired radiation pattern. The current distributions on all elements are assumed to be identical. Actually, the distributions of current are not the same and their differences affect the radiation pattern of the array. Differences in the amplitudes and phases of the currents along the elements in an array affect its field pattern in much the same way in which they affect the pattern of a single element.

Under the assumption of identical distributions of current on all elements, the far-zone electric field for the symmetrical array of Fig. 1.10 is

$$E_\Theta^r(\Theta) = \frac{j\zeta_0 I_0(0)}{2\pi} F_0(\Theta, \beta_0 h) \sum_{i=1}^{N} \left[1 + 2\frac{I_i(0)}{I_0(0)} \cos (\beta_0 ib \sin \Theta \cos \Phi) \right]$$

$$(1.94)$$

where it is assumed that t_1, the time delay between elements, is zero.

In the H-plane ($\Theta = \pi/2$) the field factor $F_0(\Theta, \beta_0 h)$ is a constant and (1.94) reduces to

$$E_\Theta^r(\Theta) = E_0 \sum_{i=1}^{N} \left[1 + 2\frac{I_i(0)}{I_0(0)} \cos (iu) \right] \qquad (1.95)$$

where

$$u = 2\pi \frac{b}{\lambda} \cos \Phi \qquad (1.96)$$

and

$$E_0 = \frac{j\zeta_0 I_0(0)}{2\pi} F_0\left(\frac{\pi}{2}, \beta_0 h\right). \qquad (1.97)$$

The synthesis problem for this array may be phrased as follows: Given a prescribed radiation pattern $E_{\Theta p}^r(u)$, what are the values of the parameters of an array which will best produce this pattern? Usually all but one of the parameters are fixed, and a single parameter is adjusted. For example, the number of elements N and the spacing b/λ might be fixed. The driving currents $I_i(0)/I_0(0)$ are then to be determined so as to produce the best approximation of the field. If this is defined in terms of the least mean-square error between $E_{\Theta p}^r(u)$ and $E_\Theta^r(u)$, the following integral must be minimized:

$$\min \int_{u_1}^{u_2} \{[E_{\Theta p}^r(u)] - [E_\Theta^r(u)]\}^2 \, du \qquad (1.98)$$

where $E_\Theta^r(\Theta)$ is given by (1.95). The integral (1.98) is minimized with respect to the coefficients $I_1(0)/I_0(0)$. Note that, since trigonometric functions are involved, the optimum range of u is $|u| \leqslant \pi$. Note also that the integral is written in terms of the variable u and not $\cos \Phi$. This serves to reduce the complexity of the results. With the interval in (1.98) equal to the range $-\pi \leqslant u \leqslant \pi$, the coefficients for the

current may be related to the Fourier trigonometric coefficients for
the representation of $E'_{\Theta p}(u)$, given by

$$E'_{\Theta p}(u) = \frac{a_0}{2} + \sum_{n=1}^{\infty} a_n \cos nu \qquad (1.99)$$

where

$$a_n = \frac{2}{\pi} \int_0^{\pi} E'_{\Theta p}(u) \cos nu \, du. \qquad (1.100)$$

When (1.95) and (1.99) are equated term by term, the following
values of the coefficients for the current are obtained:

$$2I_0(0)E_0 = \frac{a_0}{2} \qquad (1.101)$$

$$\frac{2I_n(0)}{I_0(0)} E_0 = a_n, \qquad n = 1, 2, 3, \dots . \qquad (1.102)$$

With the driving-point current on element No. 1 normalized to
unity, the relative currents on the elements are given by (1.101) and
(1.102). It follows that

$$I_n(0) = \left| \frac{a_n}{a_0} \right|. \qquad (1.103)$$

Thus, the relative driving-point currents are proportional to the
ratio of the Fourier coefficients for the prescribed radiation pattern.

Consider the particular field pattern shown in Fig. 1.19. It is
desired to build a three-element array to produce this prescribed
pattern. The inter-element spacing must also be specified. A value
of $n = b/\lambda = 1/2$ is chosen for this example. Note that the synthesis

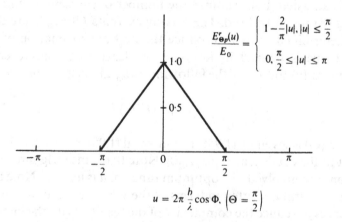

Fig. 1.19. Prescribed field pattern in H-plane.

is limited to the range $-\pi \leqslant u \leqslant \pi$. This means that for certain values of the inter-element spacing n, the range $-\pi \leqslant u \leqslant \pi$ does not correspond to the visable range of the angle Φ. For example, with $b/\lambda = 1$, the pattern may only be synthesized for the range $|\Phi| \leqslant \cos^{-1}\frac{1}{2} = 60°$. The mathematical formulation of the desired pattern is given below:

$$\frac{E'_{\Theta p}(u)}{E_0} = \begin{cases} 1 - \dfrac{2}{\pi}|u|, & |u| \leqslant \dfrac{\pi}{2} & (1.104) \\[2ex] 0, & \dfrac{\pi}{2} \leqslant |u| \leqslant \pi. & (1.105) \end{cases}$$

The coefficients for the current are found from (1.104) and (1.101) to be

$$\left.\begin{aligned} a_0 &= \frac{2}{\pi}\int_0^{\pi/2}\left(1-\frac{2u}{\pi}\right)du = \tfrac{1}{4}, \qquad \tfrac{1}{2}a_0 = \tfrac{1}{8} \\[2ex] a_n &= \frac{2}{\pi}\int_0^{\pi/2}\left(1-\frac{2u}{\pi}\right)\cos nu\,du = \frac{8}{n^2\pi^2}\sin^2\frac{n\pi}{4} \end{aligned}\right\} \quad (1.106)$$

$$n = 1, 2, 3, \dots.$$

The ratios of the driving-point currents are given by (1.103):

$$\frac{I_n}{I_0} = \frac{32}{n^2\pi^2}\sin^2\frac{n\pi}{4}, \qquad n \neq 0. \tag{1.107}$$

The realizable radiation pattern for the three-element array is compared to the prescribed pattern in Fig. 1.20. A better agreement

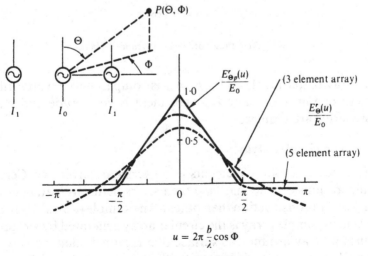

Fig. 1.20. Prescribed and actual field patterns.

between the desired and actual pattern may be achieved by increasing the number of elements. This is analogous to increasing the number of terms used in approximating a function by a Fourier series. The pattern realized with $N = 5$ is also shown in Fig. 1.20 and is a very close approximation of the triangular distribution of Fig. 1.19. Note that these patterns have been obtained for the H-plane; in other planes the field factor must be taken into account.

The synthesis of the radiation pattern is not always successful if the number of elements is limited. Consider the pattern of Fig. 1.21 for which the coefficients for the current are

$$\frac{I_n(0)}{I_0} = \frac{8}{\pi^2}\left[\frac{\pi}{2n}\sin\frac{n\pi}{2} + \frac{1}{n^2}\left(\cos\frac{n\pi}{2} - 1\right)\right], \qquad n \neq 0. \quad (1.108)$$

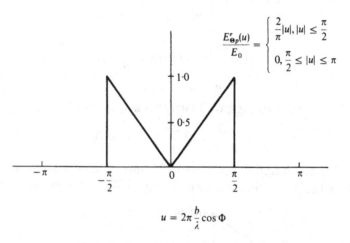

$$\frac{E_{\Theta p}^r(u)}{E_0} = \begin{cases} \dfrac{2}{\pi}|u|, |u| \leq \dfrac{\pi}{2} \\[2mm] 0, \dfrac{\pi}{2} \leq |u| \leq \pi \end{cases}$$

$$u = 2\pi\frac{b}{\lambda}\cos\Phi$$

Fig. 1.21. Prescribed field pattern in H-plane.

The array pattern realized for $N = 3$ is compared to the prescribed pattern in Fig. 1.22. Here the agreement is very crude and many more terms are required.

1.10   The circular array

The locations of the elements in an array are arbitrary. Certain configurations of antennas lend themselves better to mathematical analysis and construction than others. An example of an analytically convenient simple array is the circular array generated by wrapping a curtain array around the surface of a right cylinder. To simplify the analysis, only circular arrays with elements that are located at

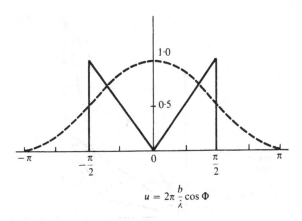

$$u = 2\pi \frac{b}{\lambda} \cos \Phi$$

Fig. 1.22.   Prescribed and actual field patterns.

the vertices of a polygon which is inscribed in a circle are considered. The case $N = 2$ is the two-element array discussed in section 1.5; the case $N = 3$ consists of three parallel elements located at the vertices of an equilateral triangle. The centres of the elements are located on a circle of radius a in the x–y plane as shown in Figs. 1.23a and 1.23b. The far-zone electric field is related to the z-component of the vector potential by

$$E_\Theta^r \doteq j\omega \sin \Theta A_z \qquad (1.109)$$

$$E_\Theta^r = \frac{j\omega\mu_0}{4\pi} \sin \Theta \int_{-h}^{h} I(z') \frac{e^{-j\beta_0 R_i}}{R_i} dz'. \qquad (1.110)$$

The distance R_1 to the centre of the element located at $(a, \pi/2, \phi_i)$ is given by

$$R_{0i}^2 = z^2 + r_i^2 = z^2 + r^2 + a^2 - 2ar \cos (\Phi - \phi_i) \qquad (1.111)$$

but $\qquad R_0^2 = z^2 + r^2 \qquad$ and $\qquad r = R_0 \sin \Theta \qquad (1.112)$

thus $\qquad R_{0i}^2 = R_0^2 \left[1 - \frac{2a}{R_0} \sin \Theta \cos (\Phi - \phi_i) \right]. \qquad (1.113)$

When higher-order terms in (a/R_0) are neglected, an application of the binomial expansion for (1.113) yields

$$R_{0i} \doteq R_0 - a \sin \Theta \cos (\Phi - \phi_i). \qquad (1.114)$$

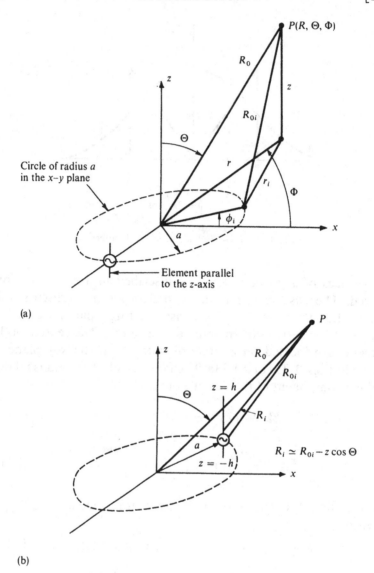

(a)

(b)

Fig. 1.23. (a) Geometry for the circular array. (b) Single-element geometry in a circular array.

Since it is the distance R_i which appears in (1.110), it follows from Fig. 1.23b and (1.114) that

$$R_i \doteq R_{0i} - z \cos \Theta \quad \text{(phase)}$$
$$R_i \doteq R_{0i} \doteq R_0 \quad \text{(amplitude).}$$
$$\left. \right\} \qquad (1.115)$$

If the array has N elements of equal length and the current distributions are identical, the total field is

$$E_\Theta^r = j\frac{\zeta_0}{2\pi}\frac{e^{-j\beta_0 R_0}}{R_0}F_0(\Theta, \beta_0 h)\sum_{i=1}^{N} I_{0i}\, e^{-j\beta_0 S_i} \qquad (1.116)$$

where $S_i = a\sin\Theta\cos(\Phi - \phi_i) - \delta_{0i}.$

δ is the phase delay between elements in fractions of a wavelength. The driving-point currents I_{0i}, $i = 1, 2, 3, ..., N$, and relative phases δ_{0i} are arbitrary. The positions of the elements are fixed by their number. Thus

$$\phi_i = \frac{2\pi}{N}(i-1), \qquad i = 1, 2, 3, ..., N. \qquad (1.117)$$

Two special cases provide examples of the far-field patterns of typical circular arrays.

Consider an electrically small array with equal driving-point currents and all phase differences δ_{0i} set equal to zero. Thus

$$I_{0i} = I_0, \qquad \delta_{0i} = 0 \qquad \text{and} \qquad (\beta_0 a)^2 \ll 1. \qquad (1.118)$$

The far-zone electric field reduces to

$$E_\Theta^r(\Theta) \doteq j\frac{\zeta_0}{2\pi}\frac{e^{-j\beta_0 R_0}}{R_0}F_0(\Theta, \beta_0 h)NI_0. \qquad (1.119)$$

The field of (1.119) is identical to that of a single element. In this case the array factor is omnidirectional. Another case of interest is when N is large. Here the finite sum may be approximated by an integral. The Euler–Maclaurin sum formula provides the necessary representation:

$$\sum_{i=1}^{N} E_{\Theta i} = \int_1^N E_{\Theta i}\, di + \tfrac{1}{2}E_{\Theta 1} + \tfrac{1}{2}E_{\Theta N} + O\!\left(\frac{1}{N^2}\right). \qquad (1.120)$$

Since $E_{\Theta 1}$ and $E_{\Theta N}$ are of the order N^{-1}, equation (1.120) may be applied to (1.116) to give the following simplified form:

$$E_\Theta^r \doteq j\frac{\zeta_0}{2\pi}\frac{e^{-j\beta_0 R_0}}{R_0}F_0(\Theta, \beta_0 h)\left(\frac{N}{2\pi}\right)\int_{1/N \doteq 0}^{2\pi} I_{0i}\, e^{j\beta_0 a\sin\Theta\cos(\Phi - u)}\, du$$

$$+ \text{ terms of } O\!\left(\frac{1}{N}\right) \text{ and higher} \qquad (1.121)$$

where $u = \dfrac{2\pi}{N}i$ and $\delta_{0i} = 0.$ \qquad (1.122)

The integral appearing in (1.121) is the Bessel function of the first kind and order zero. Hence,

$$E_\Theta^r \doteq NF_0(\Theta, \beta_0 h)J_0(\beta_0 a \sin \Theta). \qquad (1.123)$$

Since $J_0(\beta_0 a \sin \Theta)$ has a maximum at $\Theta = 0$, it represents an endfire array. However, since the array factor is multiplied by the field factor $F_0(\Theta, \beta_0 h)$ which has a null at $\Theta = 0$, the resultant maximum of the total field is shifted. The field representations (1.120) and (1.123) show that a wide range of array factors are obtainable from a circular array.

1.11 Limitations of conventional array theory

The validity of the conventional analysis of arrays, which is outlined in this chapter beginning with section 1.5, depends on the substitution of the simple approximation

$$E_\Theta^r = \frac{j\zeta_0 I_{z0}(0)}{2\pi} \frac{e^{-j\beta_0 R_1}}{R_1} F_0(\Theta, \beta_0 h)A(\Theta, \Phi) \qquad (1.124)$$

for the rigorous expression

$$E_\Theta^r = \frac{j\omega\mu_0}{4\pi} \frac{e^{-j\beta_0 R_1}}{R_1} \sum_{i=1}^{N} e^{j\beta_0 S_i} \int_{-h}^{h} I_{zi}(z') e^{j\beta_0 z' \cos \Theta} \sin \Theta \, dz' \qquad (1.125)$$

where $S_i = R_1 - R_i$ and R_i is the distance from the centre of element i to the point of calculation. The first step in the derivation of (1.124) from (1.125) is the normalization of the currents in the form

$$I_{zi}(z') = I_{zi}(0)f_i(z') = I_{z1}(0)k_{1i} e^{-j\delta_{1i}} f_i(z') \qquad (1.126)$$

where k_{1i} and δ_{1i} are the normalized real amplitude and relative phase angle of the current at the centre of element i as referred to the current at the centre of the reference element 1. The next step is the definition of the field factor for element 1,

$$F_1(\Theta, \beta_0 h) = \int_{-h}^{h} f_1(z') e^{j\beta_0 z' \cos \Theta} \sin \Theta \, \beta_0 \, dz' \qquad (1.127)$$

and the array factor

$$A(\Theta, \Phi) = \sum_{i=1}^{N} k_{1i} e^{j(\beta_0 S_i - \delta_{1i})}. \qquad (1.128)$$

With (1.126)–(1.128), (1.124) can be expressed as follows;

$$E_\Theta^r = \frac{j\zeta_0 I_{z1}(0)}{2\pi} \frac{e^{-j\beta_0 R_1}}{R_1} \left[F_1(\Theta, \beta_0 h)A(\Theta, \Phi) + D_1(\Theta, \Phi; \beta_0 h) \right]$$

$$(1.129)$$

where

$$D_1(\Theta, \Phi; \beta_0 h)$$

$$= \sum_{i=2}^{N} k_{1i} e^{j(\beta_0 S_i - \delta_{1i})} \int_{-h}^{h} [f_i(z') - f_1(z')] e^{j\beta_0 z' \cos \Theta} \sin \Theta \beta_0 \, dz'. \tag{1.130}$$

Conventional array theory *always neglects* $D_1(\Theta, \Phi; \beta_0 h)$ as defined in (1.130). This is usually justified by statements such as 'Because of the identical configuration and orientation of *all* the elements, this contribution (viz. $F_1(\Theta, \beta_0 h)$) is the same for all other elements. These latter differences are accounted for by the array factor (1.128)'. Unfortunately, when applied to arrays of centre-driven antennas, this statement is correct only for circular arrays of identical, equally spaced, non-staggered elements driven by equal voltages with constant and progressive phase differences from element to element around the circle. For such an array the geometrical and electrical environments of all elements are indeed the same, so that $f_i(z) = f_1(z)$, $F_i(\Theta, \beta_0 h) = F_1(\Theta, \beta_0 h)$ for $i = 1, 2, ..., N$ and $D_1(\Theta, \Phi; \beta_0 h) = 0$. In all other arrays, $f_i(z)$ is a complex function of z that differs both in its magnitude and phase from $f_1(z)$ for all $|z| > 0$. It cannot be assumed without explicit verification that $D_1(\Theta, \Phi; \beta_0 h)$ is negligible.

A major purpose of this book is to develop a theory of arrays that does not neglect $D_1(\Theta, \Phi; \beta_0 h)$ in (1.129).

CHAPTER 2

AN APPROXIMATE ANALYSIS OF THE CYLINDRICAL ANTENNA

2.1 The sinusoidal current

The distribution of current along a thin centre-driven antenna of length $2h$ (or along a base-driven antenna of length h over an ideal ground plane) is assumed to have the sinusoidal form

$$I_z(z) = I_z(0)\frac{\sin \beta_0(h-|z|)}{\sin \beta_0 h} \qquad (2.1)$$

throughout chapter 1. Actually, this is the correct distribution along a section of lossless coaxial line of length h that is short-circuited at $z = 0$ and terminated at $z = h$ in an infinite impedance. This is illustrated in Fig. 2.1a where the infinite impedance is obtained by means of an additional short-circuited quarter-wave section of coaxial line. In this case the current is entirely reactive, the electromagnetic field is completely confined within the coaxial shield in the form of axial standing waves and there is no radiation. When the ideal 'open' end at $z = h$ is replaced by an actual one as shown in Fig. 2.1b, the distribution of current and charges are changed in a manner that resembles a crowding of the entire pattern toward the open end. In addition to a large reactive component, the current now also includes a very small resistive part. The associated electromagnetic field is still primarily a standing wave within the coaxial sleeve, but it does extend outside especially near the open end and there is some radiation. From the point of view of the transmission line the differences between currents and fields for Figs. 2.1a and 2.1b are interpreted as end-effects. If the outside shield is removed as in Fig. 2.1c these 'end-effects' extend all the way to the generator and the distributions of current and charges are significantly changed over the entire length. The resistive component is now comparable in magnitude to the reactive part and the associated electromagnetic field includes a large radiation field that extends to infinity in the form of outward travelling waves. It is, of course, not at all surprising that the distributions of current along the conductors of radius a and length h are not the same in the three quite different situations represented in Figs. 2.1a, b, c. The boundary conditions are not alike except at $r = a, 0 \leqslant z \leqslant h$, where the tangential electric field vanishes.

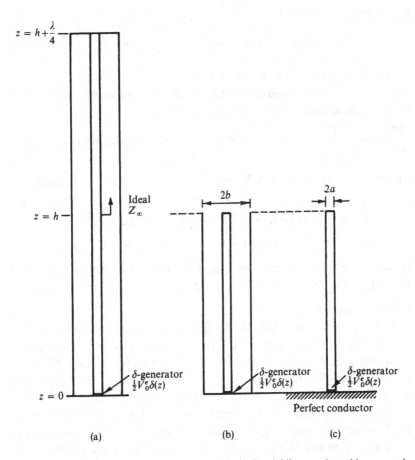

Fig. 2.1.   (a) Coaxial line terminated in Z_∞ at $z = h$. (b) Coaxial line terminated in open end. (c) Base-driven monopole over perfectly conducting ground screen.

2.2   The equation for the current

Instead of simply assuming a convenient current along the antenna in Fig. 2.1c, a more scientific and more difficult procedure is actually to *determine* the distribution of current by setting up and solving the appropriate integral equation. This is readily obtained from the boundary condition $E_z(z) = 0$ on the surface $\rho = a$ of the perfectly conducting antenna. With (1.5c) and (1.6) the vector potential is seen to satisfy the equation

$$\left(\frac{d^2}{dz^2}+\beta_0^2\right)A_z(z) = 0 \tag{2.2}$$

which has the general solution

$$A_z(z) = \frac{-j}{c}(C_1 \cos \beta_0 z + C_2 \sin \beta_0|z|) \tag{2.3}$$

if the symmetry conditions, $I_z(-z) = I_z(z)$, $A_z(-z) = A_z(z)$ are imposed. C_1 and C_2 are arbitrary constants of integration. With (1.8a) the integral equation for the current is

$$\frac{4\pi}{\mu_0}A_z(z) = \int_{-h}^{h} I_z(z')\frac{e^{-j\beta_0 R}}{R}dz' = \frac{-j4\pi}{\zeta_0}[C_1 \cos \beta_0 z + \tfrac{1}{2}V_0^e \sin \beta_0|z|] \tag{2.4}$$

where V_0^e is the EMF of the delta-function generator, $\zeta_0 = \sqrt{\mu_0/\varepsilon_0} \doteq 120\pi$ ohms, and $R = \sqrt{(z-z')^2+a^2}$. The value $C_2 = \tfrac{1}{2}V_0^e$ is obtained from (2.3) with (1.9) in the form

$$\phi(z) = j(\omega/\beta_0^2)(\partial A_z/\partial z) = -\beta_0 C_1 \sin \beta_0 z + \beta_0 C_2 \cos \beta_0 z.$$

By definition the driving voltage of the delta-function generator is $\lim_{z\to 0}[\phi(z)-\phi(-z)] = V_0^e$. The second constant C_1 must be evaluated from the condition $I_z(\pm h) = 0$.

Although it is not difficult to derive the integral equation (2.4), the problem of solving it for the current is very complicated. It has been carried out approximately in a variety of ways,† but the solutions so obtained are a very cumbersome series of terms that have proved invaluable in the accurate determination of the self-impedance but are not very useful for calculating electromagnetic fields or for determining the mutual interaction of the currents in coupled antennas. What is needed is an approximate solution that is both sufficiently simple to be useful in the evaluation of the electromagnetic field and sufficiently accurate to provide quantitatively acceptable values not only of the details of the field but of the driving-point impedance. (In anticipation, it is well to note that a generalization of the method in order to make it useful in the solution of the simultaneous integral equations that occur in the analysis of arrays is also going to be required.)

The procedure to be followed in obtaining a useful approximate solution of (2.4) is straightforward and simple. It involves the

† Many of these are described or referred to in [1], chapter 2.

replacement of the integral equation (2.4) by an approximately equivalent algebraic equation. In order to accomplish this a careful study must be made of the integral in (2.4).

2.3 Properties of integrals

The integrand in (2.4) consists of two parts: (1) the current $I_z(z)$ which is to be determined and about which nothing is known except that it vanishes at the ends $z = \pm h$, is continuous through the generator at $z = 0$, and satisfies the symmetry condition $I_z(-z) = I_z(z)$; (2) the kernel

$$K(z, z') = \frac{e^{-j\beta_0 R}}{R}, \qquad R = \sqrt{(z-z')^2 + a^2} \qquad (2.5)$$

which may be separated into its real and imaginary parts,

$$K_R(z, z') = \frac{\cos \beta_0 R}{R}; \qquad K_I(z, z') = -\frac{\sin \beta_0 R}{R}. \qquad (2.6)$$

The dimensionless quantities $K_R(z, z')/\beta_0$ and $K_I(z, z')/\beta_0$ are shown graphically in Fig. 2.2 as functions of $\beta_0|z-z'|$. A comparison in the lower figure shows that their behaviours are quite different. $K_R(z, z')/\beta_0$ has a sharp high peak precisely at $z' = z$; its magnitude $1/\beta_0 a$ is very large compared with 1 since it has been postulated that $\beta_0 a \ll 1$. On the other hand, $K_I(z, z')/\beta_0$ varies only slowly with $\beta_0|z-z'|$ and never exceeds the value 1. It is seen in the upper part of Fig. 2.2 that $\sin \beta_0 R/\beta_0 R$ is very well approximated by $\cos(\beta_0 R/2)$ in the range $0 \leqslant \beta_0|z-z'| \leqslant \pi$. Moreover, the value of $\cos(\beta_0 R/2)$ is hardly affected if the small quantity $k_0 a$ is neglected and $\beta_0 R$ is approximated by $\beta_0|z-z'|$.

These facts suggest the following approximations for the two parts of the integral in (2.4):

$$J_R(h, z) = \int_{-h}^{h} I_z(z') \frac{\cos \beta_0 R}{R} \, dz' = \Psi_1(z)I(z) \doteq \Psi_1 I(z) \qquad (2.7)$$

$$J_I(h, z) = -\int_{-h}^{h} I_z(z') \frac{\sin \beta_0 R}{R} \, dz' = -\beta_0 \int_{-h}^{h} I_z(z') \cos \tfrac{1}{2}\beta_0(z-z') \, dz'. \qquad (2.8)$$

The reasoning behind the approximation in (2.7) is simple. Since the kernel is quite small except at and very near $z' = z$, where it rises to a very large value, it is clear that the current near $z' = z$ is primarily significant in determining the value of the integral at z. In other words, the integral is approximately proportional to $I(z)$.

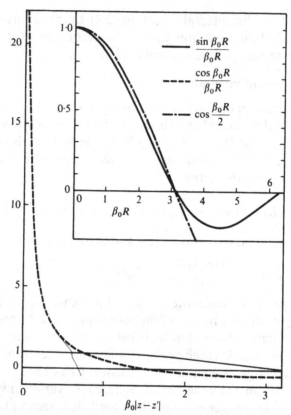

Fig. 2.2.   The functions $\dfrac{\sin \beta_0 R}{\beta_0 R}$, $\dfrac{\cos \beta_0 R}{\beta_0 R}$, and $\cos \dfrac{\beta_0 R}{2}$.

The proportionality constant Ψ_1 is best determined where $I_z(z)$ is a maximum.

The integral in (2.8) may be transformed as follows:

$$J_I(h, z) = -\beta_0 \int_{-h}^{h} I_z(z') \cos \tfrac{1}{2}\beta_0(z-z') \, dz'$$

$$= -\beta_0 \int_{0}^{h} I_z(z')[\cos \tfrac{1}{2}\beta_0(z-z') + \cos \tfrac{1}{2}\beta_0(z+z')] \, dz'$$

$$= -2\beta_0 \cos \tfrac{1}{2}\beta_0 z \int_{0}^{h} I_z(z') \cos \tfrac{1}{2}\beta_0 z' \, dz'.$$

It follows that for antennas that do not greatly exceed $\beta_0 h = \pi$ in electrical half-length, specifically, $\beta_0 h \leqslant 5\pi/4$,

$$J_I(h, z) \doteq J_I(h, 0) \cos \tfrac{1}{2}\beta_0 z; \quad J_I(h, 0) = -2\beta_0 \int_{0}^{h} I_z(z') \cos \tfrac{1}{2}\beta_0 z' \, dz'.$$

$$(2.9)$$

A further refinement in the approximation (2.7) is suggested by the fact that, while the integral on the left becomes quite small at the ends of the antenna where $z = \pm h$, the right-hand side vanishes identically at these points since $I_z(\pm h) = 0$. Evidently a better approximation than (2.7) is the following:

$$4\pi\mu_0^{-1}[A_z(z) - A_z(h)] = \int_{-h}^{h} I_z(z')[K_R(z, z') - K_R(h, z')] \, dz' \doteq \Psi_2 I_z(z)$$
(2.10)

where the left side is simply the vector potential difference between the point (a, z) and the end (a, h) of the antenna; Ψ_2 is a new constant.

2.4 Rearranged equation for the current

In order to make use of (2.10), the integral equation (2.4) may be modified by subtracting $4\pi\mu_0^{-1}A_z(h)$ from both sides. The result is

$$4\pi\mu_0^{-1}[A_z(z) - A_z(h)] = \int_{-h}^{h} I_z(z')K_d(z, z') \, dz'$$

$$= \frac{-j4\pi}{\zeta_0}[C_1 \cos \beta_0 z + \tfrac{1}{2}V_0^e \sin \beta_0|z| + U]$$
(2.11)

where
$$U = \frac{-j\zeta_0}{4\pi}\int_{-h}^{h} I_z(z')K(h, z') \, dz'$$
(2.12)

and the difference kernel is
$$K_d(z, z') = K(z, z') - K(h, z').$$
(2.13)

The constant C_1 can now be expressed in terms of U and V_0^e by setting $z = h$. Since the left side of (2.11) then vanishes, the right side can be solved for C_1 to give

$$C_1 = -\frac{\tfrac{1}{2}V_0^e \sin \beta_0 h + U}{\cos \beta_0 h}.$$
(2.14)

If this value of C_1 is substituted into (2.11) the following equation is obtained:

$$\int_{-h}^{h} I_z(z')K_d(z, z') \, dz'$$

$$= \frac{j4\pi}{\zeta_0 \cos \beta_0 h}[\tfrac{1}{2}V_0^e \sin \beta_0(h - |z|) + U(\cos \beta_0 z - \cos \beta_0 h)].$$
(2.15)

The integral equation (2.15) with (2.12) is a rearrangement of the original equation (2.4). No approximations are involved.

2.5   Reduction of integral equation to algebraic equation

The next and most important step is to make use of the information contained in (2.9) and (2.10) in order to reduce (2.15) to an approximately equivalent algebraic equation. The procedure is simple and straightforward. With (2.9) and (2.10) it is clear that the integral in (2.15) may be approximated as follows:

$$\int_{-h}^{h} I_z(z')K_d(z, z')\,dz' \doteq I_z(z)\Psi_2 + jJ_1(h, 0)(\cos \tfrac{1}{2}\beta_0 z - \cos \tfrac{1}{2}\beta_0 h).$$

(2.16)

If this is substituted in (2.15), the resulting equation can be solved explicitly for $I_z(z)$. It is seen to have the following zero-order form:

$$I_z(z) \doteq [I_z(z)]_0 = I_V[\sin \beta_0(h - |z|) + T_U(\cos \beta_0 z - \cos \beta_0 h)$$

$$+ T_D(\cos \tfrac{1}{2}\beta_0 z - \cos \tfrac{1}{2}\beta_0 h)]$$

(2.17)

where I_V, T_U and T_D are complex coefficients.

This is a very significant result. It shows that an approximation of the current consists of three terms of which each represents a different distribution. One of the terms is the simple sinusoid. As for the completely shielded transmission line, the sinusoidal component of the current is maintained directly by the generator; it does not include the components that are induced by coupling between different parts of the antenna. The currents induced by the interaction between charges moving in the more or less widely separated sections of the antenna appear in two parts. One of these, the shifted cosine, is maintained by that part of the interaction that is equivalent to a constant field acting in phase at all points along the antenna. The other part, the shifted cosine with half-angle arguments, is the correction that takes account of the phase lag introduced by the retarded instead of instantaneous interaction.

Thus, the new three-term approximation augments the conventionally assumed sinusoidal distribution with components represented by a shifted cosine and a shifted cosine with half-angle arguments, each with a complex coefficient.

It is quite possible to evaluate the coefficients Ψ_2, $J_1(h, 0)$ and U that are involved in I_V, T_U, and T_D—obtained when (2.16) is substituted in (2.15). However, it is preferable to use the arguments and approximations introduced up to this point merely to determine the *form* of the distribution of current. The three new coefficients, I_V, T_U and T_D, may be evaluated directly if (2.17) is substituted in the integral equation (2.15) and the principles involved in (2.9) and (2.10) are invoked.

The substitution of (2.17) in the integral in (2.15) involves the following parts obtained from the real part $K_{dR}(z, z')$ of the difference kernel $K_d(z, z')$ defined in (2.13):

$$\int_{-h}^{h} \sin \beta_0(h-|z'|)K_{dR}(z, z')\,dz' \doteq \Psi_{dR} \sin \beta_0(h-|z|) \qquad (2.18a)$$

$$\int_{-h}^{h} [\cos \beta_0 z' - \cos \beta_0 h]K_{dR}(z, z')\,dz' \doteq \Psi_{dUR}(\cos \beta_0 z - \cos \beta_0 h) \qquad (2.18b)$$

$$\int_{-h}^{h} [\cos \tfrac{1}{2}\beta_0 z' - \cos \tfrac{1}{2}\beta_0 h]K_{dR}(z, z')\,dz' \doteq \Psi_{dDR}(\cos \tfrac{1}{2}\beta_0 z - \cos \tfrac{1}{2}\beta_0 h). \qquad (2.18c)$$

These expressions follow from (2.10). In order to enhance the accuracy, each part of the current is separately treated and supplied with its own coefficient. The evaluation of these coefficients is considered below.

The integrals obtained with the imaginary part $K_{dI}(z, z')$ of the difference kernel are easily appproximated by the application of (2.9). Thus,

$$\int_{-h}^{h} \sin \beta_0(h-|z'|)K_{dI}(z, z')\,dz' \doteq \Psi_{dI}(\cos \tfrac{1}{2}\beta_0 z - \cos \tfrac{1}{2}\beta_0 h) \qquad (2.19a)$$

$$\int_{-h}^{h} (\cos \beta_0 z' - \cos \beta_0 h)K_{dI}(z, z')\,dz' \doteq \Psi_{dUI}(\cos \tfrac{1}{2}\beta_0 z - \cos \tfrac{1}{2}\beta_0 h) \qquad (2.19b)$$

$$\int_{-h}^{h} (\cos \tfrac{1}{2}\beta_0 z' - \cos \tfrac{1}{2}\beta_0 h)K_{dI}(z, z')\,dz' \doteq \Psi_{dDI}(\cos \tfrac{1}{2}\beta_0 z - \cos \tfrac{1}{2}\beta_0 h). \qquad (2.19c)$$

The three constants Ψ_{dI}, Ψ_{dUI} and Ψ_{dDI} are evaluated later. Finally, if the distribution (2.17) is substituted in (2.12), the result is

$$U = \frac{-j\zeta_0 I_V}{4\pi}[\Psi_V(h) + T_U \Psi_U(h) + T_D \Psi_D(h)] \qquad (2.20)$$

where
$$\Psi_V(h) = \int_{-h}^{h} \sin \beta_0(h-|z'|)K(h, z')\,dz' \qquad (2.21a)$$

$$\Psi_U(h) = \int_{-h}^{h} (\cos \beta_0 z' - \cos \beta_0 h)K(h, z')\,dz' \qquad (2.21b)$$

$$\Psi_D(h) = \int_{-h}^{h} (\cos \tfrac{1}{2}\beta_0 z' - \cos \tfrac{1}{2}\beta_0 h)K(h, z')\,dz'. \qquad (2.21c)$$

With (2.18a–c) and (2.19a–c) the integral on the left is reduced to a mere sum of terms with suitable coefficients. And the integral equation as a whole has been replaced by an algebraic equation that involves the three distributions $\sin \beta_0(h-|z|)$, $\cos \beta_0 z - \cos \beta_0 h$, and $\cos \frac{1}{2}\beta_0 z - \cos \frac{1}{2}\beta_0 h$. It is

$$\left(I_V \Psi_{dR} - \frac{j2\pi V_0^e}{\zeta_0 \cos \beta_0 h} \right) \sin \beta_0(h-|z|)$$

$$+ \left(I_V T_U \Psi_{dUR} - \frac{j4\pi U}{\zeta_0 \cos \beta_0 h} \right)(\cos \beta_0 z - \cos \beta_0 h)$$

$$+ I_V(j\Psi_{dI} + j\Psi_{dUI} T_U + \Psi_{dD} T_D)(\cos \tfrac{1}{2}\beta_0 z - \cos \tfrac{1}{2}\beta_0 h) = 0 \qquad (2.22)$$

where $\Psi_{dD} = \Psi_{dDR} + j\Psi_{dDI}$.

2.6 Evaluation of coefficients

The algebraic equation (2.22) is satisfied for all values of z when the coefficient of each of the three distributions vanishes. This step yields three equations for the determination of the coefficients I_V, T_U and T_D in (2.17). They are:

$$I_V = \frac{j2\pi V_0^e}{\zeta_0 \Psi_{dR} \cos \beta_0 h} \qquad (2.23a)$$

$$T_U[\Psi_{dUR} \cos \beta_0 h - \Psi_U(h)] - T_D \Psi_D(h) = \Psi_V(h) \qquad (2.23b)$$

$$T_U \Psi_{dUI} - j T_D \Psi_{dD} = - \Psi_{dI}. \qquad (2.23c)$$

The last two equations are easily solved for T_U and T_D. The results are

$$T_U = Q^{-1}[\Psi_V(h)\Psi_{dD} - j\Psi_D(h)\Psi_{dI}] \qquad (2.24a)$$

$$T_D = -jQ^{-1}\{\Psi_{dI}[\Psi_{dUR} \cos \beta_0 h - \Psi_U(h)] + \Psi_V(h)\Psi_{dUI}\} \quad (2.24b)$$

$$Q = \Psi_{dD}[\Psi_{dUR} \cos \beta_0 h - \Psi_U(h)] + j\Psi_D(h)\Psi_{dUI}. \qquad (2.25)$$

The several Ψ functions in (2.24)–(2.25) are defined with (2.18a–c) and (2.19a–c) at the value of z that gives the maximum of the current distribution function. Since, in the range of interest, $k_0 h < 3\pi/2$, the maximum of $\sin \beta_0(h-|z|)$ is at $z = 0$ when $\beta_0 h \leqslant \pi/2$ but at $z = h-\lambda/4$ when $\beta_0 h \geqslant \pi/2$, whereas the maxima of $(\cos \beta_0 z - \cos \beta_0 h)$ and $(\cos \frac{1}{2}\beta_0 z - \cos \frac{1}{2}\beta_0 h)$ are at $z = 0$, the following definitions are appropriate:

$$\Psi_{dR} = \Psi_{dR}(z_m), \quad \begin{cases} z_m = 0, & \beta_0 h \leqslant \pi/2 \\ z_m = h-\lambda/4, & \beta_0 h > \pi/2 \end{cases} \qquad (2.26)$$

$$\Psi_{dR}(z) = \csc \beta_0(h-|z|) \int_{-h}^{h} \sin \beta_0(h-|z'|)[K_R(z, z') - K_R(h, z')] \, dz'$$

(2.27)

$$\Psi_{dUR} = (1 - \cos \beta_0 h)^{-1} \int_{-h}^{h} (\cos \beta_0 z' - \cos \beta_0 h)$$
$$\times [K_R(0, z') - K_R(h, z')] \, dz'$$

(2.28)

$$\Psi_{dD} = (1 - \cos \tfrac{1}{2}\beta_0 h)^{-1} \int_{-h}^{h} (\cos \tfrac{1}{2}\beta_0 z' - \cos \tfrac{1}{2}\beta_0 h)$$
$$\times [K(0, z') - K(h, z')] \, dz'$$

(2.29)

$$\Psi_{dI} = (1 - \cos \tfrac{1}{2}\beta_0 h)^{-1} \int_{-h}^{h} \sin \beta_0(h-|z'|)$$
$$\times [K_I(0, z') - K_I(h, z')] \, dz'$$

(2.30)

$$\Psi_{dUI} = (1 - \cos \tfrac{1}{2}\beta_0 h)^{-1} \int_{-h}^{h} (\cos \beta_0 z' - \cos \beta_0 h)$$
$$\times [K_I(0, z') - K_I(h, z')] \, dz'.$$

(2.31)

These integrals may be evaluated directly by high-speed computer or reduced to the tabulated generalized sine and cosine integral functions given by (1.56a, b) and the exponential integral

$$E_a(h, z) = \int_{-h}^{h} \frac{e^{-j\beta_0 R_1}}{R_1} \, dz' = \int_{0}^{h} \left[\frac{e^{-j\beta_0 R_1}}{R_1} + \frac{e^{-j\beta_0 R_2}}{R_2} \right] dz'. \quad (2.32)$$

2.7 The approximate current and admittance

The final approximate expression for the current in an isolated cylindrical antenna for which $\beta_0 h < 3\pi/2$ and $\beta_0 a \ll 1$ is

$$I_z(z) = \frac{j2\pi V_0^e}{\zeta_0 \Psi_{dR} \cos \beta_0 h}[\sin \beta_0(h-|z|) + T_U(\cos \beta_0 z - \cos \beta_0 h)$$
$$+ T_D(\cos \tfrac{1}{2}\beta_0 z - \cos \tfrac{1}{2}\beta_0 h)].$$

(2.33)

The associated driving-point admittance is

$$Y_0 = \frac{j2\pi}{\zeta_0 \Psi_{dR} \cos \beta_0 h}[\sin \beta_0 h + T_U(1 - \cos \beta_0 h) + T_D(1 - \cos \tfrac{1}{2}\beta_0 h)].$$

(2.34)

When $\beta_0 h = \pi/2$, these formulas become indeterminate so that alternative expressions are needed. They are readily obtained from (2.33) and (2.34) by a simple rearrangement. The following new forms are useful near or at $\beta_0 h = \pi/2$ where both numerators and denominators in (2.33) and (2.34) are very small or zero:

$$I_z(z) = \frac{-j2\pi V_0^e}{\zeta_0 \Psi_{dR}}[(\sin \beta_0|z| - \sin \beta_0 h) + T'_U(\cos \beta_0 z - \cos \beta_0 h)$$

$$- T'_D(\cos \tfrac{1}{2}\beta_0 z - \cos \tfrac{1}{2}\beta_0 h)] \tag{2.35}$$

$$Y_0 = \frac{j2\pi}{\zeta_0 \Psi_{dR}}[\sin \beta_0 h - T'_U(1 - \cos \beta_0 h) + T'_D(1 - \cos \tfrac{1}{2}\beta_0 h)] \tag{2.36}$$

where
$$T'_U = -\frac{T_U + \sin \beta_0 h}{\cos \beta_0 h}, \qquad T'_D = \frac{T_D}{\cos \beta_0 h}. \tag{2.37}$$

T'_U and T'_D are both finite when $\beta_0 h = \pi/2$.

When the antenna is electrically short, so that $\beta_0 h < 1$, the trigonometric functions can be expanded in series and the leading terms retained. The current is then given by

$$I_z(z) = \frac{j2\pi V_0^e}{\zeta_0 \Psi_{dR}}\left[\beta_0 h\left(1 - \frac{|z|}{h}\right) + \tfrac{1}{2}\beta_0^2 h^2 T\left(1 - \frac{z^2}{h^2}\right)\right]. \tag{2.38}$$

This distribution includes triangular and parabolic components. The admittance is

$$Y_0 = \frac{j2\pi}{\zeta_0 \Psi_{dR}}[\beta_0 h + \tfrac{1}{2}\beta_0^2 h^2 T] \tag{2.39}$$

where $T = T_U + T_D/4$.

2.8 Numerical examples; comparison with experiment

Numerical computations have been made for typical antennas for which extensive measurements are available. For these antennas $a/\lambda = 7.022 \times 10^{-3}$. The parameters for the two critical lengths, $\beta_0 h = \pi/2$ with $\Omega = 2 \ln 2h/a = 8.54$ and $\beta_0 h = \pi$ with $\Omega = 9.92$ are listed below:

$$\beta_0 h = \frac{\pi}{2}, \qquad \Psi_{dR} = 6.218, \qquad T'_U = 3.085 + j3.581,$$

$$T'_D = 1.061 + j0.025 \tag{2.40a}$$

$$\beta_0 h = \pi, \qquad \Psi_{dR} = 5.737, \qquad T_U = -0.117 + j0.114,$$

$$T_D = -0.106 + j0.108. \tag{2.40b}$$

The corresponding currents in amperes per volt, admittances in mhos and impedances in ohms are as follows:

For $\beta_0 h = \pi/2$,

$$\frac{I_z(z)}{V_0^e} = \{9{\cdot}597 \cos \beta_0 z - 0{\cdot}067 \cos \tfrac{1}{2}\beta_0 z + 0{\cdot}047$$

$$-j[2{\cdot}680(\sin \beta_0|z| - 1) + 8{\cdot}269 \cos \beta_0 z - 2{\cdot}843 \cos \tfrac{1}{2}\beta_0 z$$

$$+ 2{\cdot}010]\} \times 10^{-3} \tag{2.41a}$$

$$Y_0 = (9{\cdot}577 - j4{\cdot}756) \times 10^{-3}, \quad Z_0 = 83{\cdot}76 + j41{\cdot}60. \tag{2.41b}$$

For $\beta_0 h = \pi$,

$$\frac{I_z(z)}{V_0^e} = \{0{\cdot}331(\cos \beta_0 z + 1) + 0{\cdot}314 \cos \tfrac{1}{2}\beta_0 z$$

$$-j[2{\cdot}905 \sin \beta_0|z| - 0{\cdot}340(\cos \beta_0 z + 1)$$

$$- 0{\cdot}308 \cos \tfrac{1}{2}\beta_0 z]\} \times 10^{-3} \tag{2.42a}$$

$$Y_0 = (0{\cdot}976 + j0{\cdot}988) \times 10^{-3}, \quad Z_0 = 506{\cdot}0 - j512{\cdot}2. \tag{2.42b}$$

Note that when a sinusoidal distribution of current is assumed the corresponding impedances are for $\beta_0 h = \pi/2$, $Z_0 = 73{\cdot}1 + j42{\cdot}5$ (see eq. 1.59); and for $\beta_0 h = \pi$, $Z_0 = \infty$.

Graphs of $I_z(z)/V_0^e = [I_z''(z) + jI_z'(z)]/V_0^e$ are in Figs. 2.3 and 2.4 for $\beta_0 h = \pi/2$ and π together with measured values. The approximate theoretical curves are seen to agree very well with measured values not only for $\beta_0 h = \pi/2$, but also for $\beta_0 h = \pi$.

As can be seen from Figs. 2.3 and 2.4, and especially from the latter, the theoretical currents at the driving point and, hence, the admittances differ somewhat from the measured values. In order to achieve a more accurate admittance, higher-order terms are required in the expressions for the current. Simple trigonometric functions cannot take adequate account of the rapid change in the current near the driving point when the antenna is not near resonance. Since higher-order terms are necessarily complicated, their introduction would defeat the primary purpose of this formulation, namely, to maintain a reasonably simple representation. Fortunately, there is a useful alternative. Since the only large error in the current occurs in the quadrature component of the current very near the driving point, it is possible to introduce a lumped susceptance B_c across the terminals which will correct the driving-point current and the susceptance while leaving the otherwise well-approximated current unchanged. Actually, since the use of a lumped corrective network is required in any case to take account of the local geometry of the junction between the feeding line and the antenna if quantitative accuracy is desired, the addition of B_c to the susceptance B_T of the terminal-zone network is no significant complication. In practice, it may be convenient to measure the apparent driving-point susceptance at $\beta_0 h = \pi$ and use the difference between this and the approximate theoretical value as the total lumped susceptance $B_T + B_c$ to be used with all theoretical values based on the approximate theory for any given ratio of a/λ.

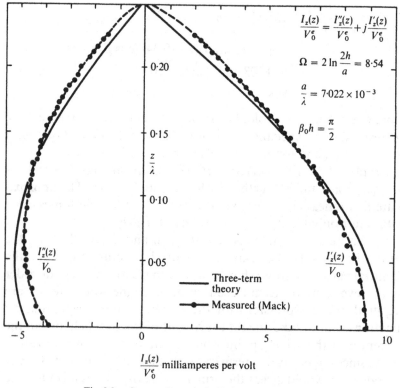

Fig. 2.3.   Current in upper half of half-wave dipole.

2.9   The radiation field

The electric field in the radiation zone of an antenna with a distribution of current $I_z(z)$ is given by the integral

$$\mathbf{E}^r = \hat{\Theta}E_\Theta^r; \quad E_\Theta^r = \frac{j\omega\mu}{4\pi}\sin\Theta\,\frac{e^{-j\beta_0 R_0}}{R_0}\int_{-h}^{h} I_z(z')\,e^{j\beta_0 z'\cos\Theta}\,dz'.$$

$$(2.43)$$

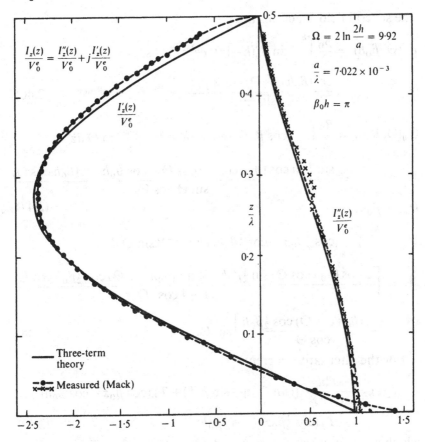

$$\frac{I_z(z)}{V_0^e} = \frac{I_z''(z)}{V_0^e} + j\frac{I_z'(z)}{V_0^e}$$

$$\Omega = 2\ln\frac{2h}{a} = 9.92$$

$$\frac{a}{\lambda} = 7.022 \times 10^{-3}$$

$$\beta_0 h = \pi$$

$$\frac{I_z'(z)}{V_0^e}$$

$$\frac{z}{\lambda}$$

$$\frac{I_z''(z)}{V_0^e}$$

——— Three-term theory

●━●━● Measured (Mack)
✕━✕━✕

$I_z(z)/V_0^e$ milliamperes per volt

Fig. 2.4.   Current in upper half of full-wave dipole.

The far field maintained by the distribution (2.33) is obtained when

$$I_z(z) = \frac{j2\pi V_0^e}{\zeta_0 \Psi_{dR} \cos\beta_0 h}[\sin\beta_0(h-|z|) + T_U(\cos\beta_0 z - \cos\beta_0 h)$$
$$+ T_D(\cos\tfrac{1}{2}\beta_0 z - \cos\tfrac{1}{2}\beta_0 h)] \tag{2.44}$$

is substituted in (2.43). The result may be expressed as follows:

$$E_\Theta^r = \frac{-V_0^e}{\Psi_{dR}}\frac{e^{-j\beta_0 R_0}}{R_0}f(\Theta, \beta_0 h) \tag{2.45a}$$

where

$$f(\Theta, \beta_0 h) = [F_m(\Theta, \beta_0 h) + T_U G_m(\Theta, \beta_0 h) + T_D D_m(\Theta, \beta_0 h)]\sec\beta_0 h. \tag{2.45b}$$

The several field functions are

$$F_m(\Theta, \beta_0 h) = \frac{\beta_0}{2} \int_{-h}^{h} \sin \beta_0(h - |z'|) \, e^{j\beta_0 z' \cos \Theta} \sin \Theta \, dz'$$

$$= \frac{\cos(\beta_0 h \cos \Theta) - \cos \beta_0 h}{\sin \Theta} \tag{2.46}$$

$$G_m(\Theta, \beta_0 h) = \frac{\beta_0}{2} \int_{-h}^{h} (\cos \beta_0 z' - \cos \beta_0 h) \, e^{j\beta_0 z' \cos \Theta} \sin \Theta \, dz'$$

$$= \frac{\sin \beta_0 h \cos (\beta_0 h \cos \Theta) \cos \Theta - \cos \beta_0 h \sin (\beta_0 h \cos \Theta)}{\sin \Theta \cos \Theta}$$

$$D_m(\Theta, \beta_0 h) \tag{2.47}$$

$$= \frac{\beta_0}{2} \int_{-h}^{h} (\cos \tfrac{1}{2}\beta_0 z' - \cos \tfrac{1}{2}\beta_0 h) \, e^{j\beta_0 z' \cos \Theta} \sin \Theta \, dz'$$

$$= \left[\frac{2 \cos (\beta_0 h \cos \Theta) \sin \tfrac{1}{2}\beta_0 h - 4 \sin (\beta_0 h \cos \Theta) \cos \tfrac{1}{2}\beta_0 h \cos \Theta}{1 - 4 \cos^2 \Theta} \right.$$

$$\left. - \frac{\sin (\beta_0 h \cos \Theta) \cos \tfrac{1}{2}\beta_0 h}{\cos \Theta} \right] \sin \Theta. \tag{2.48}$$

For the alternative current

$$I_z(z) = \frac{-j2\pi V_0^e}{\zeta_0 \Psi_{dR}} [(\sin \beta_0|z| - \sin \beta_0 h) + T_U'(\cos \beta_0 z - \cos \beta_0 h)$$

$$- T_D'(\cos \tfrac{1}{2}\beta_0 z - \cos \tfrac{1}{2}\beta_0 h)] \tag{2.49}$$

which is useful when $\beta_0 h$ is at and near $\pi/2$, the far field is

$$E_\Theta^r = \frac{V_0^e}{\Psi_{dR}} \frac{e^{-j\beta_0 R_0}}{R_0} f'(\Theta, \beta_0 h) \tag{2.50a}$$

where

$$f'(\Theta, \beta_0 h) = H_m(\Theta, \beta_0 h) + T_U' G_m(\Theta, \beta_0 h) - T_D' D_m(\Theta, \beta_0 h). \tag{2.50b}$$

The new field function is

$$H_m(\Theta, \beta_0 h) = \frac{\beta_0}{2} \int_{-h}^{h} (\sin \beta_0|z'| - \sin \beta_0 h) \, e^{j\beta_0 z' \cos \Theta} \sin \Theta \, dz'$$

$$= \frac{[1 - \cos \beta_0 h \cos (\beta_0 h \cos \Theta)] \cos \Theta - \sin \beta_0 h \sin (\beta_0 h \cos \Theta)}{\sin \Theta \cos \Theta}.$$

$$\tag{2.51}$$

$G_m(\Theta, \beta_0 h)$ and $D_m(\Theta, \beta_0 h)$ are as in (2.47) and (2.48).

For the specific cases considered above, the coefficients are:

For $\beta_0 h = \pi/2$,

$$H_m\left(\Theta, \frac{\pi}{2}\right) = \frac{\cos\Theta - \sin\left(\dfrac{\pi}{2}\cos\Theta\right)}{\sin\Theta\cos\Theta} \tag{2.52a}$$

$$G_m\left(\Theta, \frac{\pi}{2}\right) = \frac{\cos\left(\dfrac{\pi}{2}\cos\Theta\right)}{\sin\Theta} = F_m\left(\Theta, \frac{\pi}{2}\right) \tag{2.52b}$$

$$D_m\left(\Theta, \frac{\pi}{2}\right) = \frac{\sqrt{2}}{2}\left\{\frac{2\cos\left(\dfrac{\pi}{2}\cos\Theta\right) - 4\sin\left(\dfrac{\pi}{2}\cos\Theta\right)\cos\Theta}{1 - 4\cos^2\Theta}\right.$$

$$\left. - \frac{\sin\left(\dfrac{\pi}{2}\cos\Theta\right)}{\cos\Theta}\right\}\sin\Theta. \tag{2.52c}$$

For $\beta_0 h = \pi$,

$$F_m(\Theta, \pi) = \frac{\cos(\pi\cos\Theta) + 1}{\sin\Theta} \tag{2.53a}$$

$$G_m(\Theta, \pi) = \frac{\sin(\pi\cos\Theta)}{\sin\Theta\cos\Theta} \tag{2.53b}$$

$$D_m(\Theta, \pi) = \frac{2\cos(\pi\cos\Theta)\sin\Theta}{1 - 4\cos^2\Theta}. \tag{2.53c}$$

In the formulas (2.45a) and (2.50a), the field is referred to the driving voltage V_0^e. It can be referred to the current $I_z(0)$ at the driving point with the simple substitution of $I_z(0)/Y_0$ for V_0^e where Y_0 is the admittance given by (2.34) or (2.36). The field in (2.45a) is then expressed as follows:

$$E_\Theta^r = \frac{j\zeta_0 I_z(0)}{2\pi}\frac{e^{-j\beta_0 R_0}}{R_0}f_I(\Theta, \beta_0 h) \tag{2.54a}$$

where

$$f_I(\Theta, \beta_0 h) = \frac{F_m(\Theta, \beta_0 h) + T_U G_m(\Theta, \beta_0 h) + T_D D_m(\Theta, \beta_0 h)}{\sin\beta_0 h + T_U(1 - \cos\beta_0 h) + T_D(1 - \cos\frac{1}{2}\beta_0 h)}. \tag{2.54b}$$

The alternative form (2.50a) becomes

$$E_\Theta^r = \frac{j\zeta_0 I_z(0)}{2\pi}\frac{e^{-j\beta_0 R_0}}{R_0}f_I'(\Theta, \beta_0 h) \tag{2.55a}$$

where

$$f'_I(\Theta, \beta_0 h) = -\frac{H_m(\Theta, \beta_0 h) + T'_U G_m(\Theta, \beta_0 h) - T'_D D_m(\Theta, \beta_0 h)}{\sin \beta_0 h - T'_U(1 - \cos \beta_0 h) + T'_D(1 - \cos \frac{1}{2}\beta_0 h)}.$$

(2.55b)

As a numerical illustration, the three functions $F_m(\Theta, \pi)$, $G_m(\Theta, \pi)$ and $D_m(\Theta, \pi)$ are shown graphically in Fig. 2.5 for a full-wave antenna. They all have nulls at $\Theta = 0$ and maxima at $\Theta = 90°$. However, $G_m(\Theta, \pi)$ and $D_m(\Theta, \pi)$ have relatively much greater values at small values of Θ than $F_m(\Theta, \pi)$.

If use is made of the numerical values of T_U and T_D given in (2.40b) for a cylindrical antenna with $a/\lambda = 7.022 \times 10^{-3}$ (for which the distribution of current is given in (2.42a) and the admittance and impedance in (2.42b)) the field factor

$$f(\Theta, \pi) = f_r(\Theta, \pi) + jf_i(\Theta, \pi)$$

(2.56)

may be evaluated. The real and imaginary parts $f_r(\Theta, \pi)$ and $f_i(\Theta, \pi)$ are shown in Fig. 2.5 together with the magnitude $|f(\Theta, \pi)|$. This last is seen to resemble $F_m(\Theta, \pi)$ quite closely except for $\Theta < 30°$ where it is significantly greater. However, since the field is quite small when $\Theta < 30°$, no serious error is made in calculating the far field if the following approximations are used when $\beta_0 h \leq \pi$:

$$G_m(\Theta, \beta_0 h) \doteq \frac{G_m\left(\frac{\pi}{2}, \beta_0 h\right)}{F_m\left(\frac{\pi}{2}, \beta_0 h\right)} F_m(\Theta, \beta_0 h)$$

$$= \left(\frac{\sin \beta_0 h - \beta_0 h \cos \beta_0 h}{1 - \cos \beta_0 h}\right) F_m(\Theta, \beta_0 h) \quad (2.57a)$$

$$D_m(\Theta, \beta_0 h) \doteq \frac{D_m\left(\frac{\pi}{2}, \beta_0 h\right)}{F_m\left(\frac{\pi}{2}, \beta_0 h\right)} F_m(\Theta, \beta_0 h)$$

$$= \left(\frac{2\sin \frac{1}{2}\beta_0 h - \beta_0 h \cos \frac{1}{2}\beta_0 h}{1 - \cos \beta_0 h}\right) F_m(\Theta, \beta_0 h) \quad (2.57b)$$

$$H_m(\Theta, \beta_0 h) \doteq \frac{H_m\left(\frac{\pi}{2}, \beta_0 h\right)}{F_m\left(\frac{\pi}{2}, \beta_0 h\right)} F_m(\Theta, \beta_0 h)$$

$$= \left(\frac{1 - \cos \beta_0 h - \beta_0 h \sin \beta_0 h}{1 - \cos \beta_0 h} \right) F_m(\Theta, \beta_0 h). \quad (2.57c)$$

These approximations are equivalent to the use of the far-field distribution associated with a sinusoidal current, but normalizing this to the value at $\Theta = \pi/2$ obtained from the three-term form of the current.

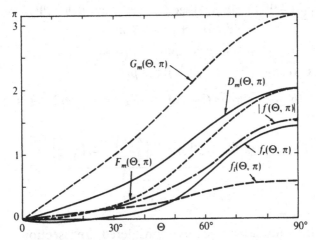

Fig. 2.5. The functions $F_m(\Theta, \pi)$, $G_m(\Theta, \pi)$, $D_m(\Theta, \pi)$ and the field components $f(\Theta, \pi) = f_r(\Theta, \pi) + j f_i(\Theta, \pi)$ when $a/\lambda = 7 \cdot 022 \times 10^{-3}$.

2.10   An approximate two-term theory

When interest is entirely in the far-field and in the driving-point impedances, the difference between the distribution functions $F_{0z} = \cos \beta_0 z - \cos \beta_0 h$ and $H_{0z} = \cos \frac{1}{2}\beta_0 z - \cos \frac{1}{2}\beta_0 h$ is small and the formulation may be simplified further by consolidating the two terms. If F_{0z} is substituted everywhere for H_{0z}, the current is well approximated as follows when $\beta_0 h \leqslant 5\pi/4$:

$$I_z(z) = \frac{j2\pi V_0^e}{\zeta_0 \Psi_{dR} \cos \beta_0 h} [\sin \beta_0(h - |z|) + T(\cos \beta_0 z - \cos \beta_0 h)] \quad (2.58)$$

or, in the form useful near $\beta_0 h = \pi/2$,

$$I_z(z) = \frac{-j2\pi V_0^e}{\zeta_0 \Psi_{dR}} [\sin \beta_0 |z| - \sin \beta_0 h + T'(\cos \beta_0 z - \cos \beta_0 h)] \quad (2.59)$$

where T and T' are obtained by forming $T_U + T_D$ and $T_U' - T_D'$ but with the substitution $\Psi_{dD} = \Psi_{dU}$, $\Psi_D(h) = \Psi_U(h)$. The function T

is simply

$$T = \frac{\Psi_V(h) - j\Psi_{dI}\cos\beta_0 h}{\Psi_{dU}\cos\beta_0 h - \Psi_U(h)}.$$

(2.60)

T' is given by

$$T' = -\frac{T + \sin\beta_0 h}{\cos\beta_0 h}$$

$$= \frac{[\Psi_V(h) - \Psi_U(h)\sin\beta_0 h]\sec\beta_0 h + \Psi_{dU}\sin\beta_0 h - j\Psi_{dI}}{\Psi_U(h) - \Psi_{dU}\cos\beta_0 h}$$

(2.61a)

$$= \frac{[\Psi_{dU} + E_a(h, h)]\sin\beta_0 h - j\Psi_{dI} - S_a(h, h)}{C_a(h, h) - [\Psi_{dU} + E_a(h, h)]\cos\beta_0 h}.$$

(2.61b)

Since $\Psi_U(h) = \Psi_V(h) = C_a(h, h)$ when $\beta_0 h = \pi/2$, this reduces simply to

$$T' = \frac{\Psi_{dU} - j\Psi_{dI} - S_a\left(\frac{\lambda}{4}, \frac{\lambda}{4}\right) + E_a\left(\frac{\lambda}{4}, \frac{\lambda}{4}\right)}{C_a\left(\frac{\lambda}{4}, \frac{\lambda}{4}\right)}$$

(2.62)

when $\beta_0 h = \pi/2$.

For the numerical cases considered in section 2.8 for $a/\lambda = 7\cdot022 \times 10^{-3}$, the results for the two-term theory are:

$$\beta_0 h = \pi/2, \quad \Psi_{dR} = 6\cdot218, \quad T' = 2\cdot65 + j3\cdot79; \\ Y_0 = (10\cdot17 - j4\cdot43) \times 10^{-3}\text{mhos}$$

(2.63)

$$\beta_0 h = \pi, \quad \Psi_{dR} = 5\cdot737, \quad T = -0\cdot172 + j0\cdot175; \\ Y_0 = (1\cdot021 + j1\cdot000) \times 10^{-3}\text{mhos}.$$

(2.64)

These are seen to be in good agreement with the values obtained with the more accurate three-term theory. A more extensive list of numerical values of Ψ_{dR}, T, T' and $Y_0 = G_0 + jB_0$ is in Table 1 of appendix 1.

As with the three-term theory, the quadrature component of the current near the driving point is not adequately represented by simple trigonometric functions so that the same expedient previously described must be used in order to obtain quantitative agreement with measured values of the susceptance. The lumped value of B_c to be used with the two-term theory differs only slightly from that for the three-term theory. For $a/\lambda = 7\cdot022 \times 10^{-3}$, it is $B_c = 0\cdot72$ millimhos. This value must be added to the two-term susceptance $B_0 + B_T$ (where B_T is the terminal-zone correction for a particular transmission-line-antenna junction) in order to obtain the measurable apparent susceptance $B_{sa} = B_0 + 0.72 + B_T$. It is seen in Fig. 2.6 that $B_0 + 0.72$ is in good agreement with the King–Middleton second-order values of B_0 and the apparent measured values corrected for the terminal-zone effects, $B_{sa} - B_T$.

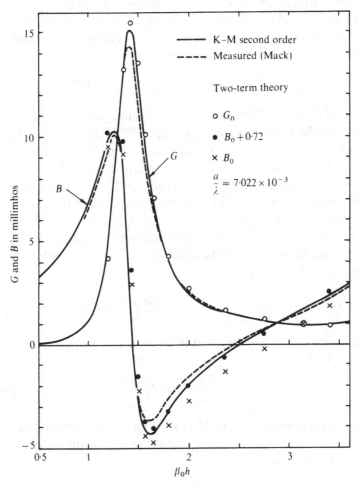

Fig. 2.6. King–Middleton second-order admittance $Y = G + jB$. Two-term zero-order admittance $Y_0 = G_0 + jB_0$, and measured.

2.11 The receiving antenna

The general method of analysis introduced in this chapter as a means of analyzing the centre-driven cylindrical antenna can be extended readily to the centre-loaded receiving antenna in an incident plane-wave field. For the purposes of this book†—which includes the properties of receiving arrays—it is sufficient to treat only the simple case of normal incidence with the electric vector parallel to the z-axis which is the axis of the antenna. The antenna is, therefore, in the plane wave front of the incident wave which may be assumed to travel in the positive x direction. That is $E_z^{inc}(x) = E_z^{inc} e^{-j\beta_0 x}$ where E_z^{inc} is the constant amplitude. The boundary condition that requires the total tangential electric field to vanish on the surface of the antenna gives

$$\left(\frac{d^2}{dz^2}+\beta_0^2\right)A_z(z) = -\frac{j\beta_0^2}{\omega}E_z^{inc} = -\beta_0^2 A_z^{inc} \qquad (2.65)$$

instead of (2.2). In (2.65), $A_z(z)$ is the vector potential due to the currents in the receiving antenna itself. A_z^{inc} is the constant amplitude of the vector potential maintained on the surface of the antenna by the distant transmitter. Note that $E_z^{inc} = -j\omega A_z^{inc}$. Since the axis of the antenna lies in the wave front, even symmetry obtains with respect to z for both the current and the associated vector potential so that $A_z(-z) = A_z(z)$, $I_z(-z) = I_z(z)$. It follows that, on an unloaded receiving or scattering antenna, the vector potential on the surface of the antenna due to the currents in the antenna satisfies the equations

$$4\pi\mu_0^{-1}A_z(z) = \int_{-h}^{h} I(z')\frac{e^{-j\beta_0 R}}{R}dz' = \frac{-j4\pi}{\zeta_0}[C_1 \cos \beta_0 z + U^{inc}] \quad (2.66)$$

where C_1 is an arbitrary constant to be evaluated from the condition $I_z(h) = 0$ and

$$U^{inc} = \frac{E_z^{inc}}{\beta_0} = -\frac{j\omega A_z^{inc}}{\beta_0}. \qquad (2.67)$$

This integral equation is like (2.4) with an added constant term on the right and with $V_0^e = 0$. If the same rearrangement is carried out as for (2.11), the result is

$$4\pi\mu_0^{-1}[A_z(z) - A_z(h)] = \int_{-h}^{h} I_z(z')K_d(z, z')\,dz'$$

$$= \frac{-j4\pi}{\zeta_0}[C_1 \cos \beta_0 z + U + U^{inc}] \qquad (2.68)$$

where U, as defined in (2.12), is proportional to the vector potential

† A more detailed analysis of the receiving antenna is in [1], chapter 4.

at $z = h$, $\rho = a$ due to the currents in the antenna; U^{inc}, as defined in (2.67), is proportional to the vector potential maintained on the surface of the antenna by the distant transmitter. The sum $U + U^{\text{inc}}$ is proportional to the total vector potential on the surface of the antenna.

Since the integral equation (2.68) is just like (2.11) with $V_0^e = 0$, it follows that the rearranged equation corresponding to (2.15) is

$$\int_{-h}^{h} I_z(z')K_d(z, z')\,dz' = \frac{j4\pi(U + U^{\text{inc}})}{\zeta_0 \cos \beta_0 h}(\cos \beta_0 z - \cos \beta_0 h). \quad (2.69)$$

The approximate solution of this equation is like (2.33) with $V_0^e = 0$. It is

$$I_z(z) = \frac{j4\pi U^{\text{inc}}}{\zeta_0 Q}[\Psi_{dD}(\cos \beta_0 z - \cos \beta_0 h)$$

$$- j\Psi_{dUI}(\cos \tfrac{1}{2}\beta_0 z - \cos \tfrac{1}{2}\beta_0 h)] \quad (2.70)$$

where Q is defined in (2.25), Ψ_{dD} in (2.29) and Ψ_{dUI} in (2.31). This solution for the unloaded receiving antenna corresponds to the three-term form (2.33) for the driven antenna. Corresponding to the simpler two-term approximation (2.58) for the driven antenna, is the expression

$$I_z(z) \doteq \frac{j4\pi U^{\text{inc}}(\Psi_{dD} - j\Psi_{dUI})}{Q}(\cos \beta_0 z - \cos \beta_0 h) = U^{\text{inc}}u(z) \quad (2.71)$$

for the unloaded receiving antenna. Note, in particular, that the shifted cosine distribution $(\cos \beta_0 z - \cos \beta_0 h)$ is characteristic of the unloaded receiving antenna when its axis is parallel to the electric vector in an incident plane-wave field. When the axis of the antenna is oriented at an arbitrary angle with respect to the incident E-vector, the distribution of current is much more complicated. In particular, if the antenna does not lie in the plane wave front (surface of constant phase) of the incident field, the current and the vector potential have components with both even and odd symmetries with respect to z.

If the antenna is cut at $z = 0$ and a load Z_L is connected in series with the halves of the antenna, the current in the antenna is readily obtained. Note first that, if a generator with voltage V_0^e is connected across the terminals at $z = 0$ instead of the load, the resulting current in the antenna is

$$I_z(z) = V_0^e v(z) + U^{\text{inc}}u(z) \quad (2.72)$$

where $v(z)$ is $I_z(z)/V_0^e$ as obtained from (2.44) and $u(z)$ is $I_z(z)/U^{inc}$ as obtained from (2.71). If now V_0^e is replaced by the negative of the voltage drop across a load Z_L that is connected across the terminals of the antenna, that is,

$$V_0^e = -I_z(0)Z_L, \qquad (2.73)$$

V_0^e is readily eliminated between (2.73) and (2.72). With Z_0, the driving-point impedance of the same antenna when driven, the result can be expressed as follows:

$$I_z(z) = U^{inc}\left[u(z) - v(z)u(0)\frac{Z_L Z_0}{Z_L + Z_0}\right]. \qquad (2.74)$$

This is the current at any point z in the centre-loaded receiving antenna. The current in the load at $z = 0$ is simply

$$I_z(0) = U^{inc}u(0)\frac{Z_0}{Z_0 + Z_L} \qquad (2.75)$$

since $v(0) = 1/Z_0$. When $Z_L = 0$, this gives $I_z(0) = U^{inc}u(0)$. The voltage drop across the load is

$$I_z(0)Z_L = U^{inc}u(0)\frac{Z_0 Z_L}{Z_0 + Z_L}. \qquad (2.76)$$

When $Z_L \to \infty$, this is the open-circuit voltage across the terminals at $z = 0$. That is

$$V(Z_L \to \infty) = \lim_{Z_L \to \infty} I_z(0)Z_L = U^{inc}u(0)Z_0 = [I_z(0)Z_0]_{Z_L=0}. \qquad (2.77)$$

It is now clear that with (2.67) and (2.75) the current in the load is given by

$$I_z(0) = \frac{V(Z_L \to \infty)}{Z_0 + Z_L} = \frac{E_z^{inc}u(0)Z_0}{\beta_0(Z_0 + Z_L)}. \qquad (2.78)$$

This is the current in a simple series circuit that consists of a generator with EMF equal to the open-circuit voltage across the terminals of the receiving antenna in series with the impedance of the antenna and the impedance of the load. The same conclusion is readily obtained by the application of Thévenin's theorem.

The quantity

$$u(0)Z_0/\beta_0 = 2h_e\left(\frac{\pi}{2}\right) \qquad (2.79)$$

which occurs in (2.78) and is dimensionally a length, is known as the complex effective length of the receiving antenna with actual length $2h$. With (2.79), the current in the load is

$$I_z(0) = \frac{2h_e\left(\frac{\pi}{2}\right)E_z^{\text{inc}}}{Z_0 + Z_L}. \qquad (2.80)$$

Note that (2.78), (2.79) and (2.80) apply only when the axis of the receiving antenna is parallel to the incident electric vector and, therefore, also perpendicular to the direction of propagation of the incident wave. Similar but more general expressions that involve the orientation of the antenna relative to the incident wave and the direction of the electric vector in the plane wave front are available in the literature.†

† See, for example, [1], chapter 4, section 4.

CHAPTER 3

THE TWO-ELEMENT ARRAY

3.1 The method of symmetrical components

An array is a configuration of two or more antennas so arranged that the superposition of the electromagnetic fields maintained at distant points by the currents in the individual elements yields a resultant field that fulfils certain desirable directional properties. Since the individual elements in an array are quite close together—the distance between adjacent elements is usually a half-wavelength or less—the currents in them necessarily interact. It follows that the distributions of both the amplitude and the phase of the current along each element depend not only on the length, radius, and driving voltage of that element, but also on the distributions in amplitude and phase of the currents along all elements in the array. Since these currents are the primary unknowns from which the radiation field is computed, a highly complicated situation arises if they are to be determined analytically, and not arbitrarily assumed known, as in conventional array theory.

In order to introduce the properties of arrays in a simple and direct manner, it is advantageous to study first the two-element array in some detail. The integral equation (2.11) for the current in a single isolated antenna is readily generalized to apply to the two identical parallel and non-staggered elements shown in Fig. 3.1. It is merely necessary to add to the vector potential on the surface of each element the contributions by the current in the other element. Thus, for element 1, the vector potential difference is

$$4\pi\mu_0^{-1}[A_{1z}(z) - A_{1z}(h)]$$

$$= \int_{-h}^{h} [I_{1z}(z')K_{11d}(z, z') + I_{2z}(z')K_{12d}(z, z')]\, dz'$$

$$= \frac{j4\pi}{\zeta_0 \cos \beta_0 h}[\tfrac{1}{2}V_{10} \sin \beta_0(h - |z|) + U_1(\cos \beta_0 z - \cos \beta_0 h)]. \quad (3.1)$$

Similarly, for element 2:

$$4\pi\mu_0^{-1}[A_{2z}(z) - A_{2z}(h)]$$

$$= \int_{-h}^{h} [I_{1z}(z')K_{21d}(z, z') + I_{2z}(z')K_{22d}(z, z')] \, dz'$$

$$= \frac{j4\pi}{\zeta_0 \cos \beta_0 h} [\tfrac{1}{2}V_{20} \sin \beta_0(h - |z|) + U_2(\cos \beta_0 z - \cos \beta_0 h)]. \quad (3.2)$$

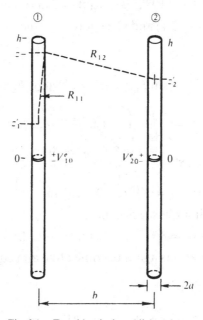

Fig. 3.1. Two identical parallel antennas.

In these expressions

$$K_{11d}(z, z') = \frac{e^{-j\beta_0 R_{11}}}{R_{11}} - \frac{e^{-j\beta_0 R_{11h}}}{R_{11h}} = K_{11}(z, z') - K_{11}(h, z') \quad (3.3a)$$

$$K_{12d}(z, z') = \frac{e^{-j\beta_0 R_{12}}}{R_{12}} - \frac{e^{-j\beta_0 R_{12h}}}{R_{12h}} = K_{12}(z, z') - K_{12}(h, z') \quad (3.3b)$$

with

$$R_{11} = \sqrt{(z - z')^2 + a^2}, \qquad R_{11h} = \sqrt{(h - z')^2 + a^2} \quad (3.4a)$$

$$R_{12} = \sqrt{(z - z')^2 + b^2}, \qquad R_{12h} = \sqrt{(h - z')^2 + b^2}. \quad (3.4b)$$

$K_{22d}(z, z')$ and $K_{21d}(z, z')$ are obtained from the above formulas when 1 is substituted for 2 and 2 for 1 in the subscripts.

The two simultaneous integral equations (3.1) and (3.2) can be reduced to a *single* equation in two special cases. These are (a) the so-called *zero-phase sequence* when the two driving voltages are identical so that the two currents are the same and (b) the *first-phase sequence* when the two driving voltages and the resulting two currents are equal in magnitude but 180° out of phase. Specifically, for the zero-phase sequence,

$$V_{10} = V_{20} = V^{(0)}, \qquad I_{1z}(z) = I_{2z}(z) = I_z^{(0)}(z), \qquad (3.5)$$

so that the equations (3.1) and (3.2) both become

$$\int_{-h}^{h} I_z^{(0)}(z')K_d^{(0)}(z, z')\, dz'$$
$$= \frac{j4\pi}{\zeta_0 \cos \beta_0 h}[\tfrac{1}{2}V^{(0)} \sin \beta_0(h-|z|) + U^{(0)}(\cos \beta_0 z - \cos \beta_0 h)] \quad (3.6)$$

where
$$U^{(0)} = \frac{-j\zeta_0}{4\pi} \int_{-h}^{h} I_z(z')K^{(0)}(h, z')\, dz' \qquad (3.7)$$

and
$$\left.\begin{aligned} K^{(0)}(z, z') &= \frac{e^{-j\beta_0 R_{11}}}{R_{11}} + \frac{e^{-j\beta_0 R_{12}}}{R_{12}}, \\ K_d^{(0)}(z, z') &= K^{(0)}(z, z') - K^{(0)}(h, z'). \end{aligned}\right\} \qquad (3.8)$$

Similarly, for the first phase sequence,

$$V_{10} = -V_{20} = V^{(1)}, \qquad I_{1z}(z) = -I_{2z}(z) = I_z^{(1)}(z) \qquad (3.9)$$

so that the two equations again become alike and equal to

$$\int_{-h}^{h} I_z^{(1)}(z')K_d^{(1)}(z, z')\, dz'$$
$$= \frac{j4\pi}{\zeta_0 \cos \beta_0 h}[\tfrac{1}{2}V^{(1)} \sin \beta_0(h-|z|) + U^{(1)}(\cos \beta_0 z - \cos \beta_0 h)] \quad (3.10)$$

where
$$U^{(1)} = \frac{-j\zeta_0}{4\pi} \int_{-h}^{h} I_z(z')K^{(1)}(h, z')\, dz' \qquad (3.11)$$

and
$$\left.\begin{aligned} K^{(1)}(z, z') &= \frac{e^{-j\beta_0 R_{11}}}{R_{11}} - \frac{e^{-j\beta_0 R_{12}}}{R_{12}}, \\ K_d^{(1)}(z, z') &= K^{(1)}(z, z') - K^{(1)}(h, z'). \end{aligned}\right\} \qquad (3.12)$$

Note that the two phase sequences differ only in the sign in $K^{(0)}(z, z')$ and $K^{(1)}(z, z')$.

If (3.6) can be solved for the zero-phase-sequence current $I_z^{(0)}(z)$ and (3.10) for the first-phase-sequence current $I_z^{(1)}(z)$, the currents $I_{1z}(z)$ and $I_{2z}(z)$ maintained by the arbitrary voltages V_{10} and V_{20} can be obtained simply by superposition. This follows directly if $V^{(0)}$ and $V^{(1)}$ are so chosen that

$$V^{(0)} = \tfrac{1}{2}[V_{10} + V_{20}], \qquad V^{(1)} = \tfrac{1}{2}[V_{10} - V_{20}]. \qquad (3.13)$$

In this case,

$$V_{10} = V^{(0)} + V^{(1)}, \qquad V_{20} = V^{(0)} - V^{(1)} \qquad (3.14)$$

so that

$$I_{1z}(z) = I_z^{(0)}(z) + I_z^{(1)}(z), \qquad I_{2z}(z) = I_z^{(0)}(z) - I_z^{(1)}(z). \qquad (3.15)$$

3.2   Properties of the integrals

The two integral equations (3.6) and (3.10) for the phase-sequence currents are formally exactly like the equation (2.15) for the isolated antenna. They differ only in the kernels of the integrals on the left and in the definitions (3.7) and (3.11) of the functions U. Each of these is now the algebraic sum of two terms that are identical except that the radius a appears in the first term, the distance b between the elements in the second term. In order to determine the effect of this difference on the current it is convenient to consider first the two extreme cases when the elements are very close together and when they are very far apart.

The two elements may be considered close together when $\beta_0 b < 1$ and $b < h$. In this case, since b satisfies substantially the same conditions as a, the behaviour of the integrals that contain b corresponds closely to that of the integrals that contain a. These are discussed in the preceding chapter. When the antennas are so far apart ($\beta_0 b \gg 1$, $b \gg h$) that $(\beta_0\sqrt{b^2 + h^2} - \beta_0 b) \ll 1$, the contribution to the difference kernels $K_d^{(0)}(z, z')$ and $K_d^{(1)}(z, z')$ by the term $K_{12d}(z, z') = (e^{-j\beta_0 R_{12}}/R_{12}) - (e^{-j\beta_0 R_{12h}}/R_{12h})$ is very small since R_{12} and R_{12h} differ only slightly. In this case, the principal part of the interaction between the currents in the two antennas is included in the function $U^{(0)}$ or $U^{(1)}$ and the integrals on the left in (3.6) and (3.10) are only slightly different from the corresponding integral for the single antenna. The interaction between the currents in the two antennas is approximately as if each maintained along the other a vector potential that is uniform in amplitude and phase. Accordingly, the current induced in each element by the other is distributed in a first approximation as a shifted cosine. This conclusion follows directly from the fact that the component of current associated with the constant part of the vector potential along the surface of the isolated antenna is distributed in this manner.

When the separation of the two elements is such that $\beta_0 b > 1$ but not so great that $\beta_0 \sqrt{b^2 + h^2}$ differs negligibly from $\beta_0 b$, the vector potentials maintained by the currents on the one antenna at points along the surface of the other differ significantly from one another in phase due to retarded action. The induced currents should then have two components, the one distributed approximately as the shifted cosine with half-angle arguments, the other as the shifted cosine.

In order to verify the correctness of these conclusions the difference integral

$$S_b(h, z) - S_b(h, h) = \int_{-h}^{h} \sin \beta_0 |z'| K_d(z, z') \, dz' \qquad (3.16)$$

has been evaluated for $\beta_0 h = \pi$ over a range of values of $\beta_0 b$ extending from 0·04 to 4·5. The real and imaginary parts are shown in Fig. 3.2 together with the three trigonometric functions, $\sin \beta_0 z$, $(\cos \beta_0 z + 1)$ and $\cos \frac{1}{2} \beta_0 z$, to which the sine, shifted cosine and shifted cosine with half-angle arguments reduce when $\beta_0 h = \pi$. For convenience in the graphical comparison, $-(\cos \beta_0 z + 1)$ and $-\cos \frac{1}{2} \beta_0 z$ are shown. It is evident from Fig. 3.2 that the real part of the difference integral approximates $\sin \beta_0 z$ when $\beta_0 b < 1$,

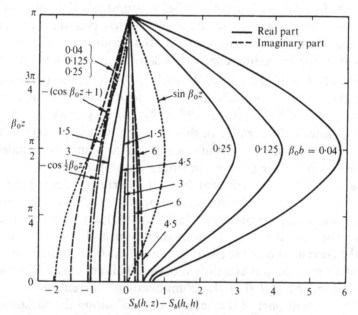

Fig. 3.2. The functions $S_b(h, z) - S_b(h, h)$ compared with three trigonometric functions.

$1 + \cos \beta_0 z$ when $\beta_0 b \geqslant 1$. On the other hand, the imaginary part resembles the shifted cosine with half-angle arguments, in this case $\cos \frac{1}{2}\beta_0 z$, for all values of $\beta_0 b$.

As a consequence of these observations, the following approximate representation of the integrals in (3.6) and (3.10) is indicated:

For $\beta_0 b < 1$,

$$\int_{-h}^{h} I_z(z') \left(\frac{\cos \beta_0 R_{12}}{R_{12}} - \frac{\cos \beta_0 R_{12h}}{R_{12h}} \right) dz' = \Psi_{12}(z)I_z(z) \doteq \Psi_{12}I_z(z)$$

$$(3.17a)$$

where Ψ_{12} is a constant.

For $\beta_0 b \geqslant 1$,

$$\int_{-h}^{h} I_z(z') \left(\frac{\cos \beta_0 R_{12}}{R_{12}} - \frac{\cos \beta_0 R_{12h}}{R_{12h}} \right) dz' \sim \cos \beta_0 z - \cos \beta_0 h. \quad (3.17b)$$

For all values of $\beta_0 b$

$$\int_{-h}^{h} I_z(z') \left(\frac{\sin \beta_0 R_{12}}{R_{12}} - \frac{\sin \beta_0 R_{12h}}{R_{12h}} \right) dz' \sim \cos \tfrac{1}{2}\beta_0 z - \cos \tfrac{1}{2}\beta_0 h.$$

$$(3.17c)$$

3.3 Reduction of integral equations for phase sequences to algebraic equations

The relations (3.17a, b, c), combined with the results of chapter 2, indicate that the current in each of the two coupled elements in both phase sequences must have leading terms that are well approximated by the following zero-order, three-term formula:

$$I_z^{(m)}(z) \doteq [I_z^{(m)}(z)]_0 = I_V^{(m)}[\sin \beta_0(h - |z|) + T_U^{(m)}(\cos \beta_0 z - \cos \beta_0 h)$$

$$+ T_D^{(m)}(\cos \tfrac{1}{2}\beta_0 z - \cos \tfrac{1}{2}\beta_0 h)] \quad (3.18)$$

where $m = 0$ or 1 and $I_V^{(m)}$, $T_U^{(m)}$ and $T_D^{(m)}$ are complex coefficients that must be determined.

The substitution of (3.18) into the integral in (3.6) and (3.10) involves the following parts obtained from the real part $K_{dR}^{(m)}(z, z')$ of the difference kernel $K_d^{(m)}(z, z')$ defined in (3.8) and (3.12) for $m = 0$ and 1:

$$\beta_0 b < 1, \qquad \int_{-h}^{h} \sin \beta_0(h - |z'|)K_{dR}^{(m)}(z, z') dz' \doteq \Psi_{dR}^{(m)} \sin \beta_0(h - |z|)$$

$$(3.19a)$$

$$\beta_0 b \geqslant 1, \quad \int_{-h}^{h} \sin \beta_0(h-|z'|)K_{dR}^{(m)}(z,z')\,dz' \doteq \Psi_{dR}\sin\beta_0(h-|z|)$$
$$+\Psi_{d\Sigma R}^{(m)}(\cos\beta_0 z-\cos\beta_0 h). \tag{3.19b}$$

For all values of $\beta_0 b$,†

$$\int_{-h}^{h}(\cos\beta_0 z'-\cos\beta_0 h)K_{dR}^{(m)}(z,z')\,dz' \doteq \Psi_{dUR}^{(m)}(\cos\beta_0 z-\cos\beta_0 h) \tag{3.20a}$$

$$\int_{-h}^{h}(\cos\tfrac{1}{2}\beta_0 z'-\cos\tfrac{1}{2}\beta_0 h)K_{dR}^{(m)}(z,z')\,dz' \doteq \Psi_{dDR}^{(m)}(\cos\tfrac{1}{2}\beta_0 z-\cos\tfrac{1}{2}\beta_0 h). \tag{3.20b}$$

The corresponding integrals obtained with the imaginary part, $K_{dI}^{(m)}(z,z')$ of the difference kernel $K_d^{(m)}(z,z')$, are valid for all values of $\beta_0 b$. They are:

$$\int_{-h}^{h}\sin\beta_0(h-|z'|)K_{dI}^{(m)}(z,z')dz' \doteq \Psi_{dI}^{(m)}(\cos\tfrac{1}{2}\beta_0 z-\cos\tfrac{1}{2}\beta_0 h) \tag{3.21}$$

$$\int_{-h}^{h}(\cos\beta_0 z'-\cos\beta_0 h)K_{dI}^{(m)}(z,z')\,dz' \doteq \Psi_{dUI}^{(m)}(\cos\tfrac{1}{2}\beta_0 z-\cos\tfrac{1}{2}\beta_0 h) \tag{3.22}$$

$$\int_{-h}^{h}(\cos\tfrac{1}{2}\beta_0 z'-\cos\tfrac{1}{2}\beta_0 h)K_{dI}^{(m)}(z,z')dz' \doteq \Psi_{dDI}^{(m)}(\cos\tfrac{1}{2}\beta_0 z-\cos\tfrac{1}{2}\beta_0 h). \tag{3.23}$$

The several Ψ functions introduced in the above expressions are defined as follows:

$$\beta_0 b < 1: \begin{cases} \Psi_{dR}^{(m)} = \Psi_{dR}^{(m)}(z_m); \begin{cases} z_m = 0, & \beta_0 h \leqslant \pi/2 \\ z_m = h-\lambda/4, & \beta_0 h > \pi/2 \end{cases} \tag{3.24a} \\ \Psi_{dR}^{(m)}(z) = \csc\beta_0(h-|z|)\int_{-h}^{h}\sin\beta_0(h-|z'|)K_{dR}^{(m)}(z,z')\,dz' \end{cases}$$

$$\tag{3.24b}$$

† Strictly according to (3.17b) the integral in (3.20b) should be treated separately with different behaviours when $\beta_0 b < 1$ and $\beta_0 b \geqslant 1$. However, since the distributions $\cos\beta_0 z-\cos\beta_0 h$ and $\cos(\beta_0 z/2)-\cos(\beta_0 h/2)$ are quite similar when $\beta_0 h \leqslant 5\pi/4$, and since considerable complication is avoided by not making this separation, the relation (3.20b) is used for both real and imaginary parts of the kernel and for all spacings.

$$\beta_0 b \geqslant 1: \begin{cases} \Psi_{dR} \text{ is defined in (2.27) and (2.28)} & \text{(3.25a)} \\ \Psi_{d\Sigma R}^{(m)} = (-1)^m (1 - \cos \beta_0 h)^{-1} \int_{-h}^{h} \sin \beta_0 (h - |z'|) \\ \qquad \times \left[\dfrac{\cos \beta_0 R_{12}}{R_{12}} - \dfrac{\cos \beta_0 R_{12h}}{R_{12h}} \right] dz'. & \text{(3.25b)} \end{cases}$$

The following apply for all values of $\beta_0 b$:

$$\Psi_{dI}^{(m)} = (1 - \cos \tfrac{1}{2}\beta_0 h)^{-1} \int_{-h}^{h} \sin \beta_0 (h - |z'|) K_{dI}^{(m)}(0, z') \, dz' \qquad (3.26)$$

$$\Psi_{dUR}^{(m)} = (1 - \cos \beta_0 h)^{-1} \int_{-h}^{h} (\cos \beta_0 z' - \cos \beta_0 h) K_{dR}^{(m)}(0, z') \, dz' \quad (3.27a)$$

$$\Psi_{dUI}^{(m)} = (1 - \cos \tfrac{1}{2}\beta_0 h)^{-1} \int_{-h}^{h} (\cos \beta_0 z' - \cos \beta_0 h) K_{dI}^{(m)}(0, z') \, dz' \quad (3.27b)$$

$$\Psi_{dD}^{(m)} = (1 - \cos \tfrac{1}{2}\beta_0 h)^{-1} \int_{-h}^{h} (\cos \tfrac{1}{2} \beta_0 z' - \cos \tfrac{1}{2}\beta_0 h) K_{d}^{(m)}(0, z') \, dz'. \tag{3.28}$$

For each pair of real and imaginary parts, the notation $\Psi_d = \Psi_{dR} + \Psi_{dI}$ will be used.

When (3.18) is substituted in $U^{(m)}$ as defined in (3.7) and (3.11), the notation of (2.20)–(2.21c) applies in the form

$$U^{(m)} = \frac{-j\zeta_0}{4\pi} I_V^{(m)} [\Psi_V^{(m)}(h) + T_U^{(m)} \Psi_U^{(m)}(h) + T_D^{(m)} \Psi_D^{(m)}(h)] \qquad (3.29)$$

where $\quad \Psi_V^{(m)}(h) = \displaystyle\int_{-h}^{h} \sin \beta_0 (h - |z'|) K^{(m)}(h, z') \, dz' \qquad (3.30)$

$$\Psi_U^{(m)}(h) = \int_{-h}^{h} (\cos \beta_0 z' - \cos \beta_0 h) K^{(m)}(h, z') \, dz' \qquad (3.31)$$

$$\Psi_D^{(m)}(h) = \int_{-h}^{h} (\cos \tfrac{1}{2}\beta_0 z' - \cos \tfrac{1}{2}\beta_0 h) K^{(m)}(h, z') \, dz' \qquad (3.32)$$

with $m = 0, 1$.

If the approximate formulas for the several parts of the integrals when $\beta_0 b < 1$ are substituted in (3.6) and (3.11), an algebraic equation is obtained that is just like (2.22) for the single antenna but with superscripts m on I, T, Ψ, V and U. It follows that (2.23a), (2.24), (2.25) and (2.26) give the solutions for $I_V^{(m)}$, $T_U^{(m)}$ and $T_D^{(m)}$ if superscripts m are affixed to V_0 and to all Ψ's.

When $\beta_0 b \geqslant 1$, the equation corresponding to (2.22) has the following slightly different form:

$$\left(I_V^{(m)}\Psi_{dR} - \frac{j2\pi V^{(m)}}{\zeta_0 \cos \beta_0 h}\right) \sin \beta_0(h - |z|)$$

$$+ \left(I_V^{(m)}\Psi_{d\Sigma R}^{(m)} + I_V^{(m)} T_U^{(m)}\Psi_{dUR}^{(m)} - \frac{j4\pi U^{(m)}}{\zeta_0 \cos \beta_0 h}\right)$$

$$\times (\cos \beta_0 z - \cos \beta_0 h) + I_V^{(m)}(j\Psi_{dI}^{(m)} + j\Psi_{dUI}^{(m)} T_U^{(m)}$$

$$+ \Psi_{dD}^{(m)} T_D^{(m)})(\cos \tfrac{1}{2}\beta_0 z - \cos \tfrac{1}{2}\beta_0 h) = 0. \qquad (3.33)$$

If the coefficients of the trigonometric functions are individually equated to zero and (3.29) is substituted for $U^{(m)}$, three relations corresponding to (2.23a–c) are obtained. They are readily solved to give

$$I_V^{(m)} = \frac{j2\pi V_0^{(m)}}{\zeta_0 \Psi_{dR} \cos \beta_0 h} \qquad (3.34)$$

$$T_U^{(m)} = \{\Psi_{dD}^{(m)}[\Psi_V^{(m)}(h) - \Psi_{d\Sigma R}^{(m)} \cos \beta_0 h] - j\Psi_D^{(m)}(h)\Psi_{dI}^{(m)}\}/Q^{(m)} \qquad (3.35)$$

$$T_D^{(m)} = -j\{\Psi_{dI}^{(m)}[\Psi_{dUR}^{(m)} \cos \beta_0 h - \Psi_U^{(m)}(h)]$$

$$+ \Psi_{dUI}^{(m)}[\Psi_V^{(m)}(h) - \Psi_{d\Sigma R}^{(m)} \cos \beta_0 h]\}/Q^{(m)} \qquad (3.36)$$

$$Q^{(m)} = \Psi_{dD}^{(m)}[\Psi_{dUR}^{(m)} \cos \beta_0 h - \Psi_U^{(m)}(h)] + j\Psi_D^{(m)}(h)\Psi_{dUI}^{(m)}. \qquad (3.37)$$

As throughout this chapter, $m = 0, 1$.

3.4 The phase-sequence currents and admittances

With the three coefficients $I_V^{(m)}$, $T_U^{(m)}$ and $T_D^{(m)}$ determined, the phase-sequence currents and the admittances may be written down directly. When $\beta_0 b < 1$, they are

$$I_z^{(m)}(z) = \frac{j2\pi V_0^{(m)}}{\zeta_0 \Psi_{dR}^{(m)} \cos \beta_0 h}[\sin \beta_0(h - |z|) + T_U^{(m)}(\cos \beta_0 z - \cos \beta_0 h)$$

$$+ T_D^{(m)}(\cos \tfrac{1}{2}\beta_0 z - \cos \tfrac{1}{2}\beta_0 h)] \qquad (3.38)$$

$$Y^{(m)} = \frac{j2\pi}{\zeta_0 \Psi_{dR}^{(m)} \cos \beta_0 h}[\sin \beta_0 h + T_U^{(m)}(1 - \cos \beta_0 h)$$

$$+ T_D^{(m)}(1 - \cos \tfrac{1}{2}\beta_0 h)] \qquad (3.39)$$

with

$$T_U^{(m)} = [\Psi_V^{(m)}(h)\Psi_{dD}^{(m)} - j\Psi_D^{(m)}(h)\Psi_{dI}^{(m)}]/Q^{(m)} \qquad (3.40)$$

$$T_D^{(m)} = -j\{\Psi_{dI}^{(m)}[\Psi_{dUR}^{(m)} \cos \beta_0 h - \Psi_U^{(m)}(h)] + \Psi_V^{(m)}(h)\Psi_{dUI}^{(m)}\}/Q^{(m)} \qquad (3.41)$$

$$Q^{(m)} = \Psi_{dD}^{(m)}[\Psi_{dUR}^{(m)} \cos \beta_0 h - \Psi_U^{(m)}(h)] + j\Psi_D^{(m)}(h)\Psi_{dUI}^{(m)}. \qquad (3.42)$$

The functions $\Psi^{(m)}$ are defined in (3.24), (3.26)–(3.28) and (3.30)–(3.32). For the zero-phase sequence, $m = 0$; for the first-phase sequence, $m = 1$.

As for the single antenna, these formulas for $I_z^{(m)}(z)$ and $Y_0^{(m)}$ become indeterminate when $\beta_0 h = \pi/2$. Convenient alternative forms when $\beta_0 h$ is at or near $\pi/2$ are

$$I_z^{(m)}(z) = \frac{-j2\pi V_0^{(m)}}{\zeta_0 \Psi_{dR}^{(m)}}[(\sin \beta_0|z| - \sin \beta_0 h) + T_U'^{(m)}(\cos \beta_0 z - \cos \beta_0 h)$$
$$- T_D'^{(m)}(\cos \tfrac{1}{2}\beta_0 z - \cos \tfrac{1}{2}\beta_0 h)] \qquad (3.43)$$

$$Y^{(m)} = \frac{j2\pi}{\zeta_0 \Psi_{dR}^{(m)}}[\sin \beta_0 h - T_U'^{(m)}(1 - \cos \beta_0 h) + T_D'^{(m)}(1 - \cos \tfrac{1}{2}\beta_0 h)] \qquad (3.44)$$

where, as in (2.37),

$$T_U'^{(m)} = -\frac{T_U^{(m)} + \sin \beta_0 h}{\cos \beta_0 h}, \qquad T_D'^{(m)} = \frac{T_D^{(m)}}{\cos \beta_0 h}. \qquad (3.45)$$

$T_U^{(m)}$ and $T_D^{(m)}$ are in (3.40) and (3.41).

When $\beta_0 b \geqslant 1$, the general form of the expressions for the phase-sequence current and admittance are similar to those for $\beta_0 b < 1$. They are

$$I_z^{(m)}(z) = \frac{j2\pi V_0^{(m)}}{\zeta_0 \Psi_{dR} \cos \beta_0 h}[\sin \beta_0(h - |z|) + T_U^{(m)}(\cos \beta_0 z - \cos \beta_0 h)$$
$$+ T_D^{(m)}(\cos \tfrac{1}{2}\beta_0 z - \cos \tfrac{1}{2}\beta_0 h)] \qquad (3.46)$$

$$Y^{(m)} = \frac{j2\pi}{\zeta_0 \Psi_{dR} \cos \beta_0 h}[\sin \beta_0 h + T_U^{(m)}(1 - \cos \beta_0 h)$$
$$+ T_D^{(m)}(1 - \cos \tfrac{1}{2}\beta_0 h)] \qquad (3.47)$$

with $T_U^{(m)}$ and $T_D^{(m)}$ given by (3.35) and (3.36) with (3.37). Similarly, when $\beta_0 h$ is near $\pi/2$ and $\beta_0 b \geqslant 1$,

$$I_z^{(m)}(z) = \frac{-j2\pi V_0^{(m)}}{\zeta_0 \Psi_{dR}}[(\sin \beta_0|z| - \sin \beta_0 h) + T_U'^{(m)}(\cos \beta_0 z - \cos \beta_0 h)$$
$$- T_D'^{(m)}(\cos \tfrac{1}{2}\beta_0 z - \cos \tfrac{1}{2}\beta_0 h)] \qquad (3.48)$$

$$Y^{(m)} = \frac{j2\pi}{\zeta_0 \Psi_{dR}}[\sin \beta_0 h - T_U'^{(m)}(1 - \cos \beta_0 h) + T_D'^{(m)}(1 - \cos \tfrac{1}{2}\beta_0 h)]. \qquad (3.49)$$

The parameters $T_U'^{(m)}$ and $T_D'^{(m)}$ are defined as in (3.45); $T_U^{(m)}$ and $T_D^{(m)}$ are given by (3.35) and (3.36).

Note that the currents and admittances when $\beta_0 b \geqslant 1$ differ from those when $\beta_0 b < 1$ not only in the T (or T') parameters but also in the appearance of Ψ_{dR} for the isolated antenna instead of $\Psi_{dR}^{(m)}$ for the coupled pair.

3.5 Currents for arbitrarily driven antennas; self- and mutual admittances and impedances

With the phase-sequence currents $I_z^{(0)}(z)$ and $I_z^{(1)}(z)$ determined, it is straightforward to obtain the expressions for the currents $I_{1z}(z)$ and $I_{2z}(z)$ in the two antennas when they are driven by the arbitrary voltages V_{10} and V_{20}. If

$$V_0^{(0)} = \tfrac{1}{2}(V_{10} + V_{20}), \qquad V_0^{(1)} = \tfrac{1}{2}(V_{10} - V_{20}) \qquad (3.50)$$

it follows that, when $\beta_0 b \geqslant 1$,

$$I_{1z}(z) = I_z^{(0)}(z) + I_z^{(1)}(z) = V_{10}v(z) + V_{20}w(z) \qquad (3.51a)$$

$$I_{2z}(z) = I_z^{(0)}(z) - I_z^{(1)}(z) = V_{10}w(z) + V_{20}v(z) \qquad (3.51b)$$

where

$$v(z) = \frac{j2\pi}{\zeta_0 \Psi_{dR} \cos \beta_0 h}[\sin \beta_0(h-|z|) + \tfrac{1}{2}(T_U^{(0)} + T_U^{(1)})(\cos \beta_0 z - \cos \beta_0 h)$$
$$+ \tfrac{1}{2}(T_D^{(0)} + T_D^{(1)})(\cos \tfrac{1}{2}\beta_0 z - \cos \tfrac{1}{2}\beta_0 h)] \qquad (3.52a)$$

$$w(z) = \frac{j2\pi}{\zeta_0 \Psi_{dR} \cos \beta_0 h}[\tfrac{1}{2}(T_U^{(0)} - T_U^{(1)})(\cos \beta_0 z - \cos \beta_0 h)$$
$$+ \tfrac{1}{2}(T_D^{(0)} - T_D^{(1)})(\cos \tfrac{1}{2}\beta_0 z - \cos \tfrac{1}{2}\beta_0 h)]. \qquad (3.52b)$$

Alternatively, when $\beta_0 h$ is near $\pi/2$,

$$v(z) = \frac{-j2\pi}{\zeta_0 \Psi_{dR}}[(\sin \beta_0|z| - \sin \beta_0 h) + \tfrac{1}{2}(T_U'^{(0)} + T_U'^{(1)})(\cos \beta_0 z - \cos \beta_0 h)$$
$$- \tfrac{1}{2}(T_D'^{(0)} + T_D'^{(1)})(\cos \tfrac{1}{2}\beta_0 z - \cos \tfrac{1}{2}\beta_0 h)] \qquad (3.52c)$$

$$w(z) = \frac{-j2\pi}{\zeta_0 \Psi_{dR}}[\tfrac{1}{2}(T_U'^{(0)} - T_U'^{(1)})(\cos \beta_0 z - \cos \beta_0 h)$$
$$- \tfrac{1}{2}(T_D'^{(0)} - T_D'^{(1)})(\cos \tfrac{1}{2}\beta_0 z - \cos \tfrac{1}{2}\beta_0 h)]. \qquad (3.52d)$$

The corresponding expressions when $\beta_0 b < 1$ are easily obtained from (3.38). The driving-point currents may be expressed in the form

$$I_{1z}(0) = V_{10}Y_{s1} + V_{20}Y_{12} \qquad (3.53a)$$

$$I_{2z}(0) = V_{10}Y_{21} + V_{20}Y_{s2} \qquad (3.53b)$$

where Y_{s1} and Y_{s2} are the self-admittances, Y_{12} and Y_{21} are the mutual admittances. They are given by

$$Y_{s1} = Y_{s2} = v(0) = \tfrac{1}{2}(Y^{(0)} + Y^{(1)}) \tag{3.54}$$

$$Y_{21} = Y_{12} = w(0) = \tfrac{1}{2}(Y^{(0)} - Y^{(1)}). \tag{3.55}$$

Specifically,

$$Y_{s1} = Y_{s2} = v(0) = \frac{j\pi}{\zeta_0 \Psi_{dR} \cos \beta_0 h}[2 \sin \beta_0 h + (T_U^{(0)} + T_U^{(1)})(1 - \cos \beta_0 h)$$

$$+ (T_D^{(0)} + T_D^{(1)})(1 - \cos \tfrac{1}{2}\beta_0 h)] \tag{3.56}$$

$$Y_{12} = Y_{21} = w(0) = \frac{j\pi}{\zeta_0 \Psi_{dR} \cos \beta_0 h}[(T_U^{(0)} - T_U^{(1)})(1 - \cos \beta_0 h)$$

$$+ (T_D^{(0)} - T_D^{(1)})(1 - \cos \tfrac{1}{2}\beta_0 h)]. \tag{3.57}$$

When $\beta_0 h = \pi/2$, the self- and mutual admittances are

$$Y_{s2} = Y_{s1} = \frac{j\pi}{\zeta_0 \Psi_{dR}}[2 \sin \beta_0 h - (T_U''^{(0)} + T_U''^{(1)})(1 - \cos \beta_0 h)$$

$$+ (T_D''^{(0)} + T_D''^{(1)})(1 - \cos \tfrac{1}{2}\beta_0 h)] \tag{3.58}$$

$$Y_{21} = Y_{12} = \frac{-j\pi}{\zeta_0 \Psi_{dR}}[(T_U''^{(0)} - T_U''^{(1)})(1 - \cos \beta_0 h)$$

$$- (T_D''^{(0)} - T_D''^{(1)})(1 - \cos \tfrac{1}{2}\beta_0 h)]. \tag{3.59}$$

The associated self- and mutual impedances are the coefficients of the currents in the equations

$$V_{10} = I_{1z}(0)Z_{s1} + I_{2z}(0)Z_{12} \tag{3.60a}$$

$$V_{20} = I_{1z}(0)Z_{21} + I_{2z}(0)Z_{s2}. \tag{3.60b}$$

It is readily shown that

$$Z_{s1} = Y_{s2}/D = \tfrac{1}{2}(Z^{(0)} + Z^{(1)}); \quad Z_{s2} = Y_{s1}/D = \tfrac{1}{2}(Z^{(0)} + Z^{(1)}) \tag{3.61a}$$

$$Z_{12} = -Y_{21}/D = \tfrac{1}{2}(Z^{(0)} - Z^{(1)}); \quad Z_{21} = -Y_{12}/D = \tfrac{1}{2}(Z^{(0)} - Z^{(1)}) \tag{3.61b}$$

where $D = Y_{s1}Y_{s2} - Y_{12}Y_{21} = (Z_{s1}Z_{s2} - Z_{12}Z_{21})^{-1}$. If lumped impedances Z_1 and Z_2 are connected in series with V_{10} and V_{20}, Z_{s1} and Z_{s2} in (3.53a, b) are replaced by $Z_{11} = Z_{s1} + Z_1$ and $Z_{22} = Z_{s2} + Z_2$.

This completes the general formulation for the currents and admittances of two parallel antennas driven by the arbitrary voltages V_{10} and V_{20}.

3.6   Currents for one driven, one parasitic antenna

If antenna 2 is parasitic instead of driven and is centre-loaded by an arbitrary impedance Z_2, the driving voltage V_{20} may be replaced by the negative of the voltage drop across the load. Thus,

$$V_{20} = -I_{2z}(0)Z_2 = -I_{2z}(0)/Y_2. \qquad (3.62)$$

If (3.62) is substituted in (3.53b), the result is

$$I_{2z}(0) = V_{10}\frac{Y_{21}}{1+Y_{s2}Z_2} = -V_{10}\frac{Z_{21}}{Z_{s2}(Z_{s1}+Z_2)+Z_{12}Z_{21}} \qquad (3.63)$$

so that

$$V_{20} = -V_{10}\left(\frac{Y_{21}}{Y_2+Y_{s2}}\right) = V_{10}\frac{Z_{21}Z_2}{Z_{s2}(Z_{s1}+Z_2)+Z_{12}Z_{21}}. \qquad (3.64)$$

It follows from (3.51a, b) that

$$I_{1z}(z) = V_{10}\left[v(z) - \left(\frac{Y_{21}}{Y_2+Y_{s2}}\right)w(z)\right] \qquad (3.65a)$$

$$I_{2z}(z) = V_{10}\left[w(z) - \left(\frac{Y_{21}}{Y_2+Y_{s2}}\right)v(z)\right]. \qquad (3.65b)$$

The driving-point admittance and impedance are

$$Y_{1in} = \frac{I_{1z}(0)}{V_{10}} = Y_{s1} - \frac{Y_{21}Y_{12}}{Y_2+Y_{s2}} \qquad (3.66a)$$

$$Z_{1in} = \frac{1}{Y_{1in}} = \frac{Z_{s1}(Z_{s2}+Z_2)+Z_{12}Z_{21}}{Z_{s2}+Z_2}. \qquad (3.66b)$$

Note that when $Z_2 = 0$ or $Y_2 = \infty$,

$$Y_{1in} = Y_{s1}; \quad I_{1z}(z) = V_{10}v(z); \quad I_{2z}(z) = V_{10}w(z). \qquad (3.67a)$$

Alternatively, when $Z_2 = \infty$ or $Y_2 = 0$,

$$\left. \begin{array}{l} Z_{1in} = Z_{s1}; \quad I_{1z}(z) = V_{10}\left[v(z) - \dfrac{Y_{21}}{Y_{s2}}w(z)\right]; \\[3mm] \qquad\qquad I_{2z}(z) = V_{10}\left[w(z) - \dfrac{Y_{21}}{Y_{s2}}v(z)\right]. \end{array} \right\} \qquad (3.67b)$$

The parasitic element is tuned to resonance when $Y_2 = jB_2$ and $B_2 = -B_{s2}$ in $Y_{s2} = G_{s2}+jB_{s2}$. With this choice, $Y_{21}/(Y_2+Y_{s2})$ is maximized so that

$$I_{1z}(z) = V_{10}\left[v(z) - \frac{Y_{21}}{G_{s2}}w(z)\right] \qquad (3.68a)$$

$$I_{2z}(z) = V_{10}\left[w(z) - \frac{Y_{21}}{G_{s2}}v(z)\right]. \qquad (3.68b)$$

Since the coefficient Y_{21}/G_{s2} is of the order of magnitude of one, the coefficients of $v(z)$ and $w(z)$ are comparable. It follows that the distributions of $I_{1z}(z)$ and $I_{2z}(z)$ are roughly similar, whereas when $Z_2 = 0$ as in (3.67a), they are quite different unless $\beta_0 h$ is near $\pi/2$.

3.7 The couplet

Perhaps the most interesting two-element array is the couplet in which the distance between the elements is $\lambda/4$ and the currents at the driving points are equal in amplitude but differ in phase by a quarter period. That is

$$I_{2z}(0) = jI_{1z}(0). \tag{3.69}$$

It follows from (3.60a, b) that with $Z_{12} = Z_{21}$ and $Z_{s2} = Z_{s1}$,

$$V_{10} = I_{1z}(0)[Z_{s1} + jZ_{12}] \tag{3.70a}$$

$$V_{20} = I_{2z}(0)[Z_{s1} - jZ_{12}] = I_{1z}(0)[Z_{12} + jZ_{s1}]. \tag{3.70b}$$

Hence, $Z_{1in} = Z_{s1} + jZ_{12}, \qquad Z_{2in} = Z_{s1} - jZ_{12}. \tag{3.70c}$

The distributions of current are obtained from (3.51a, b). Thus

$$I_{1z}(z) = V_{10}\left[v(z) + \frac{Z_{12} + jZ_{s1}}{Z_{s1} + jZ_{12}}w(z)\right] \tag{3.71a}$$

$$I_{2z}(z) = V_{10}\left[w(z) + \frac{Z_{12} + jZ_{s1}}{Z_{s1} + jZ_{12}}v(z)\right]. \tag{3.71b}$$

Instead of specifying the driving-point currents $I_{1z}(0)$ and $I_{2z}(0)$ as in (3.69), the driving voltages may be assigned so that

$$V_{20} = jV_{10}. \tag{3.72}$$

It then follows from (3.53a, b) that

$$I_{1z}(0) = V_{10}(Y_{s1} + jY_{12}) \tag{3.73a}$$

$$I_{2z}(0) = V_{20}(Y_{s1} - jY_{12}) = V_{10}(Y_{12} + jY_{s1}). \tag{3.73b}$$

The driving-point admittances are

$$Y_{1in} = Y_{s1} + jY_{12}, \qquad Y_{2in} = Y_{s1} - jY_{12}. \tag{3.74}$$

The currents are obtained from (3.51a, b) with (3.61a, b). Thus,

$$I_{1z}(z) = V_{10}\left[v(z) + \frac{Y_{12} + jY_{s1}}{Y_{s1} - jY_{12}}w(z)\right]$$

$$= V_{10}\left[v(z) - \frac{Z_{12} - jZ_{s1}}{Z_{s1} + jZ_{12}}w(z)\right] \tag{3.75a}$$

$$I_{2z}(z) = V_{10}\left[w(z) + \frac{Y_{12}+jY_{s1}}{Y_{s1}-jY_{12}}v(z)\right]$$

$$= V_{10}\left[w(z) - \frac{Z_{12}-jZ_{s1}}{Z_{s1}+jZ_{12}}v(z)\right]. \tag{3.75b}$$

The currents are not the same when $I_{1z}(0)$ and $I_{2z}(0)$ are specified as when V_{10} and V_{20} are assigned. Note that

$$[I_{1z}(z)]_I - [I_{1z}(z)]_V = 2V_{10}Z_{12}w(z) \tag{3.76a}$$

$$[I_{2z}(z)]_I - [I_{2z}(z)]_V = 2V_{10}Z_{12}v(z). \tag{3.76b}$$

If the currents differ significantly, the field patterns cannot be the same.

3.8   Field patterns

The radiation field of an array of two parallel elements is the vector sum of the fields maintained by the currents in the individual elements. In terms of the spherical coordinates R, Θ, Φ, that have their origin midway between the centres of the two elements, the individual electric fields are readily expressed in the form (2.45a, b) for the currents (3.51a, b). Thus

$$E^r_{\Theta 1} = -\frac{1}{\Psi_{dR}}\frac{e^{-j\beta_0 R_1}}{R}[V_{10}f(\Theta,\beta_0 h)+V_{20}g(\Theta,\beta_0 h)] \tag{3.77a}$$

$$E^r_{\Theta 2} = -\frac{1}{\Psi_{dR}}\frac{e^{-j\beta_0 R_2}}{R}[V_{10}g(\Theta,\beta_0 h)+V_{20}f(\Theta,\beta_0 h)] \tag{3.77b}$$

where
$$R_1 = R + \frac{b}{2}\cos\Phi\sin\Theta \tag{3.78a}$$

$$R_2 = R - \frac{b}{2}\cos\Phi\sin\Theta \tag{3.78b}$$

$$f(\Theta,\beta_0 h) = [F_m(\Theta,\beta_0 h)+\tfrac{1}{2}(T_U^{(0)}+T_U^{(1)})G_m(\Theta,\beta_0 h)$$
$$+\tfrac{1}{2}(T_D^{(0)}+T_D^{(1)})D_m(\Theta,\beta_0 h)]\sec\beta_0 h \tag{3.79}$$
$$g(\Theta,\beta_0 h) = [\tfrac{1}{2}(T_U^{(0)}-T_U^{(1)})G_m(\Theta,\beta_0 h)$$
$$+\tfrac{1}{2}(T_D^{(0)}-T_D^{(1)})D_m(\Theta,\beta_0 h)]\sec\beta_0 h. \tag{3.80}$$

The field functions $F_m(\Theta,\beta_0 h)$, $G_m(\Theta,\beta_0 h)$ and $D_m(\Theta,\beta_0 h)$ are defined in (2.46), (2.47) and (2.48). Alternatively, when $\beta_0 h$ is near $\pi/2$, the fields for the currents (3.52c, d) are:

$$E^r_{\Theta 1} = \frac{1}{\Psi_{dR}}\frac{e^{-j\beta_0 R_1}}{R}[V_{10}f'(\Theta,\beta_0 h)+V_{20}g'(\Theta,\beta_0 h)] \tag{3.81a}$$

$$E^r_{\Theta 2} = \frac{1}{\Psi_{dR}} \frac{e^{-j\beta_0 R_2}}{R} [V_{10} g'(\Theta, \beta_0 h) + V_{20} f'(\Theta, \beta_0 h)] \quad (3.81b)$$

$$f'(\Theta, \beta_0 h) = H_m(\Theta, \beta_0 h) + \tfrac{1}{2}(T_U'^{(0)} + T_U'^{(1)}) G_m(\Theta, \beta_0 h)$$
$$- \tfrac{1}{2}(T_D'^{(0)} + T_D'^{(1)}) D_m(\Theta, \beta_0 h) \qquad (3.82)$$

$$g'(\Theta, \beta_0 h) = \tfrac{1}{2}(T_U'^{(0)} - T_U'^{(1)}) G_m(\Theta, \beta_0 h)$$
$$- \tfrac{1}{2}(T_D'^{(0)} - T_D'^{(1)}) D_m(\Theta, \beta_0 h). \qquad (3.83)$$

The function $H_m(\Theta, \beta_0 h)$ is defined in (2.51).

The resultant radiation field of the arbitrarily driven two-element array is

$$E^r_{\Theta} = E^r_{\Theta 1} + E^r_{\Theta 2} = \frac{-1}{\Psi_{dR}} \frac{e^{-j\beta_0 R}}{R} \{[V_{10} f(\Theta, \beta_0 h)$$
$$+ V_{20} g(\Theta, \beta_0 h)] e^{-j(\beta_0 b/2)\cos\Phi \sin\Theta}$$
$$+ [V_{10} g(\Theta, \beta_0 h) + V_{20} f(\Theta, \beta_0 h)] e^{j(\beta_0 b/2)\cos\Phi \sin\Theta} \}. \quad (3.84)$$

When $\beta_0 h$ is near $\pi/2$, $-f'(\Theta, \beta_0 h)$ and $-g'(\Theta, \beta_0 h)$ may be substituted, respectively, for $f(\Theta, \beta_0 h)$ and $g(\Theta, \beta_0 h)$.

3.9 The two-term approximation

As pointed out in section 10 of the preceding chapter, the difference between the distribution functions $F_{0z} = \cos \beta_0 z - \cos \beta_0 h$ and $H_{0z} = \cos \tfrac{1}{2}\beta_0 z - \cos \tfrac{1}{2}\beta_0 h$ is relatively unimportant in the determination of the far-field and the driving-point admittance of an isolated antenna when $\beta_0 h \leqslant 5\pi/4$. This is also true of the far-field and driving-point admittances of two coupled antennas provided the interaction between them is not sensitive to small changes in the current distributions. When both elements are driven by comparable voltages and when the distance between them is sufficiently great so that $\beta_0 b \geqslant 1$, it may be assumed that the substitution of $\cos \beta_0 z - \cos \beta_0 h$ for $\cos \tfrac{1}{2}\beta_0 z - \cos \tfrac{1}{2}\beta_0 h$ can produce no important change in the admittances or the far-field. When one element is parasitic and unloaded, the three-term approximation is automatically reduced to two terms since the distribution $\sin \beta_0(h - |z|)$ is excited only by a generator or an equivalent voltage drop across a load. Correspondingly, the two-term approximation is reduced to a single term. However, this is quite adequate for many purposes. In anticipation, it may be added at this point that when an array consists of one driven antenna and many parasitic elements, at least two terms are desirable in the representation of the current distributions. This is considered in a later chapter.

As for the single antenna, the two-term approximations are quickly obtained from the three-term formulas by the simple substitution of $\cos \beta_0 z - \cos \beta_0 h$ for $\cos \frac{1}{2}\beta_0 z - \cos \frac{1}{2} \beta_0 h$ and the representation of the resulting coefficient $(T_U + T_D)$ by T. It is implicit that $\Psi_{dD} \rightarrow \Psi_{dU}$, $\Psi_D(h) \rightarrow \Psi_U(h)$. Thus, the phase-sequence currents and admittances (3.46) and (3.47) become, for $\beta_0 b \geqslant 1$,

$$I_z^{(m)}(z) = \frac{j2\pi V_0^{(m)}}{\zeta_0 \Psi_{dR} \cos \beta_0 h}[\sin \beta_0(h-|z|) + T^{(m)}(\cos \beta_0 z - \cos \beta_0 h)]$$

(3.85)

$$Y^{(m)} = \frac{j2\pi}{\zeta_0 \Psi_{dR} \cos \beta_0 h}[\sin \beta_0 h + T^{(m)}(1 - \cos \beta_0 h)] \qquad (3.86)$$

where

$$T^{(m)} = T_U^{(m)} + T_D^{(m)} = -\frac{\Psi_V^{(m)}(h) - (\Psi_{d\Sigma R}^{(m)} + j\Psi_{dI}^{(m)}) \cos \beta_0 h}{\Psi_U^{(m)}(h) - \Psi_{dU}^{(m)} \cos \beta_0 h}. \quad (3.87)$$

Similarly, when $\beta_0 h$ is near $\pi/2$, (3.48) and (3.49) reduce to

$$I_z^{(m)}(z) = \frac{-j2\pi V_0^{(m)}}{\zeta_0 \Psi_{dR}}[(\sin \beta_0|z| - \sin \beta_0 h) + T'^{(m)}(\cos \beta_0 z - \cos \beta_0 h)]$$

(3.88)

$$Y^{(m)} = \frac{j2\pi}{\zeta_0 \Psi_{dR}}[\sin \beta_0 h - T'^{(m)}(1 - \cos \beta_0 h)] \qquad (3.89)$$

where

$$
\begin{aligned}
T'^{(m)} &= -\frac{T^{(m)} + \sin \beta_0 h}{\cos \beta_0 h} \\
&= \frac{[\Psi_V^{(m)}(h) - \Psi_U^{(m)}(h)\sin \beta_0 h]\sec \beta_0 h + \Psi_{dU}^{(m)}\sin \beta_0 h - j\Psi_{dI}^{(m)} - \Psi_{d\Sigma R}^{(m)}}{\Psi_U^{(m)}(h) - \Psi_{dU}^{(m)}\cos \beta_0 h} \\
&= \frac{[\Psi_{dU}^{(m)} + E_a(h, h)]\sin \beta_0 h - j\Psi_{dI}^{(m)} - S_a(h, h) - \Psi_{d\Sigma R}^{(m)}}{C_a(h, h) - [\Psi_{dU} + E_a(h, h)]\cos \beta_0 h}.
\end{aligned}
$$

(3.90)

Note that when $\beta_0 h = \pi/2$, $\Psi_U^{(m)}\left(\frac{\lambda}{4}\right) = \Psi_V^{(m)}\left(\frac{\lambda}{4}\right)$.

As an example, the phase-sequence currents have been evaluated specifically for two antennas for which $\Omega = 2\ln(2h/a) = 10$, $\beta_0 h = \pi$ and $\beta_0 b = 1\cdot5$. For these

$$\Psi_{dR} = 5\cdot834, \qquad \Psi_{d\Sigma R}^{(0)} = -0\cdot245, \qquad \Psi_{d\Sigma R}^{(1)} = 0\cdot245 \quad (3.91a)$$

$$\left. \begin{aligned} \Psi_{dI}^{(0)} &= -0\cdot633 - 0\cdot524 = -1\cdot157; \\ \Psi_{dI}^{(1)} &= -0\cdot633 + 0\cdot524 = -0\cdot109 \end{aligned} \right\} \qquad (3.91b)$$

$$\Psi_{dU}^{(0)} = 7\cdot848 - j3\cdot939, \qquad \Psi_{dU}^{(1)} = 7\cdot352 - j0\cdot661. \quad (3.91c)$$

The amplitude functions are

$$T^{(0)}\left(\frac{\lambda}{2}\right) = -0\cdot216 + j0\cdot274, \qquad T^{(1)}\left(\frac{\lambda}{2}\right) = -0\cdot177 + j0\cdot066.$$

With these values the two-term zero-phase-sequence and first-phase-sequence currents (in amperes when V_0 is in volts) in the two antennas are

$$I_{2z}^{(0)}(z) = I_{1z}^{(0)}(z) = V^{(0)}\{0\cdot783(\cos\beta_0 z + 1)$$
$$-j[2\cdot805 \sin\beta_0|z| - 0\cdot617(\cos\beta_0 z + 1)]\} \times 10^{-3} \quad (3.92a)$$

$$-I_{2z}^{(1)}(z) = I_{1z}^{(1)}(z) = V^{(1)}\{0\cdot189(\cos\beta_0 z + 1)$$
$$-j[2\cdot805 \sin\beta_0|z| - 0\cdot506(\cos\beta_0 z + 1)]\} \times 10^{-3}. \quad (3.92b)$$

These currents are shown graphically in Fig. 3.3 in the form $I_z = I_z'' + jI_z'$, where I_z'' is in phase, I_z' in phase quadrature with V_0. The corresponding driving-point admittances and impedances are

$$\left.\begin{aligned} Y^{(0)} &= (1\cdot566 + j1\cdot234) \text{ millimhos,} \\ Y^{(1)} &= (0\cdot378 + j1\cdot012) \text{ millimhos,} \end{aligned}\right\} \quad (3.93a)$$

$$Z^{(0)} = 394 - j310 \text{ ohms,} \qquad Z^{(1)} = 324 - j867 \text{ ohms.} \quad (3.93b)$$

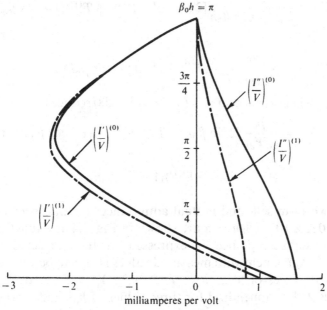

Fig. 3.3.   Zero- and first-phase-sequence currents on two-element array. $\Omega = 10, \beta_0 b = 1.5$.

The two-term approximation of the general formulas (3.51a, b) are

$$I_{1z}(z) = V_{10}v(z) + V_{20}w(z) \tag{3.94a}$$
$$I_{2z}(z) = V_{10}w(z) + V_{20}v(z) \tag{3.94b}$$

where now

$$v(z) = \frac{j2\pi}{\zeta_0 \Psi_{dR}\cos \beta_0 h}[\sin \beta_0(h-|z|)$$
$$+ \tfrac{1}{2}(T^{(0)} + T^{(1)})(\cos \beta_0 z - \cos \beta_0 h)] \tag{3.95a}$$

$$w(z) = \frac{j\pi}{\zeta_0 \Psi_{dR}\cos \beta_0 h}(T^{(0)} - T^{(1)})(\cos \beta_0 z - \cos \beta_0 h). \tag{3.95b}$$

When $\beta_0 h$ is near $\pi/2$,

$$v(z) = \frac{-j2\pi}{\zeta_0 \Psi_{dR}}[(\sin \beta_0|z| - \sin \beta_0 h) + \tfrac{1}{2}(T'^{(0)} + T'^{(1)})(\cos \beta_0 z - \cos \beta_0 h)] \tag{3.95c}$$

$$w(z) = \frac{-j\pi}{\zeta_0 \Psi_{dR}}(T'^{(0)} - T'^{(1)})(\cos \beta_0 z - \cos \beta_0 h). \tag{3.95d}$$

The self- and mutual admittances (3.55) and (3.56) become

$$Y_{s1} = Y_{s2} = \frac{j\pi}{\zeta_0 \Psi_{dR} \cos \beta_0 h}[2\sin \beta_0 h + (T^{(0)} + T^{(1)})(1 - \cos \beta_0 h)] \tag{3.96}$$

$$Y_{21} = Y_{12} = \frac{j\pi}{\zeta_0 \Psi_{dR} \cos \beta_0 h}(T^{(0)} - T^{(1)})(1 - \cos \beta_0 h). \tag{3.97}$$

Similarly, when $\beta_0 h$ is near $\pi/2$, (3.57) and (3.58) reduce to

$$Y_{s1} = Y_{s2} = \frac{j\pi}{\zeta_0 \Psi_{dR}}[2\sin \beta_0 h - (T'^{(0)} + T'^{(1)})(1 - \cos \beta_0 h)] \tag{3.98}$$

$$Y_{21} = Y_{12} = \frac{-j\pi}{\zeta_0 \Psi_{dR}}(T'^{(0)} - T'^{(1)})(1 - \cos \beta_0 h). \tag{3.99}$$

The two-term self- and mutual admittances for the special case $a/\lambda = 7\cdot022 \times 10^{-3}$, $\beta_0 h = \pi$ are shown in Fig. 3.4 as a function of b/λ. The self-susceptance is expressed in the corrected form $B_{11} + 0\cdot72$. Agreement with measured values is seen to be very good. Numerical values of Ψ_{dR}, $T^{(m)}$, $T'^{(m)}$, $Y^{(m)}$, $Y_{si} = Y_{11}$ and Y_{12} are in Tables 2–4 of appendix I for three values of $\beta_0 h$ and a range of $b/\lambda = d/\lambda$.

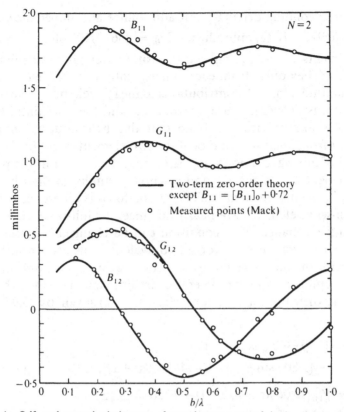

Fig. 3.4. Self- and mutual admittances of two-element array; b is the distance between elements; $\beta_0 h = \pi$.

For the special case $\Omega = 10$, $\beta_0 h = \pi$, $\beta_0 b = 1.5$, the two-term self- and mutual impedances defined in (3.61) with the two-term expressions (3.96) and (3.97) are

$$Z_{s2} = Z_{s1} = \tfrac{1}{2}(Z^{(0)} + Z^{(1)}) = 359 - j588 \text{ ohms} \qquad (3.100a)$$

$$Z_{21} = Z_{12} = \tfrac{1}{2}(Z^{(0)} - Z^{(1)}) = 35 + j278 \text{ ohms}. \qquad (3.100b)$$

If antenna 1 is driven and antenna 2 is an unloaded parasitic element, (3.67a) applies. The two-term formulas for the currents may be obtained directly from (3.94a, b) with $V_{20} = 0$. Then, in the special case $\Omega = 10$, $\beta_0 h = \pi$, $\beta_0 b = 1.5$,

$$I_{1z}(z) = V_{10}\{0.486(\cos \beta_0 z + 1)$$
$$- j[2.805 \sin \beta_0 |z| - 0.566(\cos \beta_0 z + 1)]\} \times 10^{-3}$$
$$(3.101a)$$

$$I_{2z}(z) = V_{10}(0.287 + j0.055)(\cos \beta_0 z + 1) \times 10^{-3}. \qquad (3.101b)$$

The corresponding driving-point admittance and impedance are

$$Y_{1in} = (0.972 + j1.33) \text{ millimhos}, \quad Z_{1in} = 436 - j508 \text{ ohms}. \quad (3.102)$$

The currents in the driven and parasitic antennas are shown in Fig. 3.5a. They differ from each other greatly in both distribution and amplitude. Indeed, contributions to the far-field by the currents in the parasitic element are insignificant and the horizontal field pattern is almost circular. Note that this behaviour is entirely different from what it would be if the two elements were half-wave instead of full-wave dipoles. In the former, the current in the parasitic element is comparable and essentially similar in distribution to that in the driven element. The reason for this difference is that the half-wave elements are near resonance, the full-wave elements near anti-resonance. This condition can be changed by inserting a lumped susceptance B_2 (or an equivalent transmission-line) in series with the full-wave parasitic element at its centre and tuning this susceptance to make the entire circuit resonant. When this is done the distribution functions $v(z)$ and $w(z)$ given by (3.95a, b) give

$$I_{1z}(z) = V_{10}\{0.369(\cos \beta_0 z + 1)$$
$$-j[2.805 \sin \beta_0|z| - 0.494(\cos \beta_0 z + 1)]\} \times 10^{-3} \quad (3.103a)$$

$$I_{2z}(z) = V_{10}\{[0.064(\cos \beta_0 z + 1) - 0.320 \sin \beta_0|z|]$$
$$+j[1.712 \sin \beta_0|z| - 0.343(\cos \beta_0 z + 1)]\} \times 10^{-3}. \quad (3.103b)$$

These currents are shown in Fig. 3.5b. They are very nearly alike in both distribution and amplitude, so that the horizontal field pattern of the tuned full-wave parasitic array must correspond closely to that of the half-wave array with an unloaded parasitic element.

The two-term formulas for the currents in the couplet are given by (3.71a, b) with $v(z)$ and $w(z)$ as in (3.95a, b). For the special case $\Omega = 10$, $\beta_0 h = \pi$, $\beta_0 b = 1.5$, (3.70a, b) give

$$V_{20} = V_{10}\left(\frac{Z_{12} + jZ_{s1}}{Z_{s1} + jZ_{12}}\right) = (-0.966 + j1.267)V_{10} = 1.59 V_{10} e^{-j146.3°}. \quad (3.104)$$

With this value, the explicit formulas for the current in an array are

$$I_{1z} = V_{10}\{0.129(\cos \beta_0 z + 1)$$
$$-j[2.805 \sin \beta_0|z| - 0.884(\cos \beta_0 z + 1)]\} \times 10^{-3} \quad (3.105a)$$

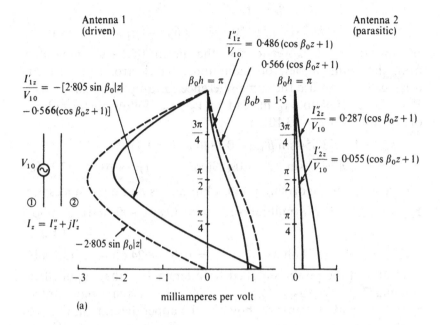

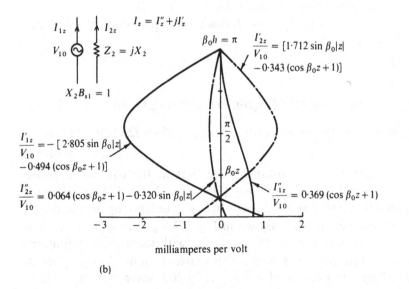

Fig. 3.5.   Currents on full-wave antenna with (a) $a/\lambda = 7 \cdot 022 \times 10^{-3}$ untuned parasite, $\Omega = 10$; (b) tuned parasite, $\Omega = 10$.

$$I_{2z} = V_{20}\{0.400(\cos \beta_0 z + 1)$$
$$-j[2.805 \sin \beta_0|z| - 0.397(\cos \beta_0 z + 1)]\} \times 10^{-3}. \quad (3.105\mathrm{b})$$

In order to obtain expressions for the current that are comparable from the point of view of maintaining an electromagnetic field, it is necessary to use the same reference for amplitude and phase. If I_{2z} is referred to V_{10} instead of V_{20}, the following formula is obtained in place of (3.105b):

$$I_{2z} = V_{10}\{[3.554 \sin \beta_0|z| - 0.884(\cos \beta_0 z + 1)]$$
$$+ j[2.710 \sin \beta_0|z| + 0.129(\cos \beta_0 z + 1)]\} \times 10^{-3}. \quad (3.105\mathrm{c})$$

The corresponding driving-point admittances and impedances are

$$Y_{10} = (0.258 + j1.768)\,\text{millimhos}, \quad Y_{20} = (0.801 + j0.784)\,\text{millimhos}$$
$$(3.106\mathrm{a})$$

$$Z_{10} = 80.8 - j554\,\text{ohms}, \quad Z_{20} = 638 - j624\,\text{ohms}. \quad (3.106\mathrm{b})$$

The ratio of the power supplied to antenna 1 to that supplied to antenna 2 is $|V_{20}|^2 G_{20}/|V_{10}|^2 G_{10} = 7.9$. The currents represented by (3.105a) and (3.105b) are shown in the upper diagram in Fig. 3.6 in the form I_{1z}/V_{10} and I_{2z}/V_{20}. The distribution of I_{2z}/V_{10} is shown in the bottom diagram in Fig. 3.6. It differs greatly from I_{1z}/V_{10} (shown in the upper graph) even though the input currents at $z = 0$ satisfy the assigned relation, $I_{20} = jI_{10}$.

The radiation field of the full-wave couplet may be expressed as follows:

$$E_\Theta^r = E_{\Theta 1}^r + E_{\Theta 2}^r = K[A_1 \, e^{j(\beta_0 b/2)\cos\Phi} + A_2 \, e^{-j(\beta_0 b/2)\cos\Phi}] \quad (3.107)$$

where

$$A_1 = V_{10}[(0.129 + j0.884)G_m(\Theta, \pi) - j2.805 F_m(\Theta, \pi)] \quad (3.108\mathrm{a})$$

$$A_2 = V_{10}[(-0.884 + j0.129)G_m(\Theta, \pi) + (3.554 + j2.710)F_m(\Theta, \pi)] \quad (3.108\mathrm{b})$$

and where K is a constant. Note that in the equatorial plane, $\Theta = \pi/2$, and $G_m(\pi/2, \pi) = \pi$, $F_m(\pi/2, \pi) = 2$. The field pattern calculated from the magnitude of (3.107) with (3.108a, b) for the couplet of full-wave elements is shown in Fig. 3.7 together with the corresponding pattern for the ideal couplet with identical distributions of current in the two elements. (This latter is quite closely approximated by the pattern of a couplet of half-wave elements.) Both patterns are normalized to unity at $\Phi = 0$. It is seen that the deep minimum at $\Phi = 180°$ in the ideal pattern (this would be a null if

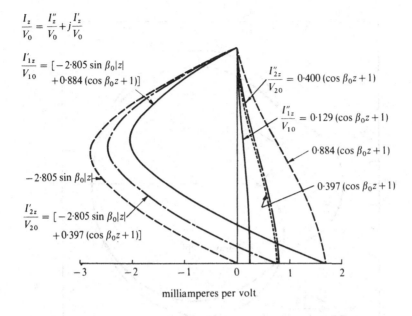

$$\frac{I_z}{V_0} = \frac{I_z''}{V_0} + j\frac{I_z'}{V_0}$$

$$\frac{I_{1z}'}{V_{10}} = [-2\cdot805 \sin \beta_0|z| \\ +0\cdot884 (\cos \beta_0 z + 1)]$$

$$\frac{I_{2z}''}{V_{20}} = 0\cdot400 (\cos \beta_0 z + 1)$$

$$\frac{I_{1z}''}{V_{10}} = 0\cdot129 (\cos \beta_0 z + 1)$$

$$0\cdot884 (\cos \beta_0 z + 1)$$

$$-2\cdot805 \sin \beta_0|z|$$

$$0\cdot397 (\cos \beta_0 z + 1)$$

$$\frac{I_{2z}'}{V_{20}} = [-2\cdot805 \sin \beta_0|z| \\ +0\cdot397 (\cos \beta_0 z + 1)]$$

milliamperes per volt

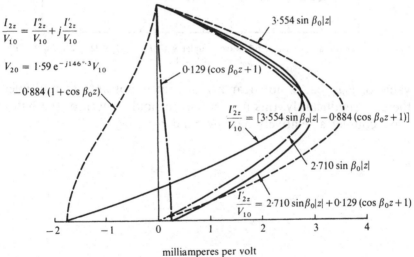

$$\frac{I_{2z}}{V_{10}} = \frac{I_{2z}''}{V_{10}} + j\frac{I_{2z}'}{V_{10}}$$

$$V_{20} = 1\cdot59 \, e^{-j146^\circ\cdot3} V_{10}$$

$$-0\cdot884 (1 + \cos \beta_0 z)$$

$$3\cdot554 \sin \beta_0|z|$$

$$0\cdot129 (\cos \beta_0 z + 1)$$

$$\frac{I_{2z}''}{V_{10}} = [3\cdot554 \sin \beta_0|z| - 0\cdot884 (\cos \beta_0 z + 1)]$$

$$2\cdot710 \sin \beta_0|z|$$

$$\frac{I_{2z}'}{V_{10}} = 2\cdot710 \sin\beta_0|z| + 0\cdot129 (\cos \beta_0 z + 1)$$

milliamperes per volt

Fig. 3.6. Currents in full-wave couplet; $I_{20} = jI_{10}$; $\beta_0 b = 2\pi b/\lambda = 1\cdot5$; $\Omega = 10$.

$\beta_0 b = \pi/2$ had been used instead of $\beta_0 b = 1\cdot5$) is replaced by a minor maximum with an amplitude that is about one-half that of the principal maximum at $\Phi = 0$. Thus, the characteristic property of the ideal couplet of providing a null in one direction does not exist in actual couplets when $\beta_0 h = \pi$ or, in fact, for any other

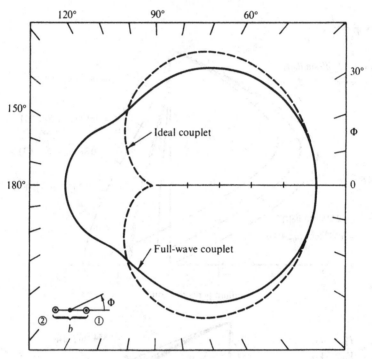

Fig. 3.7. Horizontal pattern of full-wave couplet with $I_{20} = jI_{10}$; $\Omega = 2\ln(2h/a) = 10$, $\beta_0 h = \pi$, $\beta_0 b = 1{\cdot}5$.

value of $\beta_0 h$ that is not near $\pi/2$ or that is not an odd multiple thereof. Significantly, this makes the cardioid pattern of the half-wave couplet a relatively narrow-band property!

CHAPTER 4

THE CIRCULAR ARRAY

The two-element array, which is investigated in the preceding chapter, may be regarded as the special case $N = 2$ of an array of N elements arranged either at the vertices of a regular polygon inscribed in a circle, or along a straight line to form a curtain. Owing to its greater geometrical symmetry, the circular array is advantageously treated next. Indeed, the basic assumptions which underlie the subsequent study of the curtain array (chapter 5) depend for their justification on the prior analysis of the circular array.

The real difficulty in analysing an array of N arbitrarily located elements is that the solution of N simultaneous integral equations for N unknown distributions of current is involved. Although the same set of equations applies to the circular array, they may be replaced by an equivalent set of N independent integral equations in the manner illustrated in chapter 3 for the two-element array. Since the N elements are geometrically indistinguishable, it is only necessary to make them electrically identical as well. One way is to drive them all with generators that maintain voltages that are equal in amplitude and in phase. When this is done all N currents must also be equal in amplitude and in phase at corresponding points. But this is only one of N possibilities. If the N voltages are all equal in magnitude but made to increase equally and pro-gressively in phase from element 1 to element N, a condition may be achieved such that each element is in exactly the same environ-ment as every other element. There are N such possibilities since the phase sequence closes around the circle when the phase shift from element to element is an integral multiple of $2\pi/N$. Any increment in phase given by $2\pi m/N$ with $m = 0, 1, 2, \ldots N - 1$ may be used. Specifically, when $N = 2$, the two possibilities are 0 and π. This means that the two driving voltages and the two currents may be equal in magnitude and in phase $(0, 0)$ or equal in magnitude and 180° out of phase $(0, 180°)$. Similarly, when $N = 3$, there are three possibilities, $0, 2\pi/3$ and $4\pi/3$. The voltages and currents around the circle may now be equal in magnitude with phases, $(0, 0, 0)$, $(0, 120°, 240°)$ or $(0, 240°, 480°)$.

The analysis of the circular array involves the solution of N simultaneous equations similar in form to (2.15). The case $N = 2$ is solved in chapter 3 by rearranging the two simultaneous equations for the currents $I_{1z}(z)$ and $I_{2z}(z)$ into two independent equations. These were derived by adding and subtracting the two original equations. When the elements were driven by voltages which were equal in magnitude and either in phase or 180° out of phase, the resulting currents were independent and named, respectively, the zero and first-phase-sequence currents. The solution for the phase-sequence currents was then carried out and, after a simple algebraic transformation, the actual currents in the elements were derived. The solution for arbitrary driving conditions could also be obtained from the two phase-sequence solutions. A generalization of this procedure is followed in the analysis of the circular array.

The arrays considered here consist of N identical, parallel, non-staggered, centre-driven elements that are equally spaced about the circumference of a circle. This means that the elements are at the vertices of an N-sided regular polygon. Arrays of this type are frequently called single-ring arrays in the literature. Their analysis formally parallels step-by-step the analysis of the two-element array in chapter 3. However, contributions to the vector potential on the surface of each antenna by the currents in all of the elements must be included and this leads to a set of N coupled integral equations for the N currents in the elements. The complete geometrical symmetry of the array permits the use of the method of symmetrical components to reduce the coupled set of integral equations to a single integral equation for each of N possible phase-sequence currents. All other quantities that are required to design and describe the array can be calculated from the solution of essentially one equation with N somewhat different kernels.

The coordinate system and parameters that are used to specify an array are shown in Fig. 4.1 for five elements. The diameter of each element is $2a$, its length is $2h$, the distance between the kth and the ith elements is b_{ki}, the distance between adjacent elements— the length of the side of a regular polygon with the elements at its vertices—is d, and the radius of the circle is ρ.

As indicated in sections 3.9 and 2.10, the two-term approximation is generally adequate when $h \leqslant 5\lambda/8$ and $b \geqslant \lambda/2\pi$. Since it is much simpler and has been used to compute the theoretical results discussed in this chapter, the currents, admittances, and fields in the following sections are determined in the two-term form. Later when matrix notation is introduced, both the two- and the three-term forms of the theory are presented in compact form. This serves both as a summary of the theory of circular arrays and as an introduction to the analysis of more general arrays in chapters 5 and 6.

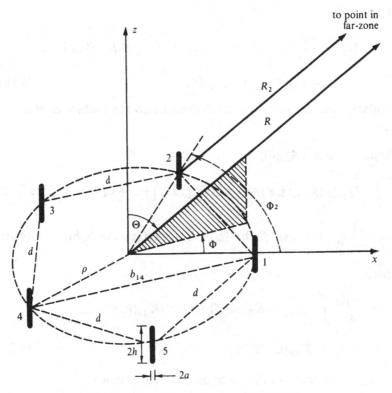

Fig. 4.1. Coordinate system for circular arrays.

4.1 Integral equations for the sequence currents

The vector potential difference at the surface of each element in a circular array of N elements is easily obtained as a generalization of (3.1). Since all elements are thin and parallel to the z-axis, only z-components of the current and the associated vector potential at the surface of each element are significant. Thus, the vector potential difference on the surface of element 1 is

$$4\pi\mu_0^{-1}[A_{1z}(z)-A_{1z}(h)]$$

$$=\int_{-h}^{h}[I_{1z}(z')K_{11d}(z,z')+I_{2z}(z')K_{12d}(z,z')+\dots+I_{Nz}(z')K_{1Nd}(z,z')]\,dz'$$

$$=\frac{j4\pi}{\zeta_0\cos\beta_0 h}[\tfrac{1}{2}V_{10}\sin\beta_0(h-|z|)+U_1(\cos\beta_0 z-\cos\beta_0 h)]. \qquad (4.1\text{a})$$

where

$$U_1=\frac{-j\zeta_0}{4\pi}\int_{-h}^{h}[I_{1z}(z')K_{11}(h,z')+I_{2z}(z')K_{12}(h,z')+\dots$$

$$+I_{Nz}(z')K_{1N}(h,z')]\,dz'. \qquad (4.1\text{b})$$

Similarly, the vector potential difference on the surface of element 2 is

$$4\pi\mu_0^{-1}[A_{2z}(z)-A_{2z}(h)]$$

$$=\int_{-h}^{h}[I_{1z}(z')K_{21d}(z,z')+I_{2z}(z')K_{22d}(z,z')+\dots+I_{Nz}(z')K_{2Nd}(z,z')]\,dz'$$

$$=\frac{j4\pi}{\zeta_0\cos\beta_0 h}[\tfrac{1}{2}V_{20}\sin\beta_0(h-|z|)+U_2(\cos\beta_0 z-\cos\beta_0 h)] \qquad (4.1\text{c})$$

where

$$U_2=\frac{-j\zeta_0}{4\pi}\int_{-h}^{h}[I_{1z}(z')K_{21}(h,z')+I_{2z}(z')K_{22}(h,z')+\dots$$

$$+I_{Nz}(z')K_{2N}(h,z')]\,dz'. \qquad (4.1\text{d})$$

The vector potential difference on the kth element is

$$4\pi\mu_0^{-1}[A_{kz}(z)-A_{kz}(h)]=\int_{-h}^{h}\sum_{i=1}^{N}I_{iz}(z')K_{kid}(z,z')\,dz'$$

$$=\frac{j4\pi}{\zeta_0\cos\beta_0 h}[\tfrac{1}{2}V_{k0}\sin\beta_0(h-|z|)+U_k(\cos\beta_0 z-\cos\beta_0 h)] \qquad (4.1\text{e})$$

$$k=1,2,\dots,N$$

where $\qquad U_k=\frac{-j\zeta_0}{4\pi}\int_{-h}^{h}\sum_{i=1}^{N}I_{iz}(z')K_{ki}(h,z')\,dz'. \qquad (4.1\text{f})$

In these expressions the kernels are

$$K_{kid}(z, z') = K_{ki}(z, z') - K_{ki}(h, z') = \frac{e^{-j\beta_0 R_{ki}}}{R_{ki}} - \frac{e^{-j\beta_0 R_{kih}}}{R_{kih}} \quad (4.1\text{g})$$

with

$$R_{ki} = \sqrt{(z_k - z_i')^2 + b_{ki}^2} \quad (4.1\text{h})$$

$$R_{kih} = \sqrt{(h - z_i')^2 + b_{ki}^2}, \qquad b_{kk} = a. \quad (4.1\text{i})$$

V_{k0} is the applied driving voltage at the centre of element k (or the voltage of an equivalent generator if the element is parasitic with an impedance connected across its terminals), and U_k is the effective driving function characteristic of that part of all the currents that maintains a vector potential of constant amplitude equal to that at $z = h$ along the entire length of the antenna. To reduce the set of N simultaneous equations in (4.1e) to N independent equations, the symmetry conditions characteristic of a circular array must be imposed and the phase-sequence voltages and currents introduced.

Assume that all of the driving voltages are equal in magnitude and have a uniformly progressive phase such that the total phase change around the circle is an integral multiple of 2π. Each multiple of 2π is one of the N phase sequences designated by a superscript (m); these range from zero to $N-1$. In the zero phase sequence, all driving voltages are the same; in the first phase sequence, the driving voltages of adjacent elements differ by $\exp(j2\pi/N)$; in the mth phase sequence, the driving voltages of adjacent elements differ by $\exp(j2\pi m/N)$, and the voltages of the kth and the ith elements are related by

$$V_i = V_k e^{j2\pi(i-k)m/N}. \quad (4.2\text{a})$$

Because of the symmetry of a circular array, the sequence currents in the elements must be related in the same manner as are the sequence driving voltages. That is,

$$I_i(z') = I_k(z') e^{j2\pi(i-k)m/N}. \quad (4.2\text{b})$$

Note that with these driving voltages both the geometric and the electrical environments of each element in the array are identical. Therefore, when (4.2a) and (4.2b) are substituted into the set of coupled integral equations, $I_i(z')$ can be removed from the summation, the remaining kernel is the same regardless of the element to which it is referred, and each equation in the set reduces to

$$\int_{-h}^{h} I^{(m)}(z')K_d^{(m)}(z, z')\,dz'$$

$$= \frac{j4\pi}{\zeta_0 \cos \beta_0 h}[\tfrac{1}{2}V_0^{(m)} \sin \beta_0(h-|z|) + U^{(m)}(\cos \beta_0 z - \cos \beta_0 h)] \quad (4.3)$$

where $m = 0, 1, \ldots N-1$ and

$$U^{(m)} = \frac{-j\zeta_0}{4\pi}\int_{-h}^{h} I^{(m)}(z')K^{(m)}(h, z')\,dz' \quad (4.3a)$$

$$K^{(m)}(h, z') = \sum_{i=1}^{N} e^{j2\pi(i-1)m/N}\frac{e^{-j\beta_0 R_{1ih}}}{R_{1ih}} \quad (4.3b)$$

$$K_d^{(m)}(z, z') = \sum_{i=1}^{N} e^{j2\pi(i-1)m/N}\left[\frac{e^{-j\beta_0 R_{1i}}}{R_{1i}} - \frac{e^{-j\beta_0 R_{1ih}}}{R_{1ih}}\right]. \quad (4.3c)$$

For later use, it is convenient to separate this difference kernel into two parts that depend, respectively, on the real and imaginary parts of the exponential functions. That is

$$K_d^{(m)}(z, z') = K_{dR}^{(m)}(z, z') + jK_{dI}^{(m)}(z, z') \quad (4.3d)$$

where $K_{dR}^{(m)}(z, z') = \sum_{i=1}^{N} e^{j2\pi(i-1)m/N}\,\mathrm{Re}\left[\frac{e^{-j\beta_0 R_{1i}}}{R_{1i}} - \frac{e^{-j\beta_0 R_{1ih}}}{R_{1ih}}\right] \quad (4.3e)$

$$K_{dI}^{(m)}(z, z') = \sum_{i=1}^{N} e^{j2\pi(i-1)m/N}\,\mathrm{Im}\left[\frac{e^{-j\beta_0 R_{1i}}}{R_{1i}} - \frac{e^{-j\beta_0 R_{1ih}}}{R_{1ih}}\right]. \quad (4.3f)$$

The method of solution for (4.3) parallels that of (3.6) and (3.10); the discussion of section 3.2 and the steps of section 3.3 are applicable if note is taken of section 3.9 which relates the two-term to the three-term theory. In fact, the solution is formally given by (3.85) and (3.86) with $m = 0, 1, 2, \ldots, N-1$. This is discussed in somewhat greater detail in a later section (section 4.6) rather than at this point in order to avoid complications in these initial stages of the analysis. Thus, the mth phase-sequence current in the two-term form is given by

$$I^{(m)}(z) = \frac{j2\pi V_0^{(m)}}{\zeta_0 \Psi_{dR} \cos \beta_0 h}[\sin \beta_0(h-|z|) + T^{(m)}(\cos \beta_0 z - \cos \beta_0 h)],$$

$$\beta_0 h \neq \frac{\pi}{2} \quad (4\cdot4a)$$

$$I^{(m)}(z) = \frac{j2\pi V_0^{(m)}}{\zeta_0 \Psi_{dR}}[1 - \sin \beta_0|z| - T'^{(m)} \cos \beta_0 z], \quad \beta_0 h = \frac{\pi}{2}.$$

$$(4.4b)$$

The Ψ and T functions which occur in (4.4a, b) are defined as follows when $\beta_0 d \geqslant 1$:

$$T^{(m)} = \frac{\Psi_V^{(m)}(h) - [\Psi_{d\Sigma}^{(m)} + j\Psi_{dI}^{(m)}]\cos \beta_0 h}{\Psi_{dU}^{(m)} \cos \beta_0 h - \Psi_U^{(m)}(h)} \tag{4.5a}$$

$$T'^{(m)} = \frac{\Psi_{dR} + E_\Sigma\left(\dfrac{\lambda}{4}, \dfrac{\lambda}{4}\right) - S_\Sigma\left(\dfrac{\lambda}{4}, \dfrac{\lambda}{4}\right)}{C_\Sigma\left(\dfrac{\lambda}{4}, \dfrac{\lambda}{4}\right)} \tag{4.5b}$$

$$\Psi_{dR} = \mathrm{Re}\left[\sin \beta_0 h C_{d\Sigma 1}\left(h, h - \frac{\lambda}{4}\right)\right.$$

$$\left. - \cos \beta_0 h S_{d\Sigma 1}\left(h, h - \frac{\lambda}{4}\right)\right], \quad h \geqslant \frac{\lambda}{4} \tag{4.6a}$$

$$= \mathrm{Re}[C_{d\Sigma 1}(h, 0) - \cot \beta_0 h S_{d\Sigma 1}(h, 0)], \quad h < \frac{\lambda}{4} \tag{4.6b}$$

$$\Psi_V^{(m)}(h) = \sin \beta_0 h C_\Sigma^{(m)}(h, h) - \cos \beta_0 h S_\Sigma^{(m)}(h, h) \tag{4.7}$$

$$\Psi_U^{(m)}(h) = C_\Sigma^{(m)}(h, h) - \cos \beta_0 h E_\Sigma^{(m)}(h, h) \tag{4.8}$$

$$\Psi_{d\Sigma}^{(m)} = (1 - \cos \beta_0 h)^{-1}[\sin \beta_0 h C_{d\Sigma 2}^{(m)}(h, 0) - \cos \beta_0 h S_{d\Sigma 2}^{(m)}(h, 0)] \tag{4.9}$$

$$\Psi_{dI}^{(m)} = \mathrm{Im}\{(1 - \cos \beta_0 h)^{-1}[\sin \beta_0 h C_{d\Sigma 1}^{(m)}(h, 0) - \cos \beta_0 h S_{d\Sigma 1}^{(m)}(h, 0)]\} \tag{4.10}$$

$$\Psi_{dU}^{(m)} = (1 - \cos \beta_0 h)^{-1}[C_{d\Sigma}^{(m)}(h, 0) - \cos \beta_0 h E_{d\Sigma}^{(m)}(h, 0)] \tag{4.11}$$

$$C_{d\Sigma}^{(m)}(h, z) = C_\Sigma^{(m)}(h, z) - C_\Sigma^{(m)}(h, h), \quad S_{d\Sigma}^{(m)}(h, z) = S_\Sigma^{(m)}(h, z) - S_\Sigma^{(m)}(h, h)$$

$$E_{d\Sigma}^{(m)}(h, z) = E_\Sigma^{(m)}(h, z) - E_\Sigma^{(m)}(h, h) \tag{4.12}$$

$$C_\Sigma^{(m)}(h, z) = \sum_{i=1}^N e^{j2\pi(i-1)m/N} C_{bi}, \quad C_{bi} = \int_{-h}^h \cos \beta_0 z' \frac{e^{-j\beta_0 R_{bi}}}{R_{bi}} dz' \tag{4.13a}$$

$$S_\Sigma^{(m)}(h, z) = \sum_{i=1}^N e^{j2\pi(i-1)m/N} S_{bi}, \quad S_{bi} = \int_{-h}^h \sin \beta_0 |z'| \frac{e^{-j\beta_0 R_{bi}}}{R_{bi}} dz' \tag{4.13b}$$

$$E_\Sigma^{(m)}(h, z) = \sum_{i=1}^N e^{j2\pi(i-1)m/N} E_{bi}, \quad E_{bi} = \int_{-h}^h \frac{e^{-j\beta_0 R_{bi}}}{R_{bi}} dz' \tag{4.13c}$$

$$R_{bi} = \sqrt{(z - z')^2 + b_i^2}, \quad b_i = a \text{ for } i = 1. \tag{4.14}$$

The subscript $d\Sigma 1$ indicates that only element number 1 ($i = 1$) is to be included and effects of all other elements are ignored; the subscript $d\Sigma2$ indicates that only the effects of elements other than element number 1 are to be included ($i = 1, ..., N$).

In order to evaluate (4.4a, b) it is convenient to lump the various coefficients into new parameters defined as follows:

$$s^{(m)} = \frac{j2\pi}{\zeta_0 \Psi_{dR} \cos \beta_0 h}, \qquad c^{(m)} = s^{(m)} T^{(m)} \qquad (4.15a)$$

$$s'^{(m)} = \frac{j2\pi}{\zeta_0 \Psi_{dR}}, \qquad c'^{(m)} = s'^{(m)} T'^{(m)} \qquad (4.15b)$$

so that (4.4a, b) become

$$\frac{I^{(m)}(z)}{V^{(m)}} = s^{(m)} \sin \beta_0(h - |z|) + c^{(m)}(\cos \beta_0 z - \cos \beta_0 h), \qquad \beta_0 h \neq \frac{\pi}{2}$$
$$(4.16a)$$

$$= s'^{(m)}(1 - \sin \beta_0|z|) - c'^{(m)} \cos \beta_0 z, \qquad \beta_0 h = \frac{\pi}{2}. \qquad (4.16b)$$

The sequence admittances are given by the normalized sequence currents in amperes per volt evaluated at $z = 0$. Thus,

$$Y^{(m)} = \frac{I^{(m)}(0)}{V^{(m)}} = s^{(m)} \sin \beta_0 h + c^{(m)}(1 - \cos \beta_0 h), \qquad \beta_0 h \neq \frac{\pi}{2} \quad (4.17a)$$

$$= s'^{(m)} - c'^{(m)}, \qquad \beta_0 h = \frac{\pi}{2}. \qquad (4.17b)$$

For a circular array of N elements there are N sequences but only $(N+1)/2$ are different if N is odd or $(N/2)+1$ if N is even. This is the same as the number of different self- and mutual admittances.

The sequence currents form a set of functions that are characteristic of the geometrical and electrical properties of the array. Thus, Ψ_{dR} and the $T^{(m)}$ or $T'^{(m)}$ function depend upon the number of elements in the array, their spacing, and the length and thickness of the elements. Once these parameters have been specified, the set of sequence currents can be calculated. Distributions of current in the elements, their driving-point admittances, and the far-zone fields of the arrays with arbitrary driving conditions can be determined from the set of sequence currents with the relations given in section 4.2. Short tables of Ψ_{dR} and $T^{(m)}$ or $T'^{(m)}$ are given in appendix I; additional values are available [1]. It may be noted parenthetically that in the notation of [1], the terms 'quasi-zeroth-order' and 'zeroth-order admittances' refer identically to what is called the 'two-term approximation' in this book.

4.2 Sequence functions and array properties

Imagine the array to be excited simultaneously with currents in all of the N possible phase sequences. Then the driving voltage and current for the kth element are

$$V_k = \sum_{m=0}^{N-1} V^{(m)} e^{j2\pi(k-1)m/N} \qquad (4.18a)$$

$$I_k(z) = \sum_{m=0}^{N-1} I^{(m)}(z) e^{j2\pi(k-1)m/N} \qquad (4.18b)$$

where $V^{(m)}$ is the mth phase-sequence voltage and $I^{(m)}(z)$ is the corresponding phase-sequence current. Similarly, from (4.18b) and (4.17), the self- and mutual admittances are

$$Y_{1k} = \frac{1}{N} \sum_{m=0}^{N-1} Y^{(m)} e^{j2\pi(k-1)m/N}. \qquad (4.19)$$

If the elements of the array are driven by arbitrary voltages V_i which produce corresponding currents $I_i(z)$ along the elements, the sequence voltages and currents are readily obtained from the relations

$$V^{(m)} = \frac{1}{N} \sum_{i=1}^{N} V_i e^{-j2\pi(i-1)m/N} \qquad (4.20a)$$

$$I^{(m)}(z) = \frac{1}{N} \sum_{i=1}^{N} I_i(z) e^{-j2\pi(i-1)m/N}. \qquad (4.20b)$$

With (4.18b) and (4.16) the normalized current distribution along the kth element can conveniently be expressed as follows:

$$\frac{I_k(z)}{V_1} = s_k \sin \beta_0(h-|z|) + c_k(\cos \beta_0 z - \cos \beta_0 h), \qquad \beta_0 h \neq \frac{\pi}{2} \quad (4.21a)$$

$$= s_k'[1 - \sin \beta_0|z|] - c_k' \cos \beta_0 z, \qquad \beta_0 h = \frac{\pi}{2} \qquad (4.21b)$$

where the complex amplitude functions s_k and c_k are

$$s_k = \sum_{m=0}^{N-1} \frac{V^{(m)}}{V_1} s^{(m)} e^{j2\pi(k-1)m/N} \qquad (4.22a)$$

$$c_k = \sum_{m=0}^{N-1} \frac{V^{(m)}}{V_1} c^{(m)} e^{j2\pi(k-1)m/N}. \qquad (4.22b)$$

The corresponding expressions for s_k' and c_k' are similar. The radiation-zone electric field for each element is given by (2.43); the total field is a superposition of the fields maintained by each element. When the currents in the form (4.4a, b) are substituted in (2.43), the resulting expressions for the field are

$$\frac{E_\Theta^r}{KK_1V} = F(\Theta, \beta_0 h) \sum_{i=1}^{N} s_i\, e^{j\rho \cos(\phi_i - \Phi)}$$

$$+ G(\Theta, \beta_0 h) \sum_{i=1}^{N} c_i\, e^{j\rho \cos(\phi_i - \Phi)}, \qquad \beta_0 h \neq \frac{\pi}{2} \qquad (4.23a)$$

$$= H\left(\Theta, \frac{\pi}{2}\right) \sum_{i=1}^{N} s_i'\, e^{j\rho \cos(\phi_i - \Phi)}$$

$$+ F\left(\Theta, \frac{\pi}{2}\right) \sum_{i=1}^{N} c_i'\, e^{j\rho \cos(\phi_i - \Phi)}, \qquad \beta_0 h = \frac{\pi}{2} \qquad (4.23b)$$

with

$$K_1 = \frac{e^{-j\beta_0 R}}{R}, \quad K = \frac{j\zeta_0}{2\pi}, \quad \rho = \frac{\pi\, d/\lambda}{\sin(\pi/N)}, \quad \phi_i = (i-1)2\pi/N.$$

$F(\Theta, \beta_0 h)$, $G(\Theta, \beta_0 h)$ and $H(\Theta, \pi/2)$ are given by (2.46), (2.47), and (2.52a), respectively. These are the so-called element factors and there is one for each type of current distribution. The sums in (4.23a, b) are the array factors. The complex amplitude coefficients are not simply related to one another and the array factors generally cannot be summed in a closed form to yield something equivalent to the familiar $\sin Nx/\sin x$ patterns. In (4.23) the driving voltage V_1 appears since the other driving voltages have been referred to the voltage of element 1. Any other element could have been used for this normalization.

The steps required to make use of this theory in the analysis of a particular array can now be summarized. If the driving voltages are specified, sequence voltages are computed from (4.20a), s_k and c_k from (4.22a, b), the current distributions from (4.21a), and far-zone fields from (4.23a, b). Driving-point admittances are found either from the current evaluated at $z = 0$, viz.,

$$Y_{kin} = \frac{I_k(0)}{V_k} = \frac{I_k(0)}{V_1}\frac{V_1}{V_k} \qquad (4.24a)$$

or from the coupled circuit equations and the self- and mutual admittances

$$Y_{kin} = \sum_{i=1}^{N} \frac{V_i}{V_k} Y_{ki}. \qquad (4.24b)$$

If the driving-point currents are specified, sequence currents can be found from (4.20b), (4.16a) or (4.16b) solved for $V^{(m)}$, and the remaining steps carried out as when the driving voltages are specified. Numerical results for a particular array can be obtained from the tables of appendix I or [1], or from the program outlined in appendix II.

4.3 Self- and mutual admittances

For a circular array with uniformly-spaced elements, self- and mutual admittances are defined in terms of the sequence admittances by (4.19). The more general definition (discussed in chapter 8) of self- and mutual admittances as the coefficients of the driving-point voltages in the coupled circuit equations also applies. For the pth element,

$$I_p(0) = \sum_{i=1}^{N} V_i Y_{pi} \qquad (4.25)$$

from which it follows that the self-admittance Y_{pp} of the pth element is the driving-point admittance of that element when all other elements are present and short-circuited at their driving points. The mutual admittance $Y_{pk}(p \neq k)$ between element p and element k is the driving-point current of element p per unit driving voltage of element k with all other elements present and short-circuited at their driving points. Thus, the mutual admittances characterize the degree in which power that is fed to one element of the array is transferred to the remaining elements.

Among the properties of circular arrays that are revealed by a study of their self- and mutual admittances are resonant spacings at which all of the elements interact vigorously and, in larger arrays, spacings at which at least some of the mutual admittances are very small compared with the self-admittance. In arrays containing only a few elements, the resonant spacings are most important for elements with lengths near $h = \lambda/4$; in larger arrays they are most important for elements with somewhat greater lengths. When the elements in an array are at the resonant spacings, their currents are essentially all in phase and their properties are very sensitive to small changes in frequency. Although calculations of the driving-point admittances generally must include all of the mutual admittances when the array consists of only a few elements, there are ranges of spacings in larger arrays over which at least some of the mutual admittances are much smaller than the self-admittance. In larger arrays there is also a range of spacings over which many of the mutual admittances are nearly the same in magnitude and phase. These properties are illustrated in Figs. 4.2–4.7, which show graphically examples of self- and mutual admittances in millimhos for a range of values of d/λ, the distance between adjacent elements.

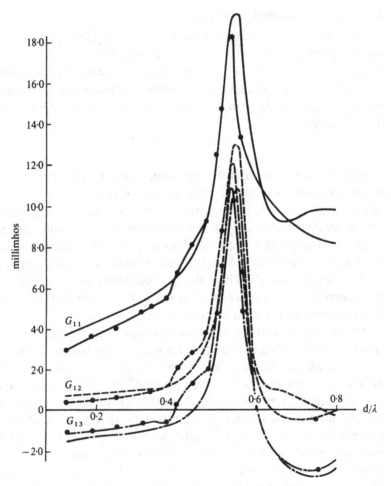

Fig. 4.2a. Measured and theoretical self- and mutual conductances for circular array: $N = 4, h/\lambda = 1/4, a/\lambda = 0{\cdot}007022$.

Except for the self-susceptance shown in Fig. 4.3b, the theoretical results are all evaluated from the two-term theory and were computed from (4.19), (4.17b), and the functions in (4.5)–(4.14) with the program outlined in appendix II. The theoretical self-susceptance

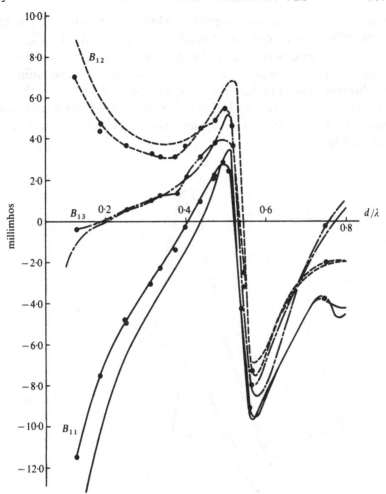

Fig. 4.2*b*.   Like Fig. 4.2*a* but for susceptances.

in Fig. 4.3*b* is shown in the corrected form $B_{11} + j1\cdot16$ with B_{11} calculated from the two-term theory. The correcting susceptance $1\cdot16$ includes the term $0\cdot72$ needed to correct the two-term susceptance and an additional susceptance that takes account of the particular end-effects of the coaxial measuring line. The measured results in Figs. 4.2, 4.3 and 4.4 were obtained from load admittances apparently terminating the coaxial line. They were measured by the distribution-curve method discussed in chapter 8. The experimental apparatus consisted of combined slotted measuring lines and monopoles driven over a ground-screen as shown in Fig. 8.27. The actual measured results have been divided by two and an approximate terminal-zone correction of $Y_T = j0.286$ millimhos as obtained from Fig. 8.3b has been combined with B_{11} so that the final results apply to an ideal centre-driven dipole with all contributions to the admittance by an associated driving mechanism eliminated.

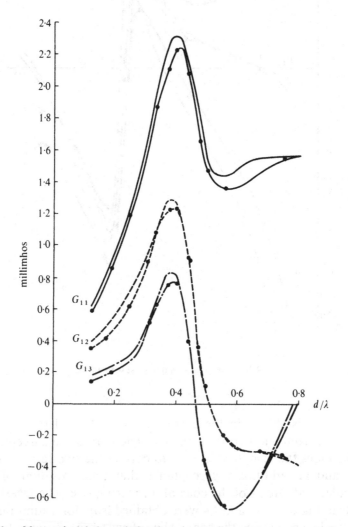

Fig. 4.3a.   Measured and theoretical self- and mutual conductances for circular array
$N = 4, h/\lambda = 3/8, a/\lambda = 0.007022.$

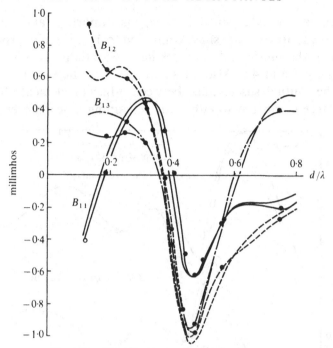

Fig. 4.3*b*.   Like Fig. 4.3*a* but for the susceptances.

An array of four elements of length $h = \lambda/4$ (Fig. 4.2) has a resonant spacing near $d/\lambda = 0.54$. At this spacing all conductances have sharp positive maxima while the susceptances are all essentially zero. If the length of the elements is increased to $h = 3\lambda/8$, a similar resonance occurs in the range between $d/\lambda = 0.37$ and 0.40, but the maxima are not as sharp. With eight elements (Fig. 4.5) there are several resonances but only the first two, which occur near $d/\lambda = 0.35$ and 0.50, are sharply defined. Also, from Fig. 4.5 it is seen that the conductances all have the same sign at the first resonance but not at the second. For twenty elements with length $h = \lambda/4$ it is seen from Fig. 4.6 that a number of resonances occur, but that they no longer have large amplitudes. On the other hand, when the length of the elements is near $h = 3\lambda/8$, the resonances are sharply defined and a small change in spacing (or frequency) produces large changes in the admittances as shown in Fig. 4.7.

Note also that, whereas the four- and eight-element arrays have only one spacing each at which some of the mutual conductances or susceptances are small compared to the self-conductance or susceptance, there is a considerable range of spacings for a twenty-

element array over which only Y_{12} is important and all other
mutual admittances are small compared to Y_{11}. For close spacings,
many of the mutual admittances have essentially the same value
in Figs. 4.5 and 4.6. Also, at small spacings the self-susceptance
and the mutual susceptance between adjacent elements become
very large compared to either the remaining susceptances or the

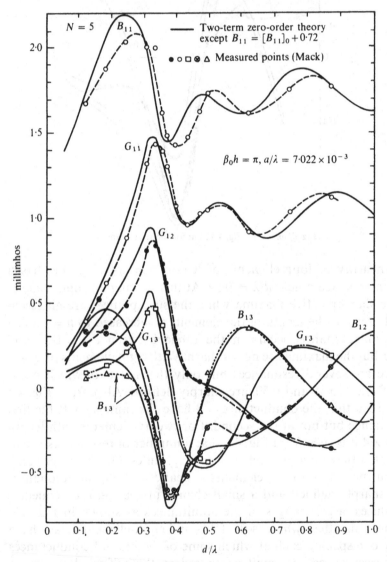

Fig. 4.4. Measured and theoretical self- and mutual admittances, 5-element circular array,
$h = \lambda/2$.

conductances. This indicates that it is these quantities which cause difficulties in matching arrays of closely-spaced elements. These susceptances can be controlled at least partially by an adjustment of the lengths of the elements. Additional, more extensive graphs and tables of self- and mutual admittances are in the literature [2]. All of the results discussed here are for elements with the radius $a/\lambda = 0.007$, but since the parameters of an array change quite slowly with the thickness of the elements, the qualitative behaviour should be the same for thicknesses that do not violate the requirement of 'thin', i.e. $\beta_0 a \leqslant 0.10$. Note, however, that the self-impedances of the individual elements change significantly with their radius—especially when h is not near $\lambda/4$.

Figs. 4.2 and 4.3 indicate that, except near the sharp resonances, the results from the two-term theory are in good agreement with the measured values for all conductances and mutual susceptances. Part of the differences between measured and computed conductances at the sharp resonant maxima for $h = \lambda/4$ may be due to the difficulty encountered in obtaining accurate measurements over this region. The self-susceptance and its correction has been discussed in chapter 2. In Fig. 4.2, no correction has been applied to B_{11}; the use of the correction 0·72 millimhos that was indicated in chapter 2 would yield better agreement for $d/\lambda < 0.40$. In Fig. 4.3, the correction applied to B_{11} is 1·16 millimhos. As previously discussed, this includes both the term 0·72 and an additional empirically determined susceptance that takes account of the end-correction for the coaxial measuring line actually used. It was determined from a comparison of theoretical and measured results for a single element (Fig. 2.6). Since the correction to B_{11} is a constant, it is evident that the correct variation of B_{11} with d/λ is given by the theory. In a practical application, the characteristics of a given array are determined from the theory, a single model of the elements of the array is constructed, and its driving-point admittance measured. The difference between theoretical and measured driving-point susceptances for the single element may be used as a correction for the computed driving-point admittances in the array.

It is sometimes convenient to characterize element intercoupling by self- and mutual impedances instead of admittances. For a general array, the conversion from an admittance basis to an impedance basis requires an inversion of the admittance matrix.

For a circular array, the sequence admittances and impedances are reciprocals, that is,

$$Z^{(m)} = 1/Y^{(m)} \tag{4.26}$$

$$Z_{1i} = \frac{1}{N} \sum_{m=0}^{N-1} e^{j2\pi(i-1)m/N} Z^{(m)} \tag{4.27}$$

so that the reciprocal of only one complex number is required for each sequence.

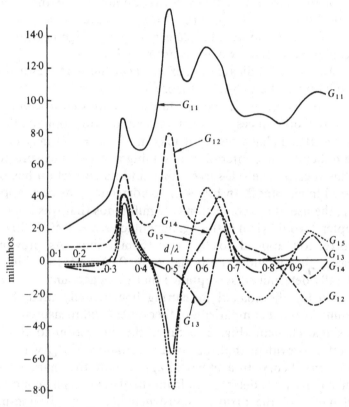

Fig. 4.5a.   Theoretical self- and mutual conductances for circular array; $N = 8$, $h = \lambda/4$, $a/\lambda = 0.007022$.

4.4   Currents and fields; arrays with one driven element

One of the simplest examples of the application of the two-term theory is provided by ring arrays with one element driven and the remaining elements short-circuited at their driving points. In the following examples, the radius of the elements is taken to be

$a/\lambda = 0.007$ and the radiation patterns are all measured or computed in the equatorial plane, $\Theta = \pi/2$. The relative radiation patterns are computed from

$$P_{dB} = 10 \log_{10} \left| \frac{E^r_\Theta(\Theta, \Phi) \cdot E^{r*}_\Theta(\Theta, \Phi)}{E^r_{\Theta m} \cdot E^{r*}_{\Theta m}} \right| \qquad (4.28)$$

when $E^r_{\Theta m}$ is the maximum value of the field in the plane $\Theta = \pi/2$. An asterisk indicates the complex conjugate, $E^r_\Theta(\Theta, \Phi)$ is given by (4.23), and P_{dB} is the relative magnitude of the Poynting vector in decibels.

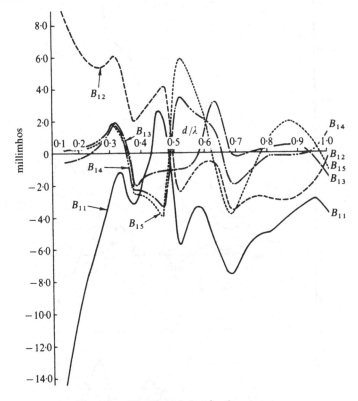

Fig. 4.5b.   Like Fig. 4.5a but for the susceptances.

Fig. 4.8a contains two examples of the radiation patterns of five-element arrays. One pattern is for $d = \lambda/4$ and $h = \lambda/4$ and has a back-to-front ratio of about -14 db with half-power beam widths of about 100°. The second pattern is also for $d = \lambda/4$ but $h = 3\lambda/8$; it has a very smooth angular variation with a back-to-front ratio of about -20 db and beam widths of about $140°$. Agreement between the theoretical and measured results is well within 1 db except near the deeper minimum in the backward direction near $\Phi = 180°$. Similar patterns for $h/\lambda = 0.5$ and two values of d/λ are in Fig. 4.8b.

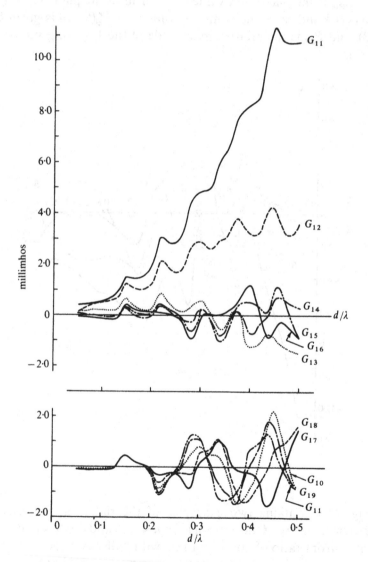

Fig. 4.6a.   Theoretical self- and mutual conductances for circular array; $N = 20$, $h = \lambda/4$, $a/\lambda = 0.007022$.

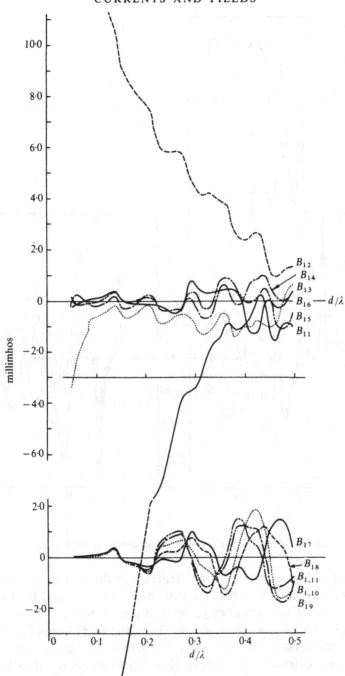

Fig. 4.6b. Like Fig. 4.6a but for the susceptances.

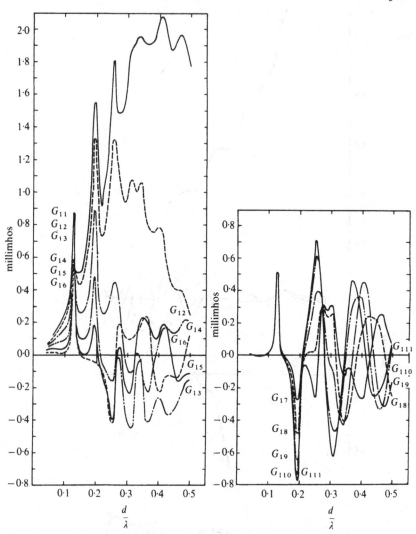

Fig. 4.7a. Theoretical self- and mutual conductances for circular array; $N = 20$, $h = 3\lambda/8$, $a/\lambda = 0.007022$.

Corresponding currents in the elements of the two arrays with the patterns given in Fig. 4.8a are shown in Figs. 4.9 and 4.10. As a consequence of the symmetry, only three of the currents are different for each five-element array. The radiation patterns depend only on the relative distributions of current. If the currents in Figs. 4.9 and 4.10 were simply normalized to their maximum values, it is evident that agreement between theoretical and measured results would be very good and, therefore, measured patterns well

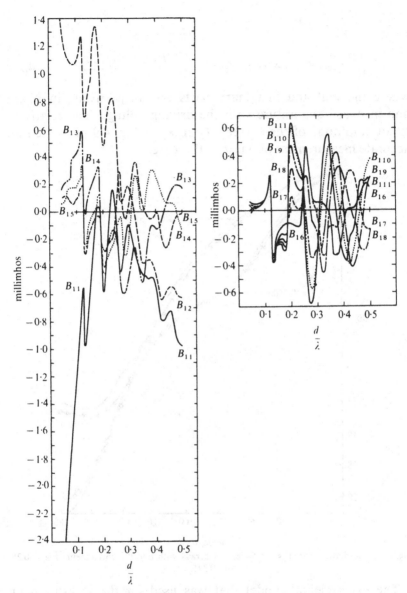

Fig. 4.7b. Like Fig. 4.7a but for the susceptances.

represented by the theory. In order to permit detailed comparison of the experimental and theoretical models, the relative amplitude and phase of the current along each element were measured and normalized to the measured self- and mutual admittances. Thus,

$$\frac{I_k(z)}{V_1} = \frac{|I_k(z)|}{V_1} e^{j\Psi_k(z)}$$

$$= \frac{|I_k(z)|}{V_1}[\cos \Psi_k(z) + j \sin \Psi_k(z)] = \frac{\mathrm{Re}I_k(z) + j\mathrm{Im}I_k(z)}{V_1} \qquad (4.29)$$

where the real and imaginary parts are, respectively, in phase and in phase quadrature with the driving voltage. The relatively small amplitude of the current $|I_3(z)|/V_1$ in Fig. 4.9 prevented an accurate measurement of phase in this case.

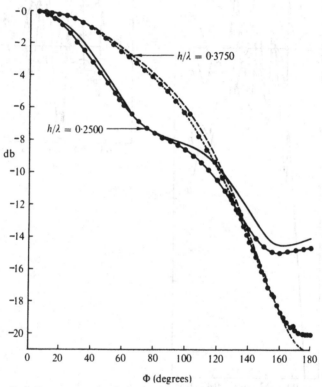

Fig. 4.8a.   Radiation patterns for 5-element arrays with one driven element; $h/\lambda = 0.25$ and 0.375.

The experimental model that was used for the measurement of both field patterns and currents consisted of five monopoles over a ground plane combined with a measuring line for each. Use was made of the image-plane technique described in section 8.9. The equipment and procedures for measuring amplitude and phase are discussed in chapter 8. The s_k and c_k coefficients for use in (4.21) and (4.23) can be computed from the Ψ_{dR}'s and T's in the tables of appendix 1 with the use of (4.15a, b) and (4.22a, b). Numerical data for the two five-element arrays under discussion are

$N = 5, \quad d = \lambda/4, \quad h = \lambda/4:$

$$s_1' = j2\cdot6824; \qquad s_2' = s_3' = s_4' = s_5' = 0;$$
$$c_1' = -4\cdot2084 + j9\cdot2159; \qquad c_2' = c_5' = -0\cdot8072 - j4\cdot6906;$$
$$c_3' = c_4' = 0\cdot6835 - j0\cdot3656; \tag{4.30}$$

$N = 5, \quad d = \lambda/4, \quad h = 3\lambda/8:$

$$s_1 = -j3\cdot6571; \qquad s_2 = s_3 = s_4 = s_5 = 0;$$
$$c_1 = 0\cdot7504 + j1\cdot1391; \qquad c_2 = c_5 = 0\cdot4492 + j0\cdot4419;$$
$$c_3 = c_4 = 0\cdot1890 + j0\cdot1678. \tag{4.31}$$

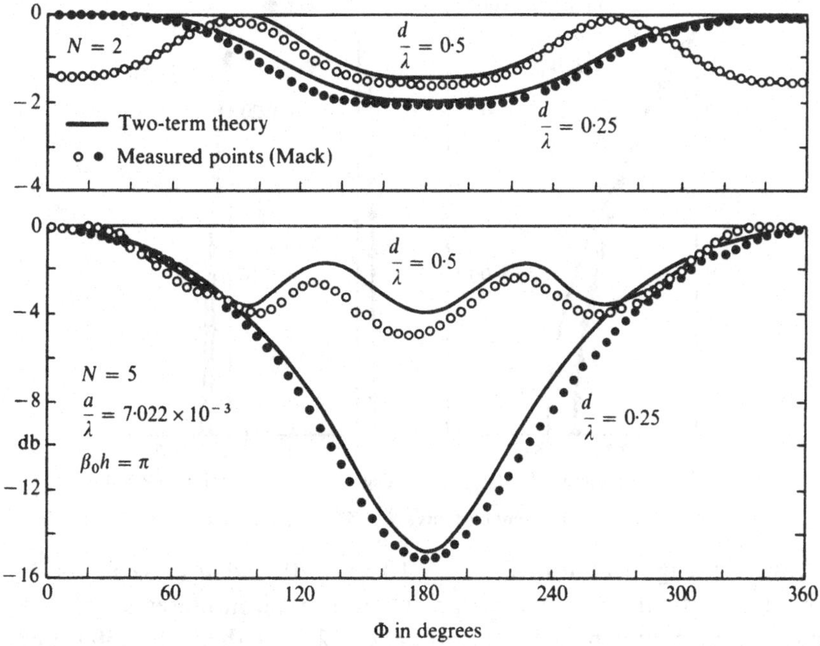

Fig. 4.8b.   Horizontal pattern of parasitic arrays of 2 and 5 elements in a circle; $h/\lambda = 0\cdot50$; d is the distance between adjacent elements.

Note that the currents in the parasitic elements are represented by shifted cosine components only.

The radiation patterns in Fig. 4.8a suggest that spacings can be found at which the pattern is a smooth function of Φ and has a deep minimum near $\Phi = 180°$. Examples of such patterns are

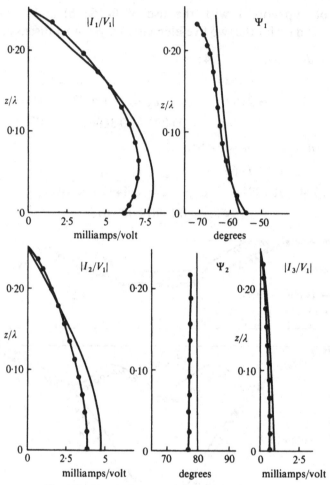

Fig. 4.9. Element currents; $N = 5$, $h = \lambda/4$, $d = \lambda/4$.

shown in Fig. 4.11 for $N = 4, 5, 10$ and 20 and $h = 3\lambda/8$. As N increases, such patterns occur when the circumference of the circle containing the array approaches 2λ. For them, the phase of the electric field is also a smooth slowly changing function of the azimuth angle ϕ as shown in Fig. 4.11. The phase was computed from (4.23a) in the form

$$\frac{E_\Theta^r(\pi/2, \Phi)}{KK_1V_1} = \mathrm{Re}\left[\frac{E_\Theta^r(\pi/2, \Phi)}{KK_1V_1}\right] + j\,\mathrm{Im}\left[\frac{E_\Theta^r(\pi/2, \Phi)}{KK_1V_1}\right]$$

$$= \left|\frac{E_\Theta^r(\pi/2, \Phi)}{KK_1V_1}\right| e^{j\Psi(\pi/2,\Phi)} \qquad (4.32a)$$

$$\Psi(\pi/2, \Phi) = \tan^{-1} \frac{\text{Im } E'_\Theta(\pi/2, \Phi)}{\text{Re } E'_\Theta(\pi/2, \Phi)}. \tag{4.32b}$$

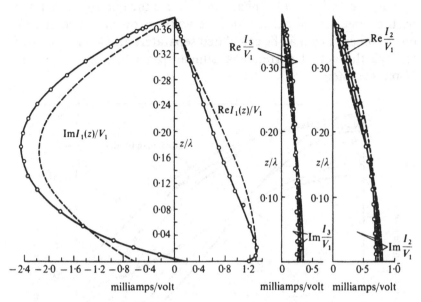

Fig. 4.10. Element currents; $N = 5$, $h = 3\lambda/8$, $d = \lambda/4$.

4.5 Matrix notation and the method of symmetrical components

In the preceding sections the N simultaneous integral equations for the N currents in a circular array were replaced by N independent integral equations by a procedure known as the method of symmetrical components. This procedure was introduced as a generalization of the corresponding treatment of the two coupled equations analysed in chapter 3. It is now appropriate to systematize the general formulation with the compact notation of matrices.

The general method of symmetrical components became well known in its application to problems in multi-phase electric circuits. Loads on three-phase power systems, for example, must generally be balanced to give equal currents in all three branches. Under some conditions unequal loads are placed across the supply lines. The calculation of the resulting branch currents is usually made in terms of three phase-sequence components. The zero phase-sequence currents are all in phase. The first sequence contains three equal phasors which have 120° progressive phase shifts. These phasors rotate in the counter-clockwise direction in the complex plane as time increases. The angular velocity is ω. The second phase sequence has three phasors with equal magnitude and a progressive $-120°$ phase shift. Since the currents generated by the three sets of phase-sequence voltages do not interact with one another, they may be calculated separately and later combined to give the actual currents. A similar procedure applies to an N-phase system.

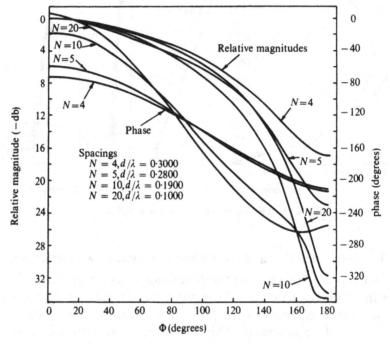

Fig. 4.11.   Power pattern of circular arrays with one driven element; $h = 3\lambda/8$, $a/\lambda = 0.007022$.

The equations which relate the currents and voltages in N coupled circuits have the following matrix form:

$$[Z]\{I\} = \{V\} \tag{4.33}$$

where
$$\{I\} = \begin{Bmatrix} I_1 \\ I_2 \\ \vdots \\ I_N \end{Bmatrix}, \qquad \{V\} = \begin{Bmatrix} V_1 \\ V_2 \\ \vdots \\ V_N \end{Bmatrix} \tag{4.34}$$

$$[Z] = \begin{bmatrix} Z_{11}Z_{12}Z_{13}\cdots Z_{1N} \\ Z_{21}Z_{22} \quad\;\; \cdots Z_{2N} \\ \vdots \\ Z_{N1} \qquad\quad \cdots Z_{NN} \end{bmatrix}. \tag{4.35}$$

The usual reciprocity of off-diagonal impedances is assumed, i.e., $Z_{ij} = Z_{ji}$.

In order to illustrate the application of the method of symmetrical components, this set of equations will not be solved in the usual manner by setting $\{I\} = [Z]^{-1}\{V\}$. Instead, the phase-sequence voltages and impedances will be calculated first by means of the following transformation matrices:

$$\begin{Bmatrix} V^{(0)} \\ V^{(1)} \\ \vdots \\ \vdots \\ V^{(N-1)} \end{Bmatrix} = \frac{1}{N} \begin{bmatrix} 1 & 1 & 1 & \cdots & 1 \\ 1 & p^{-1} & p^{-2} & \cdots & p^{-(N-1)} \\ 1 & p^{-2} & p^{-4} & \cdots & p^{-2(N-1)} \\ \vdots & \vdots & \vdots & & \vdots \\ 1 & p^{-(N-1)} & p^{-2(N-1)} & \cdots & p^{-(N-1)(N-1)} \end{bmatrix} \begin{Bmatrix} V_1 \\ V_2 \\ V_3 \\ \vdots \\ V_N \end{Bmatrix} \tag{4.36}$$

where $p = e^{j2\pi/N}$, or $\quad \{V^{(m)}\} = [P]^{-1}\{V\}$. $\tag{4.37}$

Similarly, for the impedances

$$\{Z^{(m)}\} = [P]^{-1}\{Z\} \tag{4.38}$$

where $\qquad \{Z^{(m)}\} = \begin{Bmatrix} Z^{(0)} \\ Z^{(1)} \\ \vdots \\ Z^{(N-1)} \end{Bmatrix} \tag{4.39}$

$$\{Z\} = \begin{Bmatrix} Z_{11} \\ Z_{12} \\ Z_{13} \\ \vdots \\ Z_{1N} \end{Bmatrix} \tag{4.40}$$

$$[P] = \begin{bmatrix} 1 & 1 & 1 & \dots & 1 \\ 1 & p & p^2 & \dots & p^{(N-1)} \\ 1 & p^2 & p^4 & \dots & p^{2(N-1)} \\ \vdots & \vdots & & & \\ 1 & p^{(N-1)} & p^{2(N-1)} & \dots & p^{(N-1)(N-1)} \end{bmatrix} \qquad (4.41)$$

The phase-sequence currents are given by the algebraic equations

$$I^{(m)} = V^{(m)}/Z^{(m)}, \qquad m = 0, 1, \dots (N-1). \qquad (4.42)$$

The original currents I_i, $i = 1, 2, 3, \dots N$ are given by

$$\{I\} = [P]\{I^{(m)}\} \qquad (4.43)$$

where

$$\{I\} = \begin{Bmatrix} I_1 \\ I_2 \\ \vdots \\ I_N \end{Bmatrix}. \qquad (4.44)$$

As a trivial example of the method, consider two coupled circuits with the same self-impedances. The matrix equation is

$$\begin{bmatrix} Z_{11} & Z_{12} \\ Z_{12} & Z_{11} \end{bmatrix} \begin{Bmatrix} I_1 \\ I_2 \end{Bmatrix} = \begin{Bmatrix} V_1 \\ V_2 \end{Bmatrix}. \qquad (4.45)$$

With $p = e^{j2\pi/N} = e^{j\pi} = -1$, the matrix P^{-1} defined by (4.36) is

$$[P]^{-1} = \tfrac{1}{2}\begin{bmatrix} 1 & 1 \\ 1 & -1 \end{bmatrix}. \qquad (4.46)$$

The phase-sequence voltages and impedances as obtained from (4.37) and (4.38) are

$$\begin{Bmatrix} V^{(0)} \\ V^{(1)} \end{Bmatrix} = \tfrac{1}{2}\begin{bmatrix} 1 & 1 \\ 1 & -1 \end{bmatrix}\begin{Bmatrix} V_1 \\ V_2 \end{Bmatrix} = \tfrac{1}{2}\begin{Bmatrix} V_1 + V_2 \\ V_1 - V_2 \end{Bmatrix} \qquad (4.47)$$

and

$$\begin{Bmatrix} Z^{(0)} \\ Z^{(1)} \end{Bmatrix} = \begin{bmatrix} 1 & 1 \\ 1 & -1 \end{bmatrix}\begin{Bmatrix} Z_{11} \\ Z_{12} \end{Bmatrix} = \begin{Bmatrix} Z_{11} + Z_{12} \\ Z_{11} - Z_{12} \end{Bmatrix}. \qquad (4.48)$$

The resulting phase-sequence currents $I^{(m)}$, $m = 1, 2$, are given by

$$I^{(0)} = \tfrac{1}{2}(V_1 + V_2)/(Z_{11} + Z_{12}) \qquad (4.49)$$

and

$$I^{(1)} = \tfrac{1}{2}(V_1 - V_2)/(Z_{11} - Z_{12}). \qquad (4.50)$$

The desired currents I_i, $i = 1, 2$ (which are generated by the actual driving voltages V_1 and V_2) are given by (4.43) with (4.49) and (4.50). They are

$$\begin{Bmatrix} I_1 \\ I_2 \end{Bmatrix} = \begin{bmatrix} 1 & 1 \\ 1 & -1 \end{bmatrix}\begin{Bmatrix} I^{(0)} \\ I^{(1)} \end{Bmatrix} = \begin{Bmatrix} (V_1 Z_{11} - V_2 Z_{12})/(Z_{11}^2 - Z_{12}^2) \\ (V_1 Z_{11} + V_2 Z_{12})/(Z_{11}^2 - Z_{12}^2) \end{Bmatrix}. \qquad (4.51)$$

These equations are, of course, the same as those obtained directly from (4.45). Note that in the method of symmetrical components the matrix inversion is performed in a number of straightforward steps. In the analysis of circular arrays it allows a large matrix to be inverted for each phase sequence by obtaining the reciprocal of one complex number.

4.6 General formulation and solution

In section 4.1 the solutions for the N independent integral equations for the phase-sequence currents in a circular array of identical elements were obtained by a logical generalization of the parallel analysis for the two-element array in chapter 3. A more complete formulation and solution with special reference to the complications of the N-element array is now in order.

With the matrix notation introduced in section 4.5, the integral equations (4.3) for the N phase-sequence currents may be expressed as follows

$$\int_{-h}^{h} I_z^{(m)}(z')K_d^{(m)}(z,z')\,dz'$$

$$= \frac{j4\pi}{\zeta_0 \cos \beta_0 h}\left[\tfrac{1}{2}V^{(m)}\sin \beta_0(h-|z|)+U^{(m)}(\cos \beta_0 z - \cos \beta_0 h)\right] \quad (4.52)$$

where
$$\{I_z^{(m)}\} = [P]^{-1}\{I_z\} \quad (4.53a)$$
$$\{V^{(m)}\} = [P]^{-1}\{V\} \quad (4.53b)$$
$$\{K_d^{(m)}(z,z')\} = [P]^{-1}\{K_d(z_i,z')\} \quad (4.54)$$

and
$$U^{(m)} = \frac{-j\zeta_0}{4\pi}\int_{-h}^{h} I_z^{(m)}(z')K^{(m)}(h,z')\,dz'. \quad (4.55)$$

In order to reduce the integral equation (4.52) to an approximately equivalent algebraic equation in the manner described in chapter 3, it is necessary to introduce approximate expressions for the several parts of the integral. The procedure and the reasoning behind it is in principle the same as described in sections 3.2 and 3.3 for two elements. However, for N elements in a circle the kernel consists of a sum of N instead of two terms. In the interest of simplicity, the introductory discussion in section 4.1 assumed that all elements are separated by distances sufficiently great so that $\beta_0 b_{ki} \geqslant 1$ for all values of k and i. Although this condition is satisfied in most circular arrays, there are exceptions. One is the cage antenna in which the N parallel elements are distributed around an electrically small circle so that the condition $\beta_0 b_{ki} < 1$

is satisfied for all k and i. An intermediate case arises when the circle is electrically large, but the elements are quite closely spaced so that one or more on each side of every element satisfies the inequality $\beta_0 b_{ki} < 1$, but all of the others are far enough away so that $\beta_0 b_{ki} \geqslant 1$. Since the behaviour of the parts of the integrals that relate closely spaced elements is different from the parts that represent widely spaced ones, it is necessary to treat them separately. Since for each phase sequence all elements are in identical environments, element no. 1 is conveniently selected for reference. Let it be assumed that n elements on each side of element 1 are sufficiently near so that for them $\beta_0 b_{1i} < 1$, $1 \leqslant i \leqslant n$, $N - n + 1 \leqslant i \leqslant N$ and that for all other elements, $\beta_0 b_{1i} \geqslant 1$, $n + 1 < i < N - n + 1$. Let the sum over all the $2n + 1$ elements for which $\beta_0 b_{1i} < 1$ be denoted by $\Sigma 1$, the sum over all other elements in the circle by $\Sigma 2$. Similarly, let $K_{d\Sigma 1}^{(m)}(z, z')$ be the part of the sum in (4.3c) which includes the $2n + 1$ elements for which $\beta_0 b_{1i} < 1$, $K_{d\Sigma 2}^{(m)}(z, z')$ the rest of the sum. It now follows by analogy with (3.19a, b) that

$$\int_{-h}^{h} \sin \beta_0(h - |z'|) K_{d\Sigma 1 R}^{(m)}(z, z')\, dz' \doteq \Psi_{d\Sigma 1 R}^{(m)} \sin \beta_0(h - |z|) \qquad (4.56a)$$

$$\int_{-h}^{h} \sin \beta_0(h - |z'|) K_{d\Sigma 2 R}^{(m)}(z, z')\, dz' \doteq \Psi_{d\Sigma 2 R}^{(m)} (\cos \beta_0 z - \cos \beta_0 h) \qquad (4.56b)$$

where $K_{d\Sigma 1 R}^{(m)}(z, z')$ and $K_{d\Sigma 2 R}^{(m)}(z, z')$ are the appropriate parts of $K_{dR}^{(m)}(z, z')$ as defined in (4.3e). On the other hand, all remaining parts of the integral are independent of $\beta_0 b_{1i}$, so that they are the same as in (3.20a)–(3.23) but with $K_{dR}^{(m)}(z, z')$ and $K_{dI}^{(m)}(z, z')$ as given in (4.3e) and (4.3f).

The Ψ-functions introduced in (4.56a, b) are defined as follows:

$$\Psi_{dR}^{(m)} \equiv \Psi_{d\Sigma 1 R}^{(m)} = \Psi_{d\Sigma 1 R}^{(m)}(z_m); \begin{cases} z_m = 0 & \beta_0 h \leqslant \pi/2 \\ z_m = h - \lambda/4, & \beta_0 h > \pi/2 \end{cases} \qquad (4.57a)$$

$$\Psi_{d\Sigma 1 R}^{(m)}(z) = \csc \beta_0(h - |z|) \int_{-h}^{h} \sin \beta_0(h - |z'|) K_{d\Sigma 1 R}^{(m)}(z, z')\, dz' \qquad (4.57b)$$

$$\Psi_{d\Sigma R}^{(m)} \equiv \Psi_{d\Sigma 2 R}^{(m)} = (1 - \cos \beta_0 h)^{-1} \int_{-h}^{h} \sin \beta_0(h - |z'|) K_{d\Sigma 2 R}^{(m)}(0, z')\, dz'.$$

$$(4.58)$$

These are generalizations of (3.24a, b) and (3.25a, b). The other Ψ-functions, specifically $\Psi_{dU}^{(m)} = \Psi_{dUR}^{(m)} + j\Psi_{dUI}^{(m)}$, $\Psi_{dD}^{(m)} = \Psi_{dDR}^{(m)} + j\Psi_{dDI}^{(m)}$ and $\Psi_{dI}^{(m)}$ are the same as defined in (3.26)–(3.28) but with the N-

term kernel given in (4.3d). Note that when all elements are sufficiently far apart to satisfy the inequality $\beta_0 b_{1i} > 1, 1 < i \leqslant N$, only
$i = 1$, with $b_{11} = a$ contributes to $\Psi_{dR}^{(m)}$ which is then equal to Ψ_{dR}
for the isolated element.

With the notation introduced in (4.57a) and (4.58), the equation
(3.33) applies directly to the N-element array. The same equation
with $\Psi_{dR}^{(m)}$ substituted for Ψ_{dR} is correct when some elements are
sufficiently close together so that $\beta_0 b_{1i} < 1, i > 1$. It follows that
the entire formal solution in sections 3.3 and 3.4 is valid for the phase
sequences of the N-element array. The N independent phase-
sequence currents $I^{(m)}(z), m = 0, 1, \dots N$, may be expressed as the
solution of a column matrix equation.

A summary of the relevant equations is given below.

Phase sequence currents

$$\{I_z^{(m)}(z)\} = \frac{j2\pi}{\zeta_0 \Psi_{dR} \cos \beta_0 h}$$
$$\times [\{V_0^{(m)} M_{0z}\} + \{V_0^{(m)} T_U^{(m)} F_{0z}\} + \{V_0^{(m)} T_D^{(m)} H_{0z}\}] \quad (4.59)$$

where $M_{0z} = \sin \beta_0(h - |z|)$, $F_{0z} = \cos \beta_0 z - \cos \beta_0 h$, and
$H_{0z} = \cos(\beta_0 z/2) - \cos(\beta_0 h/2)$.

$$\begin{Bmatrix} T_U^{(m)} \\ T_D^{(m)} \end{Bmatrix} = [\Phi_T^{(m)}]^{-1} \begin{Bmatrix} \Psi_V^{(m)}(h) - \Psi_{d\Sigma R}^{(m)} \cos \beta_0 h \\ -j\Psi_{dI}^{(m)} \end{Bmatrix} \quad (4.60a)$$

$$[\Phi_T^{(m)}] = \begin{bmatrix} \Phi_{T11}^{(m)} & \Phi_{T12}^{(m)} \\ \Phi_{T21}^{(m)} & \Phi_{T22}^{(m)} \end{bmatrix} \quad (4.60b)$$

$[\Phi_T^{(m)}]^{-1}$ is the reciprocal of $[\Phi_T^{(m)}]$.

$$\left.\begin{aligned} \Phi_{T11}^{(m)} &= \Psi_{dUR}^{(m)} \cos \beta_0 h - \Psi_U^{(m)}(h) \\ \Phi_{T12}^{(m)} &= -\Psi_D^{(m)}(h) \\ \Phi_{T21}^{(m)} &= j\Psi_{dUI}^{(m)} \\ \Phi_{T22}^{(m)} &= \Psi_{dD}^{(m)}. \end{aligned}\right\} \quad (4.61)$$

The phase-sequence admittance is given by setting $z = 0$ in (4.59),
thus:

$$Y_0^{(m)} = \frac{I_z^{(m)}(0)}{V_0^{(m)}}. \quad (4.62)$$

The phase-sequence impedance is the reciprocal of the phase-
sequence admittance,

$$Z^{(m)} = \frac{1}{Y^{(m)}}. \quad (4.63)$$

The mutual impedances may be calculated from the phase-sequence admittances by multiplying by the inverse (4.36) of the phase-sequence matrix P. Thus,

$$Z_{1i} = [P]^{-1}Z^{(m)}, \qquad 1 < i \leqslant N. \qquad (4.64)$$

When the identical elements of a circular array are equally spaced around a circle, symmetry reduces the number of different admittances or impedances to $(N+1)/2$ if N is odd and $(N/2)+1$ if N is even. For example,

$$Z_{12} = Z_{1N}; \qquad Z_{13} = Z_{1(N-1)}; \qquad Z_{14} = Z_{1(N-2)} \text{ etc.} \quad (4.65)$$

When the expression for the phase-sequence currents becomes indeterminate for $\beta_0 h = \pi/2$ and for a range near this value, the alternative form given in (3.43)–(3.45) is useful. It is

$$\{I_z^{(m)}(z)\} = \frac{-j2\pi}{\zeta_0 \Psi_{dR}} \{V_0^{(m)} S_{0z} + V_0^{(m)} T_U^{\prime(m)} F_{0z}$$

$$- V_0^{(m)} T_D^{\prime(m)} H_{0z}\}, \qquad \beta_0 h \sim \frac{\pi}{2} \qquad (4.66)$$

where
$$S_{0z} = \sin \beta_0 |z| - \sin \beta_0 h$$

$$= \sin \beta_0 |z| - 1 \quad \text{when } \beta_0 h = \frac{\pi}{2} \qquad (4.67)$$

and
$$\left.\begin{array}{l} \{T_U^{\prime(m)}\} = -\{(T_U^{(m)} + \sin \beta_0 h)/\cos \beta_0 h\} \\ \{T_D^{\prime(m)}\} = \{T_D^{(m)}/\cos \beta_0 h\}. \end{array}\right\} \qquad (4.68)$$

The two-term approximation used earlier in this chapter is quickly obtained from the three-term formulas. As stated in section 3.9, the procedure involves the substitution of F_{0z} for H_{0z} and $T^{(m)}$ for $T_U^{(m)} + T_D^{(m)}$. This implies that $\Psi_{dD} \to \Psi_{dU}$, $\Psi_D(h) \to \Psi_U(h)$. The two-term forms for the phase-sequence currents and admittances (cf. (3.85)–(3.87)) are

$$\{I_z^{(m)}(z)\} = \frac{j2\pi}{\zeta_0 \Psi_{dR} \cos \beta_0 h} \{V_0^{(m)} M_{0z} + V_0^{(m)} T^{(m)} F_{0z}\} \quad (4.69)$$

and
$$\{Y_0^{(m)}\} = \left\{\frac{I_z^{(m)}(0)}{V_0^{(m)}}\right\} \qquad (4.70)$$

where
$$\{T^{(m)}\} = -\left\{\frac{\Psi_V^{(m)}(h) - (\Psi_{d\Sigma R}^{(m)} + j\Psi_{dI}^{(m)}) \cos \beta_0 h}{\Psi_U^{(m)}(h) - \Psi_{dU}^{(m)} \cos \beta_0 h}\right\}. \qquad (4.71)$$

When $\beta_0 h$ is at or near $\pi/2$ the alternative formulas (3.88)–(3.90) are applicable. They are

$$\{I_z^{(m)}(z)\} = \frac{-j2\pi}{\zeta_0 \Psi_{dR}} \{V_0^{(m)} S_{0z} + V_0^{(m)} T^{\prime(m)} F_{0z}\} \qquad (4.72)$$

$$\{Y^{(m)}\} = \left\{\frac{I_z^{(m)}(0)}{V_0^{(m)}}\right\} \tag{4.73}$$

where $$\{T'^{(m)}\} = -\left\{\frac{T^{(m)} + \sin\beta_0 h}{\cos\beta_0 h}\right\}, \qquad \beta_0 h \sim \frac{\pi}{2}. \tag{4.74}$$

Note that (4.69) and (4.72) are the same as (4.4a) and (4.4b).

CHAPTER 5

THE CIRCUIT AND RADIATING
PROPERTIES OF CURTAIN ARRAYS

In chapter 1 the conventional approach to antenna theory is reviewed and the radiation and circuit properties of the single antenna and array of antennas are presented under the conventional assumptions. Possible sources of error are also pointed out. In chapters 2 and 3 a new and more accurate theory is presented for a single antenna and for a two-element array. The present chapter is concerned with the analysis and synthesis of the general N-element curtain array. This is a linear array with the centres of all elements along a straight line but with their axes all perpendicular to and in a plane containing the line.

5.1   General comparison of conventional and new theories

The analysis of arrays is conventionally formulated under the implicit assumption that distributions of current along all elements are identical. It follows that self- and mutual impedances depend only on the geometry of the elements. Circuit equations can then be written to relate the driving-point voltages and currents through an impedance matrix. Thus,

$$\{V\} = [Z]\{I\} \qquad (5.1)$$

where

$$\{V\} = \begin{Bmatrix} V_{01} \\ V_{02} \\ \vdots \\ V_{0N} \end{Bmatrix}, \qquad \{I\} = \begin{Bmatrix} I_{01} \\ I_{02} \\ \vdots \\ I_{0N} \end{Bmatrix} \qquad (5.2)$$

and

$$[Z] = \begin{bmatrix} Z_{11} & Z_{12} & Z_{13} & \cdots & Z_{1N} \\ Z_{21} & Z_{22} & Z_{23} & \cdots & Z_{2N} \\ \vdots & & & & \\ Z_{N1} & Z_{N2} & Z_{N3} & \cdots & Z_{NN} \end{bmatrix}. \qquad (5.3)$$

The bracket terms are $N \times N$ matrices; the terms in braces are column matrices. The usual reciprocity of off-diagonal impedances in (5.3) holds (i.e. $Z_{12} = Z_{21}$ etc.). Equation (5.1) relates the quantities that can be assigned at the driving point of the antenna, namely the voltages and currents. The simple matrix relation between V's and I's shows that it is *immaterial* whether the voltages or the currents are specified, since the ratio between each voltage and current is unchanged. The phase and magnitude of the currents in the individual elements are normally specified so as to produce a particular radiation pattern. The assumption of identical distributions of current on all elements involves the tacit assumption that the phase and amplitude of the current at all points in each element are completely determined by their values at the driving point.

The preceding remarks may, at first glance, seem like a repetition of well-known facts. However, the assumptions implied in the conventional formulation are not satisfactory approximations for actual arrays except when the elements are very thin and have lengths near $\lambda/2$. Even for this special case difficulties arise when the elements are very closely spaced. Fortunately, a more realistic theory can be developed that is generally applicable to arrays with elements that are less than $3\lambda/4$ in half-length. The new theory is somewhat more complicated than the conventional approach. However, for engineering purposes it is more important that a theory agree with experiment than that it be mathematically simple. As with most new approaches, much of the complexity disappears with continued use and understanding. At the outset the fundamental processes will be explained without reference to the details of the theory.

An example of the notation of a three-element array is shown in Fig. 5.1. The conventional assumption is that regardless of the driving conditions each element has the same distribution of current. For example,

$$I_i(z) = I_{0i} \frac{\sin \beta_0(h - |z|)}{\sin \beta_0 h}, \qquad i = 1, 2, 3. \qquad (5.4)$$

Equation (5.4) shows that once the currents are assigned at any point, e.g. at $z = 0$, the entire current is completely specified. The new theory does not require the individual currents to have the same distributions. They may have the following much more general trigonometric representation:

$$I_i(z) = jA_i \sin \beta_0(h - |z_i|) + B_i(\cos \beta_0 z_i - \cos \beta_0 h) \qquad (5.5)$$

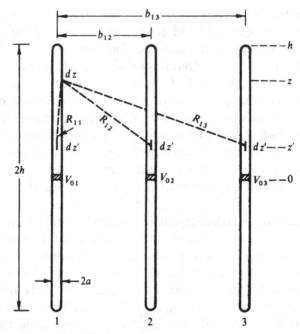

Fig. 5.1.   Three-element array.

where $i = 1, 2, 3$, and A_i is real and B_i is complex. In (5.5) the A coefficients are directly proportional to the respective driving voltages. That is,

$$A_i = CV_{0i}, \qquad i = 1, 2, 3 \qquad (5.6)$$

where C is a constant. On the other hand, the complex B coefficients depend on contributions not only from the individual element but also from all of the remaining elements. For example, there are contributions to B_1 from V_{01} and also from V_{02} and V_{03}.

Consider now the simplest circuit problem for an array of N elements. Given the N driving-point currents, what must be the driving voltages? In addition, the individual and generally different distributions of the N currents along the elements are required. By the methods of the conventional theory the specified problem is solved directly from (5.1). If the I_{0i}'s are given, the corresponding V_{0i}'s are obtained from the solution of (5.1). For the inverse problem, the I_{0i}'s that correspond to specifying the V_{0i}'s are given by the solution of the matrix equation

$$\{I\} = [Z]^{-1}\{V\} = [Y]\{V\} \qquad (5.7)$$

where $[Y] = [Z]^{-1}$ is the inverse of Z. The solution of (5.1) or (5.7) gives the relative values of the complex driving-point currents, hence also the entire current from (5.4). Clearly, the solution of this problem is more complicated in the new theory since the three quantities, A_i, and the real and imaginary parts of B_i must be determined for each element from the assigned real and imaginary parts of the driving-point currents $I_{0i} = I_i(0)$ and expressed in terms of the complex voltages V_{0i}.

5.2 New theory of curtain arrays

The theoretical solution of the general problem of the curtain array will now be examined in detail. The essential basis for this theory was given in chapter 2. Since the two-term theory described in section 2.10 yields results of sufficient accuracy, it will be used for the curtain array to reduce the complexity of the formulation. However, the more accurate three-term representation involves only added algebraic complications.

The integral equation (2.4) may be written as follows for the kth antenna of an N-element array:

$$4\pi\mu_0^{-1}A_{zk}(z) = \int_{-h}^{h} \sum_{i=1}^{N} I_{zi}(z')K_{ki}(z, z')\,dz'$$

$$= -j\frac{4\pi}{\zeta_0}(C_k \cos \beta_0 z + \tfrac{1}{2}V_{0k} \sin \beta_0|z|) \quad (5.8)$$

where
$$K_{ki}(z, z') = \frac{e^{-j\beta_0 R_{ki}}}{R_{ki}}, \quad (5.9a)$$

$$R_{ki} = \sqrt{(z_k - z_i')^2 + b_{ki}^2}, \quad b_{kk} = a, \quad \zeta_0 \doteq 120\pi \text{ ohms}. \quad (5.9b)$$

The notation is illustrated in Fig. 5.1 for a three-element array. If the array has N elements, it is necessary to solve N simultaneous integral equations of the form (5.8), where $k = 1, 2, 3, \ldots N$. Following the procedure used in (2.3)–(2.6), an approximate zero-order solution will be obtained for the general linear array. That is, given the N driving voltages, a solution will be obtained for the currents in the N elements. Alternatively, given the N driving-point currents, the N driving voltages will be determined.

As a first step in the solution, the constant part of the vector potential is removed from the right-hand side of (5.8) by the introduction of the vector potential difference

$$W_{zk}(z) = A_{zk}(z) - A_{zk}(h).$$

The result is

$$4\pi\mu_0^{-1}W_{zk}(z) = \int_{-h}^{h}\sum_{i=1}^{N}I_{zi}(z')K_{kid}(z,z')\,dz' \tag{5.10}$$

$$= -j\frac{4\pi}{\zeta_0}[C_k\cos\beta_0 z + \tfrac{1}{2}V_{0k}\sin\beta_0|z|]$$

$$-\int_{-h}^{h}\sum_{i=1}^{N}I_{zi}(z')K_{ki}(h,z')\,dz' \tag{5.11}$$

where
$$K_{kid}(z,z') = \frac{e^{-j\beta_0 R_{ki}}}{R_{ki}} - \frac{e^{-j\beta_0 R_{kih}}}{R_{kih}}. \tag{5.12}$$

The constants of integration C_k are expressed in terms of quantities U_k that are proportional to the $A_{zk}(h)$ by means of the relation $W_{zk}(h) = 0$. The final form of the integral (5.8) is [cf. (2.15)]

$$\int_{-h}^{h}\sum_{i=1}^{N}I_{zi}(z')K_{kid}(z,z')\,dz' = j\frac{4\pi}{\zeta_0 F_0(h)}(U_k F_{0z} + \tfrac{1}{2}V_{0k}M_{0z}) \tag{5.13}$$

where
$$F_{0z} = F_0(z) - F_0(h) = \cos\beta_0 z - \cos\beta_0 h \tag{5.14}$$

$$M_{0z} = F_0(z)G_0(h) - G_0(z)F_0(h) = \sin\beta_0(h-|z|) \tag{5.15}$$

$$U_k = \sum_{i=1}^{N}U_{ki} = -j\frac{\zeta_0}{4\pi}\int_{-h}^{h}\sum_{i=1}^{N}I_{zi}(z')K_{ki}(h,z')\,dz'. \tag{5.16}$$

The difference kernel (5.12) may be separated into its real and imaginary parts as follows [cf. (2.5) et seq.]:

$$K_{kidR}(z,z') + jK_{kidI}(z,z') = K_{kid}(z,z') \tag{5.17}$$

where
$$\left.\begin{array}{l}K_{kidR} = K_{kiR}(z,z') - K_{kiR}(h,z')\\K_{kidI} = K_{kiI}(z,z') - K_{kiI}(h,z')\end{array}\right\}. \tag{5.18}$$

For the single element, the integrals corresponding to those in (5.13) were separated into two groups depending on the manner in which their leading terms varied as functions of z. The same principle of separation may be applied to the integrals for the curtain array. As before, one group varies approximately as M_{0z}, the other as F_{0z}. The following functional forms for the integrals in (5.13) are important general criteria for the separation:

$$\int_{-h}^{h}I_{zi}(z')K_{kiR}(z,z')\,dz' \sim I_{zi}(z) \quad \text{when } \beta_0 b_{ki} < 1 \tag{5.19}$$

$$\int_{-h}^{h}I_{zi}(z')K_{kiR}(z,z')\,dz' \sim F_{0z} \quad \text{when } \beta_0 b_{ki} \geqslant 1 \tag{5.20}$$

$$\int_{-h}^{h} I_{zi}(z')K_{kil}(z, z')\, dz' \sim F_{0z} \quad \text{for any } I(z') \text{ and all } \beta_0 b_{ki}. \tag{5.21}$$

The current in each element can now be expressed in two parts in the form

$$I_{zi}(z) = I_{ui}(z) + I_{vi}(z) \tag{5.22}$$

where, by definition, the leading terms behave approximately as follows:

$$I_{vi}(z) \sim M_{0z}, \qquad I_{ui}(z) \sim F_{0z}. \tag{5.23}$$

Some appreciation of the importance of the general functional forms in (5.23) may be obtained from an investigation of the integral equation (5.13). If attention is directed to the right-hand side of (5.13), it is seen that the equation contains two apparent sources, the coefficients of F_{0z} and M_{0z}. The function U_k has a constant amplitude over the entire length of the kth element and is generated primarily by the distributed currents on each element in the array. The other source function is the potential difference V_{0k}, as in a transmission line or in an isolated antenna; it is localized at $z = 0$. The form of the integral equation (5.13) suggests that the current on each element can be separated into two parts, the one apparently generated by the U_k, the other by the V_{0k}. The part of the current due to U_k is closely related to the current in an unloaded receiving antenna that is located in the wave front of an incident plane-wave field that has the same amplitude and phase over the entire length of the element. For this the leading term varies as F_{0z}. Except when the elements are very closely spaced (as in an open-wire line), the sinusoidal parts of the currents (i.e. M_{0z}) are maintained primarily by the individual driving potentials V_{0k}. Thus, the current due to each of the V_{0k} is essentially the same as in an isolated antenna.

When (5.22) is substituted in (5.13), groups of integrals occur that may be expressed as follows for all k and i in the ranges 1 to N:

$$\int_{-h}^{h} I_{ui}(z')K_{kid}(z, z')\, dz' = \left(\frac{B_i}{B_k}\right)\Psi_{ki\,du}I_{uk}(z) - D_{ki\,du}(z) \tag{5.24}$$

$$\int_{-h}^{h} I_{vi}(z')K_{kid}(z, z')\, dz'$$

$$= \left(\frac{jA_i}{B_k}\right)\Psi_{ki\,dv}I_{vk}(z) - D_{ki\,dv}(z); \qquad \beta_0 b_{ki} \geqslant 1 \tag{5.25}$$

$$\int_{-h}^{h} I_{vi}(z')K_{ki\,dR}(z, z')\,dz'$$

$$= \left(\frac{A_i}{A_k}\right)\Psi_{ki\,dR}I_{vk}(z) - D_{ki\,dR}(z); \qquad \beta_0 b_{ki} < 1 \qquad (5.26)$$

$$\int_{-h}^{h} I_{vi}(z')K_{ki\,dI}(z, z')\,dz'$$

$$= \left(\frac{jA_i}{B_k}\right)\Psi_{ki\,dI}I_{uk}(z) - D_{ki\,dI}(z); \qquad \beta_0 b_{ki} < 1. \qquad (5.27)$$

It is assumed that the functions Ψ_{ki} are defined so that the difference terms $D_{ki}(z)$ are small enough to be negligible in a solution of zero order. The coefficient (jA_i/B_k) in (5.27) is the ratio of the amplitude of $I_{vi}(z)$ to that of $I_{uk}(z)$. When (5.24)–(5.27) are substituted in (5.13) and only the leading terms are retained, the following separation into two groups of equations may be carried out:

$$\sum_{i=k-m}^{k+m} \left(\frac{A_i}{A_k}\right)\Psi_{ki\,dR}I_{vk}(z) = j\frac{2\pi}{\zeta_0 F_0(h)}V_{0k}M_{0z} \qquad (5.28)$$

$$\sum_{i=1}^{N}\left(\frac{B_i}{B_k}\right)\Psi_{ki\,du}I_{vk}(z) + \left[\sum_{i=1}^{k-m-1} + \sum_{k+m+1}^{N}\right]\left(\frac{jA_i}{B_k}\right)\Psi_{ki\,dv}I_{uk}(z)$$

$$+j\sum_{i=k-m}^{k+m}\left(\frac{jA_i}{B_k}\right)\Psi_{ki\,dI}I_{uk}(z) = j\frac{4\pi}{\zeta_0 F_0(h)}U_k F_{0z}. \qquad (5.29)$$

The index m in the sums is defined by

$$\beta_0 b_{km} < 1, \qquad \beta_0 b_{k,m+1} \geqslant 1 \qquad (5.30)$$

where b_{km} is the distance between the centres of the elements m and k. In most curtain arrays the spacing of the elements is sufficiently great so that all elements with $m \neq k$ satisfy the right-hand inequality in (5.30) and only $\beta_0 b_{kk} = \beta_0 a < 1$. When this is true, (5.28) and (5.29) reduce to

$$I_{vk}(z) = j\frac{2\pi V_{0k}}{\zeta_0 \Psi_{dR} F_0(h)}M_{0z} \qquad (5.31)$$

and

$$\sum_{i=1}^{N}\left\{\left(\frac{B_i}{B_k}\right)\Psi_{ki\,du} + \left(\frac{jA_i}{B_k}\right)[\Psi_{ki\,dv}(1-\delta_{ik}) + j\Psi_{ki\,dI}\delta_{ik}]\right\}I_{uk}(z) = \frac{j4\pi U_k}{\zeta_0 F_0(h)}F_{0z}$$

$$(5.32)$$

where
$$\delta_{ik} = \begin{cases} 0, & i \neq k \\ 1, & i = k. \end{cases}$$

The notation $\Psi_{dR} = \Psi_{kkdR}$ is used, since with identical elements all the Ψ_{kkdR} are identical and equal to Ψ_{dR} for the isolated antenna.

It follows directly from (5.31) that the leading term in $I_{vk}(z)$ is always M_{0z} for each value of k. Similarly from (5.32) the leading term in $I_{uk}(z)$ is of the form F_{0z}. Hence, it is possible to set

$$I_{vi}(z) = jA_i M_{0z}, \qquad I_{ui}(z) = B_i F_{0z} \qquad (5.33)$$

or

$$I_{zi}(z) = jA_i M_{0z} + B_i F_{0z}. \qquad (5.34)$$

Since Ψ_{dR} is real, it is clear from (5.31) that A_i is real when V_{0k} is real and from (5.32) that B_i is in general complex, or

$$B_i = B_{iR} + jB_{iI}. \qquad (5.35)$$

Note that the constant (jA_i/B_k), introduced in (5.25) and (5.27), is the ratio of the coefficients of the two terms in (5.34).

With the zero-order current formally determined, the constant U_k may be obtained from the substitution of (5.34) in (5.16). It is given by

$$U_k = -j\frac{\zeta_0}{4\pi} \sum_{i=1}^{N} [jA_i \Psi_{kiv}(h) + B_i \Psi_{kiu}(h)] \qquad (5.36)$$

where

$$\Psi_{kiv}(h) = \int_{-h}^{h} M_{0z'} K_{ki}(h, z')\, dz' \qquad (5.37)$$

$$\Psi_{kiu}(h) = \int_{-h}^{h} F_{0z'} K_{ki}(h, z')\, dz'. \qquad (5.38)$$

If (5.36), (5.33), and (5.34) are substituted in (5.31) and (5.32) the result is

$$A_k = \frac{2\pi}{\zeta_0 \Psi_{dR} F_0(h)} V_{0k} \qquad (5.39)$$

$$\sum_{i=1}^{N} B_i [\Psi_{kidu} F_0(h) - \Psi_{kiu}(h)]$$

$$= j \sum_{i=1}^{N} A_i \{\Psi_{kiv}(h) - [\Psi_{kidv}(1-\delta_{ik}) + j\Psi_{kidI}\delta_{ik}]F_0(h)\} \qquad (5.40)$$

where $k = 1, 2, 3, \ldots N$. The physical significance of the zero-order solution is evident from (5.39) and (5.40). The coefficients of the 'transmitting part' of the current are given by (5.39). The N driving voltages generate the expected sinusoidal distribution of current on each element. The coefficients of the 'receiving part' of the current are given by (5.40). The N currents act as distributed sources to generate distributions of the receiving type which are present in all the elements of the array. Equation (5.40) permits the prediction in each driven element of the shifted-cosine component of the current that is due to coupling between currents distributed along the element itself and along all other elements in the array. Conventional array theory is concerned only with (5.39), since all currents are assumed to have the same sinusoidal distribution. In the special case of an array with thin half-wave elements, the real and imaginary parts of the current in each element do have approximately the same distribution. It follows that conventional array theory should work quite well for an array of very thin half-wave elements. On the other hand, in the more general case, the real and imaginary parts of the current in each element have different distributions so that (5.40) is needed along with (5.39) to determine the actual currents.

An important case to which conventional theory has no application is the array of full-wave elements in which the currents are near anti-resonance, and their real and imaginary parts have quite different distributions. Before some particular parallel arrays are analysed, (5.40) is best expressed in matrix form. A general expression will be given for the $\Psi_{ki}(z)$ functions, and rigorous expressions will be derived for the radiation field.

Equation (5.40) is a set of linear algebraic equations with N unknowns that may be solved for the B_i in terms of the A_i. The N values of the A_i are expressed in terms of the N driving voltages V_{0k} by (5.39). In order to express (5.40) in matrix form, let the following quantities be defined:

$$\Phi_{kiu} = \Psi_{ki\,du}F_0(h) - \Psi_{kiu}(h) \tag{5.41}$$

$$\Phi_{kiv} = \Psi_{kiv}(h) - \Psi_{ki\,dv}F_0(h)(1 - \delta_{ik}) - j\Psi_{ki\,dI}F_0(h)\delta_{ik}. \tag{5.42}$$

Also let
$$[\Phi_u] = \begin{bmatrix} \Phi_{11u} & \Phi_{12u} & \cdots & \Phi_{1Nu} \\ \Phi_{21u} & \Phi_{22u} & \cdots & \Phi_{2Nu} \\ \vdots & & & \\ \Phi_{N1u} & \Phi_{N2u} & \cdots & \Phi_{NNu} \end{bmatrix} \tag{5.43}$$

$$[\Phi_v] = \begin{bmatrix} \Phi_{11v} & \Phi_{12v} & \cdots & \Phi_{1Nv} \\ \Phi_{21v} & \Phi_{22v} & \cdots & \Phi_{2Nv} \\ \vdots & & & \\ \Phi_{N1v} & \Phi_{N2v} & \cdots & \Phi_{NNv} \end{bmatrix} \tag{5.44}$$

$$\{A\} = \begin{Bmatrix} A_1 \\ A_2 \\ \vdots \\ A_N \end{Bmatrix}, \qquad \{B\} = \begin{Bmatrix} B_1 \\ B_2 \\ \vdots \\ B_N \end{Bmatrix}. \tag{5.45}$$

The bracket terms are $N \times N$ matrices; the terms in braces are column matrices. From the substitution of (5.41)–(5.45) in (5.40), it follows that

$$[\Phi_u]\{B\} = [\Phi_v]\{jA\} \tag{5.46}$$

and from (5.39)

$$A_k = \frac{2\pi}{\zeta_0 \Psi_{dR} F_0(h)} V_{0k}. \tag{5.47}$$

The solutions of two important problems in linear array theory are readily obtained from (5.46) and (5.47). Case I is concerned with specifying the driving-point† currents and determining the N potentials V_{0k} required to maintain these currents. In case II the N potentials V_{0k} are specified and the corresponding driving-point currents are determined.

In the zero-order current distribution (5.34), the coefficients B_i are the amplitudes of the shifted cosine currents due to the distributed interaction of all elements of current in the array. The A_i coefficients are determined completely by the voltages of the individual generators. The distribution of the current in the ith element (5.34) may be separated into its real and imaginary parts as follows:

$$I_{zi}(z) = j\{A_i \sin \beta_0(h - |z|) + B_{iI}(\cos \beta_0 z - \cos \beta_0 h)\}$$
$$+ B_{iR}(\cos \beta_0 z - \cos \beta_0 h) \tag{5.48}$$
$$= I''_{zi}(z) + jI'_{zi}(z). \tag{5.49}$$

At $z = 0$, the real and imaginary parts of the driving-point current are

$$I''_{zi}(0) = B_{iR}(1 - \cos \beta_0 h) \tag{5.50a}$$
$$I'_{zi}(0) = A_i \sin \beta_0 h + B_{iI}(1 - \cos \beta_0 h). \tag{5.50b}$$

The driving-point impedance and admittance under the two driving conditions can be computed from the following general formulas obtained by combining (5.46)–(5.50). (Note: A special form is convenient when $\beta_0 h$ is at or near $\pi/2$.)

† The term 'base current' is also used for driving-point current.

Case I   *Input currents specified*

$$\{V_0\} = \frac{1}{c_1(1-\cos\beta_0 h)}[\Phi_u][\Phi_w]^{-1}\{I_z(0)\} \qquad (5.51)$$

where $c_1 = j2\pi/(\zeta_0\Psi_{dR}\cos\beta_0 h);$

$$[\Phi_w] = [\Phi_v + \Phi_u \sin\beta_0 h/(1-\cos\beta_0 h)]. \qquad (5.52)$$

Case II   *Driving voltages specified*

$$\{I_z(0)\} = c_1(1-\cos\beta_0 h)[\Phi_w][\Phi_u]^{-1}\{V_0\}. \qquad (5.53)$$

The matrix components in (5.51) and (5.53) as well as numerical values of the driving-point impedances and admittances under different driving conditions are given in tables in the appendix. These tables were extracted from a more complete table [1]. The forms of the current for specified driving-point voltages and currents are not generally the same since the A_i and B_i coefficients differ for the two cases.

The symmetry properties of the impedance matrix in (5.1) and its counterpart in (5.51) are not identical. The assumption of identical current distributions implies that the mutual impedance between any two elements in an array is only a function of the size and spacing of the elements. Thus, with identical elements in an array, elements with the same centre-to-centre spacing have the same value of mutual impedance. For example, in an array with elements equally spaced, $Z_{12} = Z_{23} = Z_{34}$ and $Z_{13} = Z_{24} = Z_{46}$. The mutual impedance for elements near the centre of the array is then the same as for corresponding elements near the ends of the array. The new theory correctly shows that the coupling properties of an element in the array depend on the distribution of current and the location of every element in the array. Elements near the edges of an array are coupled differently from elements near the centre.

It is important to emphasize the significance of the results in (5.51) for Case I, and in (5.53) for Case II. Here, for the first time in the theory of linear arrays it is possible to solve for the actual driving potentials in a parallel array without requiring the currents in the elements to have identical distributions. Moreover, the interaction between the currents in all of the elements has been included in a zero-order approximation. Similarly, (5.53) makes it possible for the first time to specify the N driving potentials and determine the N zero-order currents in a manner that includes the effect of their mutual interactions on their distributions.

The general $\Psi(z)$ functions, obtained from the defining integrals (5.24)–(5.27), are

$$\Psi_{ki\,du}(z) = \frac{1}{\cos\beta_0 z - \cos\beta_0 h}\{[C_b(h,z) - C_b(h,h)]$$
$$-\cos\beta_0 h[E_b(h,z) - E_b(h,h)]\} \qquad (5.54)$$

$$\Psi_{ki\,dR}(z) = \frac{1}{\sin\beta_0(h-|z|)}\mathrm{Re}\{[C_b(h,z) - C_b(h,h)]\sin\beta_0 h$$
$$-[S_b(h,z) - S_b(h,h)]\cos\beta_0 h\} \qquad (5.55)$$

$$\Psi_{ki\,dI}(z) = \frac{1}{\cos\beta_0 z - \cos\beta_0 h}\mathrm{Im}\{[C_b(h,z) - C_b(h,h)]\sin\beta_0 h$$
$$-[S_b(h,z) - S_b(h,h)]\cos\beta_0 h\} \qquad (5.56)$$

$$\Psi_{ki\,dv}(z) = \frac{1}{\cos\beta_0 z - \cos\beta_0 h}\{[C_b(h,z) - C_b(h,h)]\sin\beta_0 h$$
$$-[S_b(h,z) - S_b(h,h)]\cos\beta_0 h\} \qquad (5.57)$$

$$\Psi_{kiv}(h) = C_b(h,h)\sin\beta_0 h - S_b(h,h)\cos\beta_0 h \qquad (5.58)$$

$$\Psi_{kiu}(h) = C_h(h,h) - E_h(h,h)\cos\beta_0 h \qquad (5.59)$$

where in subscripts $b = b_{ki}$, and

$$S_b(h,z) = \int_0^h \sin\beta_0 z'\left[\frac{e^{-j\beta_0 R_1}}{R_1} + \frac{e^{-j\beta_0 R_2}}{R_2}\right]dz' \qquad (5.60)$$

$$C_b(h,z) = \int_0^h \cos\beta_0 z'\left[\frac{e^{-j\beta_0 R_1}}{R_1} + \frac{e^{-j\beta_0 R_2}}{R_2}\right]dz' \qquad (5.61)$$

$$E_b(h,z) = \int_0^h \left[\frac{e^{-j\beta_0 R_1}}{R_1} + \frac{e^{-j\beta_0 R_2}}{R_2}\right]dz' \qquad (5.62)$$

$$R_1 = \sqrt{(z-z')^2 + b_{ki}^2}, \qquad R_2 = \sqrt{(z+z')^2 + b_{ki}^2} \qquad (5.63)$$

The functions S_b, C_b and E_b are found in King† and are tabulated for a wide range of values of h, z and b by Mack [3]. In order to obtain satisfactory overall agreement, the Ψ functions are evaluated at the point of maximum current. This ensures a good approximation for the determination of both the far field and the input power. However, the input susceptance may be somewhat in error. This does not present any practical difficulty since appropriate corrections may be applied at the driving point (cf. section 2.8).

† [2] p. 94.

From the results of chapter 1, the far-zone electric field depends upon the location of each element and its current distribution. Thus, for the geometry of Fig. 5.2,

$$E_\Theta(\Theta, \Phi) = j\frac{\omega\mu_0 \sin \Theta}{4\pi} \sum_{i=1}^{N} \frac{e^{-j\beta_0 R_i}}{R_i} \int_{-h}^{h} I_i(z') e^{j\beta_0 z' \cos \Theta} dz'. \quad (5.64)$$

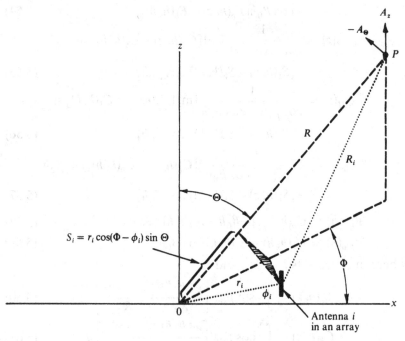

Fig. 5.2. Coordinate system locating one element with respect to the centre 0 of a parallel array.

For the conventional sinusoidal distribution of current, the electric field is given by

$$E_\Theta(\Theta, \Phi) = j\frac{\zeta_0}{2\pi} \sum_{i=1}^{N} \frac{e^{-j\beta_0 R_i}}{R_i} I_i(0) F_m(\Theta, \beta_0 h) \quad (5.65)$$

where

$$F_m(\Theta, \beta_0 h) = \frac{\cos(\beta_0 h \cos \Theta) - \cos \beta_0 h}{\sin \Theta}. \quad (5.66)$$

In contrast, the new theory yields a far-zone electric field given by

$$E_\Theta(\Theta, \Phi) = j\frac{\zeta_0}{2\pi} \sum_{i=1}^{N} \frac{e^{-j\beta_0 R_i}}{R_i} [jA_i F_m(\Theta, \beta_0 h) + B_i G_m(\Theta, \beta_0 h)] \quad (5.67)$$

where

$$G_m(\Theta, \beta_0 h) = \frac{\sin \beta_0 h \cos (\beta_0 h \cos \Theta) \cos \Theta - \cos \beta_0 h \sin (\beta_0 h \cos \Theta)}{\sin \Theta \cos \Theta}.$$

(5.68)

A comparison of (5.65) with (5.67) shows the difference between the far-field pattern for the conventional and for the new theory. It is now of interest to study the differences between the radiation patterns determined by the two theories under different driving conditions.

5.3 Examples

Consider a three-element array with elements that are a full wavelength long $(2h = \lambda)$ and separated by a quarter wavelength $(b_{i,i+1} = \lambda/4)$. The conventional approach to this problem is doomed to failure when $\beta_0 h = \pi$, since an assumed sinusoidal current (i.e. $I_z(z) \sim \sin \beta_0 |z|$) is zero at the driving point. This gives rise to a zero admittance or an infinite impedance for each element in the array. This difficulty does not exist in the new theory. When $\beta_0 h = \pi$, (5.5) gives the following form for the current:

$$I_i = jA_i \sin \beta_0 |z_i| + B_i(\cos \beta_0 z_i + 1).$$

(5.69)

At $z = 0$, the current is finite and is given by the coefficient B_i for each element in the array. In order to demonstrate the difference between the two antenna theories, the conventional approach will be used for $\beta_0 h = 3 \cdot 157$ and compared to the results of the new theory for $\beta_0 h = \pi$.

Consider now the three-element array shown in Fig. 5.1. Either the driving-point voltages or the driving-point currents may be specified. Conventionally the phases of the equal driving-point currents are specified to produce a radiation pattern. The electric field E_Θ in the far zone is given by (1.45) with (1.46). Thus,

$$E_\Theta = C \frac{\sin Nx}{\sin x}$$

(5.70)

where
$$x = n \sin \Theta \cos \Phi - t.$$

(5.71)

For the array of Fig. 5.1, the number of elements is $N = 3$. The distance between the elements in fractions of a wavelength, n, is chosen to be $\frac{1}{4}$. Now let attention be directed to the horizontal pattern in the equatorial or H-plane defined by $\Theta = \pi/2$. The three driving-point currents are equal in magnitude but the phase delay t

between elements in fractions of a period may be varied to produce a particular pattern. For example, a value of $t = \frac{1}{4}$ will produce an endfire radiation pattern with the maximum value of the directivity D toward the right in Fig. 5.1.

The actual radiation pattern of the three-element endfire array with specified base currents differs from the ideal pattern shown on the left of Fig. 5.3. The several components of current on the elements which are discussed later in this section, are equivalent to separate sources producing different radiation patterns. The additional patterns in the middle and right of Fig. 5.3 fill in the deep nulls and reduce the back-to-front ratio of the ideal pattern. Specifically, it can be shown that the actual pattern for the 3-element array is proportional to the sum

$$E_\Theta \sim a_p(\pi/2, \Phi; 3, \tfrac{1}{4}, \tfrac{1}{4}) + (-0.53 + j0.57) a_p(\pi/2, \Phi; 2, \tfrac{1}{2}, 0)$$

$$+ (-0.07 + j0.50) a_p(\pi/2, \Phi; 2, \tfrac{1}{2}, \tfrac{1}{2}).$$

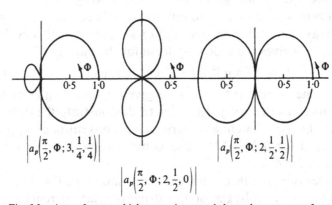

Fig. 5.3.   Array factors which comprise actual three-element array factor.

The ideal radiation pattern as determined from (5.70) depends on the vertical field factor $F_0(\Theta, \beta_0 h)$ of an isolated element. This is contained in C in (5.70). Consider now an array with full-wave elements ($\beta_0 h = \pi$). The particular value of $F_0(\Theta, \beta_0 h)$ is given by (5.66). Thus,

$$F_0(\Theta, \pi) = \frac{\cos(\pi \cos \Theta) + 1}{\sin \Theta} \qquad (5.72a)$$

$$F_0\left(\frac{\pi}{2}, \pi\right) = 2. \qquad (5.72b)$$

If the driving-point impedance for $\beta_0 h$ is computed from the

EMF method, the numerical value is infinity. This follows because the sinusoidal current is zero at the driving point. As an illustrative comparison of the results of the present theory with other methods, an examination of the driving-point impedances for the three-element array with $\beta_0 h = \pi/2$ and $\beta_0 h = \pi$ is given. The self- and mutual impedances for $\beta_0 h = \pi/2$ by the EMF method and an assumed sinusoidal current are:

$$\left.\begin{array}{ll}Z_{11} = 71 \cdot 13 + j42 \cdot 54 \text{ ohms} & \\ Z_{12} = 41 \cdot 79 - j28 \cdot 35 \text{ ohms}, & \beta_0 b = \pi/2 \\ Z_{13} = -12 \cdot 53 - j29 \cdot 94 \text{ ohms}, & \beta_0 b = \pi\end{array}\right\} \quad (5.73\text{a})$$

$$Z_{12} = Z_{21}, \quad Z_{23} = Z_{32}, \quad Z_{13} = Z_{31}, \quad Z_{11} = Z_{22} = Z_{33}.$$
$$(5.73\text{b})$$

The driving-point currents are specified in the following way to produce an endfire pattern:

$$\{I\} = \left\{\begin{array}{c}I_{01} \\ I_{02} \\ I_{03}\end{array}\right\} = I_{01}\left\{\begin{array}{c}1 \\ -j \\ -1\end{array}\right\}. \quad (5.74)$$

The substitution of (5.74) and (5.73a) into (5.1) yields the following driving-point impedances for the three elements:

$$\left.\begin{array}{lll}Z_{01} = Z_{11} - jZ_{12} - Z_{13} = 55 \cdot 31 + j30 \cdot 69 \text{ ohms} = V_{01}/I_{01} \\ Z_{02} = Z_{11} \qquad\qquad\quad = 71 \cdot 13 + j42 \cdot 54 \text{ ohms} = V_{02}/I_{02} \\ Z_{03} = Z_{11} + jZ_{12} - Z_{13} = 112 \cdot 0 + j114 \cdot 3 \text{ ohms} = V_{03}/I_{03}.\end{array}\right\} (5.75)$$

The same results are, of course, obtained when the driving voltages are assigned instead of the currents by the substitution of V for I in (5.74), since no changes are possible in the assumed distributions of current and, hence, in the mutual coupling.

An apparently obvious method for improving the accuracy of the conventional theory for the three-element endfire array is to make use of the accurate second-order iterated results for two coupled cylindrical antennas† by the simple expedient of applying them to the three elements treated as quasi-independent pairs. To be sure, this procedure implies that the distributions of current along elements 1 and 2 are the same when element 3 is present as when it is absent in the evaluation of Z_{12}, and that the currents along elements 1 and 3 are unaffected by the presence of element 2.

---

† [2], chapter 2.

Moreover, since Z_{11} is not the same for the two distances b_{12} and b_{13}, it is assumed that the value for the closer spacing is to be used. It is interesting to evaluate the driving-point impedances under these conditions. By interpolation from Fig. 8.13 on page 306 of King, *Theory of Linear Antennas*, the following values are obtained:

$$\left. \begin{aligned} Z_{11} &= 81{\cdot}7 + j39{\cdot}3 \text{ ohms}, & \beta_0 b &= \frac{\pi}{2}, & \Omega &= 2\ln\frac{2h}{a} = 10 \\[2mm] Z_{12} &= 42{\cdot}6 + j42{\cdot}7 \text{ ohms}, & \beta_0 b &= \frac{\pi}{2} \end{aligned} \right\}$$

$$(5.76a)$$

$$\left. \begin{aligned} Z_{11} &= 86{\cdot}1 + j42{\cdot}6 \text{ ohms}, & \beta_0 b &= \pi, & \Omega &= 2\ln\frac{2h}{a} = 10 \\[2mm] Z_{13} &= -15{\cdot}7 - j29{\cdot}0 \text{ ohms}, & \beta_0 b &= \pi \end{aligned} \right\}.$$

$$(5.76b)$$

Note that the two values of Z_{11} for $\beta_0 b = \pi/2$ and π are not very different when $\beta_0 h = \pi/2$. The corresponding driving-point impedances, when Z_{11} for $\beta_0 b = \pi/2$ is used, are

$$\left. \begin{aligned} Z_{01} &= 54{\cdot}7 + j25{\cdot}7 \text{ ohms}, & \Omega = 2\ln\frac{2h}{a} = 10 \\[2mm] Z_{02} &= 81{\cdot}7 + j39{\cdot}3 \text{ ohms} \\ Z_{03} &= 141{\cdot}0 + j111{\cdot}0 \text{ ohms} \end{aligned} \right\}.$$

$$(5.76c)$$

These values do not differ from those obtained from the conventional theory by more than about 25%. This is a consequence of the fact that the actual distributions of current on half-wave dipoles do not vary greatly from the sinusoidal so long as they are moderately thin.

It is now in order to examine the results obtained by the new two-term theory which takes full account of the changes in the distributions of current due to the presence of any number of coupled elements. The driving-point impedances for the three elements are readily computed.† They are

$$\left. \begin{aligned} Z_{01} &= 67{\cdot}51 + j24{\cdot}14 \text{ ohms}, & \Omega = 2\ln\frac{2h}{a} = 10 \\[2mm] Z_{02} &= 78{\cdot}47 + j31{\cdot}23 \text{ ohms} \\ Z_{03} &= 145{\cdot}61 + j96{\cdot}91 \text{ ohms} \end{aligned} \right\}.$$

$$(5.77)$$

† [1], p. 84.

These values are comparable with those in both (5.73a) and (5.75) (with differences not exceeding about 30%) simply because the current in half-wave dipoles is predominantly sinusoidal with only relatively small changes due to finite radius and mutual coupling.

The situation is quite different when the elements are not near resonance. This is well illustrated with the same three-element array but now with $\beta_0 h$ near π instead of $\pi/2$.

The conventional application of the EMF method with assumed sinusoidal currents on all elements yields meaningless results. Since the currents at all three driving points are identically zero all driving-point impedances are infinite—which is, of course, absurd.

If an attempt is made to use the second-order theory for coupled pairs, the following values are obtained from pages 297 and 298 of King, *Theory of Linear Antennas*, with $\beta_0 h = 3 \cdot 157$:

$$\left.\begin{array}{l} Z_{11} = 199 - j351 \text{ ohms}, \quad \beta_0 b = \dfrac{\pi}{2}, \quad \Omega = 2 \ln \dfrac{2h}{a} = 10 \\[2mm] Z_{12} = 116 + j187 \text{ ohms} \end{array}\right\} \quad (5.78a)$$

$$\left.\begin{array}{l} Z_{11} = 288 - j441 \text{ ohms}, \quad \beta_0 b = \pi, \quad \Omega = 2 \ln \dfrac{2h}{a} = 10 \\[2mm] Z_{13} = 122 - j15 \text{ ohms} \end{array}\right\} . \quad (5.78b)$$

Note that the difference in the self-impedances Z_{11} when the spacing is changed from $\beta_0 b = \pi/2$ to $\beta_0 b = \pi$ is very much greater when $\beta_0 h = 3 \cdot 157$ than when $\beta_0 h = \pi/2$. If it is again assumed in this rough approximation that the value for the closer elements ($\beta_0 b = \pi/2$) is to be used, the three driving-point impedances become

$$\left.\begin{array}{l} Z_{01} = 264 - j452 \text{ ohms}, \quad \Omega = 2 \ln \dfrac{2h}{a} = 10 \\[2mm] Z_{02} = 199 - j351 \text{ ohms} \\[2mm] Z_{03} = -110 - j220 \text{ ohms} \end{array}\right\} . \quad (5.79)$$

The negative input resistance for element 3 indicates that its generator supplies no power to the array but acts as a load to absorb power received by coupling from the other two elements.

Once again it is in order to introduce the results from the new two-term theory which actually determines the distributions of the currents on all three elements and the associated driving-point impedances. The following values are readily calculated[†] for the

† [1], p. 203.

driving-point currents specified in (5.74):

$$Z_{01} = 612 - j591 \text{ ohms} \qquad Y_{01} = (0\cdot845 + j0\cdot817) \times 10^{-3} \text{ mhos} \left.\vphantom{\begin{matrix}1\\1\\1\end{matrix}}\right\}$$
$$Z_{02} = 160 - j590 \text{ ohms} \qquad Y_{02} = (0\cdot429 + j1\cdot578) \times 10^{-3} \text{ mhos}$$
$$Z_{03} = 61\cdot5 - j435 \text{ ohms} \qquad Y_{03} = (0\cdot318 + j2\cdot252) \times 10^{-3} \text{ mhos}$$

$$(5.80)$$

A comparison of these values with those in (5.79) shows no agreement. Clearly, the notion that coupled antennas in an array of more than two elements may be treated as independent pairs in determining mutual impedance is unreliable. The previously shown rough agreement for half-wave dipoles is a special case that must not be assumed to have any general significance for other lengths. The coupling between any two elements depends on their relative positions in an array and only the results in (5.77) and (5.80) include this correctly.

The voltages required to maintain the currents specified in (5.74) when $\beta_0 h = \pi$ and $\beta_0 b = \pi/2$, i.e. $I_{02} = -jI_{01}$, $I_{03} = -I_{01}$, are $V_{01} = 612 - j591$, $V_{02} = -590 - j160$, $V_{03} = -61\cdot5 + j435$. The power supplied each element k by its generator is

$$P_k = |I_{0k}|^2 R_{0k} = |V_{0k}|^2 G_{0k}.$$

The ratios of the powers supplied to the three-element array are

$$P_1/P_3 = 9\cdot82, \qquad P_2/P_3 = 2\cdot51. \qquad (5.81)$$

Evidently element 1 receives almost ten times the power that is supplied to the terminals of element 3.

The new theory gives the following values† of Z_{0i} and Y_{0i}, $i = 1, 2, 3$ when the driving-point voltages are specified ($V_{02} = -jV_{01}$, $V_{03} = -V_{01}$) instead of the currents:

$$Z_{01} = 675 - j484 \text{ ohms} \qquad Y_{01} = (0\cdot979 + j0\cdot701) \times 10^{-3} \text{ mhos} \left.\vphantom{\begin{matrix}1\\1\\1\end{matrix}}\right\}$$
$$Z_{02} = 359 - j479 \text{ ohms} \qquad Y_{02} = (1\cdot003 + j1\cdot336) \times 10^{-3} \text{ mhos}$$
$$Z_{03} = 170 - j426 \text{ ohms} \qquad Y_{03} = (0\cdot808 + j2\cdot024) \times 10^{-3} \text{ mhos}$$

$$(5.82)$$

Clearly, the results for Cases I and II are not the same as seen from a comparison of (5.80) and (5.82). This difference is due to the unequal distributions of the currents in the elements which cause non-uniform coupling. This effect will become clearer when the currents in the individual elements are examined.

The conventional currents in the three-element endfire array

† [1], p. 221.

with $\beta_0 h = 3{\cdot}157$ are

$$I_1(z) = I_2(z) = I_3(z) = I_{0i}\sin{(3{\cdot}157 - \beta_0|z|)}$$

driving-point currents specified   (5.83)

$$I_i(z) = V_i Y_i \sin{(3{\cdot}157 - \beta_0|z|)}, \qquad i = 1, 2, 3$$

driving voltages specified.   (5.84)

The form of the currents in (5.83) and (5.84) is identical for each element. Both the real and imaginary parts have the same distribution. The currents in the new theory are given by (5.69) with (5.46) and (5.47). They are shown in Figs. 5.4 and 5.5 for the two different driving conditions. When the currents at $z = 0$ are specified, the distributions differ widely in form from element to element. Note that the currents are shown both with respect to the individual driving voltages and with respect to V_{02}. In the computation of radiation patterns the currents must all be normalized with respect to a single driving voltage. The large differences in the real and

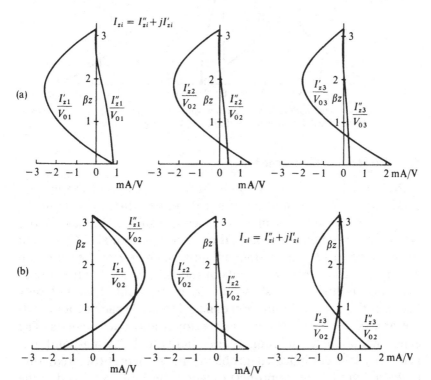

Fig. 5.4   Three-element endfire array; driving-point currents specified. Drawn with respect to (a) individual driving voltages, (b) V_{02}, ($\lambda/4$ spacing, $\beta_0 h = \pi, \Omega = 10$).

imaginary parts of the currents in Fig. 5.4 practically disappear when the driving voltages are specified in Fig. 5.5.

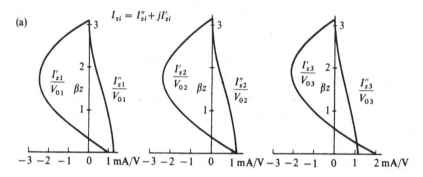

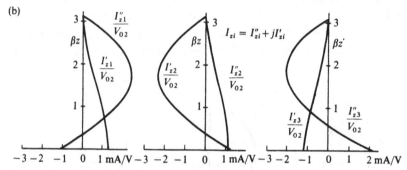

Fig. 5.5. Three-element endfire array; driving voltages specified. Drawn with respect to (a) individual driving voltages, (b) V_{02}, ($\lambda/4$ spacing, $\beta_0 h = \pi, \Omega = 10$).

5.4   Electronically scanned arrays

Previous sections of this book have demonstrated the general invalidity of the assumption of equal current distributions in the elements of an array. A most significant result of the new theory is that the expected conventional radiation pattern is not achieved since the contributions by the individual elements to the radiation pattern are different. The results of the new theory for the broadside and endfire arrays show an appreciable difference not only between the driving-point impedances for the broadside and endfire arrays, but between the conventional and new theories. The experimental determination of the individual driving-point impedances is a complicated problem and a theoretical prediction of the individual circuit properties would certainly be an aid in the efficient operation of an array.

A comparison of the corresponding expressions for the far fields based on the conventional method and the new, more accurate approach, helps to illustrate some of the problems in the theory of scanned arrays. Consider an array in which the currents at the driving points of the elements are specified in both amplitude and phase. For the present, let the amplitudes be equal and the phases required to change linearly from element to element across the array. For example, the base current might increase in phase by 30° toward one end of the array. Expressed in general terms

$$I_i = I_0 \, e^{-j\delta_i} = I_0 \, e^{-j2\pi i t} \tag{5.85}$$

where t is the time delay between elements in fractions of a period.

With the currents at $z = 0$ specified in (5.85) and under the assumption of identical distributions of current, the far-zone electric field from (1.44) has the form

$$E_\Theta^r = \left\{ \frac{j\zeta_0 I_0}{2\pi} \frac{e^{-j\beta_0 R_0}}{R_0} F_0(\Theta, \beta_0 h) \right\} \left\{ 1 + \sum_{i=1}^{\frac{1}{2}(N-1)} \left[e^{-j(\delta_i - \beta_0 S_i)} + e^{j(\delta_i - \beta_0 S_i)} \right] \right\}.$$

$$\tag{5.86}$$

The second term in braces in (5.86) is the familiar array factor given by (5.70). If the distance between the elements is small enough, the radiation pattern has only one principal maximum in the visual ranges of Θ and Φ. The first maximum of (5.70) occurs when $x = 0$. Thus, to direct the main beam in a specific direction (Θ_m, Φ_m) in space, the time delay between the currents in the elements must be set equal to a particular value t_m such that

$$n \sin \Theta_m \cos \Phi_m - t_m = 0,$$

or $$t_m = n \sin \Theta_m \cos \Theta_m. \tag{5.87}$$

For example, in an array with half-wave spacing ($n = \frac{1}{2}$) for which the main beam is to point in the direction $\Theta_m = \pi/2$ (H-plane) and $\Phi_m = 60°$, the required phase shift given by (5.87) is $t_m = \pi/4$ radians. In a single curtain array it is not possible with ordinary elements to have any control over the beam pointing in the Θ direction. The control of the main beam in the Θ direction could be achieved by a planar array formed by an array of collinear elements.

Now let the conventional requirement, that the distributions of current be equal, be removed in (1.44). Let the currents at $z = 0$

again be specified so that, on the basis of the conventional theory, the main beam will point in a desired direction. However, and this obvious fact is often overlooked, the specification of the currents at each driving point usually does not determine the entire current along each element. A variety of distributions of current may be associated with any given value at $z = 0$. In general, the radiation pattern can be considered the superposition of two parts. One part is the pattern of an array of elements with equal distributions of current; the other the pattern of the same array with dissimilar distributions of current. Conventional theories assume that the first part is the entire pattern.

The beam-pointing properties of a scanning array are affected by the interaction between the currents in the elements. The simple array factor in (5.70) characterizes an ideal array in which the exact phase and amplitude of the current are specified for each element. This specification applies not only at $z = 0$ but all along each element. The phase of the current is of primary importance in the determination of the direction of the main beam. In an actual array the variation in phase along the length of each individual element differs from element to element. In practice, this variation is responsible for a beam-pointing error of non-negligible value. Furthermore, with this phase variation perfect phase cancellation and addition is impossible. Perfect nulls in the radiation pattern will disappear and side-lobe levels will be modified significantly. The side-lobe level and the angular width of the main beam are also affected by changes in the magnitudes of the currents from element to element across the array.

As a specific example, consider the three-element array with full-wave elements ($\beta_0 h = \pi$) and half-wave spacing ($\beta_0 b = \pi/2$). The driving-point currents are specified to produce a maximum field in the direction indicated by the conventional theory. The driving-point impedance is to be calculated for each element as a function of the scanning angle. The actual position of the maximum, as given by the new theory, is to be compared with the corresponding angular position predicted by the conventional theory. The difference is the beam-pointing error Δ.

The general matrix relation (5.51) between the driving-point voltages and currents may be reduced to the following symbolic form:

$$\{V_0\} = [\Phi_T]\{I_0\} \qquad (5.88)$$

where
$$\{V_0\} = \begin{Bmatrix} V_{01} \\ V_{02} \\ \vdots \\ V_{ON} \end{Bmatrix}, \qquad \{I_0\} = \begin{Bmatrix} I_{01} \\ I_{02} \\ \vdots \\ I_{ON} \end{Bmatrix} \qquad (5.89)$$

and
$$[\Phi_T] = \begin{bmatrix} \Phi_{T11} & \Phi_{T12} & \cdots & \Phi_{T1N} \\ \Phi_{T21} & \Phi_{T22} & \cdots & \Phi_{T2N} \\ \vdots & & & \\ \Phi_{TN1} & \Phi_{TN2} & \cdots & \Phi_{TNN} \end{bmatrix} = c_1[\Phi_u][\Phi_w]^{-1} \quad (5.90)$$

[cf. (5.52)] where
$$[\Phi_w] = [\Phi_v + \Phi_u \sin \beta_0 h/(1 - \cos \beta_0 h)]$$
and
$$c_1 = j2\pi/\zeta_0 \Psi_{dR} \cos \beta_0 h.$$

The elements of Φ_T may be derived from the machine-tabulated values of Φ_u and Φ_v in the tables [appendix III]. From the tabulated values for the two different sets of driving conditions [appendix IV] and a knowledge of the symmetry properties of the Φ_u and Φ_v values, the Φ_T values may also be calculated. In the present example calculations similar to those given in appendix IV yield the following information for the case $\beta_0 h = \pi$, $\beta_0 b = \pi/2$:

When
$$\{I_0\} = \begin{Bmatrix} 1 \\ -j \\ -1 \end{Bmatrix} \qquad (5.91)$$

then
$$\{V_0\} = \begin{Bmatrix} 612 - j591 \\ -590 - j160 \\ -61 \cdot 5 + j435 \end{Bmatrix}. \qquad (5.92)$$

Also when
$$\{I_0\} = \begin{Bmatrix} 1 \\ 1 \\ 1 \end{Bmatrix} \qquad (5.93)$$

then
$$\{V_0\} = \begin{Bmatrix} 435 - j346 \\ 309 - j37 \cdot 9 \\ 435 - j346 \end{Bmatrix}. \qquad (5.94)$$

The specifications in (5.91) and (5.93) are the conventional ones for the endfire and broadside arrays. For $\beta_0 h = \pi$, $\beta_0 b = \pi/2$, the time delay between elements as given by (5.87) is

$$t_m = n \cos \Phi_m = \tfrac{1}{4} \cos \Phi_m. \tag{5.95}$$

The driving-point currents in the N elements are expressed in terms of the angle Φ_0 by the substitution of (5.95) in (5.85). The result with $I_0 = 1$ is

$$I_i = e^{-j2\pi i n \cos \Phi_m} = e^{-j(\pi/2)i \cos \Phi_m}, \qquad n = \tfrac{1}{4}, \qquad i = 1, 2, 3. \tag{5.96}$$

The following table is useful for the computation of the driving-point impedances for different values of the angle Φ_m.

Table 5.1

Φ_m	0	30°	60°	75°	90°
I_1	1	$1+j0$	$1+j0$	$1+j0$	1
I_2/I_1	$-j$	$0\cdot209-j0\cdot978$	$0\cdot707-j0\cdot707$	$0\cdot919-j0\cdot395$	1
I_3/I_1	-1	$-0\cdot913-j0\cdot409$	$-0\cdot707-j0\cdot707$	$0\cdot687-j0\cdot726$	1

Before the driving-point impedances can be computed the elements of the matrix Φ_T must be found. They can be computed directly from the basic matrix equations in terms of Φ_u and Φ_v, or they may be computed from the tables of driving-point impedances for different driving conditions. For example, from the two sets of information contained in (5.92) and (5.94), the symbolic matrix multiplication (5.88) yields

$$\left.\begin{aligned}
\Phi_{T11}+\Phi_{T12}+\Phi_{T13} &= 435-j346 \\
2\Phi_{T21}+\Phi_{T22} &= 309-j37\cdot9 \\
\Phi_{T11}-j\Phi_{T12}-\Phi_{T13} &= 612-j591 \\
-j\Phi_{T22} &= -590-j160 \\
-\Phi_{T11}-j\Phi_{T12}+\Phi_{T13} &= -61\cdot5+j435
\end{aligned}\right\} \tag{5.97}$$

where

$$[\Phi_T] = \begin{bmatrix} \Phi_{T11} & \Phi_{T12} & \Phi_{T13} \\ \Phi_{T21} & \Phi_{T22} & \Phi_{T21} \\ \Phi_{T13} & \Phi_{T12} & \Phi_{T11} \end{bmatrix}. \tag{5.98}$$

The symmetry properties of the elements of (5.98) were deduced from those of the component matrices involved in (5.88). The elements of (5.98) may be compared to the impedance matrix whose elements are the self- and mutual impedances computed under the conventional assumptions. For example, Φ_{T11} could be compared to Z_{11}, the self-impedance of the first antenna. The result shown symbolically in (5.98) indicates that the off-diagonal terms are not necessarily equal (e.g. $\Phi_{T12} \neq \Phi_{T21}$) and that the diagonal terms may differ (e.g. $\Phi_{T11} \neq \Phi_{T22}$). The numerical values of the matrix elements of Φ_T are

$$
\left.
\begin{aligned}
\Phi_{T11} &= 347 - j567 \\
\Phi_{T22} &= 160 - j590 \\
\Phi_{T12} &= 77{\cdot}9 + j275 \\
\Phi_{T21} &= 74{\cdot}3 + j276 \\
\Phi_{T13} &= 10{\cdot}4 - j53{\cdot}7.
\end{aligned}
\right\}
\tag{5.99}
$$

Consider the specific case $\Phi_m = 75°$ where the driving-point currents are given in Table 5.1 and the elements of the Φ_T matrix are given by (5.99). Thus,

$$
\begin{Bmatrix} V_{01} \\ V_{02} \\ V_{03} \end{Bmatrix} =
\begin{bmatrix} \Phi_{T11} & \Phi_{T12} & \Phi_{T13} \\ \Phi_{T21} & \Phi_{T22} & \Phi_{T21} \\ \Phi_{T13} & \Phi_{T12} & \Phi_{T11} \end{bmatrix}
\begin{Bmatrix} 1 + j0 \\ 0{\cdot}919 - j0{\cdot}395 \\ 0{\cdot}687 - j0{\cdot}726 \end{Bmatrix} I_1.
\tag{5.100}
$$

To compute, for example, $Z_{02} = (V_{02}/I_2)$, the quantity (V_{02}/I_1) is computed from (5.100) or $(V_{02}/I_1) = \Phi_{T21}(1 + j0) + \Phi_{T22}(0{\cdot}919 - j0{\cdot}395) + \Phi_{T21}(0{\cdot}687 - j0{\cdot}726) = 240 - j193$. The driving-point impedance Z_{02} is found from the substitution of the relation $I_2 = (0{\cdot}919 - j0{\cdot}395)I_1$ in this expression with the result, $Z_{02} = 297 - j82{\cdot}8$. The variation of the driving-point resistance and reactance with the beam-pointing angle Φ_m is shown in Fig. 5.6.

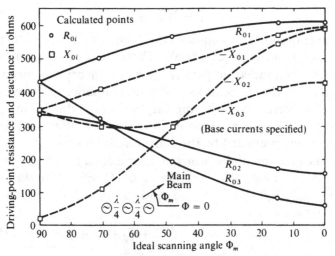

Fig. 5.6. Variation of driving-point resistance and reactance with beam-pointing angle Φ_m for 3-element array ($\lambda/4$ spacing, $\beta_0 h = \pi$, $\Omega = 10$).

It is seen that even if the beam-pointing angle is restricted to moderate departures from a normal position, significant changes in the impedance function occur. These will be apparent in the mismatch between the generator and the antenna. Note also that from symmetry, the continuation of R_{01} and X_{01} in the range $90° \leqslant \Phi_m \leqslant 180°$ is the mirror image of R_{03} and X_{03} about $\Phi_m = 90°$.

The driving-point currents have been specified according to the criteria of the conventional theory. This specification does not control the distribution of either the phase or the amplitude of the current away from the driving point. As a result, the location of the maximum of the main beam may differ from that predicted by the ideal angle in (5.87). This difference is the beam-pointing error Δ and represents the difference between the ideal scanning angle Φ_m and the actual angle Φ_a.

The far-zone electric field is given by (5.67). The computation of this field requires all currents to be normalized with respect to a single driving voltage. Thus, with the kth element as a reference, (5.67) may be rearranged to give

$$E_\Theta^r(\Theta) = \frac{j\zeta_0}{2\pi} \frac{e^{-j\beta_0 R_0}}{R_0} \sin \Theta \sum_{i=1}^{N} C_i \, e^{j\beta_0 b[(N-2i+1)/2]\cos \Phi \sin \Theta} \quad (5.101)$$

where
$$C_i = \xi_i \left[\frac{-j}{60\Psi_{dR}} F_m(\Theta, \beta_0 h) + \frac{Y_i}{2} G_m(\Theta, \beta_0 h) \right] \quad (5.102)$$

and
$$\xi_i = V_{0i}/V_{0k}. \quad (5.103)$$

The conventional theory equates the C coefficient in (5.101) to the driving-point currents [cf. (5.65)]. These, in turn, are chosen to produce a given radiation pattern. The new theory has shown that the currents in the elements as well as the radiation pattern cannot be specified merely by adjusting the currents at the driving point. Moreover, the direction of the main beam may differ considerably from the value predicted by the conventional theory.

The true location of the principal lobe is found from the location of the major maximum of $|E_\Theta^r(\Theta)|$ or of $|E_\Theta^r(\Theta)|^2$. For the special case $\Theta = \pi/2$, $\beta_0 b = \pi/2$, $N = 3$, and $\beta_0 h = \pi/2$, the electric field in the far zone is

$$E_\Theta^r(\Theta) = K(C_1 \, e^{-ju} + C_2 + C_3 \, e^{ju}) \quad (5.104)$$

where
$$K = \frac{j\zeta_0}{2\pi} \frac{e^{-j\beta_0 R_0}}{R_0}$$

$$u = \beta_0 b \cos \Phi = \frac{\pi}{2} \cos \Phi$$

$$\Theta = \frac{\pi}{2} \quad (H\text{-plane}).$$

The square of the absolute value of $E_\Theta^r(\Theta)$ is formed from (5.104) with the result

$$
\begin{aligned}
E_\Theta^r E_\Theta^{r*} &= KK^* C_2 C_2^* [C_{12}^* C_{32} \, e^{j2u} + (C_{12}^* + C_{32}) \, e^{ju} \\
&\quad + (1 + C_{12} C_{12}^* + C_{32} C_{32}^*) \\
&\quad + [(C_{12} + C_{32}^*) \, e^{-ju} + C_{12} C_{32}^* \, e^{-j2u}]
\end{aligned}
\tag{5.105}
$$

where

$$C_{12} = \frac{C_1}{C_2} \quad \text{and} \quad C_{32} = \frac{C_3}{C_2}. \tag{5.106}$$

With the substitution $x = e^{ju}$, (5.105) is seen to be an algebraic equation of fourth degree. Thus,

$$|E_\Theta^r(x)|^2 = C_a x^4 + C_b x^3 + C_c x^2 + C_d x + C_e. \tag{5.107}$$

The true location of the principal lobe is determined from the equation obtained when (5.107) is differentiated with respect to x and equated to zero. The computed beam-pointing error for the

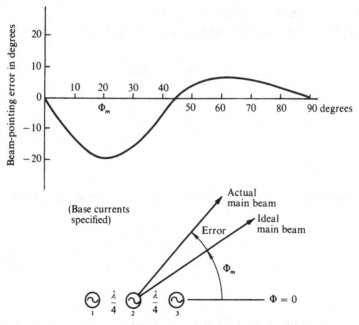

Fig. 5.7. Variation of beam-pointing error with beam-pointing angle for 3-element array ($\lambda/4$ spacing, $\beta_0 h = \pi$, $\Omega = 10$).

three-element array as determined from the conventional theory is shown in Fig. 5.7. This graph shows an appreciable plus and minus variation over most of the visible range of Φ.

The expression for the square of the absolute value of the far-field for the N-element array is

$$|E_\Theta^r(x)|^2 = KK^* \sum_i \sum_n C_i C_n^* x^{(i-n)}. \tag{5.108}$$

The location of the extremes of (5.108) is given by

$$\frac{\partial}{\partial x}|E_\Theta^r(x)|^2 = KK^*j \sum_i \sum_n (i-n)C_i C_n^* x^{(i-n)} = 0 \quad x = x_0, x_1, x_2 \dots . \tag{5.109}$$

5.5   Examples of the general theory for large arrays

Thus far the simple array with $N = 3$ and $\beta_0 b = \pi/2$ has been examined for a variety of driving-point conditions. Calculations have also been made for arrays with a larger number of elements. For these the lengths $2h$ of the elements were varied from a quarter to a full wavelength. The driving-point voltages or currents were specified according to conventional array theory to produce a broadside or endfire radiation pattern.

The driving-point currents required for an ideal broadside array are

$$I_{z1}(0) = I_{z2}(0) = I_{z3}(0), \quad \text{etc.} \tag{5.110}$$

or, in matrix form,

$$\{I_z(0)\} = I_{z1}(0) \begin{Bmatrix} 1 \\ 1 \\ \vdots \end{Bmatrix}. \tag{5.111}$$

Alternatively, the driving voltages may be assigned as follows:

$$\{V\} = V_1 \begin{Bmatrix} 1 \\ 1 \\ 1 \\ \vdots \end{Bmatrix}. \tag{5.112}$$

The relatively large sinusoidal parts of the currents on the antennas are determined directly from (5.47) by the specification of the voltages. However, the relations (5.50a) and (5.50b) between the coefficients A, B_R, B_I and the currents at $z = 0$ do not in general suffice to determine the distributions of current along the elements.

The driving-point impedances for broadside arrays are shown in Figs. 5.8 to 5.10. Driving-point currents and voltages are specified for arrays of up to 25 elements ($N \leqslant 25$) for quarter and half-wavelength spacings ($\beta_0 b = \pi/2$ and $\beta_0 b = \pi$). In Figs. 5.8a–d are shown graphs of the resistances and reactances of the individual elements of a broadside array when the driving-point currents are specified. In Figs. 5.8a and 5.8b the distance between the adjacent antennas is one-quarter wavelength; the lengths of the elements are, respectively, a quarter and a half wavelength. In Figs. 5.8c and 5.8d the spacing of the elements has been increased to a half wavelength.

Since the main beam of a broadside array is at right angles to the curtain of antennas, it is to be expected that the effect of mutual coupling will be much less than for an endfire array. However, when the elements are separated by only a quarter wavelength, differences in the interactions between the currents in differently situated elements are sufficient to produce small but significant changes in the resistances even when the elements are as short as a quarter wavelength (Fig. 5.8a). In this case there is only a very small variation in the reactance. When the length of the elements is increased to a half wavelength with the same quarter wavelength spacing, both resistance and reactance vary greatly from element to element (Fig. 5.8b). Note that the change in the reactance from the central element in the array to one at the extremities may be as large as from near 100 ohms to near zero.

As is to be expected, an increase in the spacing of the elements to a half wavelength substantially reduces the changes in resistance and reactance due to differences in mutual interaction. When $2h = \lambda/4$ both resistance and reactance are substantially constant across the array (Fig. 5.8c). When $2h = \lambda/2$ significant differences in both resistance and reactance exist, but they are much smaller than for the more closely spaced array (Fig. 5.8b). In all cases, the obviously different environment of elements at the extremities of the array is responsible for the largest differences in the impedances. For the two lengths, $2h = \lambda/4$ and $2h = \lambda/2$, there is little difference between the results obtained with specified voltages and with specified driving-point currents.

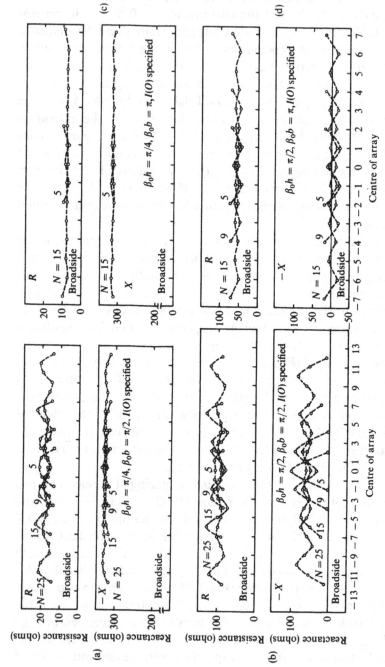

Fig. 5.8. Driving-point impedances for broadside arrays, current specified; $\beta_0 h = \pi/4$ and $\pi/2$, $N = 5$, 9, 15 and 25.

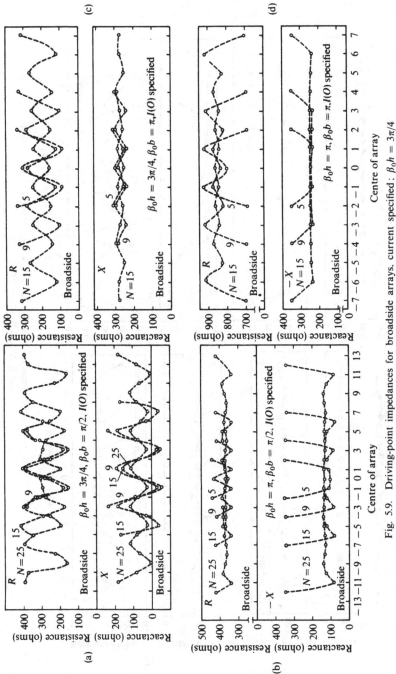

Fig. 5.9. Driving-point impedances for broadside arrays, current specified; $\beta_0 h = 3\pi/4$ and π, $N = 5, 9, 15$ and 25.

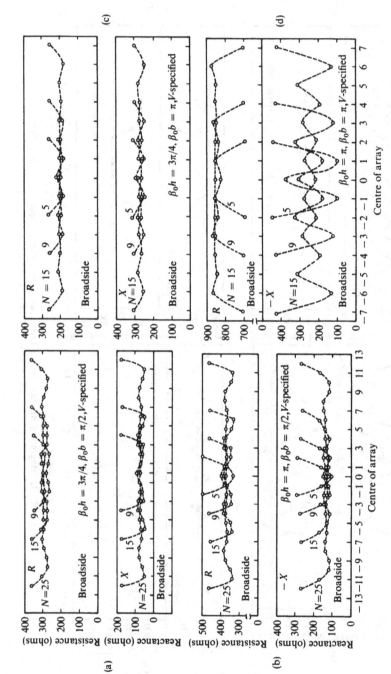

Fig. 5.10. Driving-point impedances for broadside arrays, voltage specified; $\beta_0 h = 3\pi/4$ and π, $N = 5, 9, 15$ and 25.

Graphs of the resistances of the individual antennas in a broadside array of three-quarter and full wavelength elements are shown in Figs. 5.9a–d when the driving-point currents are specified. Similar curves for the same array with the voltages specified are in Figs. 5.10a–d. Especially noteworthy when $2h = 3\lambda/4$ are the large differences between the resistances and reactances of the elements when the driving-point voltages are specified instead of the driving-point currents (Figs. 5.9a, c and 5.10a, c). When $2h = \lambda$ the reactance and to a lesser extent the resistance of the elements at the extremities of the array differ greatly from the others (Figs. 5.9b, d and 5.10b, d). As an example of typical digital results prepared for this study, a table of impedances is given in appendix IV.

The radiation patterns in the equatorial or H-plane are shown in Fig. 5.11 for a broadside array of 15 elements. The ideal patterns are fairly well approximated when the amplitude and phase of the current along each antenna are specified near the point of maximum amplitude. For the array of half-wave dipoles this occurs essentially

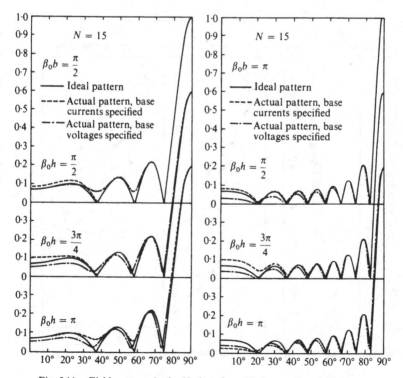

Fig. 5.11.   Field patterns in the H-plane for a 15-element broadside array.

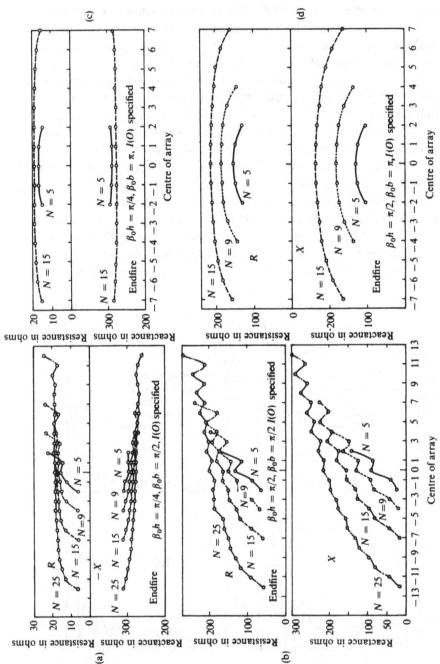

Fig. 5.12. Driving-point impedances for endfire arrays, current specified; $\beta_0 h = \pi/4$ and $\pi/2$. $N = 5, 9, 15$ and 25.

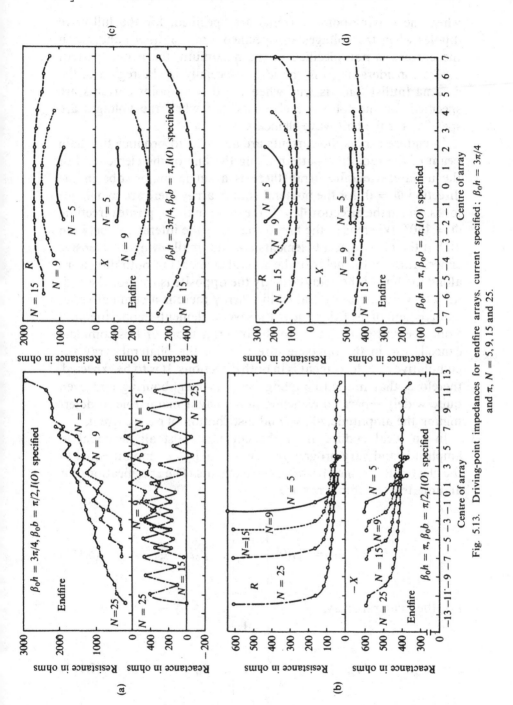

Fig. 5.13. Driving-point impedances for endfire arrays, current specified: $\beta_0 h = 3\pi/4$ and π, $N = 5, 9, 15$ and 25.

when the driving-point currents are specified, for the full-wave dipoles when the voltages are specified. On the other hand, when the current is not specified at the maximum, the actual pattern differs considerably from the ideal especially in the region of the minima (nulls). This is true when the driving-point currents are specified for the full-wave elements and when the voltages are specified for the half-wave elements.

In endfire arrays the currents are adjusted to produce the main beam of the radiation pattern along the line of the elements. For the unilateral endfire array there is a single major lobe in the direction $\Phi = 0$; for the bilateral endfire array there are two major lobes, one in the direction $\Phi = 0$, the other in the opposite direction, $\Phi = 180°$. Whereas in the broadside array the interaction between all but the next adjacent elements is quite small owing to extensive cancellation of the fields of the several elements in both directions along the line of the array; exactly the opposite is true for the endfire array. In the unilateral endfire array there is a cumulative re-enforcement of the fields due to the several elements in one direction from one end of the array to the other, a more or less complete cancellation in the opposite direction. In the bilateral array the cumulative re-enforcement is in both directions. It is to be expected, therefore, that mutual coupling between neighbouring and even quite widely separated elements must play a major role in determining the amplitude, phase, and distribution of each current.

In an ideal endfire array the currents must all be equal in amplitude and vary progressively in phase by an amount equal to the electrical distance between the elements. The specifications for an unilateral endfire array are

$$\{I_z(0)\} = I_{z1}(0) \begin{Bmatrix} 1 \\ -j \\ -1 \\ \vdots \end{Bmatrix}, \quad \beta_0 b = \frac{\pi}{2}. \tag{5.113}$$

For the bilateral array,

$$\{I_z(0)\} = I_{z1}(0) \begin{Bmatrix} 1 \\ -1 \\ 1 \\ \vdots \end{Bmatrix}, \quad \beta_0 b = \pi. \tag{5.114}$$

Alternatively, the voltages may be specified in the same manner. Thus, for the unilateral array

$$\{V\} = V_1 \left\{ \begin{matrix} 1 \\ -j \\ -1 \\ \vdots \end{matrix} \right\}, \quad \beta_0 b = \frac{\pi}{2}. \qquad (5.115)$$

For the bilateral array,

$$\{V\} = V_1 \left\{ \begin{matrix} 1 \\ -1 \\ 1 \\ \vdots \end{matrix} \right\}, \quad \beta_0 b = \pi. \qquad (5.116)$$

The resistances and reactances of the individual elements in a unilateral endfire array are shown in Figs. 5.12a and 5.12b, respectively with $2h = \lambda/4$ and $2h = \lambda/2$. The driving-point currents were specified according to (5.113). Corresponding values for the bilateral array are in Figs. 5.12c and 5.12d. Note that these are symmetrical with respect to the centre of the array. For the shorter elements ($2h = \lambda/4$), the reactances of all elements are reasonably alike; the resistances also vary little except for the two elements at the ends of the unilateral array. When the elements are a half-wavelength long, the resistances and reactances both vary greatly along the unilateral array (Fig. 5.12b), moderately along the bilateral array (Fig. 5.12d). It is interesting to note that in the unilateral array the impedance of the forward element (in the direction of the beam) is greatest, that of the rear element smallest. Since the amplitudes of the driving-point currents are all the same, the power supplied to each element is proportional to its resistance. It follows from Fig. 5.12b that the power supplied to and radiated from the forward element is approximately five times that supplied to and radiated from the rear element. Note that the resistance and the reactance of all but the last two elements in each array are significantly greater than for an isolated antenna. In effect, each element after the forward one acts partly as a driven element, partly as a parasitic reflector for the element in front of it.

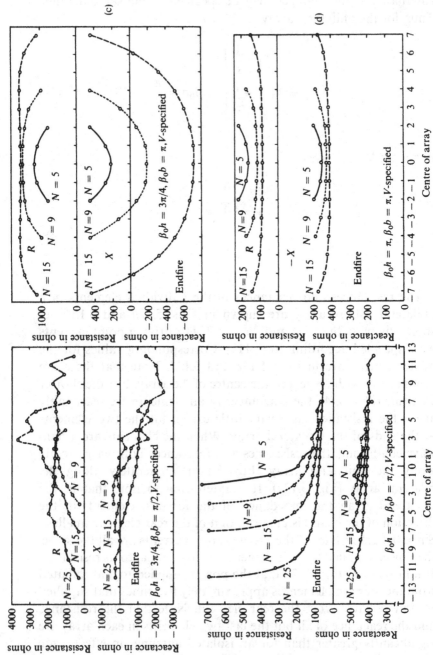

Fig. 5.14. Driving-point impedances for endfire arrays, voltage specified: $\beta_0 h = 3\pi/4$ and π, $N = 5, 9, 15$ and 25.

The resistances of the antennas in a unilateral endfire array with elements of length $2h = 3\lambda/4$ (Fig. 5.13a) decrease continually from the forward element to that in the rear in a manner resembling that for the half-wave elements (Fig. 5.12b). However, the range of magnitudes is much greater. The corresponding values for the same array but constructed of full-wave elements ($2h = \lambda$) are in Fig. 5.13b. They are startlingly different. The resistances of all elements are now reasonably alike except for that of the rear element which is much greater. Evidently, the rear element is supplied and radiates the most power—approximately four to six times as much as any other element. This suggests that all but the rear element act in part as driven radiators and in part as parasitic directors for the elements behind them, especially the rear one. Note that for the bilateral array of full-wave elements (Fig. 5.13d) the resistances of the elements increase from the centre outward, whereas for the corresponding array of half-wave elements (Fig. 5.12d) the resistances decrease from the centre outward. If the voltages are specified according to (5.115) and (5.116) instead of the

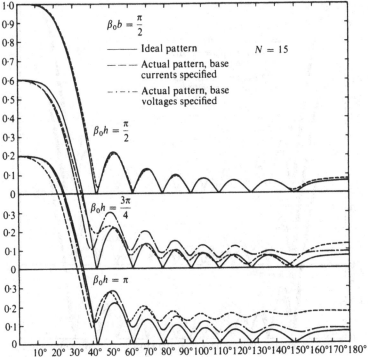

Fig. 5.15.   Field patterns in the H-plane for a 15-element unilateral endfire array.

driving-point currents, the graphs of Figs. 5.13a–d are replaced by those of Figs. 5.14a–d. The two sets are seen to differ considerably.

The radiation patterns in the equatorial or H-plane are shown in Figs. 5.15 and 5.16 respectively for the unilateral and bilateral endfire arrays. The ideal pattern is fairly well approximated when the current along each antenna is specified near its point of maximum according to the criteria for an ideal array. For the half-wave dipoles this is true essentially when the driving-point currents $I(0)$ are specified, for the full-wave dipoles when the voltages are assigned. On the other hand, when the current is not specified at its maximum value, the actual pattern differs considerably from the ideal, especially in its minor lobe structure and the region of the minima (nulls). This is true when the driving-point currents are specified for the full-wave dipoles, when the voltages are specified for the half-wave dipoles. In general, the departure from the ideal patterns is greater for the unilateral endfire array (Fig. 5.15) than for the broadside array (Fig. 5.11a, b) since the effect of mutual interaction is greater.

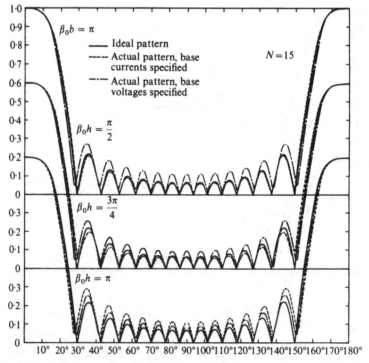

Fig. 5.16. Radiation patterns in the H-plane for a 15-element bilateral endfire array.

5.6   The special case when $\beta_0 h = \pi/2$

The general functional form for the currents in the elements given by (5.34) with (5.46) and (5.47) presents some difficulties when $\beta_0 h = \pi/2$. For both circular and curtain arrays the expression for the currents becomes indeterminate in the form $0/0$ when $\beta_0 h = \pi/2$. This behaviour is illustrated for the curtain array in the following matrix equation for the currents:

$$\{I_z(z)\} = \frac{j2\pi}{\zeta_0 \Psi_{dR} \cos \beta_0 h}\{V_0\} \sin \beta_0(h - |z|)$$

$$+ \frac{j2\pi}{\zeta_0 \Psi_{dR} \cos \beta_0 h}[\Phi_u]^{-1}[\Phi_v]\{V_0\}(\cos \beta_0 z - \cos \beta_0 h).$$

(5.117)

From the form of the Ψ functions at $\beta_0 h = \pi/2$ it follows that

$$\lim_{\beta_0 h \to \pi/2} [\Phi_u]^{-1}[\Phi_v] = -[\mathscr{I}] \qquad (5.118)$$

where $[\mathscr{I}]$ is the identity matrix. The indeterminate form for the currents in the elements follows directly when (5.118) is used in (5.117). It is

$$\{I_z(z)\} = \frac{j2\pi \cos \beta_0 z}{\zeta_0 \Psi_{dR} \cdot 0}\{V_0\} - j\frac{2\pi \cos \beta_0 z}{\zeta_0 \Psi_{dR} \cdot 0}\{V_0\} = \frac{0}{0}, \qquad \beta_0 h = \frac{\pi}{2}.$$

(5.119)

Two alternatives are available for avoiding this difficulty: (a) The formula for the currents may be rearranged as in section 2.7 or (b) a special formulation for $\beta_0 h = \pi/2$ may be used. The former method has the advantage that it is applicable over a range near $\beta_0 h = \pi/2$, whereas the latter method is valid only at $\beta_0 h = \pi/2$. Both methods are presented here, although the numerical results were calculated based on the special form for $\beta_0 h = \pi/2$. Numerical calculations have shown the results of the two approximate forms useful when $\beta_0 h = \pi/2$ to be approximately the same.

The expression for the currents when $\beta_0 h$ is near $\pi/2$ follows directly from the results of section 2.7. In matrix form

$$\{I_z(z)\} = \frac{-j2\pi}{\zeta_0 \Psi_{dR}}(\{V_0\}(\sin \beta_0|z| - \sin \beta_0 h)$$

$$+ [\Phi'_u]^{-1}[\Phi'_v]\{V_0\}(\cos \beta_0 z - \cos \beta_0 h)) \quad (5.120a)$$

where the elements of the matrices are

$$\Phi'_{kiu} = -\Phi_{kiu} \cos \beta_0 h \qquad (5.120b)$$

$$\Phi'_{kiv} = \Phi_{kiv} + \Phi_{kiu} \sin \beta_0 h. \tag{5.120c}$$

When $\beta_0 h = \pi/2$,

$$\{I_z(z)\} = \frac{-j2\pi}{\zeta_0 \Psi_{dR}} (\{V_0\}(\sin \beta_0 |z| - 1) + [\Phi'_u]^{-1}[\Phi'_v]\{V_0\} \cos \beta_0 z). \tag{5.121}$$

For $\beta_0 h = \pi/2$, the elements of the Φ'_u and Φ'_v matrices are

$$\Phi'_{kiu} = \Psi_{kiu}(h) \tag{5.122a}$$

and

$$\Phi'_{kiv} = \Psi_{kidu} - \Psi_{kidv}(1 - \delta_{ik}) - j\Psi_{kidI}\delta_{ik}. \tag{5.122b}$$

The alternative approach begins with the special form for the integral equation valid at $\beta_0 h = \pi/2$. It was this latter method which was used for the original curtain-array calculations. The final form is similar to (5.120b) with slightly different values for the constant Ψ_{dR} and the Φ'_u and Φ'_v matrices. In this method the Ψ functions are computed with the following cosine and shifted-sine currents:

$$I_{zi}(z) = -jA_i S_{0z} + B_i F_{0z}, \tag{5.123}$$

where $S_{0z} = \sin \beta_0 |z| - \sin \beta_0 h$ and $F_{0z} = \cos \beta_0 z - \cos \beta_0 h$

The final expression for the current with $\beta_0 h = \pi/2$ is

$$\{I_z(z)\} = \frac{-j2\pi}{\zeta_0 \Psi^h_{dR}}\{V_0\}(\sin \beta_0 |z| - 1) - j\frac{2\pi}{\zeta_0 \Psi^h_{dR}}[\Phi_u]^{-1}[\Phi^h_v]\{V_0\} \cos \beta_0 z \tag{5.124}$$

where

$$\Phi^h_{kiv} = \Psi^h_{ki\,dv}(1 - \delta_{ik}) + j\Psi^h_{kidI}\delta_{ik} - \Psi^h_{kiv}(0) \tag{5.125a}$$

$$\Phi_{ki\,du} = -\Psi_{ki\,du} + \Psi_{kiu}(0) \tag{5.125b}$$

and

$$\Psi^h_{dR} = -\text{Re}\{[S_b(h, 0) - E_b(h, 0)] - [S_b(h, h) - E_b(h, h)]\} \tag{5.126a}$$

$$\Psi^h_{ki\,dI} = \text{Im}\{[S_b(h, 0) - E_b(h, 0)] - [S_b(h, h) - E_b(h, h)]\} \tag{5.126b}$$

$$\Psi^h_{ki\,dv} = [S_b(h, 0) - E_b(h, 0)] - [S_b(h, h) - E_b(h, h)] \tag{5.126c}$$

$$\Psi_{ki\,du} = C_b(h, 0) - C_b(h, h) \tag{5.126d}$$

$$\Psi_{ki\,dv}(0) = S_b(h, 0) - E_b(h, 0) \tag{5.126e}$$

$$\Psi_{kiu}(0) = C_b(h, 0) \tag{5.126f}$$

$$\beta_0 h = \pi/2, \qquad b \equiv b_{ki}, \qquad b_{kk} = a.$$

Numerical calculations show that the results obtained with (5.124) are comparable with those obtained with (5.120b).

5.7   Design of curtain arrays

The antenna designer is generally confronted with both a geometrical and an electrical problem. First, the proper overall length and spacing must be chosen for a prescribed beamwidth and electrical scanning range. The currents or voltages at the driving point must then be selected for a prescribed radiation pattern. A knowledge of the driving-point impedance for each element is also necessary for properly matching the antennas to the transmission lines. If the array is operated over a large frequency range, the radiation pattern and driving-point impedances must be computed for representative points. Furthermore, these calculations must be repeated if the driving conditions are changed. With appropriate theoretical calculations available, the designer is better able to select the necessary auxiliary equipment and interpret any measurements made on the array.

The design procedure is illustrated in terms of the following somewhat simplified problem: An array is to be constructed to operate over a two-to-one frequency range. At the lowest frequency the spacing is a quarter wavelength and the elements are each a half wavelength long. The array is to have fifteen elements. Both broadside and endfire radiation patterns are desired. The problem is to find the actual impedances and radiation patterns with the currents specified at the driving point. The problem may be summarized as follows:

1. Geometry of Array
   (Number of elements, $N = 15$) ($\Omega = 2 \ln 2h/a = 10$)
   (a) at lowest frequency,   spacing $\beta_0 b = \pi/2$
                                      element half length $\beta_0 h = \pi/2$
   (b) at highest frequency,   spacing $\beta_0 b = \pi$
                                      element half length $\beta_0 h = \pi$
2. Electronic Scanning Range (in H-plane only)
   (a) at lowest frequency, $\Phi_0 = 0$ and $\Phi_0 = 90°$
   (b) at highest frequency, $\Phi_0 = 0$ and $\Phi_0 = 90°$
3. Specification of Base Currents
   (a) at lowest frequency ($\beta_0 h = \pi/2$, $\beta_0 b = \pi/2$)

$$\{I_z(0)\} = I_1(0) \begin{Bmatrix} 1 \\ 1 \\ 1 \\ \vdots \end{Bmatrix}, \qquad \Phi_0 = \pi/2 \text{ (broadside)} \quad (5.127)$$

$$\{I_z(0)\} = I_1(0) \begin{Bmatrix} 1 \\ -j \\ -1 \\ \vdots \end{Bmatrix}, \qquad \Phi_0 = 0 \text{ (endfire)} \qquad (5.128)$$

(b) at highest frequency ($\beta_0 h = \pi$, $\beta_0 b = \pi$)

$$\{I_z(0)\} = I_1(0) \begin{Bmatrix} 1 \\ 1 \\ 1 \\ \vdots \end{Bmatrix}, \qquad \Phi_0 = \pi/2 \qquad (5.129)$$

$$\{I_z(0)\} = I_1(0) \begin{Bmatrix} 1 \\ -1 \\ 1 \\ \vdots \end{Bmatrix}, \qquad \Phi_0 = 0. \qquad (5.130)$$

Calculations at lowest frequency

The driving-point impedances are calculated from (5.124) with (5.125a, b) and (5.127) with the relevant values of the Φ functions. Tables of the Φ functions along with values of Ψ_{dR} are given in appendix III based on an improved Romberg integration method. Since most of the calculations in this chapter are based on the tables of Mack [3], the calculations which follow are also based on the Mack tables.

$\beta_0 h = 1.5708 \qquad \beta_0 b = 1.5708 \qquad \Psi_{dR} = 6.9087 \qquad H_m\left(\frac{\pi}{2}, \frac{\pi}{2}\right) = -0.5708 \qquad G_m\left(\frac{\pi}{2}, \frac{\pi}{2}\right) = 1.0000$

$\Phi_{kiu} = \Phi_{um}$ with $|k - i| = m$

$\Phi_{u1} = 0.6880 - j1.2187$	$\Phi_{u2} = -0.4725 - j0.6798$	$\Phi_{u3} = -0.4988 + j0.2089$
$\Phi_{u4} = 0.1105 + j0.3750$	$\Phi_{u5} = 0.2957 - j0.0669$	$\Phi_{u6} = -0.0444 - j0.2426$
$\Phi_{u7} = -0.2051 + j0.0315$	$\Phi_{u8} = 0.0234 + j0.1773$	$\Phi_{u9} = 0.1561 - j0.0181$
$\Phi_{u10} = -0.0144 - j0.1393$	$\Phi_{u11} = -0.1257 + j0.0117$	$\Phi_{u12} = 0.0097 + j0.1146$
$\Phi_{u13} = 0.1052 - j0.0082$	$\Phi_{u14} = -0.0070 - j0.0972$	$\Phi_{u15} = -0.0904 + j0.0060$

$$(5.131)$$

$\Phi_{kir} = \Phi_{rm}$ with $|k - i| = m$.

$$\Phi_{v1} = 7{\cdot}0756 - j0{\cdot}2036 \qquad \Phi_{v2} = -0{\cdot}2864 - j0{\cdot}3970 \qquad \Phi_{v3} = -0{\cdot}2925 + j0{\cdot}1186$$

$$\Phi_{v4} = 0{\cdot}0609 + j0{\cdot}2176 \qquad \Phi_{v5} = 0{\cdot}1706 - j0{\cdot}0363 \qquad \Phi_{v6} = -0{\cdot}0239 - j0{\cdot}1395$$

$$\Phi_{v7} = -0{\cdot}1177 + j0{\cdot}0169 \qquad \Phi_{v8} = 0{\cdot}0125 + j0{\cdot}1016 \qquad \Phi_{v9} = 0{\cdot}0894 - j0{\cdot}0097$$

$$\Phi_{v10} = -0{\cdot}0077 - j0{\cdot}0797 \qquad \Phi_{v11} = -0{\cdot}0719 + j0{\cdot}0062 \qquad \Phi_{v12} = 0{\cdot}0052 + j0{\cdot}0655$$

$$\Phi_{v13} = 0{\cdot}0601 - j0{\cdot}0044 \qquad \Phi_{v14} = -0{\cdot}0037 - j0{\cdot}0556 \qquad \Phi_{v15} = -0{\cdot}0516 + j0{\cdot}0032$$

$$(5.132)$$

The driving-point currents for $\Phi_0 = 0$ are given by (5.128).

Driving-Point Currents

$I_z(0)$

$$I_1(0) = 1{\cdot}00 + j0 \qquad I_2(0) = 1{\cdot}00 + j0 \qquad I_3(0) = 1{\cdot}00 + j0 \qquad I_4(0) = 1{\cdot}00 + j0$$

$$I_5(0) = 1{\cdot}00 + j0 \qquad I_6(0) = 1{\cdot}00 + j0 \qquad I_7(0) = 1{\cdot}00 + j0 \qquad I_8(0) = 1{\cdot}00 + j0$$

$$I_9(0) = 1{\cdot}00 + j0 \qquad I_{10}(0) = 1{\cdot}00 + j0 \qquad I_{11}(0) = 1{\cdot}00 + j0 \qquad I_{12}(0) = 1{\cdot}00 + j0$$

$$I_{13}(0) = 1{\cdot}00 + j0 \qquad I_{14}(0) = 1{\cdot}00 + j0 \qquad I_{15}(0) = 1{\cdot}00 + j0$$

$$(5.133)$$

The driving-point impedances are computed from (5.124) with (5.131)–(5.133). The results are:

Driving-Point Impedances

$Z_k = V_k / I_k(0)$

$$Z_1 = 90{\cdot}816 - j21{\cdot}298 \qquad Z_2 = 129{\cdot}267 - j52{\cdot}450 \qquad Z_3 = 99{\cdot}602 - j75{\cdot}308$$

$$Z_4 = 73{\cdot}401 - j67{\cdot}853 \qquad Z_5 = 93{\cdot}734 - j54{\cdot}840 \qquad Z_6 = 114{\cdot}393 - j55{\cdot}375$$

$$Z_7 = 96{\cdot}241 - j63{\cdot}012 \qquad Z_8 = 77{\cdot}829 - j66{\cdot}659 \qquad Z_9 = 96{\cdot}241 - j63{\cdot}012$$

$$Z_{10} = 114{\cdot}393 - j55{\cdot}375 \qquad Z_{11} = 93{\cdot}734 - j54{\cdot}840 \qquad Z_{12} = 73{\cdot}401 - j67{\cdot}853$$

$$Z_{13} = 99{\cdot}602 - j75{\cdot}308 \qquad Z_{14} = 129{\cdot}267 - j52{\cdot}450 \qquad Z_{15} = 90{\cdot}816 - j21{\cdot}298$$

$$(5.134)$$

The driving-point admittances are obtained by inverting Z_{kk} in (5.134).

Driving-Point Admittances

Y_k

$$Y_1 = 0{\cdot}010437 + j0{\cdot}002448 \qquad Y_2 = 0{\cdot}006642 + j0{\cdot}002695 \qquad Y_3 = 0{\cdot}006388 + j0{\cdot}004830$$

$$Y_4 = 0{\cdot}007346 + j0{\cdot}006791 \qquad Y_5 = 0{\cdot}007948 + j0{\cdot}004650 \qquad Y_6 = 0{\cdot}007082 + j0{\cdot}003428$$

$$Y_7 = 0{\cdot}007273 + j0{\cdot}004762 \qquad Y_8 = 0{\cdot}007412 + j0{\cdot}006348 \qquad Y_9 = 0{\cdot}007273 + j0{\cdot}004762$$

$$Y_{10} = 0{\cdot}007082 + j0{\cdot}003428 \qquad Y_{11} = 0{\cdot}007948 + j0{\cdot}004650 \qquad Y_{12} = 0{\cdot}007346 + j0{\cdot}006791$$

$$Y_{13} = 0{\cdot}006388 + j0{\cdot}004830 \qquad Y_{14} = 0{\cdot}006642 + j0{\cdot}002695 \qquad Y_{15} = 0{\cdot}010437 + j0{\cdot}002448$$

$$(5.135)$$

The electric field may be computed from (5.67) which is given in the following form for ease of computation:

$$E_\Theta^r(\Theta, \Phi) = \frac{j\zeta_0}{4\pi} \frac{e^{-j\beta_0 R_0}}{R_0} \sum_{i=1}^{N} C_i \, e^{j\beta_0 b[(N-2i+1)/2]\sin\Theta\cos\Phi} \qquad (5.136)$$

where

$$C_i = \xi_i\{jA_iF_m(\Theta, \beta_0 h) + (Y_i - jA_i \sin\beta_0 h)G_m(\Theta, \beta_0 h)\} \qquad \beta_0 h \neq \pi/2 \qquad (5.137)$$

$$C_i = \xi_i\left\{H_m\left(\Theta, \frac{\pi}{2}\right) + (Y_i - jA_i)G_m\left(\Theta, \frac{\pi}{2}\right)\right\} \qquad \beta_0 h = \pi/2 \quad (5.138)$$

$$\xi_i = V_{0i}/V_{01}. \qquad (5.139)$$

For $\beta_0 h = \pi/2$, $\beta_0 b = \pi/2$, $\Theta = \pi/2$ and $\Phi_0 = 0°$, the constants C_i are:

Field-Pattern Constants

C_i

$C_1 = 0{\cdot}01147 + j0{\cdot}0014123$	$C_2 = 0{\cdot}01173 + j0{\cdot}0006786$	$C_3 = 0{\cdot}01114 + j0{\cdot}0006191$
$C_4 = 0{\cdot}01086 + j0{\cdot}0009353$	$C_5 = 0{\cdot}01123 + j0{\cdot}0009407$	$C_6 = 0{\cdot}01151 + j0{\cdot}0007627$
$C_7 = 0{\cdot}01120 + j0{\cdot}0008109$	$C_8 = 0{\cdot}01092 + j0{\cdot}0009146$	$C_9 = 0{\cdot}01120 + j0{\cdot}0008109$
$C_{10} = 0{\cdot}01151 + j0{\cdot}0007627$	$C_{11} = 0{\cdot}01123 + j0{\cdot}0009407$	$C_{12} = 0{\cdot}01086 + j0{\cdot}0009353$
$C_{13} = 0{\cdot}01114 + j0{\cdot}0006191$	$C_{14} = 0{\cdot}01173 + j0{\cdot}0006786$	$C_{15} = 0{\cdot}01147 + j0{\cdot}0001412$

$$(5.140)$$

The field patterns computed from (5.136) with (5.138)–(5.140) are shown in Fig. 5.15.

The driving-point admittances for the endfire case ($\Phi_0 = \pi/2$) are computed with (5.124) and (5.128). Thus, with $I_1(0) = 1$, the driving-point currents are specified to be

Driving-Point Currents

$I_z(0)$

$I_1 = 1{\cdot}000 + j0$	$I_2 = 0 - j1{\cdot}000$	$I_3 = -1{\cdot}000 + j0$	$I_4 = 0 + j1{\cdot}000$
$I_5 = 1{\cdot}000 + j0$	$I_6 = 0 - j1{\cdot}000$	$I_7 = -1{\cdot}000 + j0$	$I_8 = 0 + j1{\cdot}000$
$I_9 = 1{\cdot}000 + j0$	$I_{10} = 0 - j1{\cdot}000$	$I_{11} = -1{\cdot}000 + j0$	$I_{12} = 0 + j1{\cdot}000$
$I_{13} = 1{\cdot}000 + j0$	$I_{14} = 0 - j1{\cdot}000$	$I_{15} = -1{\cdot}000 + j0$	

$$(5.141)$$

The corresponding driving-point impedances and constants C_i for computing the field pattern are:

Driving-Point Impedances

Z_k

$Z_1 = 59.432 + j\ 15.260$	$Z_2 = 91.243 + j\ 43.280$	$Z_3 = 119.668 + j\ 81.021$
$Z_4 = 127.775 + j\ 94.113$	$Z_5 = 146.811 + j123.305$	$Z_6 = 148.136 + j128.287$
$Z_7 = 164.970 + j154.466$	$Z_8 = 162.206 + j153.871$	$Z_9 = 179.341 + j179.810$
$Z_{10} = 172.518 + j173.925$	$Z_{11} = 192.236 + j202.053$	$Z_{12} = 179.290 + j189.402$
$Z_{13} = 206.483 + j223.550$	$Z_{14} = 178.456 + j199.409$	$Z_{15} = 235.531 + j246.418$

$$(5.142)$$

Field-Pattern Constants

C_i

$C_1 = 0.01682 - j0.00509$	$C_2 = -0.00540 - j0.01778$	$C_3 = -0.01890 + j0.00553$
$C_4 = 0.00553 + j0.01927$	$C_5 = 0.02010 - j0.00557$	$C_6 = -0.00554 - j0.02022$
$C_7 = -0.02096 + j0.00557$	$C_8 = 0.00552 + j0.02092$	$C_9 = 0.02166 - j0.00555$
$C_{10} = -0.00548 + j0.02145$	$C_{11} = -0.02227 + j0.00555$	$C_{12} = 0.00544 + j0.02185$
$C_{13} = 0.02289 - j0.00558$	$C_{14} = -0.00530 - j0.02205$	$C_{15} = -0.02371 + j0.00590$

$$(5.143)$$

The endfire field pattern determined by (5.143) is shown in Fig. 5.11. The impedances for the endfire and broadside arrays at the lowest frequency are shown in Figs. 5.8 and 5.12.

Calculations at the highest frequency $(\beta_0 h = \pi,\ \beta_0 b = \pi)$

The relevant Φ functions from appendix III are:

$\beta_0 h = 3.1416 \quad \beta_0 b = 3.1416 \quad \Psi_{dR} = 5.8340 \quad F_m(\pi, \pi) = 2.000 \quad G_m(\pi, \pi) = 3.1416$

$\Phi_{kiu} = \Phi_{um}$ with $|k - i| = m$

$\Phi_{u1} = -6.6898 + j2.8959$	$\Phi_{u2} = 1.1030 - j0.6635$	$\Phi_{u3} = -0.7700 + j0.3211$
$\Phi_{u4} = 0.5810 - j0.1765$	$\Phi_{u5} = -0.4602 + j0.1086$	$\Phi_{u6} = 0.3786 - j0.0727$
$\Phi_{u7} = -0.3206 + j0.0518$	$\Phi_{u8} = 0.2776 - j0.0386$	$\Phi_{u9} = -0.2445 + j0.0299$
$\Phi_{u10} = 0.2183 - j0.0238$	$\Phi_{u11} = -0.1971 + j0.0194$	$\Phi_{u12} = 0.1797 - j0.0159$
$\Phi_{u13} = -0.1650 + j0.0135$	$\Phi_{u14} = 0.1525 - j0.0116$	$\Phi_{u15} = -0.1418 + j0.0100$

$$(5.144)$$

$\Phi_{kiv} = \Phi_{vm}$ with $|k - i| = m$

$\Phi_{v1} = 0.6721 - j1.6605$	$\Phi_{v2} = -0.6296 + j0.4070$	$\Phi_{v3} = 0.4441 - j0.2174$
$\Phi_{v4} = -0.3470 + j0.1287$	$\Phi_{v5} = 0.2810 - j0.0822$	$\Phi_{v6} = -0.2342 + j0.0561$
$\Phi_{v7} = 0.1999 - j0.0405$	$\Phi_{v8} = -0.1740 + j0.0304$	$\Phi_{v9} = 0.1538 - j0.0237$
$\Phi_{v10} = -0.1377 + j0.0100$	$\Phi_{v11} = 0.1245 - j0.0154$	$\Phi_{v12} = -0.1136 + j0.0128$
$\Phi_{v13} = 0.1045 - j0.0108$	$\Phi_{v14} = -0.0967 + j0.0093$	$\Phi_{v15} = 0.0899 - j0.0080$

$$(5.145)$$

The specifications of the driving-point currents for the broadside array are:

Driving-Point Currents

$I_z(0)$

$$I_1(0) = 1\cdot000 + j0 \qquad I_2(0) = 1\cdot000 + j0 \qquad I_3(0) = 1\cdot000 + j0 \qquad I_4(0) = 1\cdot000 + j0$$
$$I_5(0) = 1\cdot000 + j0 \qquad I_6(0) = 1\cdot000 + j0 \qquad I_7(0) = 1\cdot000 + j0 \qquad I_8(0) = 1\cdot000 + j0$$
$$I_9(0) = 1\cdot000 + j0 \qquad I_{10}(0) = 1\cdot000 + j0 \qquad I_{11}(0) = 1\cdot000 + j0 \qquad I_{12}(0) = 1\cdot000 + j0$$
$$I_{13}(0) = 1\cdot000 + j0 \qquad I_{14}(0) = 1\cdot000 + j0 \qquad I_{15}(0) = 1\cdot000 + j0$$

$$(5.146)$$

The driving-point impedances from (5.51) are:

Driving-Point Impedances

Z_k

$$Z_1 = 702\cdot588 - j350\cdot682 \qquad Z_2 = 915\cdot158 - j239\cdot232 \qquad Z_3 = 828\cdot211 - j245\cdot278$$
$$Z_4 = 872\cdot733 - j246\cdot841 \qquad Z_5 = 843\cdot055 - j244\cdot341 \qquad Z_6 = 866\cdot245 - j246\cdot898$$
$$Z_7 = 846\cdot108 - j244\cdot293 \qquad Z_8 = 864\cdot939 - j246\cdot801 \qquad Z_9 = 846\cdot108 - j244\cdot293$$
$$Z_{10} = 866\cdot245 - j246\cdot898 \qquad Z_{11} = 843\cdot055 - j244\cdot341 \qquad Z_{12} = 872\cdot733 - j246\cdot841$$
$$Z_{13} = 828\cdot211 - j245\cdot278 \qquad Z_{14} = 915\cdot159 - j239\cdot232 \qquad Z_{15} = 702\cdot589 - j350\cdot682$$

$$(5.147)$$

The constants C_i computed from (5.136) with (5.137) are:

Field-Pattern Constants

C_i

$$C_1 = 0\cdot00179 - j0\cdot00482 \qquad C_2 = 0\cdot00321 - j0\cdot00584 \qquad C_3 = 0\cdot00288 - j0\cdot00530$$
$$C_4 = 0\cdot00302 - j0\cdot00559 \qquad C_5 = 0\cdot00294 - j0\cdot00539 \qquad C_6 = 0\cdot00300 - j0\cdot00555$$
$$C_7 = 0\cdot00295 - j0\cdot00541 \qquad C_8 = 0\cdot00299 - j0\cdot00554 \qquad C_9 = 0\cdot00295 - j0\cdot00541$$
$$C_{10} = 0\cdot00300 - j0\cdot00555 \qquad C_{11} = 0\cdot00294 - j0\cdot00539 \qquad C_{12} = 0\cdot00302 - j0\cdot00559$$
$$C_{13} = 0\cdot00288 - j0\cdot00530 \qquad C_{14} = 0\cdot00321 - j0\cdot00584 \qquad C_{15} = 0\cdot00179 - j0\cdot00482$$
$$\Theta = \pi/2.$$

$$(5.148)$$

The field pattern determined by (5.148) is shown in Fig. 5.15. Finally, the driving-point currents, impedances, and field-pattern constants are given below for the endfire case, $\Phi_0 = \pi/2$:

Driving-Point Currents

$I_z(0)$

$$I_1 = 1\cdot000 + j0 \qquad I_2 = -1\cdot000 + j0 \qquad I_3 = 1\cdot000 + j0 \qquad I_4 = -1\cdot000 + j0$$
$$I_5 = 1\cdot000 + j0 \qquad I_6 = -1\cdot000 + j0 \qquad I_7 = 1\cdot000 + j0 \qquad I_8 = -1\cdot000 + j0$$
$$I_9 = 1\cdot000 + j0 \qquad I_{10} = -1\cdot000 + j0 \qquad I_{11} = 1\cdot000 + j0 \qquad I_{12} = -1\cdot000 + j0$$
$$I_{13} = 1\cdot000 + j0 \qquad I_{14} = -1\cdot000 + j0 \qquad I_{15} = 1\cdot000 + j0$$

$$(5.149)$$

Driving-Point Impedances

z

$$Z_1 = 196 \cdot 512 - j454 \cdot 406 \qquad Z_2 = 119 \cdot 475 - j446 \cdot 232 \qquad Z_3 = 97 \cdot 429 - j431 \cdot 258$$

$$Z_4 = 87 \cdot 725 - j423 \cdot 369 \qquad Z_5 = 82 \cdot 423 - j418 \cdot 986 \qquad Z_6 = 79 \cdot 400 - j416 \cdot 431$$

$$Z_7 = 77 \cdot 782 - j415 \cdot 071 \qquad Z_8 = 77 \cdot 300 - j414 \cdot 663 \qquad Z_9 = 77 \cdot 782 - j415 \cdot 071$$

$$Z_{10} = 79 \cdot 400 - j416 \cdot 431 \qquad Z_{11} = 82 \cdot 423 - j418 \cdot 987 \qquad Z_{12} = 87 \cdot 725 - j423 \cdot 369$$

$$Z_{13} = 97 \cdot 429 - j431 \cdot 258 \qquad Z_{14} = 119 \cdot 475 - j446 \cdot 232 \qquad Z_{15} = 196 \cdot 512 - j454 \cdot 406$$

$$(5.150)$$

Field-Pattern Constants

C_i

$$C_1 = 0 \cdot 001259 - j0 \cdot 002801 \qquad C_2 = -0 \cdot 000481 + j0 \cdot 002362$$

$$C_3 = 0 \cdot 000316 - j0 \cdot 002102 \qquad C_4 = -0 \cdot 000249 + j0 \cdot 001974$$

$$C_5 = 0 \cdot 000213 - j0 \cdot 001904 \qquad C_6 = -0 \cdot 000193 + j0 \cdot 001863$$

$$C_7 = 0 \cdot 000182 - j0 \cdot 001841 \qquad C_8 = -0 \cdot 000179 + j0 \cdot 001834$$

$$C_9 = 0 \cdot 000182 - j0 \cdot 001841 \qquad C_{10} = -0 \cdot 000193 + j0 \cdot 001863$$

$$C_{11} = 0 \cdot 000213 - j0 \cdot 001904 \qquad C_{12} = -0 \cdot 000249 + j0 \cdot 001974$$

$$C_{13} = 0 \cdot 000316 - j0 \cdot 002102 \qquad C_{14} = -0 \cdot 000481 + j0 \cdot 002362$$

$$C_{15} = 0 \cdot 001259 - j0 \cdot 002801$$

$$(5.151)$$

5.8 Summary

In this chapter a complete theory of curtain arrays of practical antennas has been presented. Mutual coupling among all elements is included in a manner that takes account of changes in the amplitudes and the phases of the currents along all elements as determined by their locations in an array. The theory is quantitatively useful for cylindrical elements with electrical half-lengths in the range $\beta_0 h \leqslant 5\pi/4$ and electrical radii with values $\beta_0 a \leqslant 0 \cdot 02$. This includes lengths over the full range in which the principal lobe in the vertical field pattern is in the equatorial plane; it provides a 5 to 1 frequency band for electrical half-lengths included in the range $\pi/4 \leqslant \beta_0 h \leqslant 5\pi/4$.

In this chapter no measurements have been cited to verify the quantitative correctness of the new theory in determining distributions of current, driving-point impedances or admittances, and field patterns of typical curtain arrays. This is due in part to the relative difficulty in carrying out accurate measurements of the self- and mutual impedances for curtain arrays owing to the lack of the symmetry which underlies the corresponding measurements with the circular array. It is due primarily to the adequacy of the experimental verification of all phases of the theory as applied to the two-element array—the simplest curtain array (chapter 3), general circular arrays (chapter 4) and to curtain arrays of parasitic elements (chapter 6). As in the case of the circular array, the most sensitive and, at the same time, the most convenient experimental verification of the theory is in its application to an array in which only one element is driven while all others are parasitic. The first section in the next chapter is concerned specifically with the application of the theory developed in this chapter to a curtain array of twenty elements of which only one is driven and a comparison of theoretically and experimentally determined currents, admittances, and field patterns.

CHAPTER 6

ARRAYS WITH UNEQUAL ELEMENTS; PARASITIC AND LOG-PERIODIC ANTENNAS

The general theory of curtain arrays which is developed in the preceding chapter requires all N elements to be identical geometrically, but allows them to be driven by arbitrary voltages or loaded by arbitrary reactors at their centres. If some of these voltages are zero, the corresponding elements are parasitic and their currents are maintained entirely by mutual interaction. In arrays of the well-known Yagi-Uda type, only one element is driven so that the importance of an accurate analytical treatment of the inter-element coupling is increased. In a long array the possible cumulative effect of a small error in the interaction between the currents in adjacent elements must not be overlooked. As an added complication, the tuning of the individual parasitic elements is accomplished by adjustments in their lengths and spacings. This introduces the important problem of arrays with elements that are different in length and that are separated by different distances. In the Yagi-Uda array the range of these differences is relatively small. On the other hand, in frequency-independent arrays of the log-periodic type the range of lengths and distances between adjacent elements is very great.

In this chapter the analytical treatment of arrays with elements that are different in length and unequally spaced is carried out successively for parasitic arrays of the conventional Yagi-Uda type and for driven log-periodic arrays. However, the formulation is sufficiently general to permit its extension to arrays of other types, both parasitic and driven, that involve geometrically different elements.

6.1 Application of the two-term theory to a simple parasitic array

The simplest parasitic array consists of N geometrically identical antennas each of length $2h$ and radius a arranged in a curtain of parallel non-staggered elements with spacing b. Element 1 is driven, all others are parasitic. Such an array is illustrated in Fig.

6.1. The directional properties of the electromagnetic field maintained by the array depend on the relative amplitudes and phases of the currents in all of the elements. The currents in the parasitic elements are all induced by their mutual interaction. The current in the driven antenna is determined in part by the driving generator, in part by the mutual interaction with the currents in the other elements. The coupling between the currents in any pair of elements of given length depends primarily on the distance between them.

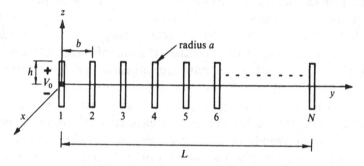

Fig. 6.1.   Parasitic array of identical elements.

The general theory of curtain arrays formulated in the preceding chapter may be applied directly by setting $V_{0i} = 0$, $1 < i \leqslant N$. The currents in the N elements are given by (5.34). They are

$$I_{z1}(z) = jA_1 \sin \beta_0(h - |z|) + B_1(\cos \beta_0 z - \cos \beta_0 h) \qquad (6.1)$$

$$I_{zi}(z) = B_i(\cos \beta_0 z - \cos \beta_0 h), \qquad i = 2, 3, \dots N \qquad (6.2)$$

where from (5.47) $\qquad A_1 = \dfrac{2\pi}{\zeta_0 \Psi_{dR} \cos \beta_0 h} V_{01} \qquad (6.3)$

and the B_i are obtained from (5.46). With V_{01} specified the currents at the centres of the elements are obtained from (5.53). The driving-point admittance of element 1 is

$$Y_{01} = I_{1z}(0)/V_{01}. \qquad (6.4)$$

The field pattern of the array is obtained from (5.67) with the appropriate values of A_i and B_i. Since only A_1 differs from zero the applicable formula is

$$E_\Theta(\Theta, \Phi) = j\frac{\omega\mu \sin \Theta}{4\pi} \left\{ jA_1 \frac{e^{-j\beta_0 R_1}}{R_1} F_m(\Theta, \beta_0 h) \right.$$

$$\left. + \sum_{i=2}^{N} B_i \frac{e^{-j\beta_0 R_i}}{R_i} G_m(\Theta, \beta_0 h) \right\} \qquad (6.5)$$

where $F_m(\Theta, \beta_0 h)$ and $G_m(\Theta, \beta_0 h)$ are defined in (5.66) and (5.68). In (6.5) the field is evaluated in the far zone of each element so that the distances R_i are measured to the centres of the elements. The far field of the array implies in addition that $R_i \doteq R_1$ in amplitudes and $R_i = R_1 - (i-1)b \sin \Theta \cos \Phi$ in phase angles.

Numerical computations have been made by Mailloux [1] for an array of 20 elements with $a/\lambda = 0.00635$ and $b/\lambda = 0.20$. Several

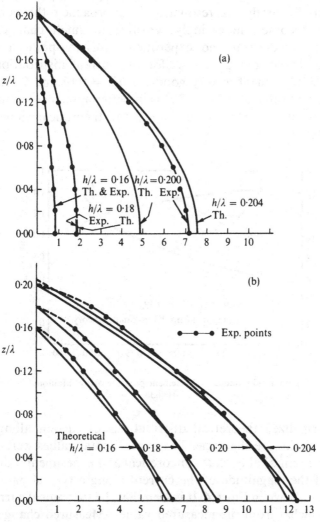

Fig. 6.2.   Components of current on driven dipole in 20-element parasitic array (Mailloux), (a) in phase with driving voltage (mA), (b) in phase quadrature with driving voltage (mA). $b/\lambda = 0.20$, $a/\lambda = 0.00635$.

values of h/λ were chosen in the range for endfire operation between 0·16 and 0·204.

The calculated distributions of current along the driven element are shown in Fig. 6.2 together with measured values. The agreement is excellent for $h/\lambda = 0·16$ and 0·18. The agreement when $h/\lambda = 0·20$ is not so close. However, the theoretical curves for antennas with h/λ increased by only 0·004—a distance of less than $a/\lambda = 0·00635$ —are in excellent agreement with the experimental data for $h/\lambda = 0·20$. Evidently, as resonance is approached the current amplitude becomes increasingly sensitive to small changes in length. The theoretical and experimental driving-point admittances are shown in Fig. 6.3. As for the current distribution in general, the agreement is very good for $h/\lambda = 0·16$ and 0·18, but the theoretical value at $h/\lambda = 0·204$ is in better agreement with the measured value for $h/\lambda = 0·20$ than is the theoretical value for $h/\lambda = 0·20$.

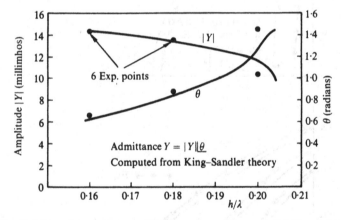

Fig. 6.3.   Driving-point admittance of 20-element parasitic array (Mailloux). $b/\lambda = 0·20$, $a/\lambda = 0·00635$.

The normalized theoretical distributions of current along all parasitic dipoles are the same. The experimental values were also found to be remarkably alike. Theoretical and experimental distributions of the magnitude of the current along a typical parasitic element are shown in Fig. 6.4. It is seen that the theoretical currents differ somewhat from the measured values. Measured changes in the phase of the current along the parasitic elements were very small.

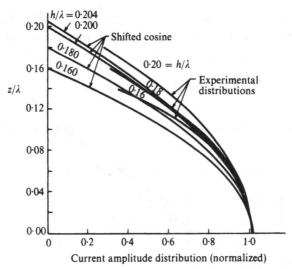

Fig. 6.4.   Normalized current amplitudes on a typical parasitic element in a 20-element array (Mailloux). $b/\lambda = 0.20$, $a/\lambda = 0.00635$.

The amplitudes of the currents at $z = 0$ along each of the twenty elements are shown in Fig. 6.5. The agreement with measured values is again excellent for $h/\lambda = 0.16$ and 0.18. As before, the theoretical curve for $h/\lambda = 0.204$ is in much better agreement

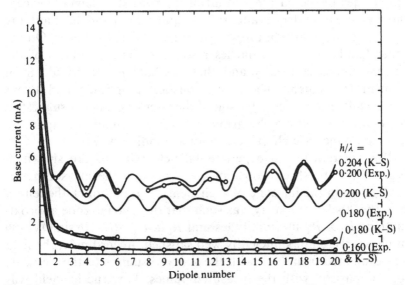

Fig. 6.5.   Amplitudes of currents at centres of the elements in a 20-element array with element No. 1 driven; comparison of King–Sandler theory with experiment (Mailloux). $b/\lambda = 0.20$, $a/\lambda = 0.00635$, frequency 600 Mc.

with the measured curve for $h/\lambda = 0.20$ than is the theoretical curve for $h/\lambda = 0.20$. The corresponding phases are shown in Fig. 6.6.

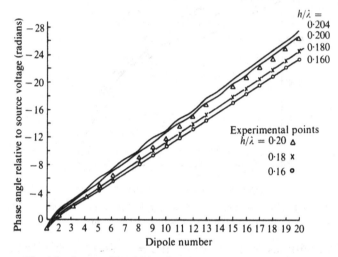

Fig. 6.6.   Same as Fig. 6.5 but for phases of currents (Mailloux).

It is interesting to note that when $h/\lambda = 0.16$ and 0.18 the amplitudes of the currents in all of the parasitic elements except those nearest the driven antenna are quite small and substantially equal and the phase shift from element to element is linear. On the other hand, as h/λ approaches resonance the amplitudes of the currents increase greatly and they oscillate in magnitude from element to element. The small constant amplitude and linear phase shift that is characteristic of the shorter elements suggests a travelling wave along the array; the large oscillating amplitudes near resonance are characteristic of a standing wave.

The theoretical and experimental field patterns are shown in Fig. 6.7 for the three values of h/λ. Although the measurements were made in the far zone of each element (E far zone), the length L of the 20-element array was such that the true far-zone approximations $R_i \doteq R_1$ in amplitudes and $R_i \doteq R_1 - (i-1)b \sin \Theta \cos \Phi$ in phases were not sufficiently well satisfied. Accordingly, the field was evaluated from (6.5) with the actual distances to the elements for comparison with the measured values. The true far-field was also computed for comparison. The former is designated 'E far zone' in the figures, the latter is labelled 'far zone'. The agreement

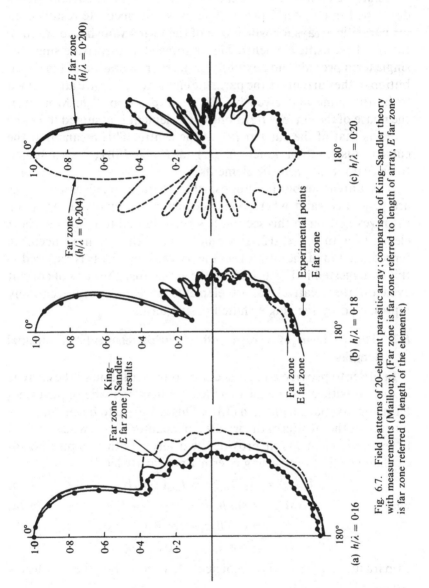

Fig. 6.7. Field patterns of 20-element parasitic array: comparison of King–Sandler theory with measurements (Mailloux). (Far zone is far zone referred to length of array, E far zone is far zone referred to length of the elements.)

between theory and experiment is seen to be quite good even in the details of the minor lobe structure.

It may be concluded that the two-term theory of curtain arrays developed in chapter 5 provides remarkably accurate results even for parasitic arrays for which one of the terms vanishes for each of the $N-1$ parasitic elements. This is somewhat surprising since the single term provides no flexibility in the representation of the distribution of the currents in the parasitic elements. They are all assumed to be the same and given by $I(z) \sim \cos \beta_0 z - \cos \beta_0 h$. Moreover, the phase of the current $I(z)$ along each element is assumed to be the same as that of the current $I(0)$ at the centre. This means that the current distribution function $f(z)$ in $I(z) = I(0)f(z)$ is assumed to be real for all parasitic elements.

It is unreasonable to suppose that these implied assumptions are generally valid when longer elements are involved. After all, the investigation in this section has been limited to relatively short elements with $h/\lambda \leqslant 0.2$. It would appear that a more accurate representation of the currents in the parasitic elements is required— this is suggested in Fig. 6.4 where the actual distributions of current even on the relatively short elements were not very accurately represented by the single shifted cosine term.

6.2   The problem of arrays with parasitic elements of unequal lengths

In order to provide a more accurate representation of the current in the parasitic elements of an array, use may be made of the three-term approximation given in (3.18). This is known to be an improvement over the two-term theory used in chapter 5 and, when applied to parasitic elements, it provides two terms with complex coefficients instead of only a single term. Specifically let

$$I_{zk}(z_k) = A_k M_{0zk} + B_k F_{0zk} + D_k H_{0zk} \qquad (6.6)$$

where
$$M_{0zk} = \sin \beta_0 (h_k - |z_k|) \qquad (6.7a)$$

$$F_{0zk} = \cos \beta_0 z_k - \cos \beta_0 h_k \qquad (6.7b)$$

$$H_{0zk} = \cos \tfrac{1}{2}\beta_0 z_k - \cos \tfrac{1}{2}\beta_0 h_k. \qquad (6.7c)$$

In parasitic elements the coefficient A_k is zero, but the two terms $B_k F_{0zk} + D_k H_{0zk}$ remain.

It is anticipated that the distribution (6.6) provides sufficient flexibility to represent the currents in elements of different lengths when each element is allowed to have its own length $2h_k$.

When the several antennas in an array are not all equal in length so that the h_i differ, the problem of solving the N simultaneous integral equations

$$\sum_{i=1}^{N} \int_{-h_i}^{h_i} I_{zi}(z_i')K_{kid}(z_k, z_i')\, dz_i' = \frac{j4\pi}{\zeta_0 \cos \beta_0 h_k}[\tfrac{1}{2}V_{0k}M_{0zk} + U_k F_{0zk}]$$

(6.8)

with $k = 1, 2, \ldots N$, is more complicated. The kernel has the form

$$K_{kid}(z_k, z_i') = K_{ki}(z_k, z_i') - K_{ki}(h_k, z_i') = \frac{e^{-j\beta_0 R_{ki}}}{R_{ki}} - \frac{e^{-j\beta_0 R_{kih}}}{R_{kih}}$$

(6.9)

where

$$R_{ki} = \sqrt{(z_k - z_i')^2 + b_{ki}^2}, \qquad R_{kih} = \sqrt{(h_k - z_i')^2 + b_{ki}^2}.$$

(6.10)

Note that $b_{kk} = a$. The function U_k is

$$U_k = \frac{-j\zeta_0}{4\pi} \sum_{i=1}^{N} \int_{-h_k}^{h_k} I_{zi}(z_i')K_{ki}(h_k, z_i')\, dz_i'.$$

(6.11)

In a parasitic antenna l the driving voltage $V_{0l} = 0$, so that

$$\sum_{i=1}^{N} \int_{-h_i}^{h_i} I_{zi}(z_i')K_{kid}(z_l, z_i')\, dz_i' = \frac{j4\pi}{\zeta_0 \cos \beta_0 h_l} U_l F_{0zl}.$$

(6.12)

In order to obtain approximate solutions of the N simultaneous integral equations (6.8) by the procedure developed in the earlier chapters, use may be made of the properties of the real and imaginary parts of the kernel. As shown in chapter 2,

$$\int_{-h_k}^{h_k} G_{0z'k}K_{kkdR}(z_k, z_k')\, dz_k' \sim G_{0zk}$$

(6.13)

where $G_{0z'k}$ stands for $M_{0z'k}$, $F_{0z'k}$ or $H_{0z'k}$ and $K_{kkdR}(z_k, z_k')$ is the real part of the kernel. On the other hand,

$$\int_{-h_k}^{h_k} G_{0z'k}K_{kkdI}(z_k, z_k')\, dz_k' \sim H_{0zk}$$

(6.14)

for any distribution G_{0zk}. It follows that

$$W_{kkV}(z_k) \equiv \int_{-h_k}^{h_k} M_{0z'k}K_{kkd}(z_k, z_k')\, dz_k' \doteq \Psi_{kkdV}^m M_{0zk} + \Psi_{kkdV}^h H_{0zk}$$

(6.15)

$$W_{kkU}(z_k) \equiv \int_{-h_k}^{h_k} F_{0z'k}K_{kkd}(z_k, z_k')\, dz_k' \doteq \Psi_{kkdU}^f F_{0zk} + \Psi_{kkdU}^h H_{0zk}$$

(6.16)

$$W_{kkD}(z_k) \equiv \int_{-h_k}^{h_k} H_{0z'k}K_{kkd}(z_k, z_k') \, dz_k' \doteq \Psi^f_{kk\,dD}F_{0zk} + \Psi^h_{kk\,dD}H_{0zk}$$

$$(6.17)$$

where the Ψ's are complex coefficients yet to be determined. Actually, (6.13) with $G = H$ and (6.14) suggest that the term $\Psi^h_{kk\,dD}H_{0zk}$ should be an adequate approximation. The term $\Psi^f_{kk\,dD}F_{0zk}$ is added in order to provide greater flexibility and symmetry.

When $i \neq k$ and $\beta_0 b \geqslant 1$, it has been shown by direct comparison in chapter 3 that

$$\int_{-h_i}^{h_i} G_{0z'i}K_{ki\,dR}(z_k, z_i') \, dz_i' \sim F_{0zk} \qquad (6.18)$$

$$\int_{-h_i}^{h_i} G_{0z'i}K_{ki\,dI}(z_k, z_i') \, dz_i' \sim H_{0zk} \qquad (6.19)$$

where $G_{0z'i}$ stands for $M_{0z'i}$, $F_{0z'i}$ or $H_{0z'i}$. It follows that with $i \neq k$,

$$W_{kiV}(z_k) \equiv \int_{-h_i}^{h_i} M_{0z'i}K_{kid}(z_k, z_i') \, dz_i' \doteq \Psi^f_{ki\,dV}F_{0zk} + \Psi^h_{ki\,dV}H_{0zk}$$

$$(6.20)$$

$$W_{kiU}(z_k) \equiv \int_{-h_i}^{h_i} F_{0z'i}K_{kid}(z_k, z_i') \, dz_i' \doteq \Psi^f_{ki\,dU}F_{0zk} + \Psi^h_{ki\,dU}H_{0zk}$$

$$(6.21)$$

$$W_{kiD}(z_k) \equiv \int_{-h_i}^{h_i} H_{0z'i}K_{kid}(z_k, z_i') \, dz_i' \doteq \Psi^f_{ki\,dD}F_{0zk} + \Psi^h_{ki\,dD}H_{0zk}$$

$$(6.22)$$

where the Ψ's are complex coefficients yet to be determined.

In the formulation developed in the earlier chapters for driven elements of equal lengths, the coefficients Ψ were defined individually in terms of the two integrals obtained from the real and imaginary parts of the kernel. In order to take account of the more varied distributions that may be obtained when the elements are neither all driven nor all equal in length, the separation into two parts is not made. Instead the entire integral is represented by a linear combination of the two distributions that best represent the parts of the integral. The complex coefficients of these distributions are to be determined by matching the integral and its approximation at two points along the antenna, instead of at only one such point.

It is anticipated that by fitting the trigonometric approximations to the integrals at $z = 0$, $h_k/2$, and h_k (where both must vanish) a good representation may be achieved in reasonably simple form of all of the different distributions which may occur along antennas of unequal lengths. It is, of course, assumed that $\beta_0 h_i \leqslant 5\pi/4$ for all h_i.

6.3   Application to the Yagi-Uda array

In order to clarify the description of the procedure used to solve the N simultaneous integral equations for a parasitic array, it will be carried out in detail for the specific and practically useful Yagi-Uda array. In general, this consists of a curtain of N antennas of which No. 1 is parasitic and adjusted in length to function as a reflector, No. 2 is driven by a voltage V_{02} and Nos. 3 to N are also parasitic and adjusted to act as directors. Such an array is shown in Fig. 6.8 for the special case (treated later) with $2h_1 = 0\cdot51\lambda$; $2h_2 = 0\cdot50\lambda$; $2h_i = 2h$, $i > 2$; $b_{21} = 0\cdot25\lambda$; $b_{i,i+1} = b, i > 2$. The details of these adjustments are examined later.

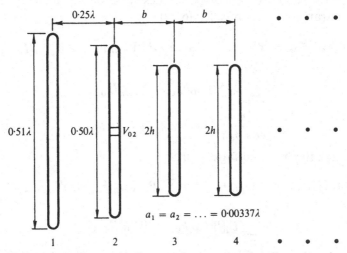

Fig. 6.8.   A Yagi array with directors of constant length, radius and spacing.

On the basis of the three-term approximation, the current in the single driven element has the form

$$I_{z2}(z_2) = A_2 M_{0z2} + B_2 F_{0z2} + D_2 H_{0z2}. \tag{6.23}$$

The currents in the parasitic elements are

$$I_{zi}(z_i) = B_i F_{0zi} + D_i H_{0zi}, \qquad i = 1, 3, 4, \dots N \tag{6.24}$$

where the constants A_2, B_i and D_i must be evaluated ultimately in terms of V_{02}. The integral equation for the driven element is

$$A_2\int_{-h_2}^{h_2} M_{0z'2}K_{22d}(z_2, z_2')\,dz_2' + \sum_{i=1}^{N} B_iF_{0z'i}K_{2id}(z_2, z_i')\,dz_i'$$

$$+ \sum_{i=1}^{N} D_iH_{0z'i}K_{2id}(z_2, z_i')\,dz_i'$$

$$= \frac{j4\pi}{\zeta_0\cos\beta_0 h_2}[\tfrac{1}{2}V_{02}M_{0z2} + U_2 F_{0z2}]. \qquad (6.25)$$

The remaining $N-1$ integral equations are

$$A_2\int_{-h_2}^{h_2} M_{0z'2}K_{k2d}(z_k, z_2')\,dz_2' + \sum_{i=1}^{N} B_iF_{0z'i}K_{kid}(z_k, z_i')\,dz_i'$$

$$+ \sum_{i=1}^{N} D_iH_{0z'i}K_{kid}(z_k, z_i')\,dz_i'$$

$$= \frac{j4\pi}{\zeta_0\cos\beta_0 h_k}U_kF_{0zk}, \qquad k = 1, 3, 4, \dots N. \qquad (6.26)$$

With (6.15)–(6.17) and (6.20)–(6.22) these may be expressed in terms of the parameters Ψ. Thus, for (6.25)

$$A_2[\Psi^m_{22dV}M_{0z2} + \Psi^h_{22dV}H_{0z2}] + \sum_{i=1}^{N} B_i[\Psi^f_{2idU}F_{0z2} + \Psi^h_{2idU}H_{0z2}]$$

$$+ \sum_{i=1}^{N} D_i[\Psi^f_{2idD}F_{0z2} + \Psi^h_{2idD}H_{0z2}]$$

$$= \frac{j4\pi}{\zeta_0\cos\beta_0 h_2}[\tfrac{1}{2}V_{02}M_{0z2} + U_2 F_{0z2}]. \qquad (6.27)$$

For (6.26), the $N-1$ equations are

$$A_2[\Psi^f_{k2dV}F_{0zk} + \Psi^h_{k2dV}H_{0zk}] + \sum_{i=1}^{N} B_i[\Psi^f_{kidU}F_{0zk} + \Psi^h_{kidU}H_{0zk}]$$

$$+ \sum_{i=1}^{N} D_i[\Psi^f_{kidD}F_{0zk} + \Psi^h_{kidD}H_{0zk}]$$

$$= \frac{j4\pi}{\zeta_0\cos\beta_0 h_k}U_kF_{0zk}, \qquad k = 1, 3, 4, \dots N. \qquad (6.28)$$

These equations will be satisfied if the coefficient of each of the three distribution functions is individually required to vanish. That is, in (6.27):

$$A_2 = \frac{j2\pi V_{02}}{\zeta_0\Psi^m_{22dV}\cos\beta_0 h_2} \qquad (6.29)$$

$$\sum_{i=1}^{N} [B_i \Psi_{2idU}^f + D_i \Psi_{2idD}^f] \cos \beta_0 h_2 - \frac{j4\pi}{\zeta_0} U_2 = 0 \qquad (6.30a)$$

$$A_2 \Psi_{22dV}^h + \sum_{i=1}^{N} [B_i \Psi_{2idU}^h + D_i \Psi_{2idD}^h] = 0. \qquad (6.30b)$$

Similarly in (6.28) with $k = 1, 3, \dots N$

$$\left\{ A_2 \Psi_{k2dV}^f + \sum_{i=1}^{N} [B_i \Psi_{kidU}^f + D_i \Psi_{kidD}^f] \right\} \cos \beta_0 h_k - \frac{j4\pi}{\zeta_0} U_k = 0 \quad (6.30c)$$

$$A_2 \Psi_{k2dV}^h + \sum_{i=1}^{N} [B_i \Psi_{kidU}^h + D_i \Psi_{kidD}^h] = 0. \qquad (6.30d)$$

Actually, the single equations in (6.30a) and (6.30b) may be combined with the $N-1$ equations in (6.30c) and (6.30d) with the aid of the Kronecker δ defined by

$$\delta_{ik} = \begin{cases} 0 & i \neq k \\ 1 & i = k \end{cases}.$$

The $2N$ equations are

$$\left\{ A_2(1 - \delta_{k2}) \Psi_{k2dV}^f + \sum_{i=1}^{N} [B_i \Psi_{kidU}^f + D_i \Psi_{kidD}^f] \right\} \cos \beta_0 h_k - \frac{j4\pi}{\zeta_0} U_k = 0$$

$$(6.31a)$$

$$A_2 \Psi_{k2dV}^h + \sum_{i=1}^{N} [B_i \Psi_{kidU}^h + D_i \Psi_{kidD}^h] = 0 \qquad (6.31b)$$

with $k = 1, 2, \dots N$. These equations, together with (6.29), determine the $2N+1$ constants A_2, B_i and D_i, $i = 1, 2, \dots N$.

Before these two sets of equations can be solved, it is necessary to evaluate the functions U_k. This is readily done in terms of the following integrals:

$$\Psi_{kiV}(h_k) = \int_{-h_i}^{h_i} M_{0z'i} K_{ki}(h_k, z_i') \, dz_i' \qquad (6.32)$$

$$\Psi_{kiU}(h_k) = \int_{-h_i}^{h_i} F_{0z'i} K_{ki}(h_k, z_i') \, dz_i' \qquad (6.33)$$

$$\Psi_{kiD}(h_k) = \int_{-h_i}^{h_i} H_{0z'i} K_{ki}(h_k, z_i') \, dz_i' \qquad (6.34)$$

where

$$K_{ki}(h_k, z_i') = \frac{e^{-j\beta_0 R_{kih}}}{R_{kih}}, \qquad R_{kih} = \sqrt{(h_k - z_i')^2 + b_{ik}}. \qquad (6.35)$$

It follows from the definition in (6.11) that

$$U_k = \frac{-j\zeta_0}{4\pi} \sum_{i=1}^{N} [A_i \Psi_{kiV}(h_k) + B_i \Psi_{kiU}(h_k) + D_i \Psi_{kiD}(h_k)]. \quad (6.36a)$$

Since only antenna 2 is driven, $A_i = 0, i \neq 2$ so that

$$U_k = \frac{-j\zeta_0}{4\pi} \left\{ A_2 \Psi_{k2V}(h_k) + \sum_{i=1}^{N} [B_i \Psi_{kiU}(h_k) + D_i \Psi_{kiD}(h_k)] \right\}. \quad (6.36b)$$

The substitution of (6.36b) in (6.31a) gives for these equations

$$A_2[\Psi_{k2V}(h_k) - (1 - \delta_{k2})\Psi_{k2dV}^f \cos \beta_0 h_k] + \sum_{i=i}^{N} B_i[\Psi_{kiU}(h_k)$$

$$- \Psi_{kidU}^f \cos \beta_0 h_k] + \sum_{i=1}^{N} D_i[\Psi_{kiD}(h_k) - \Psi_{kidD}^f \cos \beta_0 h_k] = 0$$

$$(6.37)$$

with $k = 1, 2, ... N$. These equations can be simplified formally by the introduction of the notation

$$\Phi_{k2V} = \Psi_{k2V}(h_k) - (1 - \delta_{k2})\Psi_{k2dV}^f \cos \beta_0 h_k \quad (6.38)$$

$$\Phi_{kiU} = \Psi_{kiU}(h_k) - \Psi_{kidU}^f \cos \beta_0 h_k \quad (6.39a)$$

$$\Phi_{kiD} = \Psi_{kiD}(h_k) - \Psi_{kidD}^f \cos \beta_0 h_k. \quad (6.39b)$$

With this notation, (6.37) together with (6.31b) gives the following set of 2N equations for determining the 2N coefficients B_i and D_i in terms of A_2:

$$\sum_{i=1}^{N} [\Phi_{kiU} B_i + \Phi_{kiD} D_i] = -\Phi_{k2V} A_2; \qquad k = 1, 2, ... N \quad (6.40)$$

$$\sum_{i=1}^{N} [\Psi_{kidU}^h B_i + \Psi_{kidD}^h D_i] = -\Psi_{k2dV}^h A_2; \qquad k = 1, 2, ... N. \quad (6.41)$$

These equations may be expressed in matrix form after the introduction of the following notation:

$$[\Phi_U] = \begin{bmatrix} \Phi_{11U} & \Phi_{12U} \cdots \Phi_{1NU} \\ \vdots & \\ \Phi_{N1U} & \cdots \qquad \Phi_{NNU} \end{bmatrix} \quad (6.42a)$$

$$[\Phi_D] = \begin{bmatrix} \Phi_{11D} & \Phi_{12D} \cdots \Phi_{1ND} \\ \vdots & \\ \Phi_{N1D} & \cdots \qquad \Phi_{NND} \end{bmatrix} \quad (6.42b)$$

$$[\Psi_{dU}^h] = \begin{bmatrix} \Psi_{11dU}^h & \Psi_{12dU}^h \cdots \Psi_{1NdU}^h \\ \vdots & \\ \Psi_{N1dU}^h & \cdots \quad \Psi_{NNdU}^h \end{bmatrix} \tag{6.43a}$$

$$[\Psi_{dD}^h] = \begin{bmatrix} \Psi_{11dD}^h & \Psi_{12dD}^h \cdots \Psi_{1NdD}^h \\ \vdots & \\ \Psi_{N1dD}^h & \cdots \quad \Psi_{NNdD}^h \end{bmatrix} \tag{6.43b}$$

$$\{\Phi_{2V}\} = \begin{Bmatrix} \Phi_{12V} \\ \Phi_{22V} \\ \vdots \\ \Phi_{N2V} \end{Bmatrix} \qquad \{\Psi_{2dV}^h\} = \begin{Bmatrix} \Psi_{12dV}^h \\ \Psi_{22dV}^h \\ \vdots \\ \Psi_{N2dV}^h \end{Bmatrix} \tag{6.44}$$

$$\{B\} = \begin{Bmatrix} B_1 \\ B_2 \\ \vdots \\ B_N \end{Bmatrix} \qquad \{D\} = \begin{Bmatrix} D_1 \\ D_2 \\ \vdots \\ D_N \end{Bmatrix}. \tag{6.45}$$

The matrix forms of (6.40) and (6.41) are

$$[\Phi_U]\{B\} + [\Phi_D]\{D\} = -\{\Phi_{2V}\}A_2 \tag{6.46}$$

$$[\Psi_{dU}^h]\{B\} + [\Psi_{dD}^h]\{D\} = -\{\Psi_{2dV}^h\}A_2. \tag{6.47}$$

The N coefficients B_i and the N coefficients D_i must be determined from these equations for substitution in the equations (6.23) and (6.24) for the currents in the N elements. The coefficient A_2, which is a common factor, is obtained from (6.29) in terms of the single driving voltage V_{02}.

It remains to evaluate the parameters Ψ that occur in the Φ's in (6.46) and explicitly in (6.47).

6.4 Evaluation of coefficients for the Yagi-Uda array

The equations (6.46) and (6.47) involve the elements of the $N \times N$ matrices $[\Phi_U]$, $[\Phi_D]$, $[\Psi_{dU}^h]$ and $[\Psi_{dD}^h]$. These, in turn, depend on the parameters Ψ introduced in (6.15)–(6.17) and (6.20)–(6.22) and the parameters $\Psi(h)$ defined in (6.32)–(6.34). Since each integral is approximated by a linear combination of two terms with arbitrary coefficients, these can be evaluated by equating both sides in

(6.15)–(6.17) and (6.20)–(6.22) at two values of z. The values chosen are $z = 0$, and $z = h_k/2$ in addition to $z = h_k$ where both sides must vanish.

Specific formulas for the two values of each of the integrals W defined in (6.15)–(6.17) and (6.20)–(6.22) are as follows:

$$W_{kiV}(0) \equiv A_i^{-1} \int_{-h_i}^{h_i} I_{Vi}(z_i')K_{kid}(0, z_i') \, dz_i' \doteq \int_{-h_i}^{h_i} M_{0z'i}K_{kid}(0, z_i') \, dz_i'$$

(6.48a)

$$W_{kiV}\left(\frac{h_k}{2}\right) \equiv A_i^{-1} \int_{-h_i}^{h_i} I_{Vi}(z_i')K_{kid}\left(\frac{h_k}{2}, z_i'\right) dz_i'$$

$$\doteq \int_{-h_i}^{h_i} M_{0z'i}K_{kid}\left(\frac{h_k}{2}, z_i'\right) dz_i'$$

(6.48b)

$$W_{kiU}(0) \equiv B_i^{-1} \int_{-h_i}^{h_i} I_{Ui}(z_i')K_{kid}(0, z_i') \, dz_i' \doteq \int_{-h_i}^{h_i} F_{0z'i}K_{kid}(0, z_i') \, dz_i'$$

(6.49a)

$$W_{kiU}\left(\frac{h_k}{2}\right) \equiv B_i^{-1} \int_{-h_i}^{h_i} I_{Ui}(z_i')K_{kid}\left(\frac{h_k}{2}, z_i'\right) dz_i'$$

$$\doteq \int_{-h_i}^{h_i} F_{0z'i}K_{kid}\left(\frac{h_k}{2}, z_i'\right) dz_i'$$

(6.49b)

$$W_{kiD}(0) \equiv D_i^{-1} \int_{-h_i}^{h_i} I_{Di}(z_i')K_{kid}(0, z_i') \, dz_i' \doteq \int_{-h_i}^{h_i} H_{0z'i}K_{kid}(0, z_i') \, dz_i'$$

(6.50a)

$$W_{kiD}\left(\frac{h_k}{2}\right) \equiv D_i^{-1} \int_{-h_i}^{h_i} I_{Di}(z_i')K_{kid}\left(\frac{h_k}{2}, z_i'\right) dz_i'$$

$$\doteq \int_{-h_i}^{h_i} H_{0z'i}K_{kid}\left(\frac{h_k}{2}, z_i'\right) dz_i'.$$

(6.50b)

In all of the above, $k = 1, 2, 3, \ldots N$. These are a set of complex numbers which give the values of the integrals (6.20)–(6.22) at the two points $z = 0$ and $z = h_k/2$. They are readily evaluated numerically by high-speed computer, or they may be expressed in terms of the tabulated generalized sine and cosine integral functions. Once the W's in (6.48a)–(6.50b) have been obtained for all values of i and k, the coefficients Ψ may be determined from the equations (6.15)–(6.17) and (6.20)–(6.22). At $z = 0$ these become:

$$\Psi_{kkdV}^m \sin \beta_0 h_k + \Psi_{kkdV}^h[1 - \cos(\beta_0 h_k/2)] = W_{kkV}(0) \quad (6.51a)$$

$$\Psi^f_{kidV}(1-\cos\beta_0 h_k)+\Psi^h_{kidV}[1-\cos(\beta_0 h_k/2)] = W_{kiV}(0) \qquad i \neq k$$
$$(6.51b)$$

$$\Psi^f_{kidU}(1-\cos\beta_0 h_k)+\Psi^h_{kidU}[1-\cos(\beta_0 h_k/2)] = W_{kiU}(0) \qquad (6.51c)$$

$$\Psi^f_{kidD}(1-\cos\beta_0 h_k)+\Psi^h_{kidD}[1-\cos(\beta_0 h_k/2)] = W_{kiD}(0). \qquad (6.51d)$$

At $z = h_k/2$, they are

$$\Psi^m_{kkdV}\sin(\beta_0 h_k/2)+\Psi^h_{kkdV}[\cos(\beta_0 h_k/4)-\cos(\beta_0 h_k/2)] = W_{kkV}\left(\frac{h_k}{2}\right)$$
$$(6.52a)$$

$$\Psi^f_{kidV}[\cos(\beta_0 h_k/2)-\cos\beta_0 h_k]+\Psi^h_{kidV}[\cos(\beta_0 h_k/4)-\cos(\beta_0 h_k/2)]$$
$$= W_{kiV}\left(\frac{h_k}{2}\right) \qquad i \neq k \qquad (6.52b)$$

$$\Psi^f_{kidU}[\cos(\beta_0 h_k/2)-\cos\beta_0 h_k]+\Psi^h_{kidU}[\cos(\beta_0 h_k/4)-\cos(\beta_0 h_k/2)]$$
$$= W_{kiU}\left(\frac{h_k}{2}\right) \qquad (6.52c)$$

$$\Psi^f_{kidD}[\cos(\beta_0 h_k/2)-\cos\beta_0 h_k]+\Psi^h_{kidD}[\cos(\beta_0 h_k/4)-\cos(\beta_0 h_k/2)]$$
$$= W_{kiD}\left(\frac{h_k}{2}\right). \qquad (6.52d)$$

The solutions of these equations for the Ψ's are obtained directly. They are

$$\Psi^m_{kkdV} = \Delta_1^{-1}\left\{W_{kkV}(0)\left[\cos\left(\frac{\beta_0 h_k}{4}\right)-\cos\left(\frac{\beta_0 h_k}{2}\right)\right]\right.$$
$$\left. - W_{kkV}(h_k/2)[1-\cos(\beta_0 h_k/2)]\right\} \qquad (6.53)$$

$$\Psi^h_{kkdV} = \Delta_1^{-1}\{W_{kkV}(h_k/2)\sin\beta_0 h_k - W_{kkV}(0)\sin(\beta_0 h_k/2)\} \quad (6.54)$$

$$\Psi^f_{kidV} = \Delta_2^{-1}\{W_{kiV}(0)[\cos(\beta_0 h_k/4)-\cos(\beta_0 h_k/2)]$$
$$- W_{kiV}(h_k/2)[1-\cos(\beta_0 h_k/2)]\} \qquad i \neq k \qquad (6.55)$$

$$\Psi^h_{kidV} = \Delta_2^{-1}\{W_{kiV}(h_k/2)[1-\cos\beta_0 h_k]$$
$$- W_{kiV}(0)[\cos(\beta_0 h_k/2)-\cos\beta_0 h_k]\} \qquad i \neq k \qquad (6.56)$$

$$\Psi^f_{kidU} = \Delta_2^{-1}\{W_{kiU}(0)[\cos(\beta_0 h_k/4)-\cos(\beta_0 h_k/2)]$$
$$- W_{kiU}(h_k/2)[1-\cos(\beta_0 h_k/2)]\} \qquad (6.57)$$

$$\Psi^h_{kidU} = \Delta_2^{-1}\{W_{kiU}(h_k/2)[1-\cos\beta_0 h_k]$$
$$- W_{kiU}(0)[\cos(\beta_0 h_k/2)-\cos(\beta_0 h_k)]\} \qquad (6.58)$$

$$\Psi^f_{kidD} = \Delta_2^{-1}\{W_{kiD}(0)[\cos(\beta_0 h_k/4)-\cos(\beta_0 h_k/2)]$$
$$- W_{kiD}(h_k/2)[1-\cos(\beta_0 h_k/2)]\} \qquad (6.59)$$

$$\Psi^h_{kidD} = \Delta_2^{-1}\{W_{kiD}(h_k/2)[1-\cos\beta_0 h_k]$$
$$- W_{kiD}(0)[\cos(\beta_0 h_k/2)-\cos\beta_0 h_k]\} \tag{6.60}$$

where $\Delta_1 = \sin\beta_0 h_k[\cos(\beta_0 h_k/4)-\cos(\beta_0 h_k/2)]$
$$-\sin(\beta_0 h_k/2)[1-\cos(\beta_0 h_k/2)] \tag{6.61}$$

and $\Delta_2 = [1-\cos\beta_0 h_k][\cos(\beta_0 h_k/4)-\cos(\beta_0 h_k/2)]$
$$-[\cos(\beta_0 h_k/2)-\cos\beta_0 h_k][1-\cos(\beta_0 h_k/2)]. \tag{6.62}$$

All of the Ψ's have been determined. The $\Psi(h)$ coefficients are given in (6.32)–(6.34). The elements of the Φ matrices are obtained from (6.38)–(6.39b). This completes the solution for all of the currents in the elements of the Yagi-Uda array.

6.5   Arrays with half-wave elements

When an array includes half-wave parasitic elements the formulation in sections 6.3 and 6.4 is directly applicable. Specifically, when $\beta_0 h_i = \pi/2$ and element i is parasitic, the current (6.24) has the form

$$I_{zi}(z_i) = B_i \cos\beta_0 z_i + D_i[\cos(\beta_0 z_i/2)-\sqrt{2}/2]. \tag{6.63}$$

If the length of the driven element 2 is such that $\beta_0 h_2$ is near or exactly $\pi/2$ (as in Fig. 6.8), the alternative form for the current given in (2.35) for the isolated antenna is more convenient since it does not yield an indeterminate form at $\beta_0 h_2 = \pi/2$. That is, in the notation of (6.23),

$$I_{z2}(z_2) = A'_2 S_{0z2} + B'_2 F_{0z2} + D_2 H_{0z2} \tag{6.64}$$

where $S_{0z2} = \sin\beta_0|z_2| - \sin\beta_0 h_2 \tag{6.65}$

and $A'_2 = -A_2\cos\beta_0 h_2 = -j(2\pi V_{02}/\zeta_0\Psi^m_{22dV}) \tag{6.66a}$

$$B'_2 = B_2 + A_2\sin\beta_0 h_2 = B_2 - A'_2\tan\beta_0 h_2. \tag{6.66b}$$

Note that A'_2 and B'_2 are finite when $\beta_0 h = \pi/2$. In this case

$$S_{0z2} = \sin\beta_0|z_2| - 1, \qquad B'_2 = B_2 + A_2. \tag{6.67}$$

Since (6.64) is not actually a different distribution from the original in (6.23) but merely a rearrangement that is more convenient when $\beta_0 h_2$ is at or near $\pi/2$, it is not necessary to repeat the formulation in the preceding sections with S_{0z2} substituted for M_{0z2}. A simple rearrangement of the 2N equations in (6.40) and (6.41) is all that is required. This is accomplished by the substitutions (6.66a) and (6.66b) for A_2 and B_2. Specifically, let

$$A_2 = -A'_2\sec\beta_0 h_2, \qquad B_2 = B'_2 + A'_2\tan\beta_0 h_2 \tag{6.68}$$

$$\Phi'_{k2V} = [\Phi_{k2V} - \Phi_{k2U} \sin \beta_0 h_2] \sec \beta_0 h_2 \qquad (6.69)$$

$$\Psi'^h_{k2dV} = [\Psi^h_{k2dV} - \Psi^h_{k2dU} \sin \beta_0 h_2] \sec \beta_0 h_2. \qquad (6.70)$$

Also let B'_i stand for $B_1, B'_2, B_3, \dots B_N$. With this notation, the equations (6.40) and (6.41) become:

$$\sum_{i=1}^{N} [\Phi_{kiU} B'_i + \Phi_{kiD} D_i] = \Phi'_{k2V} A'_2; \qquad k = 1, 2, \dots N \quad (6.71)$$

$$\sum_{i=1}^{N} [\Psi^h_{kidU} B'_i + \Psi^h_{kidD} D_i] = \Psi'^h_{k2dV} A'_2; \qquad k = 1, 2, \dots N. \quad (6.72)$$

In matrix form these are

$$[\Phi_U]\{B'\} + [\Phi_D]\{D\} = \{\Phi'_{2V}\} A'_2 \qquad (6.73)$$

$$[\Psi^h_{dU}]\{B'\} + [\Psi^h_{dD}]\{D\} = \{\Psi'_{2dV}\} A'_2 \qquad (6.74)$$

where the four square matrices and the column matrix $\{D\}$ are defined in (6.42a, b), (6.43a, b) and (6.45). The other column matrices are

$$\{B'\} = \begin{Bmatrix} B_1 \\ B'_2 \\ B_3 \\ \vdots \\ B_N \end{Bmatrix}, \quad \{\Phi'_{2V}\} = \begin{Bmatrix} \Phi'_{12V} \\ \Phi'_{22V} \\ \Phi'_{32V} \\ \vdots \\ \Phi'_{N2V} \end{Bmatrix}, \quad \{\Psi'_{2dV}\} = \begin{Bmatrix} \Psi'_{12dV} \\ \Psi'_{22dV} \\ \Psi'_{32dV} \\ \vdots \\ \Psi'_{N2dV} \end{Bmatrix}.$$

$$(6.75)$$

These equations are to be solved for the $2N$ coefficients B'_i and D_i in terms of $A'_2 = -j(2\pi V_{02}/\zeta_0 \Psi^m_{22dV})$. The Ψ functions that occur in these equations are defined in the same manner as in sections 6.3 and 6.4. This is illustrated below for $\beta_0 h = \pi/2$.

When $\beta_0 h = \pi/2$, $F_{0z2} = M_{0z2} = \cos \beta_0 z$. It follows from (6.48a) and (6.49) that $W_{22V}(0) = W_{22U}(0)$, $W_{22V}(h_2/2) = W_{22U}(h_2/2)$. Hence, from (6.53) and (6.57), (6.54) and (6.58), it follows $\Psi_{k2V}(z_k) = \Psi_{k2U}(z_k)$ and $\Delta_1 = \Delta_2$. This means that $\Psi^f_{k2dV} = \Psi^f_{k2dU}$ and $\Psi^h_{k2dV} = \Psi^h_{k2dU}$ when $k \neq 2$ and $\Psi^m_{22dV} = \Psi^f_{22dU}$, $\Psi^h_{22dV} = \Psi^h_{22dU}$ when $k = 2$. As a consequence, Φ'_{k2V} becomes indeterminate in the form $0/0$ when $k \neq 2$. However, the limiting value as $\beta_0 h_2 \to \pi/2$ is finite. Thus, (6.69) may be expanded as follows. When $k = 2$,

$$\Phi'_{22V} = -S_a(h_2, h_2) + E_a(h_2, h_2) + \Psi^f_{22dU}; \qquad (6.76a)$$

when $k \neq 2$,

$$
\begin{aligned}
\Phi'_{k2V} = {}& -S_{b_{k2}}(h_2, h_k) + E_{b_{k2}}(h_2, h_k) \\
& + \frac{\cos \beta_0 h_k}{\Delta_2} \Big\{ [\cos \tfrac{1}{4}\beta_0 h_k - \cos \tfrac{1}{2}\beta_0 h_k][-E_{b_{k2}}(h_2, 0) \\
& + E_{b_{k2}}(h_2, h_k) + S_{b_{k2}}(h_2, 0) - S_{b_{k2}}(h_2, h_k)] \\
& + (1 - \cos \tfrac{1}{2}\beta_0 h_k) \Big[E_{b_{k2}}\Big(h_2, \frac{h_k}{2}\Big) \\
& - E_{b_{k2}}(h_2, h_k) - S_{b_{k2}}\Big(h_2, \frac{h_k}{2}\Big) + S_{b_{k2}}(h_2, h_k)\Big] \Big\}
\end{aligned}
\tag{6.76b}
$$

where Δ_2 is defined in (6.62). Similarly, when $k = 2$,

$$
\begin{aligned}
\Psi''^{h}_{22dV} = {}& \frac{1-\sqrt{2}}{\Delta_2} \Big\{ \Big[C_a\Big(h_2, \frac{h_2}{2}\Big) - C_a(h_2, h_2)\Big]\Big[1 - \frac{1}{\sqrt{2}}\Big] \\
& - [C_a(h_2, 0) - C_a(h_2, h_2)]\Big[\cos \frac{\pi}{8} - \frac{1}{\sqrt{2}}\Big] \Big\} \\
& + \frac{1}{\Delta_2} \Big\{ \Big[-S_a\Big(h_2, \frac{h_2}{2}\Big) + S_a(h_2, h_2) \\
& + E_a\Big(h_2, \frac{h_2}{2}\Big) - E_a(h_2, h_2)\Big] \\
& + \frac{1}{\sqrt{2}}[S_a(h_2, 0) - S_a(h_2, h_2) - E_a(h_2, 0) + E_a(h_2, h_2)] \Big\};
\end{aligned}
\tag{6.76c}
$$

when $k \neq 2$,

$$
\begin{aligned}
\Psi''^{h}_{k2dV} = {}& \frac{1}{\Delta_2} \Big\{ [1 - \cos \beta_0 h_k]\Big[-S_{b_{k2}}\Big(h_2, \frac{h_k}{2}\Big) + S_{b_{k2}}(h_2, h_k) \\
& + E_{b_{k2}}\Big(h_2, \frac{h_k}{2}\Big) - E_{b_{k2}}(h_2, h_k)\Big] \\
& + [\cos \tfrac{1}{2}\beta_0 h_k - \cos \beta_0 h_k][S_{b_{k2}}(h_2, 0) - S_{b_{k2}}(h_2, h_k) \\
& - E_{b_{k2}}(h_2, 0) + E_{b_{k2}}(h_2, h_k)] \Big\}.
\end{aligned}
\tag{6.76d}
$$

The coefficients B'_i and D_i obtained for $\beta_0 h_2 = \pi/2$ from (6.73) and (6.74) with (6.76) are to be used in the current distributions

$$
I_{z2}(z_2) = A'_2 S_{0z} + B'_2 F_{0z} + D_2 H_{0z} \tag{6.77}
$$

$$
I_{zi}(z_i) = B'_i F_{0z} + D_i H_{0z}, \qquad i = 1, 3, \dots N. \tag{6.78}
$$

In the original analysis of arrays with half-wave elements [2] and in its application to arrays of the Yagi type [3], a somewhat different procedure was used. In effect, this treated the alternative form (6.64) of the distribution of current along a driven half-wave element as an independent representation. The entire procedure carried through in sections 6.3 and 6.4 was repeated with the distribution function M_{0z2} replaced by S_{0z2}. This also involved a simple rearrangement of the integral equations (6.8) so that when $k = 2$, the right-hand member is $(j4\pi/\zeta_0)[\frac{1}{2}V_{02}S_{0z2} + C_2 F_{0z2}]$.

The alternative procedure is basically equivalent to that outlined in section 6.5 but the two are not identical and involve small quantitative differences when applied to a particular array. In particular, the values of $W_{22V}(0)$ and $W_{22V}(h_2/2)$ from (6.45a, b) are necessarily somewhat different when, with $\beta_0 h = \pi/2$, $S_{0z2} = \sin \beta_{01} z_{21} - 1$ is substituted for $M_{0z2} = \cos \beta_0 z_2$ in the integrals. It follows that the two values of Ψ_{22dV}^m as defined in (6.53) are also not quite the same when S_{0z2} is used instead of M_{0z2}. These differences are small and either procedure should give satisfactory results, although in the interest of simplicity and consistency the generalization in section 6.3 is to be preferred.

Reference is here made to the alternative procedure primarily because it was used by Morris in an extensive quantitative study of the Yagi-Uda array. The results of his work, described later in this chapter, differ negligibly from those actually given.

6.6   The far field of the Yagi-Uda array; gain

The electric field maintained at distant points by the currents in the N elements of the Yagi-Uda array is readily determined. For the currents

$$I_{z2}(z_2) = A_2 \sin \beta_0(h_2 - |z_2|) + B_2(\cos \beta_0 z_2 - \cos \beta_0 h_2)$$
$$+ D_2(\cos \tfrac{1}{2}\beta_0 z_2 - \cos \tfrac{1}{2}\beta_0 h_2) \qquad (6.79a)$$

$$I_{zi}(z_i) = B_i(\cos \beta_0 z_i - \cos \beta_0 h_i) + D_i(\cos \tfrac{1}{2}\beta_0 z_i - \cos \tfrac{1}{2}\beta_0 h_i), \quad i \neq 2$$
$$(6.79b)$$

the electromagnetic field is

$$E_\Theta(R_2, \Theta, \Phi) = \frac{j\zeta_0}{2\pi}\left\{ A_2 \frac{e^{-j\beta_0 R_2}}{R_2} F_m(\Theta, \beta_0 h_2) \right.$$

$$\left. + \sum_{i=1}^{N} \frac{e^{-j\beta_0 R_i}}{R_i}[B_i G_m(\Theta, \beta_0 h_i) + D_i D_m(\Theta, \beta_0 h_i)] \right\} \qquad (6.80)$$

where $F_m(\Theta, \beta_0 h)$, $G_m(\Theta, \beta_0 h)$ and $D_m(\Theta, \beta_0 h)$ are defined in (2.46)–(2.48) and R_i is the distance from the point of calculation to the centre of element i. This may be rearranged as follows:

$$E_{\Theta N}(R_2, \Theta, \Phi) = -\frac{V_{02}}{\Psi}\frac{e^{-j\beta_0 R_2}}{R_2}f_{VN}(\Theta, \Phi). \qquad (6.81a)$$

Since no ambiguity can arise the symbol Ψ without subscripts and superscripts is used for Ψ_{22dV}^m as defined in (6.53). The field factor in (6.81a) for the N-element array is given by

$$f_{VN}(\Theta, \Phi) = \left\{ F_m(\Theta, \beta_0 h_2) + \sum_{i=1}^{N} e^{-j\beta_0(R_i - R_2)}[T_{Ui}G_m(\Theta, \beta_0 h_i) + T_{Di}D_m(\Theta, \beta_0 h_i)] \right\} \sec \beta_0 h_2. \qquad (6.81b)$$

In obtaining (6.81a, b) the far-field approximation, $R_i \doteq R_2$, in amplitudes has been made. In the spherical coordinates R_2, Θ, Φ, (Fig. 6.9), and $b_{i,i\pm1} = b$,

$$R_i - R_2 = -(i-2)b \sin \Theta \cos \Phi. \qquad (6.81c)$$

The following set of parameters has been introduced:

$$T_{Ui} = B_i/A_2, \qquad T_{Di} = D_i/A_2 \qquad (6.82)$$

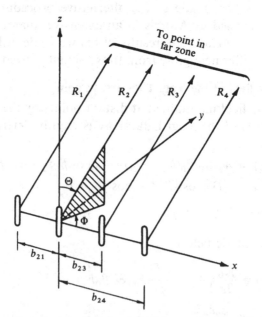

Fig. 6.9. Coordinates for 4-element array referred to origin at centre of element 2. $b_{21} = b_{23} = b_{34} = b$.

where $A_2 = j2\pi V_{02}/\zeta_0 \Psi \cos \beta_0 h$. The quantity $E_\Theta(\Theta, \Phi)/V_{02}$ is the far field per unit voltage driving element 2.

An alternative expression for the field per unit input current to the driven antenna 2, i.e. $E_\Theta(\Theta, \Phi)/I_{z2}(0)$, is readily obtained with the substitution of $V_{02} = I_{z2}(0)/Y_{2N}$ where, from (6.23), the input admittance of antenna 2 when driving the N-element array is

$$Y_{2N} = \frac{I_{z2}(0)}{V_{02}} = \frac{j2\pi}{\zeta_0 \Psi \cos \beta_0 h}[\sin \beta_0 h_2 + T_{U2}(1 - \cos \beta_0 h_2)$$

$$+ T_{D2}(1 - \cos \tfrac{1}{2}\beta_0 h_2)]. \tag{6.83}$$

The result is

$$E_{\Theta N}(R_2, \Theta, \Phi) = \frac{j\zeta_0 I_{z2}(0)}{2\pi} \frac{e^{-j\beta_0 R_2}}{R_2} f_{IN}(\Theta, \Phi) \tag{6.84a}$$

where

$$f_{IN}(\Theta, \Phi) = \left\{ \frac{F_m(\Theta, \beta_0 h_2) + \sum\limits_{i=1}^{N} e^{-j\beta_0(R_i - R_2)}[T_{Ui}G_m(\Theta, \beta_0 h_i) + T_{Di}D_m(\Theta, \beta_0 h_i)]}{\sin \beta_0 h_2 + T_{U2}(1 - \cos \beta_0 h_2) + T_{D2}(1 - \cos \tfrac{1}{2}\beta_0 h_2)} \right\}. \tag{6.84b}$$

If the driven element is near a half wavelength long, the more convenient alternative form of the current is

$$I_{z2}(z_2) = A_2'(\sin \beta_0 |z_2| - \sin \beta_0 h_2) + B_2'(\cos \beta_0 z_2 - \cos \beta_0 h_2)$$

$$+ D_2(\cos \tfrac{1}{2}\beta_0 z_2 - \cos \tfrac{1}{2}\beta_0 h_2) \tag{6.85}$$

where $A_2' = -j2\pi V_{02}/\zeta_0 \Psi$. The currents in the parasitic elements are given by (6.79b). With the notation

$$T_{Ui}' = B_i'/A_2', \qquad B_i' = B_1, B_2', B_3, \dots B_N \tag{6.86}$$

the formula for the distant field is

$$E_\Theta(R_2, \Theta, \Phi) = \frac{V_{02}}{\Psi} \frac{e^{-j\beta_0 R_2}}{R_2} f_{VN}'(\Theta, \Phi) \tag{6.87a}$$

where

$$f_{VN}'(\Theta, \Phi) = H_m(\Theta, \beta_0 h_2) + \sum_{i=1}^{N} e^{-j\beta_0(R_i - R_2)}$$

$$\times [T_{Ui}'G_m(\Theta, \beta_0 h_i) + T_{Di}D_m(\Theta, \beta_0 h_i)]. \tag{6.87b}$$

$H_m(\Theta, \beta_0 h)$ is defined in (2.51) and, specifically for $\beta_0 h = \pi/2$, in (2.52a). As before, $G_m(\Theta, \beta_0 h)$ and $D_m(\Theta, \beta_0 h)$ are given in (2.47) and (2.48). Special values for $\beta_0 h = \pi/2$ are in (2.52b, c). If desired $E_\Theta(\Theta, \Phi)$ as given in (6.87a, b) may be referred to the current $I_{z2}(0)$

instead of the voltage V_{02}. In this case

$$Y_{2N} = \frac{-j2\pi}{\zeta_0 \Psi}[(1 - \sin \beta_0 h_2) + T'_{U2}(1 - \cos \beta_0 h_2) + T_{D2}(1 - \cos \tfrac{1}{2}\beta_0 h_2)]$$

(6.88)

so that $$E_{\Theta N}(R_2, \Theta, \Phi) = \frac{j\zeta_0 I_{z2}(0)}{2\pi} \frac{e^{-j\beta_0 R_2}}{R_2} f'_{1N}(\Theta, \Phi)$$ (6.89a)

where

$$f'_{1N}(\Theta, \Phi) = \frac{H_m(\Theta, \beta_0 h_2) + \sum_{i=1}^{N} e^{-j\beta_0(R_i - R_2)}[T'_{Ui}G_m(\Theta, \beta_0 h_i) + T_{Di}D_m(\Theta, \beta_0 h_i)]}{(1 - \sin \beta_0 h_2) + T'_{U2}(1 - \cos \beta_0 h_2) + T_{D2}(1 - \cos \tfrac{1}{2}\beta_0 h_2)}$$

(6.89b)

The graphical representations of the normalized field factors $|f_N(\Theta, \Phi)|/|f_N(\pi/2, 0)|$ or $|f'_N(\Theta, \Phi)|/|f'_N(\pi/2, 0)|$ in appropriate planes are the field patterns. The field pattern in the equatorial (horizontal) plane is given by $|f_N(\pi/2, \Phi)|/|f_N(\pi/2, 0)|$ as a function of Φ. Important field patterns in planes perpendicular to the equatorial plane are with $\Phi = 0$ and π. In this case $|f_N(\Theta, \{_\pi^0\})|/|f_N(\pi/2, 0)|$ is shown graphically as a function of Θ. The ratio of the field in the forward direction (i.e. toward the directors, $\Phi = 0$) to the field in the backward direction (i.e. toward the reflector, $\Phi = \pi$) in the equatorial plane $\Theta = \pi/2$ is known as the front-to-back ratio. It is given by

$$R_{FB} = \frac{\left|f_N\left(\frac{\pi}{2}, 0\right)\right|}{\left|f_N\left(\frac{\pi}{2}, \pi\right)\right|}.$$ (6.90a)

The front-to-back ratio in decibels is

$$r_{FB} = 20 \log_{10}\left|f_N\left(\frac{\pi}{2}, 0\right)\Big/ f_N\left(\frac{\pi}{2}, \pi\right)\right|.$$ (6.90b)

Note that in all of the ratios involving $f_N(\Theta, \Phi)$ either $f_{VN}(\Theta, \Phi)$ or $f_{1N}(\Theta, \Phi)$ may be used.

Since the total power radiated by an array is given by the integral over a great sphere of the normal component of the Poynting vector

$$|S_R(R, \Theta, \Phi)| = |E_\Theta(R, \Theta, \Phi)|^2/2\zeta_0$$ (6.91)

the distribution as a function of Θ and Φ of $|S_R(R, \Theta, \Phi)|$ is of interest. The total power supplied to the N-element array is

$$P_{2N} = \tfrac{1}{2}|I_{z2}(0)|^2 R_{2N} = \tfrac{1}{2}|V(0)|^2 G_{2N}$$ (6.92)

where R_{2N} and G_{2N} are, respectively, the driving-point resistance and conductance of the element 2 when driving the N-element parasitic array. With (6.84a) and (6.92)

$$|S_R(R_2, \Theta, \Phi)| = \frac{P_{2N}}{4\pi^2 R_2^2} \frac{\zeta_0}{G_{2N}} |f_{VN}(\Theta, \Phi)|^2 \qquad (6.93a)$$

$$= \frac{P_{2N}}{4\pi^2 R_2^2} \frac{\zeta_0}{R_{2N}} |f_{IN}(\Theta, \Phi)|^2. \qquad (6.93b)$$

A graphical representation of $|f_N(\Theta, \Phi)/f_N(\pi/2, 0)|^2$ is known as a power pattern. (Note that R_2 is a distance, R_{2N} a resistance.)

If ohmic losses in the conductors of the antennas and in the surrounding dielectric medium (air) are neglected, the total power radiated by an array outside a great sphere of radius R_2 is the same as the total power supplied at the terminals of the driven element 2. That is

$$P_{2N} = \tfrac{1}{2}|V_{02}|^2 G_{2N} = \tfrac{1}{2}|I_{z2}(0)|^2 R_{2N}$$

$$= \int_0^{2\pi} \int_0^\pi |S_R(R_2, \Theta, \Phi)| R_2^2 \sin \Theta \, d\Theta \, d\Phi. \qquad (6.94)$$

With (6.93) and (6.94), formulas are obtained for R_{2N} and G_{2N} in terms of the far field. They are

$$R_{2N} = \frac{\zeta_0}{4\pi^2} \int_0^{2\pi} \int_0^\pi |f_{IN}(\Theta, \Phi)|^2 \sin \Theta \, d\Theta \, d\Phi \qquad (6.95a)$$

$$G_{2N} = \frac{\zeta_0}{4\pi^2} \int_0^{2\pi} \int_0^\pi |f_{VN}(\Theta, \Phi)|^2 \sin \Theta \, d\Theta \, d\Phi. \qquad (6.95b)$$

Actually, both R_{2N} and G_{2N} are already known from

$$I_{z2}(0)/V_{02} = G_2 + jB_2$$

when the medium in which the array is immersed is lossless.

The absolute directivity of the N-element Yagi array is defined in terms of the power radiated by a fictitious omnidirectional antenna that maintains the same field in *all* directions as the Yagi array does in the one direction of its maximum, viz., $\Theta = \pi/2$, $\Phi = 0$. This power is

$$P_{N\,omni} = 4\pi R_2^2 \left| S_R\left(R_2, \frac{\pi}{2}, 0\right) \right|$$

$$= P_{2N} \frac{\zeta_0}{\pi R_{2N}} |f_{IN}(\Theta, \Phi)|^2 = P_{2N} \frac{\zeta_0}{\pi G_{2N}} |f_{VN}(\Theta, \Phi)|^2. \qquad (6.96)$$

The ratio $P_{N\,\text{omni}}/P_{2N}$ is the absolute directivity. Thus

$$D_N\left(\frac{\pi}{2},0\right) = \frac{P_{N\,\text{omni}}}{P_{2N}} = \frac{\zeta_0}{\pi R_{2N}}\left|f_{IN}\left(\frac{\pi}{2},0\right)\right|^2 = \frac{\zeta_0}{\pi G_{2N}}\left|f_{VN}\left(\frac{\pi}{2},0\right)\right|^2. \quad (6.97)$$

This formula is often written with R_{2N} expressed explicitly as given in (6.95a). The quantity

$$G_N\left(\frac{\pi}{2},0\right) = 10\log_{10}D_N\left(\frac{\pi}{2},0\right) \quad (6.98)$$

is the absolute gain in decibels.

The absolute directivity of the driven element 2 when isolated is

$$D_1\left(\frac{\pi}{2},0\right) = \frac{P_{1\,\text{omni}}}{P_{21}} = \frac{\zeta_0}{\pi R_{21}}\left|f_{I1}\left(\frac{\pi}{2},0\right)\right|^2 = \frac{\zeta_0}{\pi G_{21}}\left|f_{V1}\left(\frac{\pi}{2},0\right)\right|^2. \quad (6.99)$$

The relative directivity at constant power of the array referred to the isolated driven element is

$$D_r(0) = \frac{D_N\left(\frac{\pi}{2},0\right)}{D_1\left(\frac{\pi}{2},0\right)} = \frac{R_{21}\left|f_{IN}\left(\frac{\pi}{2},0\right)\right|^2}{R_{2N}\left|f_{I1}\left(\frac{\pi}{2},0\right)\right|^2} = \frac{G_{21}\left|f_{VN}\left(\frac{\pi}{2},0\right)\right|^2}{G_{2N}\left|f_{V1}\left(\frac{\pi}{2},0\right)\right|^2}. \quad (6.100)$$

The corresponding relative gain in decibels is

$$G_r(0) = G_N\left(\frac{\pi}{2},0\right) - G_1\left(\frac{\pi}{2},0\right) = 10\left[\log_{10}D_N\left(\frac{\pi}{2},0\right) - \log_{10}D_1\left(\frac{\pi}{2},0\right)\right]. \quad (6.101)$$

The relative directivity (6.100) is readily expressed in terms of the electric field in (6.84a). Thus

$$D_r(0) = \frac{\left|E_{\Theta N}\left(R_2,\frac{\pi}{2},0\right)\right|^2}{\left|E_{\Theta 1}\left(R_2,\frac{\pi}{2},0\right)\right|^2}\frac{P_{21}}{P_{2N}}. \quad (6.102)$$

The relative directivity at constant power, $P_{21} = P_{2N}$, is

$$D_r(0) = \frac{\left|E_{\Theta N}\left(R_2,\frac{\pi}{2},0\right)\right|^2}{\left|E_{\Theta 1}\left(R_2,\frac{\pi}{2},0\right)\right|^2}. \quad (6.103)$$

This is equivalent to (6.100).

The relative directivity (6.100) or (6.103) is also the relative forward directivity in the direction $\Theta = \pi/2$, $\Phi = 0$. The relative directivity at constant power $P_{21} = P_{2N}$ in the backward direction $\Theta = \pi/2, \Phi = \pi$ is defined by

$$D_r(\pi) = \frac{\left|E_{\Theta N}\left(R_2, \frac{\pi}{2}, \pi\right)\right|^2}{\left|E_{\Theta 1}\left(R_2, \frac{\pi}{2}, \pi\right)\right|^2} = \frac{R_{21}}{R_{2N}} \frac{\left|f_{IN}\left(\frac{\pi}{2}, \pi\right)\right|^2}{\left|f_{I1}\left(\frac{\pi}{2}, \pi\right)\right|^2} = \frac{G_{21}}{G_{2N}} \frac{\left|f_{VN}\left(\frac{\pi}{2}, \pi\right)\right|^2}{\left|f_{V1}\left(\frac{\pi}{2}, \pi\right)\right|^2}.$$

(6.104)

The relative backward gain is

$$G_r(\pi) = 10 \log_{10} D_r(\pi).$$

(6.105)

Since for a single element rotational symmetry with respect to Φ gives $f_N(\pi/2, 0) = f_N(\pi/2, \pi)$, it follows that

$$\frac{D_r(0)}{D_r(\pi)} = \frac{\left|f_N\left(\frac{\pi}{2}, 0\right)\right|^2}{\left|f_N\left(\frac{\pi}{2}, \pi\right)\right|^2}$$

(6.106)

and

$$r_{FB} = G_r(0) - G_r(\pi)$$

(6.107)

in decibels. Note that R_2 is a distance, R_{21} and R_{2N} resistances.

6.7   Simple applications of the modified theory; comparison with experiment

The theory of arrays developed in the preceding sections is like that formulated in the earlier chapters in that the complicated simultaneous integral equations for the currents in the elements are replaced by a set of algebraic equations. This is accomplished by approximating the integrals with an appropriate combination of trigonometric functions. In dealing with arrays of driven elements of equal length it was convenient to use different trigonometric functions for different parts of the integrals and to match these to the integrals at the point of maximum current, $z = z_m$, and at the ends, $z = \pm h$. For use with parasitic elements of unequal length this procedure is modified. Each integral is approximated by a sum of trigonometric terms with coefficients matched to the integral at $z = 0, \pm h/2$ and $\pm h$. In order to illustrate the application of the modified theory and at the same time verify its accuracy it is convenient to consider the simplest cases, the isolated antenna and the two-element parasitic array. Since conventional (sinusoidal) theory fails completely when full-wave elements are involved, the examples are selected deliberately to include such elements.

In Fig. 6.10 are the distributions of current along a full-wave isolated antenna as computed from the modified theory, and as measured. They may be compared with the three-term approximation in Fig. 2.4 where the same experimental data are also shown.

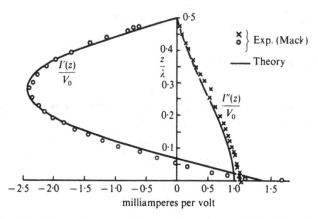

Fig. 6.10.   Distribution of current on full-wave antenna; $I(z) = I''(z) + jI'(z)$; $a/\lambda = 0.007022$, $h/\lambda = 0.5$.

The two theoretical representations, while not identical, are nevertheless both very good approximations of the current. The modified theory does not provide quite as good an overall fit, but is somewhat better in specifying the susceptance—as would be expected since all integrals are matched at $z = 0$ and not only at the maximum of current. The admittance in the modified theory is $Y_0 = (0.926 + j1.350) \times 10^{-3}$ mhos; the value obtained previously is $Y_0 = (0.976 + j0.988) \times 10^{-3}$ mhos. The measured value after correction for end effects is $(1.025 + j1.676) \times 10^{-3}$ mhos. As indicated in conjunction with Fig. 2.6 a lumped susceptance $B_0 = 0.72 \times 10^{-3}$ mhos must be added to the three-term admittance to give $Y_0 = (0.976 + j1.708) \times 10^{-3}$ mhos. A similar lumped correction is also required with the modified theory, but it is smaller, viz. $B_0 = 0.35$ mhos. It is clear that when suitably corrected to give the right susceptance either theory provides a very acceptable approximation of the current in a dipole.

The distributions of current in an array of two full-wave elements in which element 1 is centre driven and element 2 is parasitic are shown in Fig. 6.11 for four values of b, the distance between the parallel antennas. The corresponding field patterns in the equatorial plane are in Fig. 6.12. The distributions of current in Fig. 6.11 may

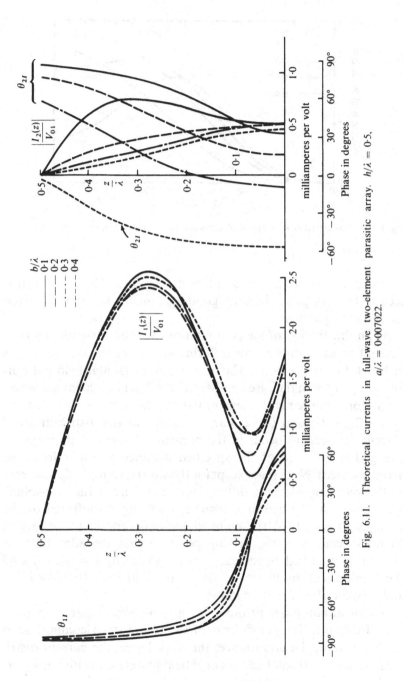

Fig. 6.11. Theoretical currents in full-wave two-element parasitic array. $h/\lambda = 0.5$, $a/\lambda = 0.007022$.

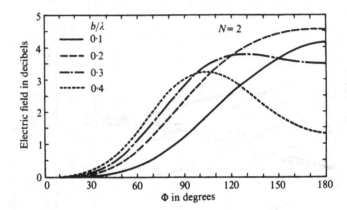

Fig. 6.12. Horizontal field patterns of full-wave two-element parasitic array. $h/\lambda = 0.5$, $a/\lambda = 0.007022$.

be compared with measured values in Fig. 6.13. The agreement is seen to be very good. Equally good agreement has been observed for the field patterns.

As an illustration of the computations for the currents in a two-element array with elements differing greatly in length, the graphs in Fig. 6.14 are provided. The associated horizontal field patterns are in Fig. 6.15. In the case considered, the driven element is a wavelength long, the parasitic element has successively the three lengths $h_2 = 0.2\lambda$, 0.4λ, and 0.65λ. Large changes in the distributions of current are seen to occur in the parasitic element as its length is changed while fixed at the specified distance $b = 0.2\lambda$ from the driven element. Note that except for the shortest length, the currents in the parasitic element differ significantly from the sinusoidal. The current in the driven antenna is only slightly affected by the changes in length of the coupled parasitic antenna, the largest changes occur near the driving point so that the admittance is noticeably modified. Specifically, for the values $h_2/\lambda = 0.2, 0.4, 0.65$ the admittances are $(0.916 + j1.041) \times 10^{-3}$, $(0.790 + j1.480) \times 10^{-3}$, and $(0.805 + j1.510) \times 10^{-3}$ mhos.

A typical computer printout for a two-element parasitic array is in Table 6.1. The coefficients of the trigonometric components of the current, the admittance, the impedance, the current distributions, the horizontal and vertical field patterns, the forward gain, the backward gain and the front-to-back ratio are all given.

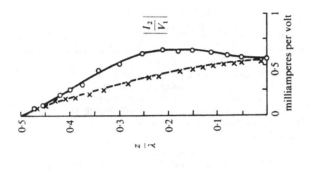

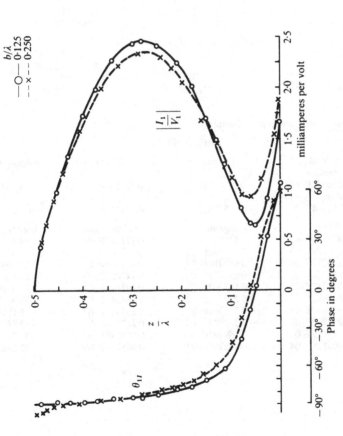

Fig. 6.13.   Measured currents in full-wave array of two elements (Mack). $h/\lambda = 0.5$, $a/\lambda = 0.007022$.

Table 6.1

No. of elements	= 2
Half-length of driving antenna	= 0·5000000E 00
Half-length of parasitic antenna	= 0·6500000E 00
Radius	= 0·7022000E–02
Element spacing	= 0·2000000E–00

Coefficients for current distributions

Element No. 1

AR	AI	BR
−0·249034E–04	−0·318019E–02	0·182919E–03
BI	DR	DI
0·441346E–03	0·439517E–03	0·627672E–03

Element No. 2

		BR
		0·707925E–04
BI	DR	DI
−0·493231E–03	0·221251E–03	0·456011E–03

Current distributions and input admittances

Element No. 1

	Real	Imaginary	Magnitude	Argument
Input admittance =				
	0·805356E–03	0·151036E–02	0·171167E–02	61·8473
Input impedance =				
	0·274884E 03	−0·515518E 03	0·584226E 03	−61·8473

Z/H	Real	Imaginary	Magnitude	Argument
0·	0·805356E–03	0·151036E–02	0·171167E–02	61·8473
0·1	0·783297E–03	0·498302E–03	0·928363E–03	32·4182
0·2	0·734272E–03	−0·473916E–03	0·873929E–03	−32·7939
0·3	0·661902E–03	−0·131281E–02	0·147023E–02	−63·1562
0·4	0·571337E–03	−0·193902E–02	0·202144E–02	−73·4809
0·5	0·468802E–03	−0·229502E–02	0·234241E–02	−78·3470
0·6	0·361052E–03	−0·235064E–02	0·237821E–02	−81·1558
0·7	0·254792E–03	−0·210594E–02	0·212130E–02	−82·9870
0·8	0·156115E–03	−0·159102E–02	0·159866E–02	−84·2797
0·9	0·700128E–04	−0·862943E–03	0·865779E–03	−85·2440

Table 6.1—*continued*

Element No. 2

Z/H	Real	Imaginary	Magnitude	Argument
0·	0·434100E–03	–0·120109E–03	0·450410E–03	–15·4447
0·1	0·423681E–03	–0·890176E–04	0·432931E–03	–11·8492
0·2	0·393571E–03	–0·202264E–05	0·393577E–03	–0·2940
0·3	0·347053E–03	0·123120E–03	0·368245E–03	19·5057
0.4	0·289068E–03	0·260242E–03	0·388955E–03	41·9382
0·5	0·225521E–03	0·379298E–03	0·441278E–03	59·1837
0·6	0·162456E–03	0·451620E–03	0·479950E–03	70·1187
0·7	0·105249E–03	0·455010E–03	0·467024E–03	76·8698
0·8	0·579298E–04	0·377819E–03	0·382234E–03	81·1709
0·9	0·227404E–04	0·221326E–03	0·222491E–03	84·0178

Horizontal field pattern

Phi	E	E DB
0·	1·000000	–0·
10·00	0·999009	–0·0086
20·00	0·997528	–0·0215
30·00	0·999831	–0·0015
40·00	1·012188	0·1052
50·00	1·041196	0·3507
60·00	1·091405	0·7597
70·00	1·163272	1·3136
80·00	1·252706	1·9570
90·00	1·352388	2·6220
100·00	1·453907	3·2507
110·00	1·549639	3·8046
120·00	1·633938	4·2647
130·00	1·703568	4·6272
140·00	1·757561	4·8982
150·00	1·796660	5·0893
160·00	1·822565	5·2137
170·00	1·837157	5·2829
180·00	1·841848	5·3051

F gain = 0·4079 DB   B gain = 5·7130 DB   FTBR = –5·3051 DB

Vertical field pattern

Theta	E	E DB
10·00	0·068390	–23·3002
20·00	0·151670	–16·3820
30·00	0·245067	–12·2143
40·00	0·345650	–9·2273
50·00	0·460228	–6·7405
60·00	0·606571	–4·3424
70·00	0·782575	–2·1295
80·00	0·937397	–0·5615

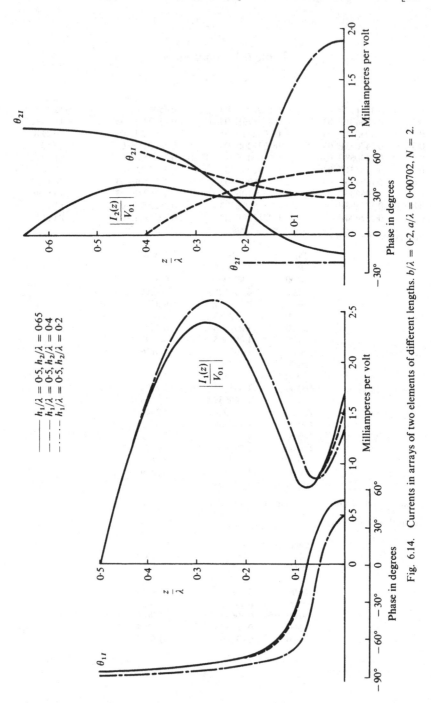

Fig. 6.14. Currents in arrays of two elements of different lengths. $b/\lambda = 0.2$, $a/\lambda = 0.00702$, $N = 2$.

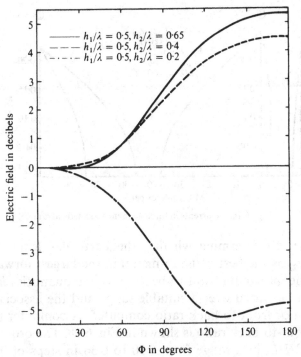

Fig. 6.15.   Horizontal field patterns of arrays of two elements of different lengths. $b/\lambda = 0.2$, $a/\lambda = 0.007022$, $N = 2$.

6.8   The three-element Yagi-Uda array†

The computed distributions of current and the field pattern for a three-element array consisting of a reflector of length $2h_1 = 0.51\lambda$, a driven element of length $2h_2 = 0.50\lambda$ and a single director of length $2h_3 = 0.45\lambda$ are shown in Figs. 6.16 and 6.17. For this array the radius of all elements was taken as $a = 0.003369\lambda$. The driving-point impedance of element 2 is $Z_2 = 27.4 + j1.27$ ohms. The computed values of the phase angle of the current along the reflector are nearly constant; it decreases from $74°.5$ at $z/h_1 = 0$ to $72°.7$ at $z/h_1 = 0.9$. The phase angle of the current along the driven element decreases from $-2°.66$ at $z/h_2 = 0$ to $-8°.47$ at $z/h_2 = 0.9$. The phase angle of the current along the director is almost exactly constant, changing only from $-154°.3$ at $z/h_3 = 0$ to $-154°.0$ at $z/h_3 = 0.9$. It is clear from Fig. 6.16 that the current in the reflector is so small that it actually contributes negligibly to the field.

† This section is based on the work of Dr I. L. Morris [3].

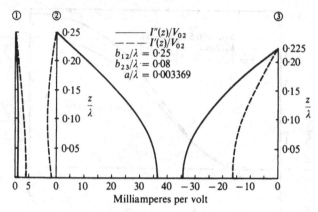

Fig. 6.16.   Currents in three-element Yagi-Uda array.

In order to determine whether the particular length h_3 and spacing b_{23} is the best value to maintain the largest forward gain $G(0)$ or the maximum front-to-back ratio, the quantities h_3/λ and b_{23}/λ can be varied over a suitable range and the associated forward gain or front-to-back ratio computed. A computer printout for the front-to-back ratio is shown in Fig. 6.18. The ordinates are $2h_3/\lambda = 2H/L$, in a range from 0·50 to 0·36 in steps of 0·01; the abscissae are $b_{23}/\lambda = B/L$ in the range from 0·02 to 0·30 in steps of 0·02. The contours are drawn along estimated lines of constant front-to-back ratio ranging from 1 to 19. It is seen that the maximum value of front-to-back ratio is close to $b_{23}/\lambda = 0·12$ with

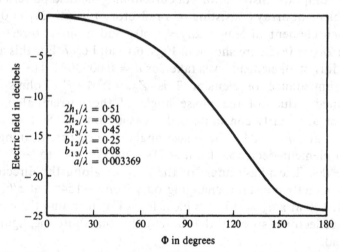

Fig. 6.17.   Field pattern of three-element Yagi-Uda array with element No. 2 driven.

1-Director Yagi antenna with constant director length and spacing
front-to-back ratio in dB

Reflector-feeder spacing = 0·250000
Radius of each element = 0·003369

Reflector length = 0·510000
Feeder length = 0·500000

$2h/\lambda$ \ b/λ	0·020000	0·040000	0·060000	0·080000	0·100000	0·120000	0·140000	0·160000	0·180000	0·200000	0·220000	0·240000	0·260000	0·280000	0·300000
0·500000	0·52	0·33	0·22	0·11	0·02	−0·07	−0·16	−0·19	−0·21	−0·21	−0·18	−0·11	0·01	0·17	0·39
0·490000	10·94	4·95	2·72	1·71	1·15	0·81	0·58	0·41	0·30	0·24	0·22	0·24	0·30	0·41	0·58
0·480000	15·19	13·23	8·72	5·98	4·31	3·24	2·52	2·02	1·66	1·40	1·23	1·12	1·08	1·10	1·18
0·470000	13·42	17·39	16·01	11·96	9·02	7·02	5·61	4·60	3·85	3·29	2·88	2·58	2·37	2·24	2·20
0·460000	12·24	15·72	20·17	19·66	15·18	11·85	9·55	7·89	6·65	5·71	5·00	4·45	4·03	3·73	3·52
0·450000	11·53	14·03	17·81	24·14	25·08	18·38	14·34	11·70	9·82	8·42	7·34	6·51	5·86	5·35	4·98
0·440000	11·08	12·93	15·54	19·66	28·40	30·70	20·39	15·93	13·14	11·16	9·68	8·54	7·63	6·95	6·40
0·430000	10·76	12·19	14·08	16·70	20·58	26·07	25·03	19·71	16·10	13·60	11·76	10·35	9·24	8·36	7·67
0·420000	10·53	11·66	13·09	14·93	17·28	20·01	21·75	20·41	17·68	15·23	13·27	11·73	10·49	9·50	8·71
0·410000	10·36	11·27	12·40	13·77	15·39	17·13	18·49	18·66	17·48	15·74	14·05	12·58	11·34	10·32	9·48
0·400000	10·22	10·98	11·88	12·94	14·14	15·38	16·40	16·82	16·43	15·42	14·17	12·93	11·81	10·84	10·01
0·390000	10·11	10·74	11·48	12·33	13·26	14·19	14·97	15·40	15·32	14·76	13·90	12·93	11·98	11·11	10·34
0·380000	10·02	10·55	11·17	11·86	12·60	13·33	13·95	14·33	14·37	14·06	13·47	12·74	11·96	11·20	10·51
0·370000	9·94	10·40	10·91	11·49	12·09	12·68	13·17	13·50	13·58	13·41	13·00	12·45	11·82	11·19	10·58
0·360000	9·88	10·27	10·71	11·19	11·69	12·16	12·57	12·85	12·95	12·85	12·56	12·14	11·64	11·10	10·58

Fig. 6.18. Typical printout for three-element Yagi-Uda array.

$2h_3/\lambda = 0.44$. Thus, the distributions of current and the field pattern in Figs. 6.16 and 6.17 do not quite correspond to those for maximum front-to-back ratio. A small readjustment in the length of the director from $2h_3 = 0.45\lambda$ to $2h_3 = 0.44\lambda$ and an increase in its spacing b_{23} from 0.08λ to 0.12λ produces an increase in front-to-back ratio from 24·14 to 30·70. If the parameters $2h_3/\lambda$ and b_{23}/λ were varied in steps smaller than 0·01 and 0·02, respectively, an even higher ratio might be obtained within the narrow ranges $2h_3 = 0.44\pm0.01$, $b_{23} = 0.12\pm0.02$. A more extended set of contours of the front-to-back ratios is shown in Fig. 6.19 in which the computed numbers have been deleted and only the contours of constant r_{FB} are shown. It is clear that a number of successive maxima in front-to-back ratio are obtained as the distance b_{23} between the director and the driven element is increased. These occur substantially at intervals of $\lambda/2$ with $2h_3$ between 0.44λ and 0·46λ. Similar computer printouts for forward gain, driving-point resistance and reactance are also shown in Fig. 6.19.

6.9   The four, eight and ten director Yagi-Uda arrays†

The theory developed and illustrated with simple examples in the preceding sections can be applied to analyse the properties of longer Yagi-Uda arrays. For such arrays the quantities of principal interest include the distributions of current along all elements (since these determine the field), the admittance or impedance of the single driven element, the far-field pattern, the forward gain and the front-to-back ratio. For many purposes the determination of conditions that yield a maximum in the forward gain or in the front-to-back ratio is important. The parameters that may be varied are the length $2h_i$ and radius a_i of each element i, the distances b_{ij} between the elements i and j and their number N. Thus, there are a total of $3N$ parameters.

Because of the large number of possible combinations, an exhaustive study of the Yagi array would be very costly in both time and money even when a high-speed digital computer is available. An investigation of reasonable proportions must be restricted to a choice and range of parameters that is appropriate to a particular purpose.

In general, the purpose of the Yagi-Uda array is to obtain a highly directive field pattern with large values of the forward gain

† This section is based on the work of Dr I. L. Morris [3].

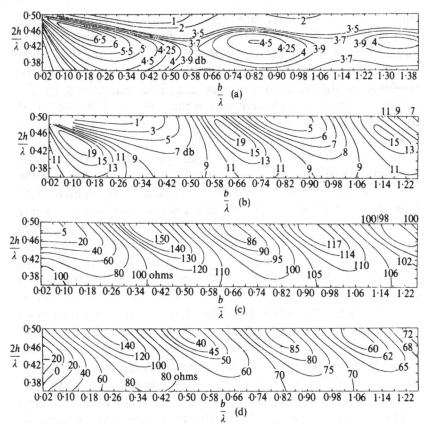

Fig. 6.19. Contour diagrams constructed with computer printouts for a 1-director Yagi array, (a) forward gain, (b) front-to-back ratio, (c) input resistance, (d) input reactance.

and front-to-back ratio. It has been shown implicitly that these desired properties can be achieved with the array pictured in Fig. 6.8. It consists of the following components: (1) A single driven element No. 2 that is a typical half-wave dipole of length $2h_2 = 0.5\lambda$ but with a finite radius a_2 and a distribution of current that is not assumed in advance to be sinusoidal, but remains to be determined. (2) A single reflecting element No. 1 that is slightly longer $(2h_1 = 0.51\lambda)$ than the driven antenna is placed at a distance $b_{12} = 0.25\lambda$ from it. The field maintained by the currents induced in a parasitic element of this length and relative location tends to reinforce the field maintained by the currents in the driven element in the forward direction (from 1 to 2) and to reduce or cancel it in the opposite or backward direction (from 2 to 1). (3) The balance of the array consists of $N-2$ directors that all have the same half-length $h_i = h$ and that are separated by the same distance $b_{i-1,i} = b$ with $3 \leqslant i \leqslant N$. In order to function as directors, the length h of the $N-2$ parasitic elements must satisfy the inequality, $h_i < h_2 = 0.5\lambda$ if the field maintained by the currents in them is to reinforce in the forward direction the field maintained by the currents in the driven element and in the reflector. If it is required that all antennas have the same radius, $a_i = a$, the $3N$ parameters have been reduced to three; h, b and N.

Contour diagrams constructed from computer printouts of the forward gain, the front-to-back ratio, the input resistance and the input reactance are shown in Figs. 6.20 and 6.21 for an array with four identical directors. The parameters are $2h/\lambda$ and b/λ where $h = h_3 = h_4 = h_5 = h_6$ and $b = b_{23} = b_{34} = b_{45} = b_{56}$. From these, combinations of h and b may be selected for which the forward gain or the front-to-back ratio is a maximum. For example, the following pairs of values are obtained from Fig. 6.20 to give a maximum front-to-back ratio: $2h/\lambda = 0.413$, 0.420, 0.426, 0.424; $b/\lambda = 0.033$, 0.139, 0.248, 0.360. These four sets all give a maximum front-to-back ratio, but the field patterns are quite different. These are shown in Fig. 6.22 together with corresponding patterns for similarly optimized one- and two-director arrays. From these it is seen that the field patterns for the most closely spaced condition for maximum front-to-back ratio are practically identical regardless of the number of directors. This is due to the fact that the directors are all so close to the driven element that no minor lobes are possible. As the distance between directors is increased, but limited to values that yield maxima in the front-to-back ratio, minor lobes appear and the beam width is reduced. The currents at the centres of the elements for the arrays that maintain the field patterns in Fig. 6.22 are represented in the form of phasor diagrams in Figs. 6.23a, b. The magnitude and angle of $I_2(0)$ in each element is shown. Note that for the very closely spaced 4-director array with $b/\lambda = 0.033$, the currents in the directors are almost equal and in phase and much smaller than the current in the driven element. On the other hand, for the largest spacing shown $b/\lambda = 0.36$, the currents in the directors are comparable in magnitude with the current in the driven element and their phase differences are close to the progressive phase difference $360°b/\lambda = 130°$ of a wave travelling with the velocity of light from element to element.

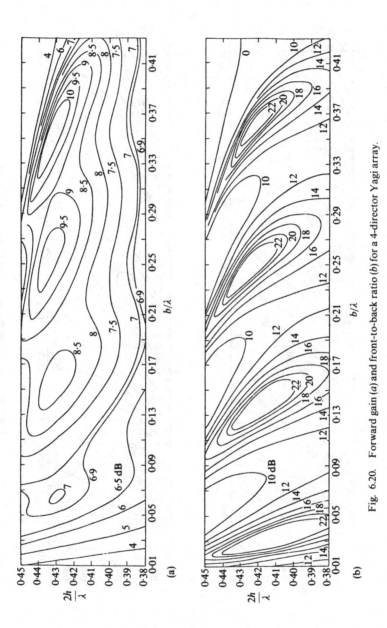

Fig. 6.20.   Forward gain (a) and front-to-back ratio (b) for a 4-director Yagi array.

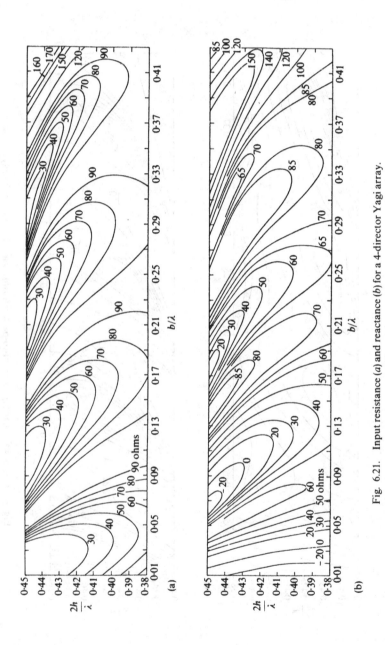

Fig. 6.21.   Input resistance (a) and reactance (b) for a 4-director Yagi array.

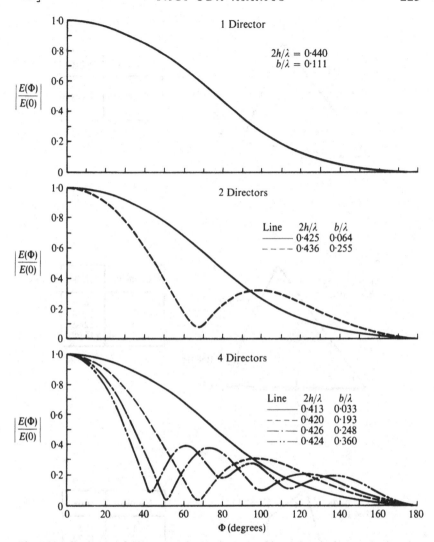

Fig. 6.22. Horizontal field patterns for Yagi arrays with maxima in front-to-back ratio.

Composite diagrams showing the forward gain, the front-to-back ratio, the input resistance and the input reactance as functions of b/λ for 1, 2, 4 and 8 director arrays with $h/\lambda = 0.43$, $a/\lambda = 0.00337$ are shown in Figs. 6.24a, b and 6.25a, b. From these the major quantities of interest are readily obtained.

A computer printout of a 10-director Yagi-Uda array† with $2h/\lambda = 0.4$ and $b/\lambda = 0.3$ is given in the accompanying Table 6.2.

† The numerical evaluation for the 10-director Yagi array was done by V. W. H. Chang.

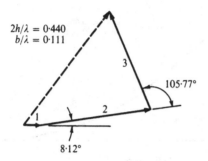

Element	$|I(0)|$
1	4·28
2	24·47
3	23·08

$2h/\lambda = 0\cdot440$
$b/\lambda = 0\cdot111$

105·77°

8·12°

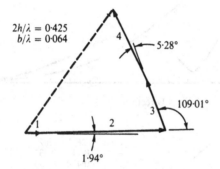

Element	$|I(0)|$
1	4·35
2	27·34
3	12·55
4	16·68

$2h/\lambda = 0\cdot425$
$b/\lambda = 0\cdot064$

5·28°

109·01°

1·94°

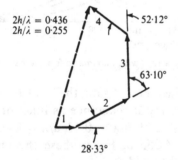

Element	$|I(0)|$
1	4·55
2	13·32
3	14·17
4	11·05

$2h/\lambda = 0\cdot436$
$2h/\lambda = 0\cdot255$

52·12°

63·10°

28·33°

Fig. 6.23a. Phasor diagrams for Yagi arrays with maxima in front-to-back ratio; 1 and 2 directors.

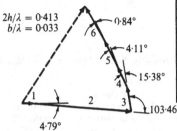

Element	$|I(0)|$
1	4·49
2	30·66
3	7·91
4	7·34
5	8·79
6	12·21

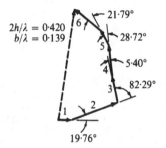

Element	$|I(0)|$
1	4·48
2	15·54
3	9·64
4	9·14
5	5·70
6	9·73

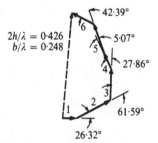

Element	$|I(0)|$
1	4·60
2	12·04
3	9·96
4	6·37
5	8·05
6	7·89

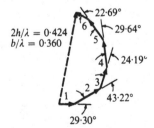

Element	$|I(0)|$
1	4·74
2	9·59
3	7·46
4	8·14
5	7·58
6	5·29

Fig. 6.23b.   Like Fig. 6.23a but for 4 directors.

Table 6.2. *Computer printout for 8-director Yagi-Uda array*

No. of elements = 10
Half-length of driving antenna = 0·2500000E–00
Half-length of parasitic antenna = 0·2000000E–00
Half-length of reflector antenna = 0·2550000E–00
Radius = 0·3369000E–02
Spacing between reflector and driving antennas = 0·2500000E–00
Spacing between parasitic antennas = 0·3000000E–00

Coefficients for current distributions

Element No.	AR	AI	BR
1	0	0	−0·261108E–03
2	0·260791E–04	−0·128598E–02	0·603188E–03
3	0	0	0·373823E–02
4	0	0	−0·247170E–02
5	0	0	−0·117657E–02
6	0	0	0·302159E–02
7	0	0	−0·130879E–02
8	0	0	−0·176549E–02
9	0	0	0·264461E–02
10	0	0	−0·391583E–03

Element No.	BI	DR	DI
1	0·443744E–03	0·626141E–02	0·121838E–01
2	0·204197E–01	0·200499E–01	−0·917549E–01
3	−0·956042E–03	−0·339069E–01	0·757068E–02
4	−0·290843E–02	0·207297E–01	0·252446E–01
5	0·282252E–02	0·102848E–01	−0·238580E–01
6	0·397791E–03	−0·256008E–01	−0·357119E–02
7	−0·284188E–02	0·109612E–01	0·241588E–01
8	0·176994E–02	0·149901E–01	−0·149562E–01
9	0·139167E–02	−0·223427E–01	−0·119466E–01
10	−0·278379E–02	0.390606E–02	0·234452E–01

Table 6.2. *Computer printout for 8-director Yagi-Uda array—cont.*

Current distributions and input admittances

Element No. 1

Z/H	Real	Imaginary	Magnitude	Argument
0·	0·163470E–02	0·416262E–02	0·447210E–02	68·4651
0·1	0·161797E–02	0·411786E–02	0·442432E–02	68·4550
0·2	0·156780E–02	0·398398E–02	0·428136E–02	68·4247
0·3	0·148433E–02	0·376216E–02	0·404439E–02	68·3743
0·4	0·136779E–02	0·345436E–02	0·371530E–02	68·3042
0·5	0·121849E–02	0·306328E–02	0·329672E–02	68·2146
0·6	0·103685E–02	0·259232E–02	0·279199E–02	68·1060
0·7	0·823442E–03	0·204556E–02	0·220508E–02	67·9789
0·8	0·578922E–03	0·142768E–02	0·154059E–02	67·8340
0·9	0·304115E–03	0·743922E–03	0·803683E–03	67·6719

Element No. 2

Z/H	Real	Imaginary	Magnitude	Argument
0·	0·644960E–02	−0·516874E–02	0·826517E–02	−38·6555
0·1	0·638444E–02	−0·533846E–02	0·832227E–02	−39·8463
0·2	0·618129E–02	−0·543588E–02	0·823147E–02	−41·2718
0·3	0·584171E–02	−0·544297E–02	0·798446E–02	−42·9171
0·4	0·536841E–02	−0·533362E–02	0·756752E–02	−44·7520
0·5	0·476516E–02	−0·507442E–02	0·696107E–02	−46.7358
0·6	0·403674E–02	−0·462572E–02	0·613943E–02	−48·8223
0·7	0·318893E–02	−0·394290E–02	0·507107E–02	−50·9645
0·8	0·222841E–02	−0·297779E–02	0·371928E–02	−53·1176
0·9	0·116268E–02	−0·168029E–02	0·204333E–02	−55·2423

Element No. 3

Z/H	Real	Imaginary	Magnitude	Argument
0·	−0·389258E–02	0·785263E–03	0·397100E–02	168·6103
0·1	−0·385515E–02	0·777863E–03	0·393285E–02	168·6082
0·2	−0·374266E–02	0·755602E–03	0·381817E–02	168·6018
0·3	−0·355451E–02	0·718302E–03	0·362636E–02	168·5912
0·4	−0·328973E–02	0·665672E–03	0·335640E–02	168·5765
0·5	−0·294700E–02	0·597315E–03	0·300693E–02	168·5580
0·6	−0·252470E–02	0·512741E–03	0·257624E–02	168·5358
0·7	−0·202096E–02	0·411377E–03	0·206240E–02	168·5102
0·8	−0·143372E–02	0·292589E–03	0·146327E–02	168·4815
0·9	−0·760799E–03	0·155698E–03	0·776567E–03	168·4500

Table 6.2. *Computer printout for 8-director Yagi-Uda array—cont.*

Element No. 4

Z/H	Real	Imaginary	Magnitude	Argument
0·	0·225112E–02	0·281162E–02	0·360177E–02	51·2469
0·1	0·222970E–02	0·278474E–02	0·356740E–02	51·2456
0·2	0·216531E–02	0·270393E–02	0·346408E–02	51·2416
0·3	0·205751E–02	0·256871E–02	0·329114E–02	51·2350
0·4	0·190559E–02	0·237827E–02	0·304753E–02	51·2259
0·5	0·170859E–02	0·213152E–02	0·273178E–02	51·2144
0·6	0·146531E–02	0·182713E–02	0·234212E–02	51·2006
0·7	0·117439E–02	0·146354E–02	0·187647E–02	51·1848
0·8	0·834289E–03	0·103904E–02	0·133254E–02	51·1671
0·9	0·443383E–03	0·551817E–03	0·707878E–03	51·1476

Element No. 5

Z/H	Real	Imaginary	Magnitude	Argument
0·	0·115124E–02	−0·260615E–02	0·284910E–02	−66·0760
0·1	0·114022E–02	−0·258133E–02	0·282195E–02	−66·0770
0·2	0·110710E–02	−0·250670E–02	0·274030E–02	−66·0799
0·3	0·105169E–02	−0·238177E–02	0·260363E–02	−66·0847
0·4	0·973653E–03	−0·220574E–02	0·241108E–02	−66·0913
0·5	0·872566E–03	−0·197751E–02	0·216147E–02	−66·0997
0·6	0·747887E–03	−0·169575E–02	0·185335E–02	−66·1096
0·7	0·598995E–03	−0·135890E–02	0·148506E–02	−66·1212
0·8	0·425205E–03	−0·965221E–03	0·105473E–02	−66·1340
0·9	0·225787E–03	−0·512883E–03	0·560383E–03	−66·1481

Element No. 6

Z/H	Real	Imaginary	Magnitude	Argument
0·	−0·280144E–02	−0·407170E–03	0·283087E–02	−171·7418
0·1	−0·277475E–02	−0·403259E–03	0·280390E–02	−171.7424
0·2	−0·269450E–02	−0·391507E–03	0·272279E–02	−171.7442
0·3	−0·256017E–02	−0·371848E–03	0·258703E–02	−171.7473
0·4	−0·237090E–02	−0·344178E–03	0·239575E–02	−171·7516
0·5	−0·212552E–02	−0·308354E–03	0·214777E–02	−171·7569
0·6	−0·182261E–02	−0·264202E–03	0·184166E–02	−171·7633
0·7	−0·146050E–02	−0·211518E–03	0·147573E–02	−171·7707
0·8	−0·103734E–02	−0·150081E–03	0·104814E–02	−171·7790
0·9	−0·551178E–03	−0·796550E–04	0·556904E–03	−171·7880

Table 6.2. *Computer printout for 8-director Yagi-Uda array—cont.*

Element No. 7

Z/H	Real	Imaginary	Magnitude	Argument
0·	0·118905E–02	0·265022E–02	0·290474E–02	65·7455
0·1	0·117774E–02	0·262496E–02	0·287706E–02	65·7450
0·2	0·114373E–02	0·254901E–02	0·279384E–02	65·7437
0·3	0·108680E–02	0·242187E–02	0·265454E–02	65·7414
0·4	0·100657E–02	0·224275E–02	0·245827E–02	65·7383
0·5	0·902524E–03	0·201056E–02	0·220384E–02	65·7344
0·6	0·774037E–03	0·172395E–02	0·188974E–02	65·7297
0·7	0·620375E–03	0·138136E–02	0·151427E–02	65·7243
0·8	0·440727E–03	0·981069E–03	0·107552E–02	65·7182
0·9	0·234232E–03	0·521244E–03	0·571454E–03	65·7116

Element No. 8

Z/H	Real	Imaginary	Magnitude	Argument
0·	0·164294E–02	−0·163338E–02	0·231672E–02	−44·7710
0·1	0·162728E–02	−0·161782E–02	0·229464E–02	−44·7712
0·2	0·158020E–02	−0·157105E–02	0·222828E–02	−44·7718
0·3	0·150140E–02	−0·149275E–02	0·211720E–02	−44·7727
0·4	0·139038E–02	−0·138243E–02	0·196068E–02	−44·7741
0·5	0·124645E–02	−0·123940E–02	0·175777E–02	−44·7757
0·6	0·106878E–02	−0·106281E–02	0·150727E–02	−44·7777
0·7	0·856408E–03	−0·851691E–03	0·120781E–02	−44·7800
0·8	0·608252E–03	−0·604957E–03	0·857871E–03	−44·7826
0·9	0·323173E–03	−0·321454E–03	0·455822E–03	−44·7854

Element No. 9

Z/H	Real	Imaginary	Magnitude	Argument
0·	−0·243969E–02	−0·131999E–02	0·277389E–02	−151·6237
0·1	−0·241646E–02	−0·130739E–02	0·274746E–02	−151·6242
0·2	−0·234660E–02	−0·126951E–02	0·266799E–02	−151·6258
0·3	−0·222965E–02	−0·120611E–02	0·253497E–02	−151·6285
0·4	−0·206487E–02	−0·111680E–02	0·234754E–02	−151·6321
0·5	−0·185124E–02	−0·100106E–02	0·210457E–02	−151·6367
0·6	−0·158748E–02	−0·858237E–03	0·180462E–02	−151·6422
0·7	−0·127214E–02	−0·687575E–03	0·144607E–02	−151·6486
0·8	−0·903608E–03	−0·488242E–03	0·102708E–02	−151·6557
0·9	−0·480150E–03	−0·259353E–03	0·545718E–03	−151·6634

Element No. 10

Z/H	Real	Imaginary	Magnitude	Argument
0·	0·475414E–03	0·255409E–02	0·259796E–02	79·3463
0·1	0·470794E–03	0·252977E–02	0·257321E–02	79·3483
0·2	0·456916E–03	0·245667E–02	0·249880E–02	79·3545
0·3	0·433725E–03	0·233429E–02	0·237425E–02	79·3647
0·4	0·401134E–03	0·216185E–02	0·219875E–02	79·3787
0·5	0·359024E–03	0·193825E–02	0·197122E–02	79·3965
0·6	0·307248E–03	0·166218E–02	0·169033E–02	79·4177
0·7	0·245641E–03	0·133207E–02	0·135453E–02	79·4421
0·8	0·174023E–03	0·946231E–03	0·962100E–03	79·4695
0·9	0·922045E–04	0·502831E–03	0·511215E–03	79·4994

Table 6.2. *Computer printout for 8-director Yagi-Uda array—cont.*

	Real	Imaginary	Magnitude	Argument
Input admittance =				
	0·644960E–02	−0·516874E–02	0·826517E–02	−38·6555
Input impedance =				
	0·944123E 02	0·756624E 02	0·120990E 03	38·6555

Horizontal field pattern

Phi	E	E DB
0·	1·000000	−0·
5·00	0·986511	−0·1180
10·00	0·944099	−0·4996
15·00	0·867684	−1·2328
20·00	0·751469	−2·4818
25·00	0·593565	−4·5306
30·00	0·404522	−7·8611
35·00	0·232860	−12·6581
40·00	0·225985	−12·9184
45·00	0·344538	−9·2553
50·00	0·415399	−7·6307
55·00	0·390921	−8·1582
60·00	0·298334	−10·5060
65·00	0·247244	−12·1375
70·00	0·299673	−10·4671
75·00	0·334813	−9·5039
80·00	0·292586	−10·6749
85·00	0·225493	−12·9373
90·00	0·229328	−12·7909
95·00	0·260556	−11·6820
100·00	0·241345	−12·3473
105·00	0·183770	−14·7145
110·00	0·156452	−16·1124
115·00	0·176931	−15·0439
120·00	0·187247	−14·5517
125·00	0·166602	−15·5664
130·00	0·130073	−17·7163
135·00	0·107503	−19·3716
140·00	0·118454	−18·5290
145·00	0·146355	−16·6919
150·00	0·171587	−15·3103
155·00	0·187308	−14·5489
160·00	0·193479	−14·2673
165·00	0·192924	−14·2923
170·00	0·189266	−14·4586
175·00	0·185684	−14·6245
180·00	0·184254	−14·6917

F gain = 11·5646 DB   B gain = −3·1270 DB   FTBR = 14·6917 DB

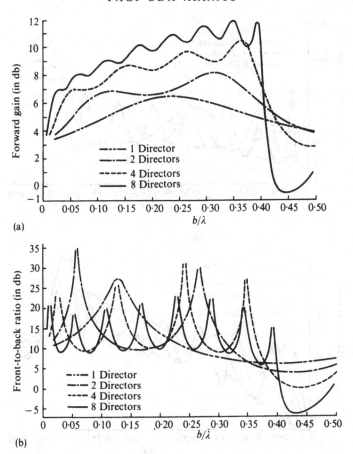

Fig. 6.24.	Forward gain (a), and front-to-back ratio (b), for a Yagi array with directors of constant length, radius and spacing (0·43λ, 0·00337λ and b, respectively).

The impedance of the driven element when isolated is $Z_0 = 88·94 + j39·11$ ohms. Graphs of the currents in all of the elements are shown in Fig. 6.26. The phase angle along each parasitic element is essentially constant. It is represented in Fig. 6.27 as a function of the distance of the element from the driven antenna No. 2. The curve drawn through the points has no physical significance; it serves merely to interrelate the discrete points and thus reveal how nearly constant the phase change from director to director actually is. The electrical separation of adjacent directors is 108°, the average phase difference of the currents is 115°·6. The horizontal field pattern maintained by the currents in the ten-element array is shown in Fig. 6.28.

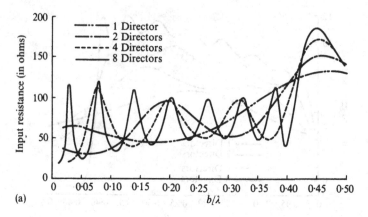

(a)

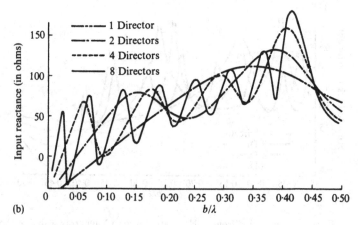

(b)

Fig. 6.25.   Input resistance (a), and reactance (b), for a Yagi array with directors of constant
length, radius and spacing (0·43λ, 0·00337λ and b, respectively).

6.10   Receiving arrays

The study of arrays of cylindrical antennas in all of the earlier
sections of the book has been directed specifically to the problem
of transmission, which involves the determination of distributions
of current, driving-point admittances and field patterns. Arrays of
antennas are also used to secure desired directional properties for
receivers.

In a transmitting array a single element may be driven, as in
parasitic arrays of the Yagi-Uda type, or all the elements may be
active as in the broadside or endfire arrays. In these latter the
driving voltage is usually supplied from a single power oscillator by

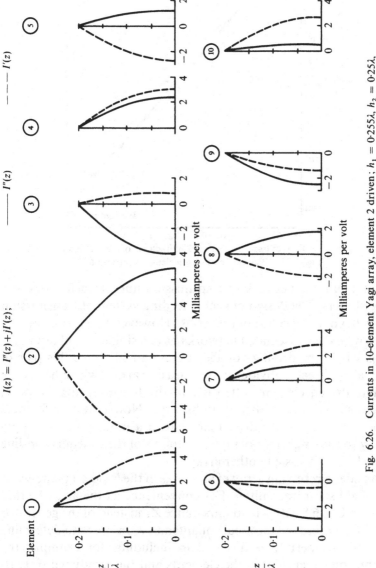

Fig. 6.26. Currents in 10-element Yagi array, element 2 driven; $h_1 = 0.255\lambda$, $h_2 = 0.25\lambda$, $h_3 = \ldots = h_{10} = 0.20\lambda$; $b_{12} = 0.25\lambda$, $b_{23} = \ldots = b_{9,10} = 0.3\lambda$; $a_1 = a_2 \ldots = a_{10} = 0.00337\lambda$; $\Omega = 10$ for $h = 0.25\lambda$.

$I(z) = I''(z) + jI'(z)$;   ———— $I''(z)$ ----- $I'(z)$

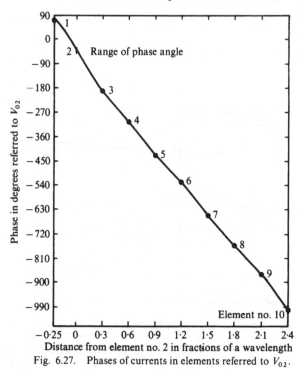

Fig. 6.27. Phases of currents in elements referred to V_{02}.

way of a suitable network of transmission lines, transformers and phase shifters. The design of such a feeding system of transmission lines is beyond the scope of this book. However, most transmitting arrays with their associated networks have a single pair of terminals across which the driving voltage is maintained. Since this pair of terminals is directly obvious in the parasitic arrays which have only a single driven element, attention in the following discussion is focused specifically on arrays of this type. Note that all references to the terminals of the driven element in a parasitic array apply equally to the single pair of input terminals of the transmission-line network that drives any other array.

Consider a receiving array of antennas in the incident plane-wave field of a distant transmitter. For convenience let the array be that shown in Fig. 6.8 with a load impedance Z_L instead of the generator connected across the terminals of antenna 2. In order to determine all of the properties of this system including, for example, the distributions of current in the elements and the reradiated or scattered field, it is necessary to formulate the coupled integral equations from the boundary condition that requires the tangential

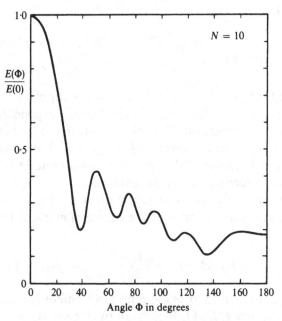

Fig. 6.28.   Field of 10-element Yagi array.

component of the total electric field to vanish on the perfectly conducting surface of each element. Fortunately, if interest is restricted to the transmission of information from a distant transmitter to the load Z_L, this elaborate analysis is unnecessary since the current in the load between the given terminals can be determined by the application of the reciprocal theorem† to the identical array when driven by the voltage V_0^e across the same terminals.

The reciprocal theorem applies to two arbitrarily located pairs of terminals, the one, for example, in an array A, the other in a simple dipole D. First, let the array be used for transmission, the dipole for reception. A generator with EMF V_0^e and internal impedance Z_g is connected across the terminals of the array; a load Z_L is connected across the terminals of the dipole. The centre of the driven element 2 in the array is located at the origin of the spherical coordinates r, Θ, Φ; the receiving dipole is used to measure the field pattern of the array. For this purpose it is moved along the surface of a great sphere so that its axis is always tangent to the electric field maintained by the transmitter. The current $I_D(\Theta, \Phi)$ in Z_L at the centre of the dipole varies as the dipole is moved. From (2.78) with

† See, for example, [4], p. 690 and [5], p. 216.

(2.79), it is given by

$$I_D(\Theta, \Phi) = \frac{2h_e\left(\frac{\pi}{2}\right)E_z^{inc}}{Z_0 + Z_L} = \frac{-2h_e\left(\frac{\pi}{2}\right)E_\Theta(R_2, \Theta, \Phi)}{Z_0 + Z_L} \qquad (6.108)$$

where $2h_e(\pi/2)$ is the effective length of the dipole when its axis is parallel to the incident electric field and perpendicular to the direction of propagation. Note that when the axis of the receiving dipole is tangent to the surface of the great sphere parallel to E^{inc}, the positive direction of the spherical coordinate Θ is opposite to the positive direction z along the antenna.

The far-zone electric field maintained by the N-element Yagi array driven by a generator at the centre of element No. 2 is given by (6.84a). It is

$$E_\Theta(R_2, \Theta, \Phi) = \frac{j\zeta_0 I_{z2}(0)}{2\pi} \frac{e^{-j\beta_0 R_2}}{R_2} f_{IN}(\Theta, \Phi) \qquad (6.109)$$

where R_2 is measured from the centre of element No. 2 and the field factor of the array, $f_{IN}(\Theta, \Phi)$ is given by (6.84b). If the driving-point impedance of the array at the terminals of element No. 2 is Z_{02} and the internal impedance of the generator is Z_g, it follows that

$$I_{z2}(0) = \frac{V_0^e}{Z_{02} + Z_g}. \qquad (6.110)$$

With (6.109) and (6.110), (6.108) becomes

$$I_D(\Theta, \Phi) = \frac{2h_e\left(\frac{\pi}{2}\right)}{Z_0 + Z_L} \cdot \frac{j\zeta_0 V_0^e}{Z_{02} + Z_g} \frac{e^{-j\beta_0 R_2}}{2\pi R_2} f_{IN}(\Theta, \Phi). \qquad (6.111)$$

Now let the generator with its emf V_0^e and internal impedance Z_g be interchanged with the load Z_L so that the dipole is the transmitter, the array the receiver. The dipole is again moved over the surface of the same great sphere; the array remains fixed at the origin of coordinates. The current $I_A(\Theta, \Phi)$ in the load Z_L in the array varies as the location of the transmitter is changed.

The reciprocal theorem states that if the same voltage V_0^e is applied successively to both antennas and provided $Z_g = Z_L$, then

$$I_D(\Theta, \Phi) = I_A(\Theta, \Phi) \qquad (6.112)$$

for all values of Θ and Φ. It follows by a rearrangement of (6.111) and with (6.112) that the current in the load Z_L of the Yagi array when used for reception is given by

$$I_A(\Theta, \Phi) = \frac{2f_{IN}(\Theta, \Phi)}{\beta_0(Z_{02} + Z_L)} \cdot \frac{j\zeta_0 V_0^e}{Z_0 + Z_g} \frac{e^{-j\beta_0 R_2}}{R_2} \beta_0 h_e\left(\frac{\pi}{2}\right) \quad (6.113)$$

provided $Z_L = Z_g$. Since it has been proved† in general that

$$\beta_0 h_e\left(\frac{\pi}{2}\right) = f_I\left(\frac{\pi}{2}, \beta_0 h\right) \quad (6.114)$$

where $f_I(\pi/2, \beta_0 h)$ is the field factor of the dipole given in (2.54b) and evaluted at $\Theta = \pi/2$, it follows that (6.113) can be expressed as follows:

$$I_A(\Theta, \Phi) = -\frac{2h_{eN}(\Theta, \Phi)E_\Theta^r}{Z_{02} + Z_L} \quad (6.115)$$

where
$$E_\Theta^r = \frac{j\zeta_0 I_z(0)}{2\pi} \frac{e^{-j\beta_0 R_2}}{R_2} f_I\left(\frac{\pi}{2}, \beta_0 h\right) \quad (6.116)$$

is the field maintained by the dipole at the centre of element No. 2 of the array and where

$$2h_{eN}(\Theta, \Phi) = f_{IN}(\Theta, \Phi)/\beta_0 \quad (6.117)$$

is by definition the effective length of the Yagi array. It follows that the directional properties of the Yagi (or any other array) are the same for reception as for transmission.

The preceding discussion has been concerned with reciprocity with constant applied voltage. If reciprocity is to be preserved with constant power somewhat different conditions must be fulfilled. This problem is considered elsewhere.‡

6.11 Driven arrays of elements that differ greatly in length

The procedure outlined in section 6.2 for approximating the integrals in the simultaneous integral equations (6.8) for the currents in a parasitic array of unequal elements is quite adequate when the elements do not differ greatly in length. In the Yagi-Uda array the lengths $2h_i$ of the individual elements $i = 1, \dots N$ always lie in a range that extends from slightly greater than $\lambda/2$ to approximately $\lambda/3$. Unfortunately, when elements have lengths that encompass the full range permitted by the present theory, viz., $0 \leqslant \beta_0 h_i \leqslant 5\pi/4$, the representations (6.20)–(6.22) for the several integrals are not adequate under certain conditions. In particular

† [4], pp. 568–570.
‡ [4], p. 694.

the two-term approximations on the right in (6.20)–(6.22) do not adequately represent the integrals $W_{ki}(z_k)$ on the left whenever element k is quite long ($\beta_0 h_k \sim \pi$) but element i is short ($\beta_0 h_i \lesssim \pi/4$). Extensive computations and measurements by W. M. Cheong [6] have shown that the two-term approximations in (6.20)–(6.22) with the two-point fitting used in (6.53)–(6.60) are especially unsatisfactory for points on the longer element in the range $|z| > h/2$.

A better representation of all of the integrals (6.15)–(6.17) and (6.20)–(6.22) is obtained when full advantage is taken of the three-term distribution of current given in (6.6) to approximate the integrals. Specifically, let

$$W_{kiV}(z_k) \equiv \int_{-h_i}^{h_i} M_{0z'i} K_{kid}(z_k, z_i')\, dz_i'$$
$$\doteq \Psi^m_{kidV} M_{0zk} + \Psi^f_{kidV} F_{0zk} + \Psi^h_{kidV} H_{0zk} \qquad (6.118)$$

$$W_{kiU}(z_k) \equiv \int_{-h_i}^{h_i} F_{0z'i} K_{kid}(z_k, z_i')\, dz_i'$$
$$\doteq \Psi^m_{kidU} M_{0zk} + \Psi^f_{kidU} F_{0zk} + \Psi^h_{kidU} H_{0zk} \qquad (6.119)$$

$$W_{kiD}(z_k) \equiv \int_{-h_i}^{h_i} H_{0z'i} K_{kid}(z_k, z_i')\, dz_i'$$
$$\doteq \Psi^m_{kidD} M_{0zk} + \Psi^f_{kidD} F_{0zk} + \Psi^h_{kidD} H_{0zk} \qquad (6.120)$$

The inclusion of the distribution M_{0z} in the approximate representation of the integrals $W_{kiU}(z_k)$ and $W_{kiD}(z_k)$ is a new departure. In all previous discussions it has been pointed out that the part of the integral that depends on the real part of the kernel is approximately proportional to the distribution in the integrand when the distance $\beta_0 b_{ki} < 1$ (which usually occurs only when $i = k$ and $b_{kk} = a$) and that otherwise the entire integral is proportional to combinations of $F_{0z} = \cos \beta_0 z - \cos \beta_0 h$ and $H_{0z} = \cos(\beta_0 z/2) - \cos(\beta_0 h/2)$. This means that the distribution $M_{0z} = \sin \beta_0(h - |z|)$ can appear on the right only when $M_{0z'}$ appears in the integrand. These statements are still correct. However, the investigations of Cheong [6] have shown that the current induced in the relatively long antenna ($h \sim \lambda/2$) by a very short one ($h < \lambda/4$) is not well represented by combinations of F_{0z} and H_{0z} alone. These distributions are excellent when the amplitude and phase of the inducing field are approximately constant along the entire length of an antenna. Clearly, this is not at all true of the field maintained, for example, along a full-wave antenna by the current in an adjacent quite short element. By including the term in M_{0z}, Cheong has obtained an improved over-all representation of the amplitudes of the currents, especially at points at some distance from the centres of the longer elements. On the other hand, since M_{0z} has a discontinuous slope at $z = 0$ (except when $\beta_0 h = (2n+1)\pi/2$) which the actual induced current cannot have, the slope of an approximate representation that makes use of M_{0z} is necessarily somewhat in error near $z = 0$ even though the amplitude is quite well described. The slope of the current is, of course, proportional to the charge per unit length. Fortunately, an incorrect slope with a discontinuity at $z = 0$ does not significantly affect the admittance or the far field. These are determined by the magnitude and phase of the current alone.

Since combinations of F_{0z} and H_{0z} are excellent approximations of the two integrals in (6.119) and (6.120) except in the special situations just described, it is to be anticipated that the coefficients Ψ^m_{kidU} and Ψ^m_{kidD} will be small except under those conditions. In any event, the three-term representation of the current for all elements including the M_{0z} terms in (6.119) and (6.120), can only serve to improve the representation of the amplitudes of the currents at the expense of a small error in their slopes near $z = 0$.

In order to determine the complex parameters Ψ in (6.118)–(6.120), the approximate expressions on the right are made exactly equal to the integrals at the three points $z_i = 0$, $z_i = h_i/3$ and $z_i = 2h_i/3$ instead of only at the two points $z_i = 0$ and $z_i = h_i/2$ used in section 6.4. That is, three equations are obtained from each of the relations (6.118)–(6.120) in the form:

$$
\begin{aligned}
W_{kiV}(0) &= \int_{-h_i}^{h_i} M_{0z'i} K_{kid}(0, z_i')\, dz_i' \\
&= \Psi^m_{kidV} \sin \beta_0 h_k + \Psi^f_{kidV}(1 - \cos \beta_0 h_k) \\
&\quad + \Psi^h_{kidV}[1 - \cos(\beta_0 h_k/2)]
\end{aligned}
\tag{6.121}
$$

$$
\begin{aligned}
W_{kiV}(h_k/3) &= \int_{-h_i}^{h_i} M_{0z'i} K_{kid}(h_k/3, z_i')\, dz_i' \\
&= \Psi^m_{kidV} \sin(2\beta_0 h_k/3) \\
&\quad + \Psi^f_{kidV}[\cos(\beta_0 h_k/3) - \cos \beta_0 h_k] \\
&\quad + \Psi^h_{kidV}[\cos(\beta_0 h_k/6) - \cos(\beta_0 h_k/2)]
\end{aligned}
\tag{6.122}
$$

$$W_{kiV}(2h_k/3) = \int_{-h_i}^{h_i} M_{0z'i} K_{kid}(2h_k/3, z'_i)\, dz'_i$$

$$= \Psi_{kidV}^m \sin(\beta_0 h_k/3)$$
$$+ \Psi_{kidV}^f [\cos(2\beta_0 h_k/3) - \cos \beta_0 h_k]$$
$$+ \Psi_{kidV}^h [\cos(\beta_0 h_k/3) - \cos(\beta_0 h_k/2)]. \quad (6.123)$$

Each integral when evaluated is a complex number. There are, then, three simultaneous complex algebraic equations to evaluate the three complex parameters Ψ_{kidV}^m, Ψ_{kidV}^f, and Ψ_{kidV}^h for each pair of values i and k. A similar second set of three equations is obtained with the different complex numbers $W_{kiU}(0)$, $W_{kiU}(h_k/3)$ and $W_{kiU}(2h_k/3)$ on the left. These are obtained from the same integrals when $M_{0z'i}$ is replaced by $F_{0z'i}$. The simultaneous solution of these three equations for each pair of values i and k yields the complex parameters Ψ_{kidU}^m, Ψ_{kidU}^f and Ψ_{kidU}^h. A third set of three equations is obtained with the quantities $W_{kiD}(0)$, $W_{kiD}(h_k/3)$ and $W_{kiD}(2h_k/3)$ appearing on the left in (6.121)–(6.123). These quantities are defined by the integrals in (6.121)–(6.123) with $M_{0z'i}$ replaced by $H_{0z'i}$. For each pair of values of i and k this third set of three equations yields Ψ_{kidD}^m, Ψ_{kidD}^f and Ψ_{kidD}^h. In this manner all values of the parameters Ψ_{kid} are determined. They have the following forms for each of the subscripts V, U and D on the Ψ's and W's:

$$\Psi_{ki}^m = \Delta^{-1} \begin{vmatrix} W_{ki}(0) & 1 - \cos \beta_0 h_k & 1 - \cos(\beta_0 h_k/2) \\ W_{ki}(h_k/3) & \cos(\beta_0 h_k/3) - \cos \beta_0 h_k & \cos(\beta_0 h_k/6) - \cos(\beta_0 h_k/2) \\ W_{ki}(2h_k/3) & \cos(2\beta_0 h_k/3) - \cos \beta_0 h_k & \cos(\beta_0 h_k/3) - \cos(\beta_0 h_k/2) \end{vmatrix}$$

$$(6.124)$$

$$\Psi_{ki}^f = \Delta^{-1} \begin{vmatrix} \sin \beta_0 h_k & W_{ki}(0) & 1 - \cos(\beta_0 h_k/2) \\ \sin(2\beta_0 h_k/3) & W_{ki}(h_k/3) & \cos(\beta_0 h_k/6) - \cos(\beta_0 h_k/2) \\ \sin(\beta_0 h_k/3) & W_{ki}(2h_k/3) & \cos(\beta_0 h_k/3) - \cos(\beta_0 h_k/2) \end{vmatrix} \quad (6.125)$$

$$\Psi_{ki}^h = \Delta^{-1} \begin{vmatrix} \sin \beta_0 h_k & 1 - \cos \beta_0 h_k & W_{ki}(0) \\ \sin(2\beta_0 h_k/3) & \cos(\beta_0 h_k/3) - \cos \beta_0 h_k & W_{ki}(h_k/3) \\ \sin(\beta_0 h_k/3) & \cos(2\beta_0 h_k/3) - \cos \beta_0 h_k & W_{ki}(2h_k/3) \end{vmatrix} \quad (6.126)$$

where

$$\Delta = \begin{vmatrix} \sin \beta_0 h_k & 1 - \cos \beta_0 h_k & 1 - \cos(\beta_0 h_k/2) \\ \sin(2\beta_0 h_k/3) & \cos(\beta_0 h_k/3) - \cos \beta_0 h_k & \cos(\beta_0 h_k/6) - \cos(\beta_0 h_k/2) \\ \sin(\beta_0 h_k/3) & \cos(2\beta_0 h_k/3) - \cos \beta_0 h_k & \cos(\beta_0 h_k/3) - \cos(\beta_0 h_k/2) \end{vmatrix}$$

$$(6.127)$$

The N simultaneous integral equations for the currents in the elements are

$$\sum_{i=1}^{N} \left\{ A_i \int_{-h_i}^{h_i} M_{0z'i} K_{kid}(z_k, z_i') \, dz_i' + B_i \int_{-h_i}^{h_i} F_{0z'i} K_{kid}(z_k, z_i') \, dz_i' \right.$$

$$\left. + D_i \int_{-h_i}^{h_i} H_{0z'i} K_{kid}(z_k, z_i') \, dz_i' \right\}$$

$$= \frac{j4\pi}{\zeta_0 \cos \beta_0 h_k} [\tfrac{1}{2} V_{0k} M_{0zk} + U_k F_{0zk}] \tag{6.128}$$

$$k = 1, 2, \ldots N$$

where $U_k = \dfrac{-j\zeta_0}{4\pi} \displaystyle\sum_{i=1}^{N} \int_{-h_k}^{h_k} I_{zi}(z_i') K_{ki}(h_k, z_i') \, dz_i'$

$$= \frac{-j\zeta_0}{4\pi} \sum_{i=1}^{N} [A_i \Psi_{kiV}(h_k) + B_i \Psi_{kiU}(h_k) + D_i \Psi_{kiD}(h_k)] \tag{6.129}$$

with $\Psi_{kiV}(h_k)$, $\Psi_{kiU}(h_k)$ and $\Psi_{kiD}(h_k)$ defined in (6.32)–(6.34). If the integrals in (6.128) are replaced by their approximate algebraic equivalents, the following set of algebraic equations for the co-efficients A_i, B_i and D_i is obtained:

$$\sum_{i=1}^{N} \left\{ A_i [\Psi_{kidV}^m M_{0zk} + \Psi_{kidV}^f F_{0zk} + \Psi_{kidV}^h H_{0zk}] \right.$$

$$+ B_i [\Psi_{kidU}^m M_{0zk} + \Psi_{kidU}^f F_{0zk} + \Psi_{kidU}^h H_{0zk}]$$

$$\left. + D_i [\Psi_{kidD}^m M_{0zk} + \Psi_{kidD}^f F_{0zk} + \Psi_{kidD}^h H_{0zk}] \right\}$$

$$= \frac{j4\pi}{\zeta_0 \cos \beta_0 h_k} [\tfrac{1}{2} V_{0k} M_{0zk} + U_k F_{0zk}]. \tag{6.130}$$

Finally, if (6.129) is substituted for U_k, the set of equations may be arranged as follows:

$$M_{0zk} \sum_{i=1}^{N} \left[(A_i \Psi_{kidV}^m + B_i \Psi_{kidU}^m + D_i \Psi_{kidD}^m) \cos \beta_0 h_k - \frac{j2\pi}{\zeta_0} V_{0k} \right]$$

$$+ F_{0zk} \sum_{i=1}^{N} [(A_i \Psi_{kidV}^f + B_i \Psi_{kidU}^f + D_i \Psi_{kidD}^f) \cos \beta_0 h_k$$

$$- A_i \Psi_{kiV}(h_k) - B_i \Psi_{kiU}(h_k) - D_i \Psi_{kiD}(h_k)]$$

$$+ H_{0zk} \sum_{i=1}^{N} [A_i \Psi_{kidV}^h + B_i \Psi_{kidU}^h + D_i \Psi_{kidD}^h] \cos \beta_0 h_k = 0 \tag{6.131}$$

with $k = 1, 2, \ldots N$. These equations are satisfied if the coefficient of each of the three distribution functions is allowed to vanish. The result is a set of $3N$ simultaneous equations for the $3N$ unknown coefficients A, B and D. They are:

$$\sum_{i=1}^{N} [A_i\Psi_{kidV}^m + B_i\Psi_{kidU}^m + D_i\Psi_{kidD}^m] = \frac{j2\pi}{\zeta_0}\frac{V_{0k}}{\cos\beta_0 h_k} \quad (6.132)$$

$$\sum_{i=1}^{N} [A_i\Phi_{kiV} + B_i\Phi_{kiU} + D_i\Phi_{kiD}] = 0 \quad (6.133)$$

$$\sum_{i=1}^{N} [A_i\Psi_{kidV}^h + B_i\Psi_{kidU}^h + D_i\Psi_{kidD}^h] = 0 \quad (6.134)$$

with $k = 1, 2, \dots N$. In (6.133) the following notation has been introduced:

$$\Phi_{kiV} \equiv \Psi_{kiV}(h_k) - \Psi_{kidV}^f \cos\beta_0 h_k \quad (6.135)$$

$$\Phi_{kiU} \equiv \Psi_{kiU}(h_k) - \Psi_{kidU}^f \cos\beta_0 h_k \quad (6.136)$$

$$\Phi_{kiD} \equiv \Psi_{kiD}(h_k) - \Psi_{kidD}^f \cos\beta_0 h_k. \quad (6.137)$$

These equations can be expressed in matrix notation. Let

$$[\Phi] = \begin{bmatrix} \Phi_{11} & \Phi_{12}\dots\Phi_{1N} \\ \vdots & \\ \Phi_{N1} & \dots\quad\Phi_{NN} \end{bmatrix} \quad (6.138)$$

where the Φ_{ki}'s are defined in (6.135)–(6.137) for each subscript V, U and D. Also let

$$[\Psi^h] = \begin{bmatrix} \Psi_{11}^h & \Psi_{12}^h\dots\Psi_{1N}^h \\ \vdots & \\ \Psi_{N1}^h & \dots\quad\Psi_{NN}^h \end{bmatrix}, \quad [\Psi^m] = \begin{bmatrix} \Psi_{11}^m & \Psi_{12}^m\dots\Psi_{1N}^m \\ \vdots & \\ \Psi_{N1}^m & \dots\quad\Psi_{NN}^m \end{bmatrix} \quad (6.139)$$

where the Ψ_{ki}^h are obtained from (6.127). The following column matrices are needed:

$$\{A\} = \begin{Bmatrix} A_1 \\ A_2 \\ \vdots \\ A_N \end{Bmatrix}, \quad \{B\} = \begin{Bmatrix} B_1 \\ B_2 \\ \vdots \\ B_N \end{Bmatrix}, \quad \{D\} = \begin{Bmatrix} D_1 \\ D_2 \\ \vdots \\ D_N \end{Bmatrix} \quad (6.140)$$

$$\left\{\frac{j2\pi}{\zeta_0}\frac{V_0}{\cos\beta_0 h}\right\} = \frac{j2\pi}{\zeta_0}\begin{Bmatrix} V_{01}/\cos\beta_0 h_1 \\ V_{02}/\cos\beta_0 h_2 \\ \vdots \\ V_{0N}/\cos\beta_0 h_N \end{Bmatrix}. \quad (6.141)$$

With this notation, the equivalent matrix equations for determining the coefficients A_i, B_i and D_i are

$$[\Psi_{dV}^m]\{A\} + [\Psi_{dU}^m]\{B\} + [\Psi_{dD}^m]\{D\} = \left\{ \frac{j2\pi}{\zeta_0} \frac{V_0}{\cos \beta_0 h} \right\} \quad (6.142a)$$

$$[\Phi_V]\{A\} + [\Phi_U]\{B\} + [\Phi_D]\{D\} = 0 \quad (6.142b)$$

$$[\Psi_{dV}^h]\{A\} + [\Psi_{dU}^h]\{B\} + [\Psi_{dD}^h]\{D\} = 0. \quad (6.142c)$$

These equations correspond to (6.29) with (6.46) and (6.47) in the simpler case of the Yagi array with two-term fitting of the integrals.

The solutions of (6.132)–(6.134) or (6.142a, b, c) express each of the coefficients A_i, B_i and D_i as a sum of terms in the N voltages $V_{0k}, k = 1, 2, \dots N$. That is

$$A_i = j \frac{2\pi}{\zeta_0} \sum_{k=1}^N \frac{V_{0k}}{\cos \beta_0 h_k} \alpha_{ik} \quad (6.143)$$

$$B_i = j \frac{2\pi}{\zeta_0} \sum_{k=1}^N \frac{V_{0k}}{\cos \beta_0 h_k} \beta_{ik} \quad (6.144)$$

$$D_i = j \frac{2\pi}{\zeta_0} \sum_{k=1}^N \frac{V_{0k}}{\cos \beta_0 h_k} \gamma_{ik} \quad (6.145)$$

where the α_{ik}, β_{ik} and γ_{ik} are the appropriate cofactors divided by the determinant of the system.

It follows that with the coefficients A_i, B_i and D_i evaluated, the currents in all elements are available in the form:

$$I_{zi}(z) = j \frac{2\pi}{\zeta_0} \sum_{k=1}^N \frac{V_{0k}}{\cos \beta_0 h_k} \{\alpha_{ik} \sin \beta_0(h_k - |z|) + \beta_{ik}(\cos \beta_0 z - \cos \beta_0 h_k)$$
$$+ \gamma_{ik}[\cos (\beta_0 z/2) - \cos (\beta_0 h_k/2)]\} \quad (6.146)$$

$$I_{zi}(0) = j \frac{2\pi}{\zeta_0} \sum_{k=1}^N \frac{V_{0k}}{\cos \beta_0 h_k} \{\alpha_{ik} \sin \beta_0 h_k + \beta_{ik}(1 - \cos \beta_0 h_k)$$
$$+ \gamma_{ik}[1 - \cos (\beta_0 h_k/2)]\}$$

$$= \sum_{k=1}^N V_{0k} Y_{ik}. \quad (6.147)$$

In these relations $i = 1, 2, \dots N$, and

$$Y_{ik} = j \frac{2\pi}{\zeta_0 \cos \beta_0 h_k} \{\alpha_{ik} \sin \beta_0 h_k + \beta_{ik}(1 - \cos \beta_0 h_k)$$
$$+ \gamma_{ik}[1 - \cos (\beta_0 h_k/2)]\}. \quad (6.148)$$

The quantities Y_{ik}, with $k = i$, are the self-admittances of the N elements in the array; the Y_{ik}, with $k \neq i$, are the mutual admittances.

They are readily determined from (6.148). Note that in general the self-admittance of an element when coupled to other antennas is not the same as the self-admittance of the same element when isolated.

In matrix form, the equations for the N driving-point currents are

$$\{I_z(0)\} = [Y_A]\{V_0\} \tag{6.149}$$

where
$$\{I_z(0)\} = \begin{Bmatrix} I_{z1}(0) \\ I_{z2}(0) \\ \vdots \\ I_{zN}(0) \end{Bmatrix}, \quad \{V_0\} = \begin{Bmatrix} V_{01} \\ V_{02} \\ \vdots \\ V_{0N} \end{Bmatrix} \tag{6.150}$$

and
$$[Y_A] = \begin{Bmatrix} Y_{11} & Y_{12} \cdots Y_{1N} \\ \vdots & \\ Y_{N1} \cdots & Y_{NN} \end{Bmatrix}. \tag{6.151}$$

The solution for the currents in the N-elements of the array is thus completed in terms of arbitrary voltages. When these are specified, the complete distributions of current are given in the form (6.146). The driving-point admittances Y_{0i} and impedances Z_{0i} are given by

$$Y_{0i} = \frac{I_{zi}(0)}{V_{0i}} = \frac{1}{Z_{0i}}. \tag{6.152}$$

6.12 The log-periodic dipole array

An interesting and important example of a curtain of driven elements that all have different lengths and radii and that are un-equally spaced is the so-called log-periodic dipole array illustrated in Fig. 6.29. In spite of the fact that in this array all elements are connected directly to an active transmission line, its operation when suitably designed is closely related to that of the Yagi-Uda antenna in which only one element is driven and all others are parasitic. However, unlike the Yagi antenna, the log-periodic array has important broad band properties. These are best intro-duced in terms of an array of an infinite number of centre-driven dipoles arranged as shown in Fig. 6.29. Let the half-length of a typical element i be h_i, let its radius be a_i. The distance between element i and the next adjacent element to the right is $b_{i,i+1}$ where

$i = 1, 2, 3, \dots$. The array is constructed so that the following parameters

$$\frac{h_i}{h_{i+1}} = \tau, \qquad \frac{h_{i+1}}{b_{i,i+1}} = \sigma, \qquad 2 \ln \frac{2h_i}{a_i} = \Omega \qquad (6.153)$$

are treated as constants independent of i. As throughout this book, it is assumed that $h_i \gg a_i$.

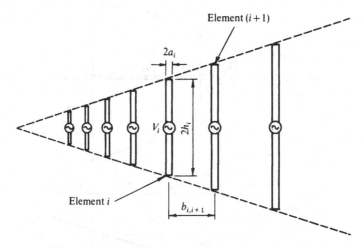

Fig. 6.29.   Seven elements of an infinite log-periodic array.

If the dipoles individually approximate perfect conductors, the electrical properties of the array (such as the driving-point admittances of the elements and the field pattern of the array) at an angular frequency ω_0 depend only on the electrical dimensions $\beta_0 h_i$, $\beta_0 b_{i,i+1}$, and $\beta_0 a_i$ where $\beta_0 = \omega_0/c = 2\pi/\lambda_0$ and c is the velocity of light. If the angular frequency is changed to $\omega_n = \tau^n \omega_0$, where n is a positive or negative integer, the original electrical properties are determined by $\tau^{-n}\beta_n h_i$, $\tau^{-n}\beta_n b_{i,i+1}$ and $\tau^{-n}\beta_n a_i$ where $\beta_n = \omega_n/c$. However, there are along the array antennas with half-lengths $h_{i+n} = \tau^{-n} h_i$ for which $(h_{i+n+1}/b_{i+n,i+n+1}) = \sigma$ and $2 \ln (2h_{i+n}/a_{i+n}) = \Omega$. Since $\beta_0 h_i = \tau^{-n}\beta_n h_i = \beta_n h_{i+n}$, it follows that all properties of the array at the angular frequency ω_0 referred to element i are repeated at the angular frequency ω_n but referred to the element $i+n$. This periodicity of the properties with respect to frequency is linear with respect to the logarithm of the frequency. That is, since $\log \omega_n = \log \omega_0 + n \log \tau$, it is clear that any property shown graphically on a logarithmic frequency scale is periodic with period log τ. Accordingly, arrays with this construction are known as log-periodic dipole arrays [7–10]. Such arrays are generally driven from a two-wire line in the manner illustrated in Figs. 6.30a, b. The arrangement with reversed connections in Fig. 6.30b is the one required for end-fire operation.

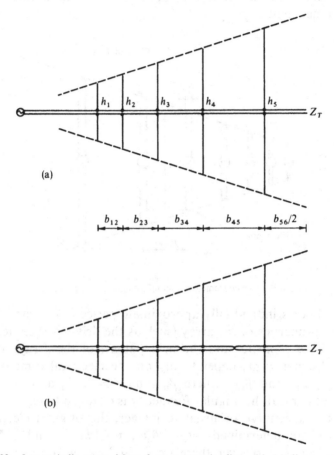

Fig. 6.30.   Log-periodic array driven from a two-wire line with (a) direct connexions, (b) reversed connexions.

Actual arrays are, of course, never infinite so that the ideal frequency-independent properties of the infinite array are modified by asymmetries near the ends. These may be modified by the use of a terminating impedance Z_T as shown in Fig. 6.30 which provides an additional parameter. The value $Z_T = Z_c$, where Z_c is the characteristic impedance of the line, is an obvious choice.

6.13 Analysis of the log-periodic dipole array

The theory developed in section 6.11 for arrays of antennas with unequal lengths, spacings and radii can be applied directly to the log-periodic dipole array. It is only necessary to specify the driving-point voltages to the elements in order to obtain a complete solution for the distributions of current along the elements and their individual input admittances. The driving-point admittance of the array and the complete field pattern are then readily obtained over any frequency range for which the condition $\beta_0 h_i \leqslant 5\pi/4$ is satisfied for all elements. Such quantities as the beam width, the directivity, front-to-back ratio and side-lobe level can, of course, be obtained from the field pattern.

Consider specifically the array shown in Fig. 6.30b. The driving voltage is applied to a transmission-line that is connected successively to all of the elements beginning with the shortest. Between each adjacent pair of elements the connexions are reversed by crossing the conductors of the transmission line in order to achieve the desired phase relations. The analysis of this circuit is conveniently carried out following the method introduced by Carrel [9]. The procedure is simply to determine first the matrix equation for the antenna circuit shown in Fig. 6.31a, then the matrix equation for the transmission-line circuit shown in Fig. 6.31b, and finally the matrix equation for the two circuits in parallel. Note that in Fig. 6.31, a generator is connected across each of the N terminals.

The matrix equation for the antenna circuit in Fig. 6.31a has already been given in (6.149). The elements of the admittance matrix $[Y_A]$ are the self- and mutual admittances of the antenna array.

The matrix equation for the transmission-line circuit in Fig. 6.31b is readily derived. Consider a typical section of the line between the terminal pairs i and $i+1$ which are separated by a length of line $b_{i,i+1}$ as shown in Fig. 6.32. The relations between the current and voltage at terminals i and those at terminals $i+1$ are readily obtained.† For temporary convenience let $d_i = b_{i,i+1}$; also let ϕ be any constant phase-shift introduced between adjacent elements in addition to the value $\beta_0 d$ which is determined by the length of line between elements i and $i+1$.

$$V_i = V_{i+1} \cos{(\beta_0 d_i + \phi)} + jI'_{i+1} R_c \sin{(\beta_0 d_i + \phi)} \quad (6.154a)$$

$$I''_i R_c = jV_{i+1} \sin{(\beta_0 d_i + \phi)} + I'_{i+1} R_c \cos{(\beta_0 d_i + \phi)} \quad (6.154b)$$

where R_c is the characteristic resistance of the lossless line.

† [11], p. 83, equations (6) and (7).

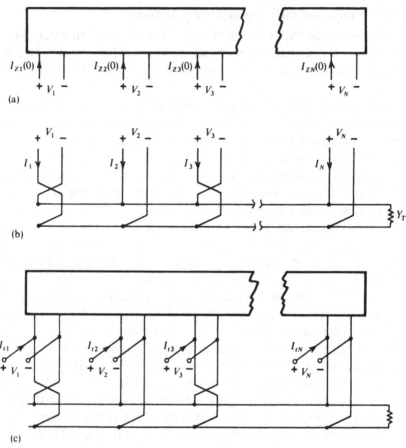

Fig. 6.31. Schematic diagram of (a) the antenna circuit, (b) the transmission-line circuit, and (c) the antenna and transmission-line circuits connected in parallel.

These equations can be rearranged in the form

$$I_i'' = -jG_c[V_i \cot(\beta_0 d_i + \phi) - V_{i+1} \csc(\beta_0 d_i + \phi)] \quad (6.155a)$$

$$I_{i+1}' = -jG_c[V_i \csc(\beta_0 d_i + \phi) - V_{i+1} \cot(\beta_0 d_i + \phi)] \quad (6.155b)$$

where $G_c = R_c^{-1}$ is the characteristic conductance of the lossless line. It follows that

$$I_{i+1}'' = -jG_c[V_{i+1} \cot(\beta_0 d_{i+1} + \phi) - V_{i+2} \csc(\beta_0 d_{i+1} + \phi)]$$
$$(6.156a)$$

$$I_{i+2}' = -jG_c[V_{i+1} \csc(\beta_0 d_{i+1} + \phi) - V_{i+2} \cot(\beta_0 d_{i+1} + \phi)].$$
$$(6.156b)$$

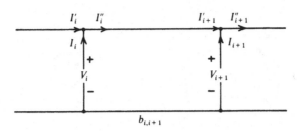

Fig. 6.32. Section of transmission line between the terminal pairs i and $i+1$ when the voltages V_i and V_{i+1} are maintained.

The total current in the generator at the terminals $i+1$ is

$$I_{i+1} = I''_{i+1} - I'_{i+1} = jG_c\{V_i \csc (\beta_0 d_i + \phi) - V_{i+1}[\cot (\beta_0 d_i + \phi)$$
$$+ \cot (\beta_0 d_{i+1} + \phi)] + V_{i+2} \csc (\beta_0 d_{i+1} + \phi)\}. \tag{6.157}$$

In particular, when $\phi = \pi$ as in Fig. 6.31b,

$$I_{i+1} = -jG_c\{V_i \csc \beta_0 b_{i,i+1}$$
$$+ V_{i+1}(\cot \beta_0 b_{i,i+1} + \cot \beta_0 b_{i+1,i+2}) + V_{i+2} \csc \beta_0 b_{i+1,i+2}\}. \tag{6.158}$$

Also $\qquad I_1 = I''_1 = -jG_c[V_1 \cot \beta_0 b_{12} + V_2 \csc \beta_0 b_{12}] \tag{6.159}$

and

$$I_N = -jG_c[V_{N-1} \csc \beta_0 b_{N-1,N} + V_N(\cot \beta_0 b_{N-1,N} + jy_N)] \tag{6.160a}$$

since $\qquad\qquad I''_N = V_N Y_N = V_N y_N G_c \tag{6.160b}$

where $\qquad y_N = Y_N/G_c = \left[\dfrac{Y_T + jG_c \tan \beta_0 b_T}{G_c + jY_T \tan \beta_0 b_T} \right] \tag{6.161}$

is the normalized admittance in parallel with element N. $Y_T = 1/Z_T$ is the admittance terminating the final section of line of length $b_T = b_{N,N+1}/2$.

With (6.158), (6.159) and (6.160), the matrix equation for the transmission line has the form

$$\{I\} = [Y_L]\{V\} \tag{6.162}$$

where $\qquad \{I\} = \left\{ \begin{array}{c} I_1 \\ I_2 \\ \vdots \\ I_N \end{array} \right\}, \qquad \{V\} = \left\{ \begin{array}{c} V_1 \\ V_2 \\ \vdots \\ V_N \end{array} \right\} \tag{6.163}$

250   ARRAYS WITH UNEQUAL ELEMENTS   [6.13

and

$$[Y_L] = -jG_c \begin{bmatrix} \cot\beta_0 b_{12} & \csc\beta_0 b_{12} & 0 & 0 & \cdots \\ \csc\beta_0 b_{12} & (\cot\beta_0 b_{12}+\cot\beta_0 b_{23}) & \csc\beta_0 b_{23} & 0 & \cdots \\ 0 & \csc\beta_0 b_{23} & (\cot\beta_0 b_{23}+\cot\beta_0 b_{34}) & \csc\beta_0 b_{34}\cdots \\ \vdots & \vdots & \vdots & \vdots \\ 0 & 0 & 0 & 0 & \cdots \\ 0 & 0 & 0 & 0 & \cdots \end{bmatrix}$$

$$\begin{bmatrix} 0 & 0 & 0 \\ 0 & 0 & 0 \\ 0 & 0 & 0 \\ \vdots & \vdots & \vdots \\ \csc\beta_0 b_{N-2,N-1} & (\cot\beta_0 b_{N-2,N-1}+\cot\beta_0 b_{N-1,N}) & \csc\beta_0 b_{N-1,N} \\ 0 & \csc\beta_0 b_{N-1,N} & (\cot\beta_0 b_{N-1,N}+jy_N) \end{bmatrix}. \quad (6.164)$$

The final step in the analysis of the array in Fig. 6.30b is to connect the transmission-line circuit in Fig. 6.31b in parallel with the antenna circuit in Fig. 6.31a as shown schematically in Fig. 6.31c. The same driving voltages are maintained across the N input terminals. Let the total currents in the generators be represented by $I_{ti} = I_{zi}(0)+I_i$ where $I_{zi}(0)$ is the current entering antenna i and I_i is the current into the transmission line at terminals i. The matrix equation for the total current is

$$\{I_t\} = ([Y_A]+[Y_L])\{V_0\} = [Y]\{V_0\}. \quad (6.165)$$

This gives the N currents supplied by N generators connected across the N sets of terminals in Fig. 6.31c. In the actual circuit in Fig. 6.30b, there is only one generator, V_{01}, and all of the total currents I_{ti} are zero except I_{t1}. Hence, in (6.165)

$$\{I_t\} = \begin{Bmatrix} I_{t1} \\ 0 \\ 0 \\ \vdots \\ 0 \end{Bmatrix}, \quad \{V_0\} = \begin{Bmatrix} V_{01} \\ \vdots \\ V_{0N} \end{Bmatrix} \quad (6.166)$$

$$[Y] = [Y_A]+[Y_L]. \quad (6.167)$$

The voltages V_{0i} driving the N elements are, therefore, given by

$$\{V_0\} = [Y]^{-1}\{I_t\} \quad (6.168)$$

in terms of the total current I_{t1}. The driving-point admittances of the N elements can be determined as follows. The substitution of

$$\{V_0\} = [Y_A]^{-1}\{I_z(0)\} \tag{6.169}$$

in (6.165) yields

$$\{I_t\} = [U + [Y_L][Y_A]^{-1}]\{I_z(0)\} \tag{6.170}$$

where U is the unit matrix. Note that $[Z_A] = [Y_A]^{-1}$ is the impedance matrix of the array. The equation (6.170) can be solved for the driving-point currents of the several elements in terms of the driving-point current in element 1. Thus,

$$\{I_z(0)\} = [U + [Y_L][Y_A]^{-1}]^{-1}\{I_t\}. \tag{6.171}$$

These currents with a common phase and amplitude reference value are convenient for calculating the field pattern and for comparing relative amplitudes. The admittances of the N elements are

$$Y_{0i} = G_{0i} + jB_{0i} = I_{zi}(0)/V_{0i}, \qquad i = 1, 2, \dots N \tag{6.172}$$

where V_{0i} and $I_{zi}(0)$ are given, respectively, by (6.168) and (6.171). The driving-point admittance of the array at the terminals $i = 1$ of the first element is

$$Y_1 = G_1 + jB_1 \doteq I_{t1}/V_{01}. \tag{6.173}$$

6.14   Characteristics of a typical log-periodic dipole array†

A complete determination of the properties of the log-periodic dipole array involves a systematic study in which the several parameters that characterize its operation are varied progressively over adequately wide ranges. These include the degree of taper of the array ($\tau = h_i/h_{i+1}$), the relative spacing of the elements ($\sigma = h_{i+1}/b_{i,i+1}$), the relative thickness of the elements ($\Omega = 2 \ln (2h_i/a_i)$), the total number of elements N, the normalized admittance ($y_T = Y_T R_c$) terminating the transmission line beyond the Nth element, and the phase shift ϕ introduced between successive elements in addition to that specified by the electrical distance $\beta_0 b_{i,i+1}$ between adjacent elements. Such an investigation could also make use of optimization procedures for the forward gain, front-to-back ratio, band width, and other properties of practical interest in a manner similar to that used earlier in this chapter for the Yagi-Uda array. Use of the formulation of sections 6.11–6.13, which takes full account of the coupling among all elements in determining the different distributions of current and the individual

† This section is based on chapter 9 of [6]. Parts of sections 6.14–6.16 were first published in *Radio Science* [12].

driving-point admittances, should lead to results of considerable quantitative accuracy to supplement those of earlier, more approximate investigations [7–10]. Although no such exhaustive numerical study is yet available, a complete analysis of a typical log-periodic dipole array has been made by Cheong [6] with a high-speed computer. The parameters for this array are $\tau = 0.93$, $\sigma = 0.70$, $\Omega = 11.4$, $N = 12$, $Y_T R_c = 1$, and $\phi = \pi$. The results obtained serve admirably to illustrate both the detailed operation of the log-periodic dipole array and the power of the theory.

Consider first the operation of the array at a frequency† such that an element k near its centre is a half wavelength long. At this frequency the admittances of the twelve elements when individually isolated lie on a curve in the complex admittance plane that is very nearly an arc of a circle that extends on both sides of the axis $B_0 = 0$ as shown in Fig. 6.33. Note that element 7 is nearest to resonance with only a small negative susceptance. The actual admittances $Y_{0i} = G_{0i} + jB_{0i}$ of the same elements when driven as parts of the log-periodic array lie on a curve that departs significantly from the circle for the isolated admittances.‡ It is roughly circular for the group of elements from No. 3 to No. 9, but the circle has a much greater radius than that for the isolated elements. Indeed, it is so great that the conductances of a number of elements (Nos. 2 and 3) are negative. This large difference in the driving-point admittances is due to coupling; it indicates a strong interaction between the currents in this group of elements. Note that element 7 is still very nearly resonant. Since the admittance curve near its ends bends inward and comes quite close to the circle for the isolated elements, it must be concluded that the elements near the ends of the array behave much as if they were individually isolated. This is possible only if their currents are relatively small and contribute little to the properties of the array.

In Fig. 6.34 are shown the magnitudes and relative phase angles‡ of the complex voltages V_{0i} that obtain across the input terminals of the elements in the array. The amplitudes are fairly constant for the shorter capacitive elements but they decrease rapidly as soon as the elements are long enough to pass through resonance and become inductive. The phase of the voltages is seen to shift continuously

---

† Designated as f_{14} in a notation described in section 6.15.

‡ Note that only the plotted points are physically meaningful; the continuous curve serves only to guide the eye.

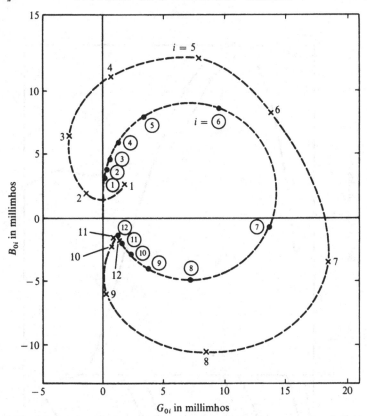

Fig. 6.33. Admittances of elements in a log-periodic dipole array when individually isolated and when in an array with $\tau = 0.93$, $\sigma = 0.7$, $\Omega = 11.4$, $Y_T R_c = 1$, $\phi = \pi$ (operating frequency f_{14}). ●, Isolated admittances; ×, admittances in array.

from element to element along the line. Corresponding curves for the driving-point currents $I_{zi}(0)$ are also in Fig. 6.34. Note particularly that elements 4, 5 and 6 all carry larger currents than element 7 which is nearest resonance. Note also that the phase curve for the current crosses that for the voltage at resonance. The shorter elements have leading (capacitive) currents, the longer elements lagging (inductive) currents. The relative powers† P_{0i} in each element and in the termination are given in Fig. 6.35. Note that in the elements 2 and 3, which have a negative input conductance, the power is negative. This means that power is transferred from the other elements to Nos. 2 and 3 by radiation coupling and then from these back to the feeder. The small rise in voltage shown in Fig. 6.34

† Note that only the plotted points are physically significant. The continuous curve serves merely to guide the eye.

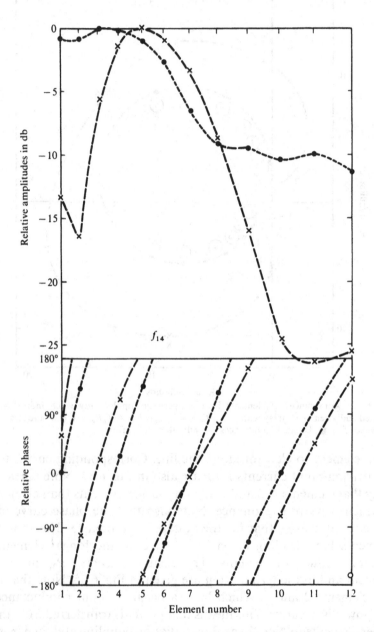

Fig. 6.34. Relative amplitudes and phases of driving-point currents and voltages for a log-periodic dipole array; $\tau = 0.93$, $\sigma = 0.7$, $\Omega = 11.4$, $Y_T R_c = 1$, $\phi = \pi$. ●, driving-point voltages, V_{0i}; ×, driving-point currents, $I_{zi}(0)$.

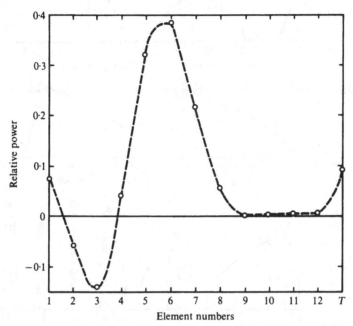

Fig. 6.35.   Relative power in the twelve elements and the termination. (Operating frequency f_{14}.)

at elements 3 and 4 may be ascribed to elements 2 and 3 acting as generators and not as loads. It is significant that the maximum power per element is not in the resonant element 7 but in the shorter elements 5 and 6 which also have larger currents. This is a consequence of the very much smaller voltage maintained across the terminals of element 7 as compared with the voltages across the terminals of elements 5 and 6.

The roles played by the several elements in the array may be seen most clearly from their currents. The distributions of current $I_{0i}(z)$ along all twelve elements are shown in Fig. 6.36a referred to the driving voltage V_{01} at the input terminals of the array. Note these distributions differ greatly from element to element—they are not simple sinusoids. The quantity $I_{0i}(z)/V_{01}$ is represented in its real and imaginary parts; it provides the relative currents that together maintain the electromagnetic field. It is seen that (as predicted from the admittance curves in Fig. 6.33) the currents in the outer elements 1, 2, 9, 10, 11, 12 are extremely small so that their contributions are negligible. Clearly, the distant electromagnetic field is determined essentially by the currents in elements 3 to 8 and of

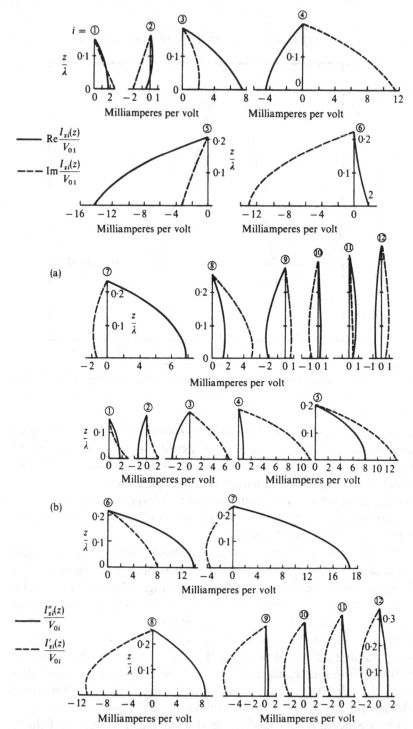

Fig. 6.36. (a) Normalized currents in the elements of a log-periodic dipole array; $I_{zi}(z)/V_{01}$; $\tau = 0.93$, $\sigma = 0.7$, $\Omega = 11.4$, $Y_T R_c = 1$, $\phi = \pi$. (b) Normalized currents in the log-periodic array; $I_{zi}(z)/V_{0i} = [I''_{zi}(z) + jI'_{zi}(z)]/V_{0i}$; $I_{zi}(0)/V_{0i} = Y_{0i} = G_{0i} + jB_{0i}$; $\tau = 0.93$, $\sigma = 0.7$, $\Omega = 11.4$, $Y_T R_c = 1$, $\phi = \pi$.

these elements 4, 5, 6 and 7 predominate. Note in particular that the currents in the shorter-than-resonant elements 4, 5 and 6 actually exceed the current in the practically resonant element 7.

The current distributions are also shown in Fig. 6.36b but each current is now referred to its own driving voltage. Thus, the quantities represented are $I_{zi}(z)/V_{0i} = [I''_{zi}(z) + jI'_{zi}(z)]/V_{0i}$ where $I''_{zi}(z)$ is the component in phase with V_{0i}, $I'_{zi}(z)$ the component in phase quadrature. Note that $I_{zi}(0)/V_{0i} = Y_{0i}$ so that $I''_{zi}(0)/V_{0i} = G_{0i}$ and $I'_{zi}(0)/V_{0i} = B_{0i}$. The power in antenna i is $P_{0i} = |V_{0i}|^2 G_{0i}$, but since the value of V_{0i} differs greatly from element to element as seen in Fig. 6.34, the relative powers in the several elements are not proportional simply to the real parts of the currents $I''_{zi}(0)$ in the terminals. However, the distributions in Fig. 6.36b are instructive since they show the negative real parts for elements 2 and 3 that transfer power to the feeding line. They also show that the imaginary parts of the currents in elements 1 to 6 are capacitive, those in elements 7 to 12 inductive. This means that each of the elements 1 to 6 acts as a director for the elements to its right, whereas each of the elements 7 to 12 acts as a reflector for all elements to its left. Actually, the capacitive components of current in elements 3, 4 and 5 exceed the conductive components so that relatively little power is supplied to them from the line, and they behave substantially like parasitic directors. The inductive component of current predominates in elements 8 to 12 and these act in major part like parasitic reflectors. However, since the amplitudes of the currents in elements 9 to 12 are quite small, it is clear that the principal reflector action comes from element 8. In summary, Figs. 6.36a, b indicate that of the twelve elements numbers 1, 2, 9, 10, 11 and 12 may be ignored since their currents are small; elements 5, 6, 7 are supplied most of the power from the feeder and behave primarily like driven antennas in an endfire array; elements 3 and 4 act predominantly like parasitic directors; and element 8 is essentially a parasitic reflector. Thus, the log-periodic antenna is very much like a somewhat generalized Yagi-Uda array when driven at a frequency for which the antenna closest to resonance is not too near the ends and the array is long enough to include relatively inactive elements at each end. A lengthening of the array by the addition of one or two or even a great many more elements at either end or at both ends cannot significantly modify the circuit or field properties of the array at the particular frequency since these are determined by the active group.

The normalized far-field pattern in the equatorial or H-plane (variable Φ with $\Theta = \pi/2$) is shown in Fig. 6.37. Note the smoothness of the pattern and the very small minor lobes. As is to be expected this low minor-lobe level is achieved at the expense of the beam width. A comparison with the field pattern in Fig. 6.28 for a 10-element Yagi-Uda array shows that the latter has larger minor lobes but a much narrower beam. However, the Yagi-Uda array does not have the important frequency-independent properties of the log-periodic dipole array.

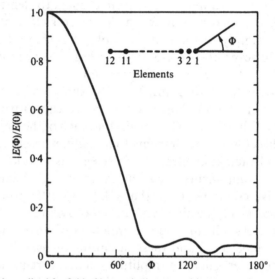

Fig. 6.37.   Normalized far field of log-periodic array with currents shown in Fig. 6.36a.

6.15   Frequency-independent properties of the log-periodic dipole array

The principle underlying the properties of the log-periodic dipole array when driven at the terminals of the shortest element as shown in Fig. 6.30b and operated as illustrated in the preceding section depends upon the following: (1) A small group of about seven dipoles constitutes the active or radiating part of the array. These may be described approximately as including (a) three strongly driven and radiating elements near resonance, (b) three shorter elements each of which combines the functions of a rather weakly driven antenna and a highly active parasitic director, and (c) one longer antenna that acts both as a weakly driven element and a strong parasitic reflector. (2) All other elements in the array and the terminating admittance Y_T have such small currents and so little power that they may be ignored both as loads on the feeding line and as contributing radiators of the far-zone field. (3) The driving-point admittance of the array at the terminals of the shortest element is approximately equal to the characteristic conductance G_c of the transmission line. (4) The currents in the active elements maintain a unilateral endfire field pattern with very small minor lobes.

The effect of a change in frequency is to shift the active group toward the terminated end with longer elements when the frequency is lowered. So long as the frequency range is bounded so that neither the shortest nor the longest element in the array is a part of the active group, there can be no significant change in either the circuit or the field properties. The array must behave substantially as if infinitely long. On the other hand, as the frequency is increased or decreased sufficiently to make the element at either end of the array a member of the active group, all of the properties of the array must begin to change. This change becomes drastic when the frequency is varied so much that none of the N elements is near resonance.

The general behaviour of the twelve-element log-periodic dipole array as a function of frequency has been investigated by Cheong [6] using a discrete set of frequencies $f_1, f_2, \ldots f_{27}$. These are chosen so that the lowest frequency f_1 is below the resonant frequency of the longest element No. 12 and the highest frequency f_{27} is above the resonant value for the shortest element No. 1 as shown in the table on page 260. In order to distribute the frequencies according to the log-periodic scheme of lengths and spacings, the ratio factor $\sqrt{0.93}$ was chosen so that $f_{j+2}/f_j = 0.93$ where j is an integer. This provides an intermediate frequency step $f_{j+1}/f_j = \sqrt{0.93}$ to achieve a closer approximation of a continuous spectrum. The properties of the array described in the preceding section and represented in Figs. 6.33–6.37 are obtained specifically at the centre frequency f_{14} in this set for which an element (No. 7) near the middle of the array is most nearly resonant.

Consider first a decrease in frequency from f_{14} to f_7 so that resonance is moved from approximately element 7 to approximately element 10. The corresponding driving-point admittances are shown in the complex admittance plane in Fig. 6.38 together with the admittances of the elements when these are individually isolated. The admittance circle for the isolated antennas and the

Relation between the relative heights of the elements, h/λ, and the frequencies f_i.

i in f_i	h_1/λ	h_{12}/λ	f_i when $h_1 = 1$ m
1	0·0962	0·2138	28.86 MHz
2	0·0998	0·2217	29.94
3	0·1035	0·2299	31·05
4	0·1073	0·2384	32·19
5	0·1113	0·2473	33·39
6	0·1154	0·2564	34·62
7	0·1197	0·2659	35·91
8	0·1241	0·2757	37·23
9	0·1287	0·2859	38·61
10	0·1335	0·2966	40·05
11	0·1384	0·3075	41·52
12	0·1435	0·3188	43·05
13	0·1488	0·3306	44·64
14	0·1543	0·3428	46·29
15	0·1600	0·3555	48·00
16	0·1659	0·3686	49·77
17	0·1721	0·3824	51·63
18	0·1785	0·3966	53·55
19	0·1850	0·4110	55·50
20	0·1918	0·4261	57·54
21	0·1989	0·4419	59·67
22	0·2063	0·4583	61·89
23	0·2139	0·4752	64·17
24	0·2218	0·4928	66·54
25	0·2300	0·5110	69·00
26	0·2385	0·5299	71·55
27	0·2473	0·5494	74·19

admittance curve† for the array resemble those in Fig. 6.33 but appear to have been moved in a counter-clockwise direction. The admittances of the short elements from 1 to 6 now form a small spiral around the values for the same elements when isolated. The previous tight little spiral of admittances for the longer elements in Fig. 6.33 is completely unwound and the admittance curve for the array no longer comes near to the circle for the admittances of the isolated elements. It is clear that in Fig. 6.38 elements 6 to 12 instead of 3 to 9 as in Fig. 6.33 form the active group. This is further confirmed in Fig. 6.39 which shows the voltages and currents at the driving points of the elements. The voltage amplitudes are quite constant from elements 1 to 8, then decrease rapidly. The associated current amplitudes are small for elements 1 to 6, large for elements 6 to 11 and again small for element 12. Evidently, with reference to

† Note that only the plotted points are physically significant. The continuous curve serves merely to guide the eye.

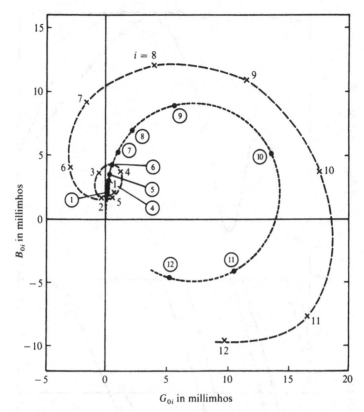

Fig. 6.38.   Like Fig. 6.33 but for lower frequency with resonance near element 10. (Operating frequency f_7.) ●, Isolated admittances; ×, admittances in array.

Fig. 6.39 (and Fig. 6.34), the group consisting of director-radiators 6, 7 and 8 (instead of 3, 4, 5), radiators 9, 10, 11 (instead of 6, 7, 8) and reflector-radiator 12 (instead of 9) are primarily responsible for the properties of the array. These conclusions may also be reached from a study of the current-distribution curves for $I_{zi}(z)/V_{01}$ in Fig. 6.40a and for $I_{zi}(z)/V_{0i}$ in Fig. 6.40b. The former show clearly that the amplitudes of the currents in elements 1 through 5 are negligibly small. The latter indicate the following: the capacitive currents dominate in elements 6, 7 and 8, in element 9 the capacitive and conductive currents are practically equal, element 10 is nearly resonant with a very small capacitive current, element 11 has large inductive and conductive components, and in element 12 the inductive current exceeds the conductive component. It may be concluded, therefore, that the decrease in frequency which moved

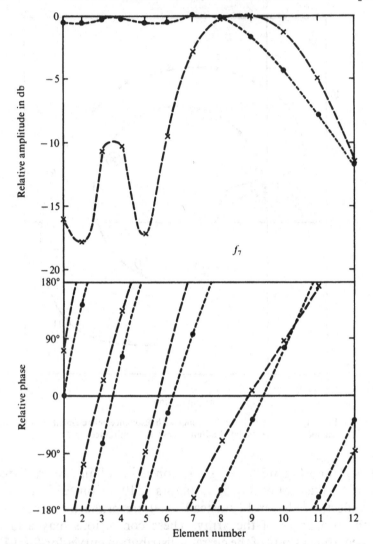

Fig. 6.39.   Like Fig. 6.34 but for lower frequency with resonance near element 10.
●, Driving-point voltages, V_{0i}; ×, driving-point currents, $I_{zi}(0)$.

resonance from near element 7 to near element 10 has not significantly changed the properties of the active group and, hence, of the array.

If the frequency is decreased still further to f_3 at which even element No. 12 is too short to be resonant, the admittance curve is that shown in Fig. 6.41. The counter-clockwise rotation of the curves has been increased beyond that in Fig. 6.38 so that now none of the

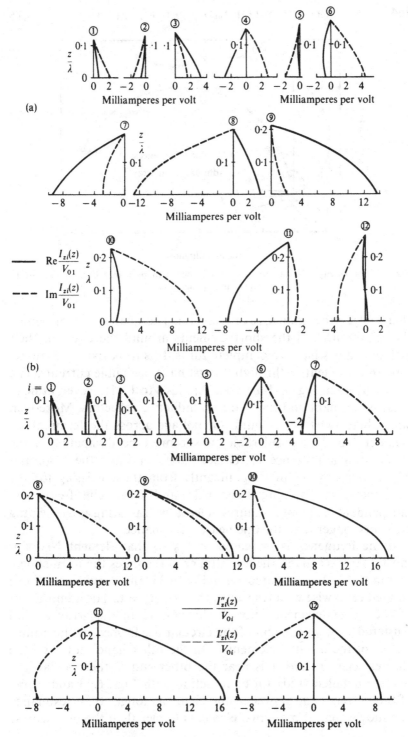

Fig. 6.40. (a) Like Fig. 6.36a but for lower frequency with resonance near element 10.
(b) Like Fig. 6.36b but for lower frequency with resonance near element 10.

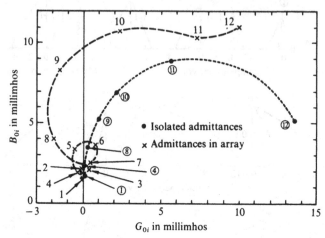

Fig. 6.41.   Like Fig. 6.38 but for lower frequency with resonance beyond element 12.
(Operating frequency f_3.)

elements is either inductive or resonant. The small spiral formed by
the admittances of the short elements around the circle of their
isolated values has two complete turns. It is to be expected, therefore, that elements 1 through 7 must have negligible currents. The
active group in Fig. 6.41 includes dipoles 8 to 12. However, none of
these is resonant and there are no inductive reflectors. Moreover,
since there must be a significant voltage across the terminals of
element No. 12, considerable power must be dissipated in the
terminating admittance Y_T. Under these conditions the properties
of the array must differ significantly from those existing for the
frequencies determining Figs. 6.33 and 6.38. The frequency-
independent behaviour requires at least two radiating and reflecting
elements longer than the one nearest resonance.

  If the frequency is increased to f_{19} so that element No. 4 is
most nearly resonant, the admittance curve takes the form shown
in Fig. 6.42. As compared with Fig. 6.33, the curves have been
rotated clockwise with respect to the axis $B_0 = 0$. The admittances
of the longer elements Nos. 8 through 12 in the array are all
clustered close to one end of the circular arc formed by the admittances of the isolated elements. On the other hand, not even the
shortest element No. 1 is near the other end of the circular arc.
Since a detailed study (in conjunction with Figs. 6.33 and 6.36a)
of the currents and power in the elements longer than resonance
has shown that at most two elements longer than the one nearest

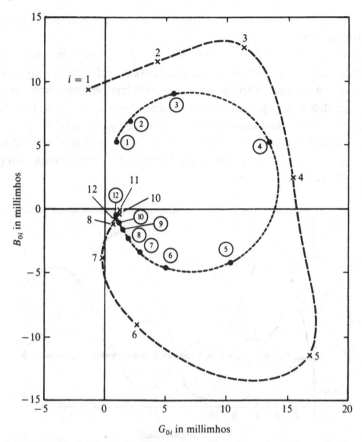

Fig. 6.42. Like Fig. 6.33 but for higher frequency with resonance near element 4. (Operating frequency f_{19}.) ●, Isolated admittances; ×, admittances in array.

resonance carry significant currents, it follows that all elements from No. 12 down through No. 7 play no significant role in the array. On the other hand, it is clear from Fig. 6.42 that the admittance of element No. 1 does not produce a curve that bends inward toward the circular arc of isolated admittances, but rather outward away from the arc. This is a consequence of the fact that the region of active elements has been moved too close to the end of the array. It is clear from Fig. 6.36a that the active region includes at least four elements shorter than the one nearest resonance. For the frequency f_{19} leading to Fig. 6.42 there are only three such elements available. This means that the frequency responsible for Fig. 6.42 is already somewhat higher than acceptable for the frequency-independent properties of the array and that the currents in element

No. 1 must differ from the expected since one of the required director-radiators is missing.

The useful range for a frequency-independent behaviour lies between the frequencies at which elements 5 and 10 (or, in general, $N-2$) are resonant. In the scale of discrete frequencies used for the twelve-element array this range is approximately $f_7 \leqslant f \leqslant f_{17}$. The power in the several elements at the frequencies f_3, f_7, f_{14}, f_{19} and f_{23} is shown in Figs. 6.35 and 6.43. Note that in these figures only the plotted points are significant. The connecting curves serve only to guide the eye.

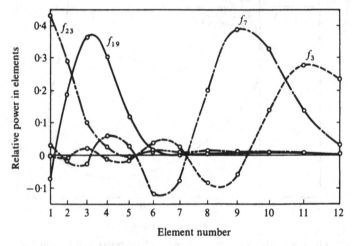

Fig. 6.43.   Like Fig. 6.35 but for frequencies f_3, f_7, f_{19}, and f_{23}.

A detailed study of the operation of the twelve-element array over the full range of frequencies from f_1 to f_{27} has been made by Cheong [6]. Important results in addition to those already discussed are contained in Figs. 6.44–6.46. They may be summarized as follows:

1. As shown in Fig. 6.44, curve T, a large fraction of the total power is dissipated in the terminating admittance $Y_T = G_c$, in the ranges $f < f_5$ and $f > f_{26}$. As a consequence only a small fraction of power appears in the dipoles so that little is radiated. It is also clear from Fig. 6.44 that in the range $f_5 \leqslant f \leqslant f_{26}$ only a small part of the power is dissipated in the terminating admittance, most of it appears in and is radiated from a relatively small group of active dipoles near resonance.

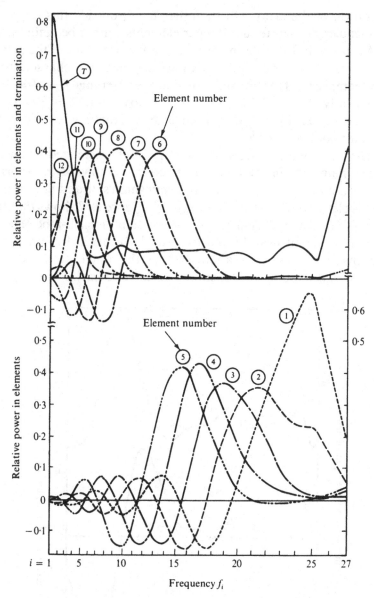

Fig. 6.44.   Relative power in the elements and the termination.

2. In the range $f_5 \leqslant f \leqslant f_{17}$ elements which have half-lengths h_i in the range $0.18 \leqslant h_i/\lambda \leqslant 0.255$ form the active group. Resonance occurs with $h_i/\lambda \doteq 0.216$. Elements which have half-lengths h_i less than 0.18λ or greater than 0.255λ play an insignificant part in the

operation of the array. On the other hand, outside this range of frequencies the shorter and longer elements cannot be ignored.

3. As shown in Fig. 6.45 the driving-point admittance of the array, Y_t, is reasonably constant at a value very near the characteristic conductance G_c of the transmission line over the range $f_5 \leqslant f \leqslant f_{17}$. Specifically $Y_T \doteq (23 \cdot 0 + j0 \cdot 0) \times 10^{-3}$ mhos with $G_c = 20 \times 10^{-3}$ mhos. Outside this range of frequencies Y_T varies widely in both real and imaginary parts.

4. The band of frequencies $f_5 \leqslant f \leqslant f_{17}$ is characterized by a very stable main lobe in the forward direction, i.e. toward the shorter elements and the driving point, and very small side and back lobes. This is clear from Fig. 6.37 and Fig. 6.46. Figure 6.47 shows that the ratio of the forward field to the largest side or back-lobe level is roughly constant near 15 and that the 3 db forward beam width remains quite stable at about 38° in the range $f_5 \leqslant f \leqslant f_{17}$. Outside this band of frequencies large side and back lobes appear.

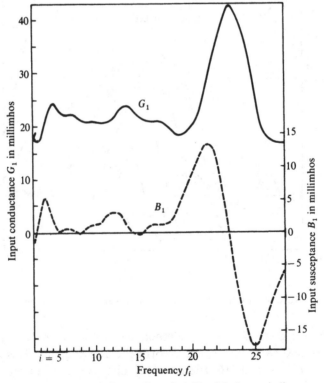

Fig. 6.45.   Input admittance $Y_1 = G_1 + jB_1$ of the log-periodic array.

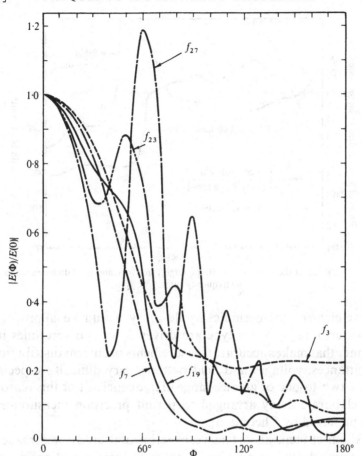

Fig. 6.46.   Like Fig. 6.37 but for a number of different frequencies.

It is important to note that all of the computed data apply to a particular array with a single set of values of the basic parameters τ, σ, Ω, Y_T, R_c and ϕ. An extensive study by high-speed computer of the effects of changes in these parameters and of optimum designs based on the three-term theory is indicated but is not available at the time of this writing. Additional information is given in Cheong [6], Cheong and King [12], and Carrel [9].

6.16   Experimental verification of the theory for arrays of unequal dipoles

In order to verify experimentally the predictions of the general theory developed in section 6.11 for arrays of dipoles with a wide range of lengths and spacings, a series of measurements on the

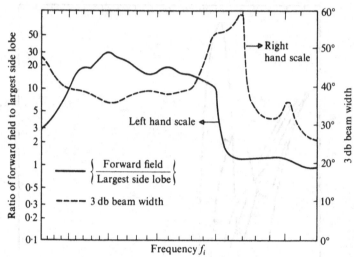

Fig. 6.47.   Ratio of the forward field to the largest side lobe and the 3 db beam width over the frequency range f_1 to f_{27}.

twelve-element log-periodic dipole array would be appropriate. However, arrays of this type are driven from two-wire lines in a manner that makes accurate measurements of current distributions, admittances, voltages and field patterns very difficult—especially over a two-to-one or greater range of frequencies. For this reason a less elaborate array arranged to permit precision measurements was preferred by Cheong [6].

As a first step, an extensive experimental study was made of two-coupled dipoles over wide ranges of lengths and spacings in order to verify the adequency of the three-term representation of the currents. When this had been established, a complete array of five elements was constructed after the log-periodic design with the longest element approximately twice as long as the shortest element. This array consisted of monopoles over a very large ground screen. Each element was the extension of the inner conductor of a coaxial line of which the outer conductor pierced the metal ground screen. In order to provide an equivalent for the reversal of the connexions between adjacent pairs of elements, provision was made to permit the insertion of an arbitrary length of coaxial line in addition to a length equal to the spacing of the elements. Since the added phase shift had to be exactly π for each different frequency, it was necessary to readjust the length of the sections of coaxial line between the elements for each frequency.

Careful measurements were made of the driving-point admittance, the currents and voltages in amplitude and phase at the base of each element, and the field pattern over a range of some 17 different frequencies that included resonance for the longest and the shortest elements. The agreement between theory and measurement was remarkable in all details, thus confirming the adequacy of the theory for use not only on the five-element array but on an array of any type that satisfies the requirements of the theory. Details and extensive graphs are in the work of Cheong [6] and Cheong and King [13].

CHAPTER 7

PLANAR AND THREE-DIMENSIONAL ARRAYS

The study of dipole arrays in chapters 3 through 6 has proceeded from simpler to more complicated configurations. In chapters 3 and 4 all elements are physically alike and arranged to be parallel with their centres uniformly spaced around a circle so that when driven in suitable phase sequences all elements are geometrically and electrically identical. Chapter 5 is also concerned with parallel elements that are structurally alike, but they lie in a curtain with their centres along a straight line of finite length; consequently the electromagnetic environments of the several elements are not all the same. In chapter 6 the requirement that the elements in a curtain array be equal in length is omitted and consideration is given first to arrays of elements that differ only moderately in length, then to arrays in which not only the lengths but also the radii of the elements and the distances between them vary widely. The lifting of each restriction introduces additional complications in the approximate representation of the currents on the elements by simple trigonometric functions and in the reduction of the integrals in the simultaneous integral equations to sums of such functions with suitably defined complex coefficients.

The final generalization which is carried out in this chapter is the omission of the requirement maintained throughout the book until this point, that all elements be non-staggered. The removal of this condition leads to the discussion of arrays of parallel elements that are arranged in a plane as in Fig. 7.1 and in three dimensions as shown in Fig. 7.2. Note that such arrays include arbitrarily staggered elements and collinear elements which do not occur in the circular and curtain arrays considered in chapters 3 through 6. When the centres of the elements are displaced from a common plane, the halves of many antennas are in different electrical environments so that an even symmetry with respect to their individual centres no longer obtains for the distributions of current. An important new complication is thus introduced:

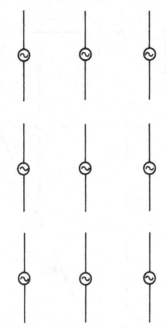

Fig. 7.1.   Planar array of nine identical elements.

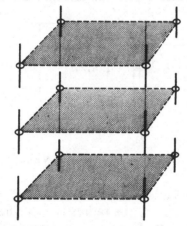

Fig. 7.2.   Three-dimensional array of twelve identical elements.

components of current with odd symmetries in addition to those with even symmetries.

7.1 Vector potentials and integral equations for the currents

Four typical elements in an array of N parallel dipoles are shown in Fig. 7.3. All antennas have their axes parallel to the Z-axis

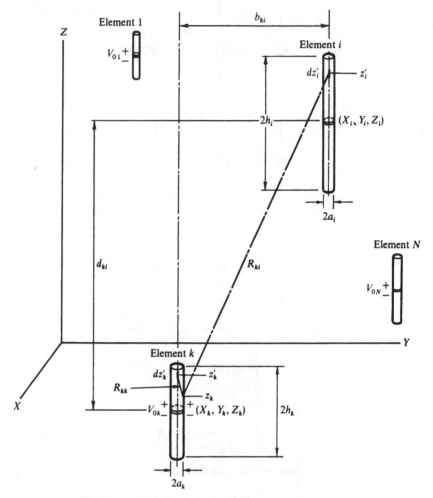

Fig. 7.3.   Typical elements in an array of N parallel antennas.

of a system of rectangular coordinates X, Y, Z. The centre of the kth element is at X_k, Y_k, Z_k; its radius is a_k, its half-length h_k, it is centre driven by a delta-function generator with EMF V_{0k}. As before, the antennas are assumed to be perfectly conducting and electrically thin so that $\beta_0 a_k \ll 1$ for $k = 1, 2, ..., N$. A local axial coordinate z_k has its origin at the centre of element k.

The vector potential on the surface of antenna k no longer has the simple form given in (2.3), since the even symmetry conditions $I_{zk}(-z_k) = I_{zk}(z_k)$ for the current and $A_{zk}(-z_k) = A_{zk}(z_k)$ for the vector potential no longer apply. However, the vector potential can be resolved into two parts, one with even symmetry, the other with odd symmetry. Thus

$$A_{zk}(z_k) = A_{zk}^{even}(z_k) + A_{zk}^{odd}(z_k) \tag{7.1}$$

where, in the range $-h_k \leqslant z_k \leqslant h_k$,

$$A_{zk}^{even}(z_k) = \frac{-j}{c}[C_{k1} \cos \beta_0 z_k + \tfrac{1}{2} V_{0k} \sin \beta_0 |z_k|] \tag{7.2}$$

as in (2.3), and

$$A_{zk}^{odd}(z_k) = \frac{-j}{c} C_{k2} \sin \beta_0 z_k. \tag{7.3}$$

The vector potential on the surface of antenna k is also given by the sum of integrals,

$$A_{zk}(z_k) = \sum_{i=1}^{N} \frac{\mu_0}{4\pi} \int_{-h_i}^{h_i} I_{zi}(z_i') G_{ki}(d_{ki}, z_k, z_i') \, dz_i' \tag{7.4}$$

where

$$G_{ki}(d_{ki}, z_k, z_i') = \frac{e^{-j\beta_0 R_{ki}}}{R_{ki}} \tag{7.5a}$$

with

$$R_{ki} = \sqrt{(d_{ki} + z_i' - z_k)^2 + b_{ki}^2}. \tag{7.5b}$$

As shown in Fig. 7.3, $d_{ki} = |Z_k - Z_i|$ is the axial distance between the plane containing the centres of elements k and i, $d_{kk} = 0$; $b_{ki} = \sqrt{(X_k - X_i)^2 + (Y_k - Y_i)^2}$, $i \neq k$, is the distance between the centre of element k and the projection of the centre of element i onto the plane $z_k = 0$; $b_{kk} = a_k$. The currents $I_{zi}(z_i)$ in the N elements that generate the vector potential on the surface of antenna k as given in (7.4) include even and odd parts with respect to the centres of the respective elements. That is,

$$I_{zi}(z_i) = I_{zi}^{even}(z_i) + I_{zi}^{odd}(z_i) \tag{7.6}$$

where

$$I_{zi}^{even}(z_i) = (1/2)[I_{zi}(z_i) + I_{zi}(-z_i)], \quad I_{zi}^{odd}(z_i) = (1/2)[I_{zi}(z_i) - I_{zi}(-z_i)].$$

In order to separate the even and the odd parts of the vector potential in (7.4), the kernel $G_{ki}(z_k, z_i')$ in (7.4) must be separated into its even and odd parts. Thus,

$$G_{ki}(d_{ki}, z_k, z_i') = G_{ki}^{even}(d_{ki}, z_k, z_i') + G_{ki}^{odd}(d_{ki}, z_k, z_i') \tag{7.7}$$

where, as is readily shown,

$$G_{ki}^{even}(d_{ki}, z_k, z_i') = \tfrac{1}{2}[K_{ki}(z_k - d_{ki}, z_i') + K_{ki}(z_k + d_{ki}, z_i')] \tag{7.8}$$

$$G_{ki}^{odd}(d_{ki}, z_k, z_i') = \tfrac{1}{2}[K_{ki}(z_k - d_{ki}, z_i') - K_{ki}(z_k + d_{ki}, z_i')]. \tag{7.9}$$

The function K occurring in (7.8) and (7.9) is the kernel previously used for non-staggered arrays, viz.,

$$K_{ki}(z_k, z_i') = K_{kiR}(z_k, z_i') + jK_{kiI}(z_k, z_i') = \frac{e^{-j\beta_0\sqrt{(z_k - z_i')^2 + b_{ki}^2}}}{\sqrt{(z_k - z_i')^2 + b_{ki}^2}}. \qquad (7.10)$$

Note that when $d_{ki} = 0$, $G_{ki}^{\text{even}}(0, z_k, z_i') = K_{ki}(z_k, z_i')$ and $G_{ki}^{\text{odd}}(0, z_k, z_i') = 0$, as required for the previously analysed non-staggered array. By means of the obvious relation,

$$K_{ki}(z_k, z_i') = K_{ki}(-z_k, -z_i'),$$

it is readily shown that, when (7.6) and (7.7) are substituted in (7.4), the parts of the integral that involve the products $I^{\text{even}}G^{\text{even}}$, $I^{\text{odd}}G^{\text{odd}}$ are themselves even in z, the parts that contain $I^{\text{even}}G^{\text{odd}}$, $I^{\text{odd}}G^{\text{even}}$ are themselves odd in z. It follows that the even part of the vector potential is given by

$$4\pi\mu_0^{-1}A_{zk}^{\text{even}}(z_k) = \int_{-h_k}^{h_k} I_{zk}^{\text{even}}(z_k')G_{kk}^{\text{even}}(0, z_k, z_k')\, dz_k'$$

$$+ \sum_{i=1}^{N}{}' \int_{-h_i}^{h_i} I_{zi}^{\text{even}}(z_i')G_{ki}^{\text{even}}(d_{ki}, z_k, z_i')\, dz_i'$$

$$+ \sum_{i=1}^{N}{}' \int_{-h_i}^{h_i} I_{zi}^{\text{odd}}(z_i')G_{ki}^{\text{odd}}(d_{ki}, z_k, z_i')\, dz_i'$$

$$= \frac{-j4\pi}{\zeta_0}[C_{k1}\cos\beta_0 z_k + (1/2)V_{0k}\sin\beta_0|z_k|] \qquad (7.11)$$

where $k = 1, 2, \dots N$; $\zeta_0 \doteq 120\pi$ ohms; and Σ' is the sum with $i = k$ omitted. The odd part of the vector potential is contained in

$$4\pi\mu_0^{-1}A_{zk}^{\text{odd}}(z_k) = \int_{-h_k}^{h_k} I_{zk}^{\text{odd}}(z_k')G_{kk}^{\text{even}}(0, z_k, z_k')\, dz_k'$$

$$+ \sum_{i=1}^{N}{}' I_{zi}^{\text{even}}(z_i')G_{ki}^{\text{odd}}(d_{ki}, z_k, z_i')\, dz_i'$$

$$+ \sum_{i=1}^{N}{}' I_{zi}^{\text{odd}}(z_i')G_{ki}^{\text{even}}(d_{ki}, z_k, z_i')\, dz_i'$$

$$= -(j4\pi/\zeta_0)C_{k2}\sin\beta_0 z_k \qquad (7.12)$$

where $k = 1, 2, \dots N$.

The relations on the right in (7.11) and (7.12) are $2N$ simultaneous integral equations for the even and odd parts of the currents in the N elements.

7.2   Vector potential differences and integral equations

In order to determine approximate distributions of current from the two sets of N simultaneous integral equations in the general manner described in earlier chapters, it is convenient to introduce the vector potential differences. This is quite straight forward for the even part of $A_{zk}(z_k)$. Thus, if $4\pi\mu_0^{-1}A_{zk}(h_k)$ is subtracted from both sides of (7.11), the result is

$$4\pi\mu_0^{-1}[A_{zk}^{\text{even}}(z_k) - A_{zk}^{\text{even}}(h_k)] = \int_{-h_k}^{h_k} I_{zk}^{\text{even}}(z_k')G_{kkd}^{\text{even}}(0, z_k, z_k')\, dz_k'$$

$$+ \sum_{i=1}^{N}{}' \int_{-h_i}^{h_i} I_{zi}^{\text{even}}(z_i')G_{kid}^{\text{even}}(d_{ki}, z_k, z_i')\, dz_i'$$

$$+ \sum_{i=1}^{N}{}' \int_{-h_i}^{h_i} I_{zi}^{\text{odd}}(z_i')G_{kid}^{\text{odd}}(d_{ki}, z_k, z_i')\, dz_i'$$

$$= -(j4\pi/\zeta_0)[\tfrac{1}{2}V_{0k}(\sin\beta_0|z_k| - \sin\beta_0 h_k)$$

$$+ C_{k1}(\cos\beta_0 z_k - \cos\beta_0 h_k)]$$

$$= (j4\pi/\zeta_0 \cos\beta_0 h_k)[\tfrac{1}{2}V_{0k}\sin\beta_0(h_k - |z_k|)$$

$$+ U_k(\cos\beta_0 z_k - \cos\beta_0 h_k)] \qquad (7.13)$$

where $k = 1, 2, \dots N$ and

$$U_k = \frac{-j\zeta_0}{\mu_0}A_{zk}^{\text{even}}(h_k), \qquad C_{k1} = \frac{-U_k + (1/2)V_{0k}\sin\beta_0 h_k}{\cos\beta_0 h_k} \qquad (7.14)$$

as in the corresponding equation with $d_{ki} = 0$ for the curtain array. The difference kernel (with extra subscript d) is defined by

$$G_{kid}^{\text{even}}(d_{ki}, z_k, z_i') = G_{ki}^{\text{even}}(d_{ki}, z_k, z_i') - G_{ki}^{\text{even}}(d_{ki}, h_k, z_i'). \qquad (7.15)$$

It is not possible to form an equation like (7.13) with $A_{zk}^{\text{odd}}(z_k)$ since this is an odd function of z_k so that if $A_{zk}^{\text{odd}}(z_k) - A_{zk}^{\text{odd}}(h_k)$ is zero at $z_k = h_k$, it is $-2A_{zk}^{\text{odd}}(h_k)$ at $z_z = -h_k$. A convenient alternative† is to subtract the odd function $(z_k/h_k)A_{zk}^{\text{odd}}(h_k)$ which is equal to the vector potential at both $z_k = h_k$ and $z_k = -h_k$.

† See [1].

Thus, with (7.12),

$$4\pi\mu_0^{-1}[A_{zk}^{odd}(z_k) - (z_k/h_k)A_{zk}^{odd}(h_k)]$$

$$= \int_{-h_k}^{h_k} I_{zk}^{odd}(z_k')\mathscr{G}_{kkd}^{even}(0, z_k, z_k')\,dz_k'$$

$$+ \sum_{i=1}^{N} \int_{-h_i}^{h_i} I_{zi}^{even}(z_i')\mathscr{G}_{kid}^{odd}(d_{ki}, z_k, z_i')\,dz_i'$$

$$+ \sum_{i=1}^{N} \int_{-h_i}^{h_i} I_{zi}^{odd}(z_i')\mathscr{G}_{kid}^{even}(d_{ki}, z_k, z_i')\,dz_i'$$

$$= -(j4\pi C_{k2}/\zeta_0)[\sin\beta_0 z_k - (z_k/h_k)\sin\beta_0 h_k] \quad (7.16)$$

where $k = 1, 2, \ldots N$ and the difference kernels are given by

$$\mathscr{G}_{kid}^{even}(d_{ki}, z_k, z_i') = G_{ki}^{even}(d_{ki}, z_k, z_i') - (z_k/h_k)G_{ki}^{even}(d_{ki}, h_k, z_i') \quad (7.17a)$$

$$\mathscr{G}_{kid}^{odd}(d_{ki}, z_k, z_i') = G_{ki}^{odd}(d_{ki}, z_k, z_i') - (z_k/h_k)G_{ki}^{odd}(d_{ki}, h_k, z_i'). \quad (7.17b)$$

For each superscript, the kernel may be expanded into its real and imaginary parts as follows:

$$\mathscr{G}_{kid}(d_{ki}, z_k, z_i') = \mathscr{G}_{kidR}(d_{ki}, z_k, z_i') + j\mathscr{G}_{kidI}(d_{ki}, z_k, z_i'). \quad (7.18)$$

The desired alternative set of $2N$ simultaneous integral equations for the even and odd parts of the currents in the N elements is contained in (7.13) and (7.16).

7.3   Approximate distribution of current

It has been shown in earlier chapters that the first integral in (7.13) is well approximated by

$$\int_{-h_k}^{h_k} I_{zk}^{even}(z_k')G_{kkdR}^{even}(0, z_k, z_k')\,dz_k' = \int_{-h_k}^{h_k} I_{zk}^{even}(z_k')K_{kkdR}(z_k, z_k')\,dz_k'$$

$$\sim I_{zk}^{even}(z_k) \quad (7.19)$$

and

$$\int_{-h_k}^{h_k} I_{zk}^{even}(z_k')G_{kkdI}^{even}(0, z_k, z_k')\,dz_k' = \int_{-h_k}^{h_k} I_{zk}^{even}(z_k')K_{kkdI}(z_k, z_k')\,dz_k'$$

$$\sim H_{0zk} \quad (7.20)$$

where

$$H_{0zk} = \cos(\beta_0 z_k/2) - \cos(\beta_0 h_k/2) \quad (7.21)$$

provided $\beta_0 h_k \leqslant 5\pi/4$. By the same procedure it is readily shown that the first integral in (7.16) can be separated into analogous parts for which the following relations are good approximations:

$$\int_{-h_k}^{h_k} I_{zk}^{odd}(z_k')\mathscr{G}_{kkdR}^{even}(0, z_k, z_k')\,dz_k' \sim I_{zk}^{odd}(z_k) \quad (7.22)$$

$$\int_{-h_k}^{h_k} I_{zk}^{odd}(z_k')\mathscr{G}_{kkdI}^{even}(0, z_k, z_k')\,dz_k' \sim E_{0zk} \quad (7.23)$$

where
$$E_{0zk} = \sin(\beta_0 z_k/2) - (z_k/h_k)\sin(\beta_0 h_k/2). \qquad (7.24)$$

As a consequence of (7.19) and (7.22) it follows that the trigonometric functions that occur on the right side of (7.13) and (7.16) must also occur as leading terms in the approximate expressions for the currents, together with (7.21) and (7.24). That is, appropriate approximate formulas for the even and odd currents in antenna k are given below. For the even currents

$$I_{zk}^{\text{even}}(z_k) = A_k M_{0zk} + B_k F_{0zk} + D_k H_{0zk} \qquad (7.25a)$$

or the alternative equivalent form:

$$I_{zk}^{\text{even}}(z_k) = A'_k S_{0zk} + B'_k F_{0zk} + D_k H_{0zk} \qquad (7.25b)$$

where A_k, A'_k, B_k, B'_k and D_k are complex coefficients and

$$M_{0zk} = \sin \beta_0(h_k - |z_k|) \qquad (7.26)$$
$$S_{0zk} = \sin \beta_0|z_k| - \sin \beta_0 h_k \qquad (7.27)$$
$$F_{0zk} = \cos \beta_0 z_k - \cos \beta_0 h_k \qquad (7.28)$$
$$H_{0zk} = \cos(\beta_0 z_k/2) - \cos(\beta_0 h_k/2). \qquad (7.29)$$

For the odd currents

$$I_{zk}^{\text{odd}}(z_k) = Q_k P_{0zk} + R_k E_{0zk} \qquad (7.30)$$

where Q_k and R_k are complex coefficients and

$$P_{0zk} = \sin \beta_0 z_k - (z_k/h_k)\sin \beta_0 h_k. \qquad (7.31)$$

E_{0zk} is defined in (7.24). The above formulas are for $k = 1, 2, \ldots N$. The approximate formulas (7.25a, b) and (7.30) are obtained specifically from the first integrals in (7.13) and (7.16). When there are no staggered elements ($d_{ik} = 0$), it is known that the induced currents are well represented by a linear combination of F_{0zk} and H_{0zk}. It may be argued that a similar linear combination must also be an acceptable representation of the even parts of the currents induced in staggered elements. This follows from the theoretical and experimental studies referred to in chapter 6 of currents in non-staggered elements that differ greatly in length. If the current induced on a relatively long element (but with $\beta_0 h_k \leqslant 5\pi/4$) by an adjacent very short antenna is well represented by (7.25a, b), it may be concluded that the same must be true of the current induced in antenna k by other coupled elements which maintain a vector potential with an even part that varies less in amplitude and phase along antenna k than the vector potential generated by the currents in a very short element. Since no measurements were available of the currents induced by coupled staggered elements, a numerical check was made of the degree in which the assumed current distributions satisfy the integral equation. The results were quite satisfactory.

It may be concluded that the current along any element k in an array of N parallel antennas is approximately

$$I_{zk}(z_k) = A_k \sin \beta_0 (h_k - |z_k|) + B_k (\cos \beta_0 z_k - \cos \beta_0 h_k) + D_k [\cos (\beta_0 z_k/2)$$
$$- \cos (\beta_0 h_k/2)] + Q_k [\sin \beta_0 z_k - (z_k/h_k) \sin \beta_0 h_k]$$
$$+ R_k [\sin (\beta_0 z_k/2) - (z_k/h_k) \sin (\beta_0 h_k/2)]. \tag{7.32}$$

If more convenient, the first two terms may be replaced by those in (7.25b). The first three terms for the even part of the current are the same in form as for arrays of parallel, non-staggered elements. They include the term $\sin \beta_0 (h_k - |z_k|)$ which represents that part of the current excited directly by the generator voltage V_{0k}. No such term is possible for the odd part of the current in a centre-driven dipole. The remaining problem is to determine the coefficients in (7.32).

7.4 Evaluation of coefficients

The coefficients in the approximate formula (7.32) for the current in a typical element k in an array of N arbitrarily located parallel elements may be evaluated in various ways. The method outlined here is the one selected by V. W. H. Chang in his study of planar and three-dimensional arrays. He preferred to use the following alternative form for the current:

$$I_{zk}(z_k) = A'_k (\sin \beta_0 |z_k| - \sin \beta_0 h_k) + B'_k (\cos \beta_0 z_k - \cos \beta_0 h_k)$$
$$+ D_k [\cos (\beta_0 z_k/2) - \cos (\beta_0 h_k/2)]$$
$$+ Q_k [\sin \beta_0 z_k - (z_k/h_k) \sin \beta_0 h_k]$$
$$+ R_k [\sin (\beta_0 z_k/2) - (z_k/h_k) \sin (\beta_0 h_k/2)] \tag{7.33}$$

where $k = 1, 2, \dots N$. Instead of substituting the even and odd parts into the integral equation (7.13) and (7.16) he used the simpler integral equation for the total current obtained when (7.4) is equated to (7.1) with (7.2) and (7.3). That is,

$$\sum_{i=1}^{N} \int_{-h_i}^{h_i} I_{zi}(z'_i) \frac{e^{-j\beta_0 R_{ki}}}{R_{ki}} \, dz'_i = -(j4\pi/\zeta_0)[C_{k1} \cos \beta_0 z_k + C_{k2} \sin \beta_0 z_k$$
$$+ (1/2) V_{0k} \sin \beta_0 |z_k|]. \tag{7.34}$$

The substitution of (7.33) in the integral in (7.34) yields N equations with $7N$ unknowns, viz. the $5N$ coefficients in (7.33) and the $2N$

constants C_{k1} and C_{k2} with $k = 1, 2, \ldots N$. The required $7N$ equations can be obtained by satisfying (7.34) exactly at seven points along each antenna. The points chosen for z_k are h_k, $2h_k/3$, $h_k/3$, 0, $-h_k/3$, $-2h_k/3$, and $-h_k$. These correspond to the values used in the evaluation of the coefficients for the array of unequal elements in the last sections of chapter 6, but since the currents are now not even functions of z_k, the negative values $-h_k$, $-2h_k/3$ and $-h_k/3$ must also be used.

The number of unknowns can be reduced by the elimination of the constants C_{k1} and C_{k2}. The former is conveniently evaluated at $z_k = h_k$ where the current vanishes; the latter can be obtained from the equation at $z_k = 2h_k/3$. Thus, with the notation

$$U_{k1} = \sum_{i=1}^{N} \int_{-h_i}^{h_i} I_{zi}(z_i')G_{ki}(d_{ki}, h_k, z_i')\,dz_i' \qquad (7.35)$$

$$U_{k2} = \sum_{i=1}^{N} \int_{-h_i}^{h_i} I_{zi}(z_i')G_{ki}(d_{ki}, 2h_k/3, z_i')\,dz_i'. \qquad (7.36)$$

(7.34) evaluated at $z_k = h_k$ and $2h_k/3$ yields

$$(j4\pi/\zeta_0)C_{k1} = [U_{k1}\sin(2\beta_0 h_k/3) - U_{k2}\sin\beta_0 h_k]\csc(\beta_0 h_k/3) \qquad (7.37)$$

$$(j4\pi/\zeta_0)(C_{k2} + V_{0k}/2) = [U_{k2}\cos\beta_0 h_k - U_{k1}\cos(2\beta_0 h_k/3)]\csc(\beta_0 h_k/3). \qquad (7.38)$$

Note that in the range $\beta_0 h_k \leqslant 5\pi/4$, these expressions remain finite.

With (7.37) and (7.38), C_{k1} and C_{k2} can be eliminated from (7.34) to obtain

$$\sum_{i=1}^{N} \sin(\beta_0 h_k/3)\int_{-h_i}^{h_i} I_{zi}(z_i')G_{ki}(d_{ki}, z_k, z_i')\,dz_i'$$

$$+ \sum_{i=1}^{N} \sin\beta_0(\tfrac{2}{3}h_k - z_k)\int_{-h}^{h} I_{zi}(z_i')G_{ki}(d_{ki}, h_k, z_i')\,dz_i'$$

$$- \sum_{i=1}^{N} \sin\beta_0(h_k - z_k)\int_{-h_i}^{h_i} I_{zi}(z_i')G_{ki}(d_{ki}, \tfrac{2}{3}h_k, z_i')\,dz_i'$$

$$= \frac{j4\pi V_{0k}}{\zeta_0}\sin(\beta_0 h_k/3)\sin\beta_0 z_k H(-z_k) \qquad (7.39)$$

where $H(-z_k)$ is the Heaviside function defined by $H(-z_k) = 0$, $z_k > 0$; $H(-z_k) = 1$, $z_k \leqslant 0$.

The next step is to substitute the current (7.33) in the integrals in (7.39). This leads to quantities of the following form:

$$\xi_{kij}(z_k) = \sin{(\beta_0 h_k/3)} \int_{-h_i}^{h_i} J_{zi}^j(z_i') G_{ki}(d_{ki}, z_k, z_i') \, dz_i'$$

$$+ \sin{\beta_0(\tfrac{2}{3}h_k - z_k)} \int_{-h_i}^{h_i} J_{zi}^j(z_i') G_{ki}(d_{ki}, h_k, z_i') \, dz_i'$$

$$+ \sin{\beta_0(h_k - z_k)} \int_{-h_i}^{h_i} J_{zi}^j(z_i') G_{ki}(d_{ki}, 2h_k/3, z_i') \, dz_i'$$

$$(7.40)$$

where $k = 1, 2, \dots N$, $i = 1, 2, \dots N$ and $j = 1, 2, \dots 5$. The notation

$$J_{zi}^1(z_i') = S_{0zi} = \sin{\beta_0|z_i|} - \sin{\beta_0 h_i} \qquad (7.41a)$$

$$J_{zi}^2(z_i') = F_{0zi} = \cos{\beta_0 z_i} - \cos{\beta_0 h_i} \qquad (7.41b)$$

$$J_{zi}^3(z_i') = H_{0zi} = \cos{(\beta_0 z_i/2)} - \cos{(\beta_0 h_i/2)} \qquad (7.41c)$$

$$J_{zi}^4(z_i') = P_{0zi} = \sin{\beta_0 z_i} - (z_i/h_i)\sin{\beta_0 h_i} \qquad (7.41d)$$

$$J_{zi}^5(z_i') = E_{0zi} = \sin{(\beta_0 z_i/2)} - (z_i/h_i)\sin{(\beta_0 h_i/2)} \qquad (7.41e)$$

is used. Note that for any specified value of z_k in a fixed array, (7.40) defines a set of N complex numbers that can be evaluated by high-speed computer. With (7.40) and (7.41a–e), (7.39) becomes:

$$\sum_{i=1}^{N} [A_i' \xi_{ki1}(z_k) + B_i' \xi_{ki2}(z_k) + D_i \xi_{ki3}(z_k) + Q_i \xi_{ki4}(z_k) + R_i \xi_{ki5}(z_k)]$$

$$= j(4\pi V_{0k}/\zeta_0)\sin{(\beta_0 h_k/3)}\sin{\beta_0 z_k H(-z_k)} \qquad (7.42)$$

with $k = 1, 2, \dots N$. Five sets of N equations can be obtained from (7.42) if z_k is successively made equal to the five values $h_k/3, 0,$ $-h_k/3$, $-2h_k/3$ and $-h_k$. These contain the $M = 5N$ unknown coefficients given by the column matrix

$$\{A\} = \text{tr}\,(A_1, \dots A_N, B_1, \dots B_N, D_1, \dots D_N, Q_1, \dots, Q_N, R_1, \dots R_N) \qquad (7.43)$$

where tr indicates the transpose. Let

$$[\Phi] = \begin{bmatrix} \Phi_{11} \cdots \Phi_{1M} \\ \vdots \qquad \vdots \\ \Phi_{M1} \cdots \Phi_{MM} \end{bmatrix} \qquad (7.44)$$

where $M = 5N$ and

$$\Phi_{k+(m-1)N,\,i+(j-1)N} = \xi_{kij}(z_k^m) \qquad (7.45)$$

with $j = 1, 2, \dots 5$; $m = 1, 2, \dots 5$; $k = 1, 2, \dots N$; $i = 1, 2, \dots N$. The notation $z_k^1 = h_k/3$, $z_k^2 = 0$, $z_k^3 = -h_k/3$, $z_k^4 = -2h_k/3$, $z_k^5 = -h_k$ is used. Also let the following column matrix of $5N$ terms be defined:

$$\{W\} = \text{tr} \,(0 \ldots 0, 0 \ldots 0, W_1 \ldots W_N, T_1 \ldots T_N, S_1 \ldots S_N) \quad (7.46)$$

where
$$W_k = -(j4\pi V_{0k}/\zeta_0)\sin^2(\beta_0 h_k/3) \quad (7.47a)$$

$$T_k = -(j4\pi V_{0k}/\zeta_0)\sin(\beta_0 h_k/3)\sin(2\beta_0 h_k/3) \quad (7.47b)$$

$$S_k = -(j4\pi V_{0k}/\zeta_0)\sin(\beta_0 h_k/3)\sin\beta_0 h_k \quad (7.47c)$$

with $k = 1, 2, \ldots N$.

With this matrix notation the $5N$ equations for the N coefficients of the currents in terms of the N driving voltages V_{0k} with $k = 1, 2, \ldots N$ are given by the single matrix equation

$$[\Phi]\{A\} = \{W\}. \quad (7.48)$$

If (7.48) is solved for the $5N$ coefficients given by (7.43), the N currents $I_{zk}(z_k)$, $k = 1, 2, \ldots N$ given in (7.33) are known in terms of the N voltages V_{0k}. The currents at the driving points are then given by the matrix equation

$$\{I_z(0)\} = [Y]\{V_0\} \quad (7.49a)$$

where
$$\{I_z(0)\} = \left\{ \begin{matrix} I_{z1}(0) \\ \vdots \\ I_{zN}(0) \end{matrix} \right\}, \qquad \{V_0\} = \left\{ \begin{matrix} V_{01} \\ \vdots \\ V_{0N} \end{matrix} \right\}. \quad (7.49b)$$

The square matrix

$$[Y] = \begin{bmatrix} Y_{11} & Y_{12} \ldots & Y_{1N} \\ \vdots & & \vdots \\ Y_{N1} & \ldots\ldots\ldots\ldots & Y_{NN} \end{bmatrix} \quad (7.49c)$$

is the admittance matrix. The terms Y_{ii} are the self-admittances, the terms Y_{ij}, $i \neq j$ are the mutual admittances.

The N driving voltages can be expressed in terms of the currents at the driving points in the form

$$\{V_0\} = [Z]\{I_z(0)\} \quad (7.49d)$$

where $[Z] = [Y]^{-1}$ is the impedance matrix.

The driving-point admittance of element k is defined by

$$Y_{0k} = I_{zk}(0)/V_{0k}. \quad (7.49e)$$

7.5 The field patterns

The radiation field of an array of arbitrarily located parallel elements is the superposition of the fields generated by the individual elements. The far field of element i in such an array is given by

$$E^r_{\Theta i} = \frac{j\omega\mu_0}{4\pi} \frac{e^{-j\beta_0 R_i}}{R_i} \int_{-h_i}^{h_i} I_{zi}(z_i') e^{j\beta_0 z_i' \cos\Theta} \sin\Theta \, dz_i' \quad (7.50a)$$

where R_i is the distance from the centre of the antenna i at X_i, Y_i, Z_i to the point of calculation P, and Θ is the angle between the Z-axis and the line $0P$ from the origin of coordinates near the centre of the array.

If the distribution of current (7.33) is substituted in (7.50a), the field of element i can be expressed in the following integrated form:

$$E^r_{\Theta i} = \frac{j\zeta_0}{4\pi} \frac{e^{-j\beta_0 R_i}}{R_i} [A_i' H_m(\Theta, \beta_0 h_i) + B_i' G_m(\Theta, \beta_0 h_i)$$

$$+ D_i D_m(\Theta, \beta_0 h_i) + Q_i Q_m(\Theta, \beta_0 h_i) + R_i R_m(\Theta, \beta_0 h_i)] \quad (7.50b)$$

where the individual field factors are as follows:

$$H_m(\Theta, \beta_0 h_i) = \frac{\beta_0 \sin\Theta}{2} \int_{-h_i}^{h_i} (\sin\beta_0|z_i'| - \sin\beta_0 h_i) e^{j\beta_0 z_i' \cos\Theta} \, dz_i'$$

$$= \{\cos\Theta - [1 - \cos\beta_0 h_i \cos(\beta_0 h_i \cos\Theta)]\}\sec\Theta \csc\Theta \quad (7.51a)$$

$$H_m(0, \beta_0 h_i) = H_m(\pi, \beta_0 h_i) = 0 \quad (7.51b)$$

$$H_m\left(\frac{\pi}{2}, \beta_0 h_i\right) = 1 - \cos\beta_0 h_i - \beta_0 h_i \sin\beta_0 h_i \quad (7.51c)$$

$$G_m(\Theta, \beta_0 h_i) = \frac{\beta_0 \sin\Theta}{2} \int_{-h_i}^{h_i} (\cos\beta_0 z_i' - \cos\beta_0 h_i) e^{j\beta_0 z_i' \cos\Theta} \, dz_i'$$

$$= [\cos\Theta \sin\beta_0 h_i \cos(\beta_0 h_i \cos\Theta)$$

$$- \cos\beta_0 h_i \sin(\beta_0 h_i \cos\Theta)] \sec\Theta \csc\Theta \quad (7.52a)$$

$$G_m(0, \beta_0 h_i) = G_m(\pi, \beta_0 h_i) = 0 \quad (7.52b)$$

$$G_m\left(\frac{\pi}{2}, \beta_0 h_i\right) = \sin\beta_0 h_i - \beta_0 h_i \cos\beta_0 h_i \quad (7.52c)$$

$$D_m(\Theta, \beta_0 h_i) = \frac{\beta_0 \sin\Theta}{2} \int_{-h_i}^{h_i} [\cos(\beta_0 z_i'/2) - \cos(\beta_0 h_i/2)] e^{j\beta_0 z_i' \cos\Theta} \, dz_i'$$

$$= \left[\frac{\sin\Theta}{\cos\Theta(1 - 4\cos^2\Theta)}\right][2\cos\Theta \sin(\beta_0 h_i/2)$$

$$\times \cos(\beta_0 h_i \cos\Theta) - \cos(\beta_0 h_i/2) \sin(\beta_0 h_i \cos\Theta)]$$

$$(7.53a)$$

$$D_m\left(\frac{\pi}{2}, \beta_0 h_i\right) = 2 \sin (\beta_0 h_i/2) - \beta_0 h_i \cos (\beta_0 h_i/2) \tag{7.53b}$$

$$D_m\left(\frac{\pi}{3}, \beta_0 h_i\right) = D_m\left(\frac{2\pi}{3}, \beta_0 h_i\right) = \frac{\sqrt{3}}{4}(\beta_0 h_i - \sin \beta_0 h_i) \tag{7.53c}$$

$$Q_m(\Theta, \beta_0 h_i) = \frac{\beta_0 \sin \Theta}{2} \int_{-h_i}^{h_i} [\sin \beta_0 z_i' - (z_i'/h_i) \sin \beta_0 h_i] \, e^{j\beta_0 z_i' \cos \Theta} \, dz_i'$$

$$= (j/\beta_0 h_i)[- \beta_0 h_i \cos^2 \Theta \cos \beta_0 h_i \sin (\beta_0 h_i \cos \Theta)$$

$$- \sin^2 \Theta \sin \beta_0 h_i \sin (\beta_0 h_i \cos \Theta)$$

$$+ \beta_0 h_i \cos \Theta \sin \beta_0 h_i \cos (\beta_0 h_i \cos \Theta)]$$

$$\times \csc \Theta \sec^2 \Theta \tag{7.54a}$$

$$Q_m(0, \beta_0 h_i) = Q_m\left(\frac{\pi}{2}, \beta_0 h_i\right) = Q_m(\pi, \beta_0 h_i) = 0 \tag{7.54b}$$

$$R_m(\Theta, \beta_0 h_i) = \frac{\beta_0 \sin \Theta}{2} \int_{-h_i}^{h_i} [\sin (\beta_0 z_i'/2) - (z_i'/h_i)$$

$$\times \sin (\beta_0 h_i/2)] e^{j\beta_0 z_i' \cos \Theta} \, dz_i'$$

$$= \left[\frac{j \sin \Theta}{\beta_0 h_i \cos^2 \Theta (1 - 4 \cos^2 \Theta)}\right]\{\sin (\beta_0 h_i \cos \Theta)$$

$$\times [-2\beta_0 h_i \cos^2 \Theta \cos (\beta_0 h_i/2) - \sin (\beta_0 h_i/2)$$

$$+ 4 \sin (\beta_0 h_i/2) \cos^2 \Theta] + \beta_0 h_i \sin (\beta_0 h_i/2)$$

$$\times \cos \Theta \cos (\beta_0 h_i \cos \Theta)\} \tag{7.55a}$$

$$R_m\left(\frac{\pi}{2}, \beta_0 h_i\right) = 0 \tag{7.55b}$$

$$R_m\left(\frac{\pi}{3}, \beta_0 h_i\right) = - R_m\left(\frac{2\pi}{3}, \beta_0 h_i\right) = j\frac{\sqrt{3}}{4}[\beta_0 h_i + \sin \beta_0 h_i$$

$$- (8/\beta_0 h_i) \sin^2 (\beta_0 h_i/2)]. \tag{7.55c}$$

The radiation field of N parallel antennas is the sum of the contributions from each element. In the far-field approximation

$$\frac{e^{-j\beta_0 R_i}}{R_i} = \frac{e^{-j\beta_0 R_0}}{R_0} e^{j\beta_0(\mathbf{r}_i \cdot \hat{\mathbf{R}}_0)} \tag{7.56}$$

where $\mathbf{r}_i$ is the vector drawn from the origin near the centre of the array to the centre of antenna i and $\hat{\mathbf{R}}_0$ is the unit vector along the line $0P$ where P is the point of calculation as shown in Fig. 7.4.

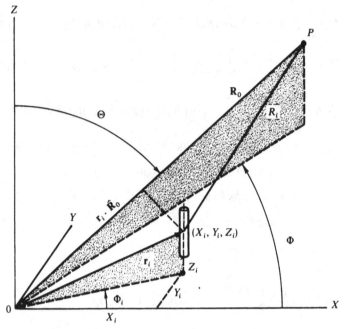

Fig. 7.4.   Point P in the far field of an array of parallel elements of which element i at X_i, Y_i, Z_i is typical.

With (7.56) the far-field of the array is

$$E_\Theta^r = \frac{j\zeta_0}{4\pi} \frac{e^{-j\beta_0 R_0}}{R_0} \sum_{i=1}^{N} e^{j\beta_0(\mathbf{r}_i \cdot \hat{\mathbf{R}}_0)}[A_i' H_m(\Theta, \beta_0 h_i) + B_i' G_m(\Theta, \beta_0 h_i)$$

$$+ D_i D_m(\Theta, \beta_0 h_i) + Q_i Q_m(\Theta, \beta_0 h_i) + R_i R_m(\Theta, \beta_0 h_i)]. \quad (7.57)$$

If the point P where the field is calculated is located by the spherical coordinates R_0, Θ, Φ and the centre of element i is at X_i, Y_i, Z_i, then

$$\mathbf{r}_i \cdot \hat{\mathbf{R}}_0 = X_i \sin \Theta \cos \Phi + Y_i \sin \Theta \sin \Phi + Z_i \cos \Theta. \quad (7.58)$$

7.6   The general two-element array†

In the introductory analysis of the two-element array in chapter 2 only parallel non-staggered antennas are considered. As a consequence of the resulting even symmetry for the currents in the elements and the vector potentials, a three- or even two-term representation is adequate. The more general five-term approximation of the currents introduced in this chapter includes the previous three terms to describe the even currents and two additional terms

† The computations in this section were planned and programmed by V. W. H. Chang.

to represent the odd currents generated by asymmetrical coupling when the elements are collinear as shown in Fig. 7.5 or staggered as in Fig. 7.6.

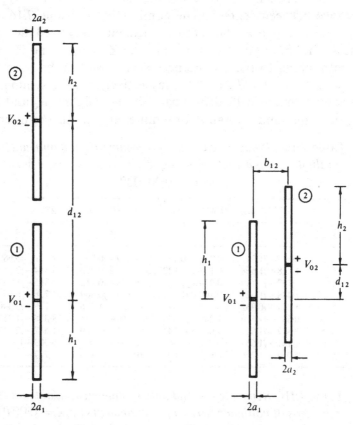

Fig. 7.5. Two-element collinear array.   Fig. 7.6. Two-element staggered array.

The general formulas for the currents, driving-point admittances and field patterns derived in the preceding sections are readily specialized for the two-element array. Comparative examples of the admittances of coupled full-wave and half-wave elements when driven symmetrically with $V_{01} = V_{02} = 1$ and anti-symmetrically with $V_{01} = -V_{02} = 1$ as a function of the distance between centres are given in Tables 7.1a, b, c for antennas with $a/\lambda = 0.007022$. These tables give the symmetrical admittance $Y^s = G^s + jB^s$ and the anti-symmetrical admittance $Y^a = G^a + jB^a$. The associated self- and mutual admittances are $Y_{1s} = (Y^s + Y^a)/2$ and $Y_{12} = -(Y^s - Y^a)/2$.

Table 7.1a applies to the non-staggered antennas considered in chapter 2; the variable parameter is b_{12}/λ, the normalized distance between centres. Table 7.1b is for the collinear pair with the distance $(d_{12}-2h)$ between the adjacent ends as the parameter (d_{12} is the distance between centres). The admittances in Table 7.1c are for the staggered pair as the centre of element 2 is moved along a 45° line so that $b_{12} = d_{12}$. The impedances $Z^s = 1/Y^s$ and $Z^a = 1/Y^a$ corresponding to the admittances in Tables 7.1a, b, c are shown graphically in Figs. 7.7 and 7.8, respectively, for the symmetrically and anti-symmetrically driven pairs. In these figures R_0 and X_0 are the resistance and reactance for infinite separation. The interaction

Table 7.1a. *Symmetrical and anti-symmetrical admittances in millimhos of parallel, non-staggered array of two elements;* $a/\lambda = 0.007022$.

b_{12}/λ	$h/\lambda = 0.50$		$h/\lambda = 0.25$	
	Y^s	Y^a	Y^s	Y^a
0·05	$0·813+j1·397$	$0·071+j0·941$	$4·939-j0·820$	$4·831-j53·897$
0·10	$1·028+j1·668$	$0·146+j1·122$	$5·572-j0·290$	$4·344-j24·150$
0·15	$1·197+j1·749$	$0·230+j1·286$	$6·258+j0·112$	$4·385-j15·386$
0·25	$1·448+j1·627$	$0·408+j1·531$	$7·853+j0·809$	$4·758-j8·655$
0·50	$1·079+j0·932$	$0·865+j1·774$	$14·321-j1·496$	$6·129-j3·340$
0·75	$0·635+j1·374$	$1·244+j1·570$	$8·960-j7·101$	$8·318-j1·318$
1·00	$0·872+j1·657$	$1·066+j1·141$	$7·211-j3·839$	$11·577-j2·517$
1·25	$1·157+j1·537$	$0·728+j1·359$	$8·543-j2·095$	$9·503-j5·707$
1·50	$1·053+j1·226$	$0·877+j1·602$	$10·725-j2·875$	$7·777-j3·918$

Table 7.1b. *Symmetrical and anti-symmetrical admittances in millimhos of collinear array of two elements;* $a/\lambda = 0.007022$.

$\dfrac{d_{12}-2h}{\lambda}$	$h/\lambda = 0.50$		$h/\lambda = 0.25$	
	Y^s	Y^a	Y^s	Y^a
0	$1·050+j1·581$	$0·816+j1·139$	$5·549-j1·953$	$15·281-j6·078$
0·05	$1·042+j1·505$	$0·808+j1·338$	$7·508-j1·868$	$11·013-j6·483$
0·10	$1·035+j1·465$	$0·844+j1·409$	$8·444-j1·913$	$9·529-j5·821$
0·15	$1·026+j1·437$	$0·875+j1·440$	$9·100-j2·104$	$8·858-j5·208$
0·25	$0·999+j1·403$	$0·919+j1·461$	$9·850-j2·764$	$8·412-j4·288$
0·50	$0·941+j1·409$	$0·966+j1·446$	$9·521-j4·026$	$8·832-j3·243$
0·75	$0·945+j1·435$	$0·965+j1·422$	$8·923-j3·800$	$9·415-j3·404$
1·00	$0·958+j1·435$	$0·950+j1·422$	$9·042-j3·445$	$9·300-j3·785$
1·25	$0·959+j1·426$	$0·950+j1·431$	$9·295-j3·518$	$9·045-j3·699$
1·50	$0·953+j1·425$	$0·956+j1·432$	$9·238-j3·708$	$9·103-j3·517$

Table 7.1c. *Symmetrical and anti-symmetrical admittances in millimhos of parallel, staggered array of two elements;* $a/\lambda = 0.007022.$

$\dfrac{b_{12}}{\lambda} = \dfrac{d_{12}}{\lambda}$	$h/\lambda = 0.50$		$h/\lambda = 0.25$	
	Y^s	Y^a	Y^s	Y^a
0.05	$0.825 + j1.318$	$0.086 + j0.608$	$4.692 - j0.775$	$11.923 - j76.274$
0.10	$1.040 + j1.720$	$0.786 + j0.801$	$5.456 - j0.470$	$8.631 - j29.120$
0.15	$1.189 + j1.832$	$0.236 + j0.984$	$6.336 - j0.297$	$7.399 - j16.489$
0.25	$1.363 + j1.681$	$0.420 + j1.379$	$8.372 - j0.097$	$6.669 - j8.084$
0.50	$1.035 + j1.110$	$0.965 + j1.673$	$11.077 - j4.641$	$7.804 - j2.856$
0.75	$0.785 + j1.474$	$1.089 + j1.414$	$8.235 - j3.925$	$10.159 - j3.106$
1.00	$1.008 + j1.489$	$0.906 + j1.350$	$9.340 - j2.908$	$8.903 - j4.302$
1.25	$0.976 + j1.380$	$0.939 + j1.484$	$9.483 - j4.115$	$8.862 - j3.166$
1.50	$0.918 + j1.400$	$0.989 + j1.419$	$8.729 - j3.561$	$9.650 - j3.650$

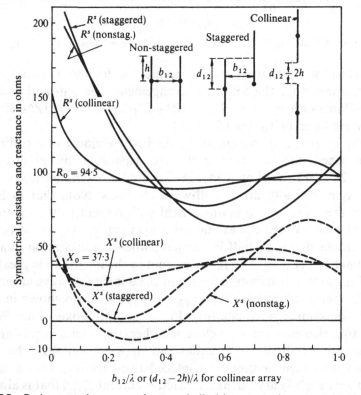

Fig. 7.7. Resistance and reactance of symmetrically driven array of two parallel half-wave dipoles when non-staggered, staggered with $b_{12} = d_{12}$, and collinear; $a/\lambda = 0.007022$, $h/\lambda = 0.25$, $V_{02} = V_{01}$.

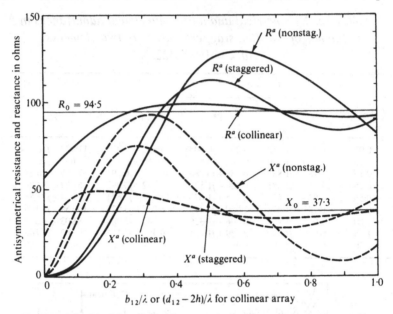

Fig. 7.8.   Like Fig. 7.7 but for anti-symmetrically driven elements, $V_{02} = -V_{01}$.

between the elements is seen to be greatest for the non-staggered pair, smallest for the collinear arrangement. The self- and mutual impedances are given by $Z_{s1} = (Z^s + Z^a)/2$ and $Z_{12} = (Z^s - Z^a)/2$; they are listed in Tables 7.2a, b.

The current distribution along the lower element of a collinear pair when the adjacent ends are separated by a distance $d_{12} - 2h = 0.1\lambda$ is shown in Fig. 7.9 for both symmetrically and anti-symmetrically driven full-wave elements. Note that in both cases the currents are asymmetrical with respect to the centre of the element. When the excitation is symmetrical ($V_{01} = V_{02}$), the current in the outer half is greatest; when the excitation is anti-symmetrical ($V_{01} = -V_{02}$), the current in the inner half is greatest.

The current distribution for a pair of coupled full-wave antennas in the staggered position with $b_{12}/\lambda = d_{12}/\lambda = 0.1$ is shown in Fig. 7.10 for symmetric excitation ($V_{01} = V_{02}$) in broken line. Since the two elements are very close together, the interaction is great. When centre driven with equal and opposite voltages, the two conductors form a slightly displaced two-wire line with a large and only slightly asymmetrical reactive current $I'_{z1}(z)$ that is almost sinusoidal, and a very small in-phase component $I''_{z1}(z)$. Since the current induced in each element by that in the other is essentially

Table 7.2a. *Self- and mutual impedances of coupled half-wave dipoles; $h/\lambda = 0.25$; $a/\lambda = 0.007022$.*

| | Parallel, non-staggered | | Staggered | | Collinear | |
| | $x = b_{12}/\lambda$ | | $x = b_{12}/\lambda = d_{12}/\lambda$ | | $x = (d_{12} - 2h)/\lambda$ | |
x	$Z_{s1} = R_{s1}+jX_{s1}$	$Z_{12} = R_{12}+jX_{12}$	$Z_{s1} = R_{s1}+jX_{s1}$	$Z_{12} = R_{12}+jX_{12}$	$Z_{s1} = R_{s1}+jX_{s1}$	$Z_{12} = R_{12}+jX_{12}$
0					$108.4+j39.5$	$51.9+j17.0$
0.05	$99.4+j25.6$	$97.7+j7.15$	$104.7+j23.5$	$102.7+j10.7$	$96.4+j35.4$	$29.0-j4.2$
0.10	$93.1+j24.7$	$85.9+j15.4$	$95.6+j23.6$	$86.3-j7.9$	$94.5+j36.1$	$18.1-j10.6$
0.15	$88.4+j28.6$	$71.3-j31.5$	$90.1+j28.9$	$67.4-j21.5$	$94.1+j36.7$	$10.2-j12.6$
0.25	$87.4+j37.9$	$38.6-j50.8$	$90.1+j37.5$	$29.4-j36.1$	$94.2+j37.3$	$-0.1-j10.8$
0.50	$97.4+j37.9$	$-28.4-j30.7$	$94.1+j39.1$	$-1.7-j30.9$	$94.4+j37.2$	$-5.3+j0.5$
0.75	$92.9+j36.4$	$-24.4+j17.9$	$94.9+j36.8$	$-18.1-j4.6$	$94.4+j37.2$	$0.5+j3.2$
1.00	$95.3+j37.7$	$12.8+j19.8$	$94.5+j37.3$	$4.5+j9.8$	$94.4+j37.2$	$2.2-j0.4$
1.25	$93.9+j37.8$	$16.5-j9.7$	$94.3+j37.2$	$3.3-j6.8$	$94.4+j37.2$	$-0.3-j1.6$
1.50	$94.8+j37.5$	$-7.8-j14.2$	$94.4+j37.1$	$-5.7+j1.4$	$94.4+j37.2$	$-1.2+j0.2$

Table 7.2b. *Self- and mutual impedances of coupled full-wave dipoles; $h/\lambda = 0.5$; $a/\lambda = 0.007022$.*

| | Parallel, non-staggered | | Staggered | | Collinear | |
| | $x = b_{12}/\lambda$ | | $x = b_{12}/\lambda = d_{12}/\lambda$ | | $x = (d_{12} - 2h)/\lambda$ | |
x	$Z_{s1} = R_{s1}+jX_{s1}$	$Z_{12} = R_{12}+jX_{12}$	$Z_{s1} = R_{s1}+jX_{s1}$	$Z_{12} = R_{12}+jX_{12}$	$Z_{s1} = R_{s1}+jX_{s1}$	$Z_{12} = R_{12}+jX_{12}$
0					$353.7-j509.5$	$-62.1+j70.6$
0.05	$195.5-j795.5$	$115.6+j260.8$	$285.1-j1078.8$	$56.2+j533.7$	$320.8-j498.4$	$-9.8+j49.2$
0.10	$190.9-j655.5$	$76.8+j221.0$	$250.5-j825.2$	$7.0+j399.4$	$317.3-j488.9$	$4.4+j33.7$
0.15	$200.5-j571.3$	$66.0+j182.0$	$239.9-j672.3$	$9.4+j288.3$	$318.6-j484.1$	$10.5+j23.1$
0.25	$233.8-j476.4$	$71.4+j133.5$	$246.6-j511.3$	$44.4+j152.4$	$322.6-j481.8$	$14.1+j8.7$
0.50	$376.4-j457.1$	$154.2-j1.6$	$354.2-j465.1$	$95.4-j16.6$	$323.5-j484.5$	$4.1-j6.4$
0.75	$293.7-j495.5$	$-16.4-j104.2$	$311.8-j486.2$	$-30.2-j42.2$	$323.3-j483.7$	$-3.4-j2.4$
1.00	$342.9-j470.3$	$-94.4-j2.3$	$327.3-j485.7$	$-15.6-j25.1$	$323.4-j484.0$	$-16+j2.1$
1.25	$309.3-j493.5$	$3.2+j78.2$	$323.1-j482.0$	$18.4-j0.8$	$323.3-j483.9$	$1.3+j1.1$
1.50	$333.0-j474.8$	$70.0+j5.4$	$322.5-j484.5$	$-8.1-j10.2$	$323.4-j484.0$	$0.8-j0.9$

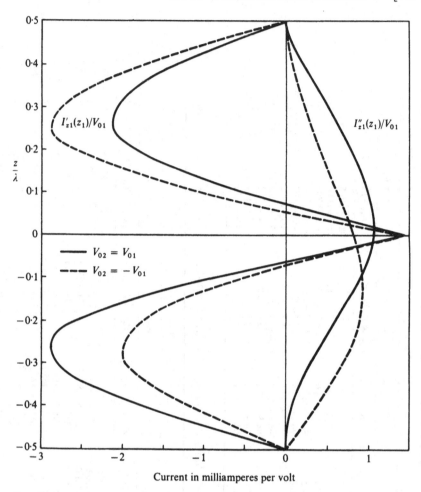

Fig. 7.9.   Currents in element No. 1 of a symmetrically and anti-symmetrically driven pair of collinear antennas. $I_{z1}(z) = I''_{z1}(z) + jI'_{z1}(z)$. $V_{02} = \pm V_{01} = \pm 1$, $a/\lambda = 0.007022$, $h/\lambda = 0.5$, $d_{12}/\lambda = 1.1$. (Element 1 is below element 2.)

180° out of phase, the coupling reinforces the currents excited by the generator voltages. When centre driven by equal and co-directional generators the distribution of current is extremely asymmetrical. The half of each element that is removed from the other has a very large approximately sinusoidal reactive current $I'_{zi}(z)$, whereas the adjacent halves have only a small and oppositely directed reactive component. The in-phase component $I''_{zi}(z)$ is much greater than when the antennas are anti-symmetrically driven and the asymmetry is in the opposite directions.

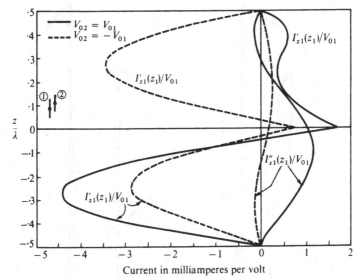

Fig. 7.10.   Like Fig. 7.9 but for two staggered elements with $b_{12}/\lambda = d_{12}/\lambda = 0.1$.

The distributions of current for the symmetrically and anti-symmetrically driven staggered pair are sketched approximately to scale in Figs. 7.11a, b. Note that the more closely coupled adjacent halves of the elements have the greater current when the excitation is asymmetrical, very much the smaller when the excitation is symmetrical. In the former case the coupling between the elements reinforces the generators, in the latter it opposes them.

It is interesting to note that the distribution along the symmetrically driven pair in Fig. 7.11a resembles that along a sleeve dipole.† This is to be expected since the two elements are very closely coupled.

7.7   A simple planar array‡

The application of the general theory developed earlier in this chapter to planar arrays is conveniently illustrated with the three by three nine-element array shown in Fig. 7.12. This involves non-staggered, staggered and collinear elements, so that the effects of the different types of coupling on otherwise identical elements can be studied.

† See, for example, R. W. P. King, [2], p. 413, Fig. 30.7e.
‡ The computations on this section are those of V. W. H. Chang. Parts of sections 7.7–7.10 were first published in *Radio Science* [3].

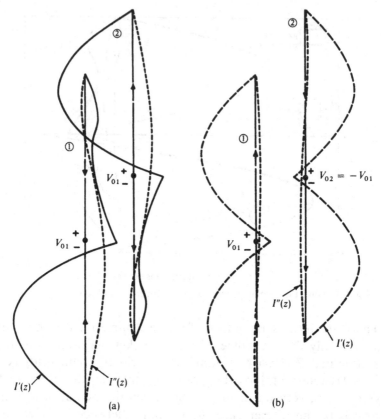

Fig. 7.11.   Currents in two-element staggered array driven (a) symmetrically with $V_{02} = V_{01}$ and (b) anti-symmetrically with $V_{02} = -V_{01}$. The distributions of current are taken from Fig. 7.10.

Consider first the broadside array in which all elements are driven with equal voltages, that is, $V_{0i} = 1$, $i = 1, 2, \ldots 9$. Since conventional theory is unable to treat full-wave elements, let an array of nine elements with $h/\lambda = 0.5$ be analysed. Let the lateral distances between elements be $b = 0.25$ and the axial distance between adjacent ends be $(d - 2h)/\lambda = 0.1$ where d is the distance between centres of adjacent collinear elements. With this symmetric excitation, elements 1, 2 and 3 are like elements 7, 8 and 9 in the even parts of these currents, but the algebraic sign of the odd parts is reversed. The coefficients of the five trigonometric functions in the current distribution given in (7.33) are listed in Table 7.3 for all of the elements. The associated driving-point admittances and impedances are also listed. Note that these differ significantly. The

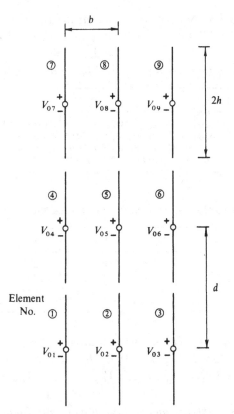

Fig. 7.12.   Planar array of 9 identical, equally-spaced elements.

four different distributions of current are shown in Fig. 7.13 in the form $I_{zi}(z_i) = I''_{zi}(z_i) + jI'_{zi}(z_i)$ where $I''_{zi}(z_i)$ is in phase with V_{0i}, $I'_{zi}(z_i)$ in phase quadrature. Note that the currents for elements 7, 8 and 9 are like those for 1, 2 and 3 but with $-z_i$ substituted for z_i. Elements 4, 5 and 6 have even currents. The contribution by the odd currents in elements 1, 2 and 3 is seen to be large.

When the same nine-element array is driven to obtain a unilateral endfire pattern with $V_{01} = V_{04} = V_{07} = 1$, $V_{02} = V_{05} = V_{08} = -j$, $V_{03} = V_{06} = V_{09} = -1$, the coefficients for the trigonometric functions in the current distribution (7.33) are listed in Table 7.4. Note that there are now six different sets of coefficients since elements 1 and 3 and their counterparts are no longer electrically identical. The driving-point admittances and impedances are also given in Table 7.4. They are seen to have a wider range of values than in the broadside array. Note that the resistances of the elements

Table 7.3. *Nine-element planar array—broadside*
$a/\lambda = 0.007022$, $h/\lambda = 0.5$, $b/\lambda = 0.25$, $d/\lambda = 1.1$,
$V_{0i} = 1$, $i = 1, 2, \ldots 9$
Coefficients of trigonometric functions in milliamperes per volt

i	A'_i	B'_i	D'_i	Q_i†	R_i†
1, 3, 7. 9	$0.006 - j3.626$	$0.763 - j0.739$	$0.287 + j2.849$	$0.197 + j0.444$	$-0.008 + j0.134$
2, 8	$0.010 - j3.580$	$1.065 - j0.643$	$0.197 + j3.164$	$0.317 + j0.604$	$0.279 + j0.199$
4, 6	$0.007 - j3.615$	$0.832 - j0.590$	$0.176 + j2.596$	$0 + j0$	$0 + j0$
5	$0.010 - j3.567$	$1.176 - j0.443$	$0.087 + j2.821$	$0 + j0$	$0 + j0$

† Reverse signs for $i = 7, 8, 9$.

Admittances in millimhos and impedances in ohms

i	$Y_{0i} = G_{0i} + jB_{0i}$	$Z_{0i} = R_{0i} + jX_{0i}$
1, 3, 7, 9	$1.759 + j1.371$	$353.7 - j275.7$
2, 8	$2.328 + j1.878$	$260.2 - j209.9$
4, 6	$1.840 + j1.416$	$341.3 - j262.8$
5	$2.440 + j1.955$	$249.6 - j200.0$

Table 7.4. *Nine-element planar array—endfire*
$a/\lambda = 0.007022$, $h/\lambda = 0.5$, $b/\lambda = 0.25$, $d/\lambda = 1.1$
$V_{01} = V_{04} = V_{07} = 1$, $V_{02} = V_{05} = V_{08} = -j$,
$V_{03} = V_{06} = V_{09} = -1$
Coefficients of trigonometric functions in milliamperes per volt

i	A'_i	B'_i	D'_i	Q_i†	R_i†
1, 7	$0.053 - j3.674$	$0.321 - j0.426$	$0.391 - j2.060$	$0.206 + j0.463$	$0.288 + j0.213$
2, 8	$-3.675 - j0.006$	$-0.282 - j0.501$	$2.375 - j0.027$	$0.500 - j0.151$	$0.196 - j0.168$
3, 9	$0.042 + j3.668$	$-0.724 - j0.152$	$0.480 - j2.201$	$-0.119 - j0.562$	$-0.223 - j0.124$
4	$0.053 - j3.664$	$0.381 - j0.299$	$0.321 + j1.845$	$0 + j0$	$0 + j0$
5	$-3.665 - j0.005$	$-0.129 - j0.546$	$2.137 + j0.010$	$0 + j0$	$0 + j0$
6	$0.044 + j3.657$	$-0.735 - j0.344$	$0.463 - j1.918$	$0 + j0$	$0 + j0$

† Reverse signs for $i = 7, 8, 9$.

Admittances in millimhos and impedances in ohms

i	$Y_{0i} = G_{0i} + jB_{0i}$	$Z_{0i} = R_{0i} + jX_{0i}$
1, 7	$1.034 + j1.208$	$409.0 - j477.8$
2, 8	$1.030 + j1.811$	$237.2 - j417.2$
3, 9	$0.966 + j2.506$	$134.0 - j347.4$
4	$1.084 + j1.250$	$396.0 - j456.7$
5	$1.083 + j1.879$	$230.2 - j399.5$
6	$1.008 + j2.607$	$129.0 - j333.7$

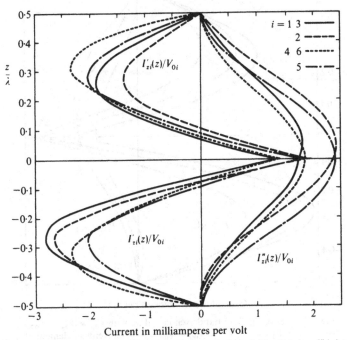

Fig. 7.13.   Currents in a planar array of 9 elements in broadside. $I_{zi}(z_i) = I''_{zi}(z_i) + jI'_{zi}(z_i)$,
$V_{0i} = 1$; $a/\lambda = 0.007022$, $h/\lambda = 0.5$, $b/\lambda = 0.25$, $d/\lambda = 1.1$.

in the collinear trio in the backward direction $(1, 4, 7)$ are much greater than the corresponding resistances in the forward trio $(3, 6, 9)$. This is characteristic of endfire arrays of full-wave elements. The six different currents are shown in Figs. 7.14 and 7.15. The currents in elements 7, 8 and 9 are like those in 1, 2 and 3 but with $-z_i$ substituted for z_i. Note that the currents in the rear collinear trio $(1, 4, 7)$ are greater and contribute more to the far field than the currents in the forward trio of the elements $(3, 6, 9)$. The far-field patterns in the horizontal or H-plane and the vertical or E-plane are shown in Fig. 7.16 for both the broadside and the endfire arrays. The horizontal pattern of the broadside array is bidirectional with maxima at $\Phi = 90°$ and $270°$, the endfire pattern is unidirectional with a broad maximum in the direction $\Phi = 0°$. The vertical patterns in the direction $\Phi = 0$ are seen to be very sharp as would be expected when three full-wave elements (which correspond to six half-wave elements) are stacked. (Note that the vertical pattern shown for the broadside array is not in the direction of the maximum at $\Phi = 90°$.)

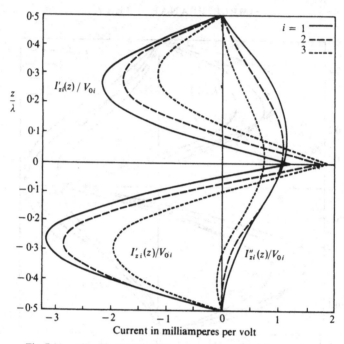

Fig. 7.14. Like Fig. 7.13 but driven in endfire with $V_{01} = V_{04} = V_{07} = 1$, $V_{02} = V_{05} = V_{08} = -j$, $V_{03} = V_{06} = V_{09} = -1$. Currents in elements 1, 2, 3.

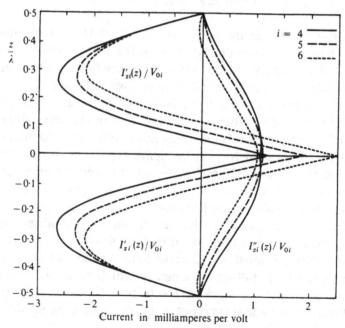

Fig. 7.15. Like Fig. 7.14 but for elements 4, 5 and 6.

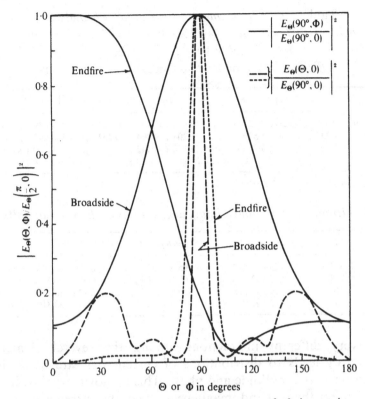

Fig. 7.16. Horizontal ($\Theta = 90°$) and vertical ($\Phi = 0°$) patterns of a 9-element planar array
of full-wave antennas; broadside and endfire excitation.

When the length of the elements is a half wavelength instead of a
full wavelength, it is usually desirable to assign the driving-point
currents $I_{zi}(0)$ instead of the voltages V_{0i}. If the array shown in
Fig. 7.12 is constructed of half-wave elements with $h/\lambda = 0.25$,
$b_x/\lambda = b_y/\lambda = 0.25, (d - 2h)/\lambda = 0.1$, and the currents are assigned for
a broadside pattern with $I_{zi}(0) = 2.5$ milliamperes for $i = 1, 2, \ldots 9$,
the coefficients for the trigonometric functions in the expression
(7.33) for the currents in the elements are those given in Table 7.5.
The required driving voltages V_{0i} are also listed together with the
associated driving-point admittances and impedances. Note that
the voltages differ considerably, as do the impedances. This is due
entirely to mutual coupling.

The fact that the driving-point currents are all maintained equal
and in phase by a suitable choice of voltages does not mean that the
several distributions of current are therefore also equal and in phase.

Table 7.5. *Nine-element planar array—broadside*
$a/\lambda = 0.007022$, $h/\lambda = 0.25$, $b/\lambda = 0.25$, $d/\lambda = 0.6$,
$I_{zi}(0) = 2.5 \times 10^{-3}$; $i = 1, 2, \dots 9$
Coefficients of trigonometric functions in milliamperes

i	A_i'	B_i'	D_i'	Q_i†	R_i†
1, 3, 7, 9	$-0.810 - j1.304$	$-4.422 - j2.913$	$20.869 + j5.492$	$-0.392 + j0.650$	$2.679 - j6.099$
2, 8	$-1.233 - j2.300$	$-5.529 - j5.054$	$23.204 + j9.404$	$-0.202 + j1.214$	$0.897 - j11.409$
4, 6	$-1.194 - j1.213$	$-5.759 - j2.193$	$24.120 + j3.342$	$0 + j0$	$0 + j0$
5	$-1.188 - j2.419$	$-7.570 - j4.620$	$27.976 + j7.513$	$0 + j0$	$0 + j0$

† Reverse signs for $i = 7, 8, 9$.

Admittance in millimhos, impedances in ohms, EMF's in volts

i	$Y_{0i} = G_{0i} + jB_{0i}$	$Z_{0i} = R_{0i} + jX_{0i}$	V_{0i}
1, 3, 7, 9	$8.203 + j4.310$	$95.5 - j50.2$	$0.239 - j0.125$
2, 8	$4.817 + j2.320$	$168.6 - j81.1$	$0.421 - j0.203$
4, 6	$6.369 + j5.538$	$89.4 - j77.7$	$0.223 - j0.194$
5	$3.713 + j2.663$	$177.8 - j127.5$	$0.445 - j0.319$

The very different interactions among the several elements necessarily leads to distributions of current that are quite dissimilar in both amplitude and phase. This is shown graphically in Fig. 7.17 for the real and imaginary parts of the currents. The real parts are seen to be more nearly triangular than cosinusoidal; the imaginary parts are quite large and distributed so differently from the real part that the phase angle is very far from constant. This means that even for half-wave elements the conventional assumption that all currents are cosinusoidally distributed and constant in phase along each element is of questionable validity for determining impedances and minor lobe structures.

The far-field pattern of the nine-element broadside planar array of half-wave elements is shown in Fig. 7.18 in the horizontal plane ($\Theta = 90°$) and the vertical plane in the direction of the maximum horizontal field ($\Phi = 90°$).

The general five-term theory is also valid for arrays that include parasitic elements. For example, in the nine-element planar array, the upper and lower rows may be parasitic with only elements 4, 5 and 6 driven and all constants the same as for the array described in Table 7.5 except that $V_{01} = V_{02} = V_{03} = V_{07} = V_{08} = V_{09} = 0$, $V_{04} = V_{05} = V_{06} = 1$. The coefficients of the trigonometric

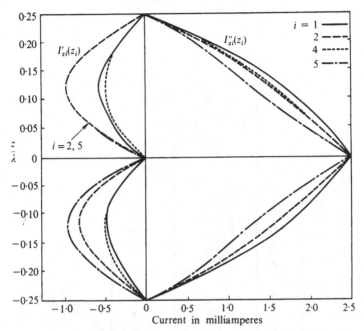

Fig. 7.17. Normalized currents in the nine elements of a planar array.
$I_{zi}(z_i) = I''_{zi}(z_i) + jI'_{zi}(z_i)$, $I_{zi}(0) = 2.5$ milliamperes with $i = 1, 2, \ldots 9$.
$a/\lambda = 0.007022$, $h/\lambda = 0.25$, $b/\lambda = 0.25$, $d/\lambda = 0.6$.

functions in the distribution of current (7.33) are as given in Table 7.6. The admittances and impedances for the three driven elements are also tabulated. The distributions of the real and imaginary parts of the current referred to the driving voltage are shown in Fig. 7.19. The currents in the collinear parasitic elements are, of course, much smaller than in the driven elements and their distributions are quite different. Note, however, that the current in the middle element No. 5 has quite a different distribution from that of the other two driven elements Nos. 4 and 6.

7.8 A three-dimensional array of twenty-seven elements†

As a final example of the application of the five-term theory consider a three-dimensional array consisting of three stacked, three-element, broadside curtains arranged in endfire as shown in Fig. 7.20. Let the lateral distances between the adjacent identical elements be $b_x/\lambda = b_y/\lambda = 0.25$, the axial distance between adjacent ends $(d-2h)/\lambda = 0.1$; also let $a/\lambda = 0.007022$. If the antennas are

† The computations in this section are those of V. W. H. Chang.

Table 7.6. *Nine-element planar array, with three elements driven*

$a/\lambda = 0.007022$, $h/\lambda = 0.25$, $b/\lambda = 0.25$, $d/\lambda = 0.6$;
$V_{01} = V_{02} = V_{03} = V_{07} = V_{08} = V_{09} = 0$; $V_{04} = V_{05} = V_{06} = 1$

Coefficients of trigonometric functions in milliamperes per volt

i	A_i'	B_i'	D_i'	Q_i†	R_i†
1, 3, 7, 9	0·020−j0·038	−0·554−j0·197	−1·305−j3·260	−2·225−j3·406	16·467−j33·106
2, 8	0·026−j0·045	−0·519−j0·340	−1·735−j4·588	−2·145−j2·471	16·336−j27·315
4, 6	−0·386−j5·485	−8·106−j14·193	54·503+j29·666	0+j0	0+j0
5	−0·309−j5·731	−7·211−j18·733	45·860+j62·562	0+j0	0+j0

† Reverse the signs for $i = 7, 8, 9$.

Admittances in millimhos and impedances in ohms

i	$Y_{0i} = G_{0i}+jB_{0i}$	$Z_{0i} = R_{0i}+jX_{0i}$
4, 6	8·244−j0·019	121·3+j0·3
5	6·530+j5·322	92·0−j75·0

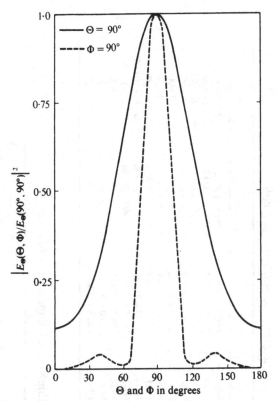

Fig. 7.18. Horizontal ($\Theta = 90°$) and vertical ($\Phi = 90°$) patterns of a planar array of nine elements with currents shown in Fig. 7.17; $a/\lambda = 0.007022, h/\lambda = 0.25, b/\lambda = 0.25, d/\lambda = 0.6$; $I_{zi}(0) = 2.5$ milliamperes, $i = 1, 2, \ldots 9$.

individually a full wavelength long ($h/\lambda = 0.5$), the desired uni-directional endfire pattern is well realized when the driving voltages (which directly excite the large sinusoidal components of the currents) are assigned the following values: $V_{1+3n} = 1, V_{2+3n} = -j$, $V_{3+3n} = -1$ with $n = 0, 1, 2, \ldots 8$. The unidirectional beam is to be in the positive x direction. With this choice of parameters the five coefficients for the trigonometric components of the current in (7.33) have been computed and listed in Table 7.7. The associated driving-point admittances and impedances are also given. These are seen to vary widely as a necessary consequence of differences in the induced currents. Since the power in each element is given by $P_{0i} = \frac{1}{2}V_{0i}^2 G_{0i}$, and $V_{0i}^2 = V_{0i}V_{0i}^* = 1$, the relative powers are proportional to the driving-point conductances. It is seen from Table 7.7 that the nine elements in the plane $x = -b_x$ (which are

Table 7.7. *Twenty-seven element three-dimensional endfire array*
Main beam in direction $\Phi = 0$

$a/\lambda = 0\cdot007022, \quad h/\lambda = 0\cdot5, \quad b_x/\lambda = b_y/\lambda = 0\cdot25, \quad d/\lambda = 1\cdot1$
$V_{1+3n} = 1, \quad V_{2+3n} = -j, \quad V_{3+3n} = -1; \quad n = 0, 1, 2, \dots 8$

Coefficients of trigonometric functions in millimhos per volt

i	A'_i	B'_i	D'_i	$Q_i\dagger$	$R_i\dagger$
1, 7, 19, 25	$0\cdot062 - j3\cdot633$	$0\cdot442 - j0\cdot687$	$1\cdot033 - j2\cdot590$	$0\cdot157 + j0\cdot529$	$0\cdot102 + j0\cdot430$
2, 8, 20, 26	$-3\cdot639 - j0\cdot017$	$-0\cdot394 - j0\cdot815$	$3\cdot419 - j0\cdot333$	$0\cdot593 - j0\cdot120$	$0\cdot225 - j0\cdot015$
3, 9, 21, 27	$0\cdot033 + j3\cdot613$	$-1\cdot355 - j0\cdot336$	$1\cdot635 - j2\cdot430$	$-0\cdot110 - j0\cdot651$	$-0\cdot309 + j0\cdot062$
4, 22	$0\cdot062 - j3\cdot623$	$0\cdot522 - j0\cdot544$	$0\cdot963 - j2\cdot358$	$0 + j0$	$0 + j0$
5, 23	$-3\cdot628 - j0\cdot017$	$-0\cdot189 - j0\cdot869$	$2\cdot314 + j0\cdot382$	$0 + j0$	$0 + j0$
6, 24	$0\cdot037 + j3\cdot599$	$-1\cdot348 - j0\cdot608$	$1\cdot601 - j2\cdot041$	$0 + j0$	$0 + j0$
10, 16	$0\cdot075 - j3\cdot590$	$0\cdot744 - j0\cdot361$	$1\cdot063 + j2\cdot997$	$0\cdot242 + j0\cdot686$	$0\cdot306 + j0\cdot474$
11, 17	$-0\cdot359 - j0\cdot025$	$-0\cdot157 - j0\cdot934$	$3\cdot637 + j0\cdot656$	$0\cdot762 + j0\cdot006$	$0\cdot494 - j0\cdot002$
12, 18	$0\cdot037 + j3\cdot559$	$-1\cdot381 - j0\cdot642$	$1\cdot915 - j2\cdot338$	$0\cdot060 - j0\cdot746$	$-0\cdot126 - j0\cdot165$
13	$0\cdot074 - j3\cdot578$	$0\cdot846 - j0\cdot422$	$0\cdot985 + j2\cdot675$	$0 + j0$	$0 + j0$
14	$-2\cdot358 - j0\cdot023$	$0\cdot097 - j0\cdot945$	$3\cdot310 - j0\cdot646$	$0 + j0$	$0 + j0$
15	$-0\cdot041 + j3\cdot545$	$-1\cdot330 - j0\cdot931$	$1\cdot828 - j1\cdot936$	$0 + j0$	$0 + j0$

† Reverse signs for $i = 7, 8, 9, 16, 17, 18, 25, 26, 27$.

Admittances in millimhos and impedances in ohms

i	$Y_{0i} = G_{0i} + jB_{0i}$	$Z_{0i} = R_{0i} + jX_{0i}$	i	$Y_{0i} = G_{0i} + jB_{0i}$	$Z_{0i} = R_{0i} + jX_{0i}$
1, 7, 19, 25	$1\cdot917 + j1\cdot217$	$371\cdot9 - j236\cdot1$	10, 16	$2\cdot552 + j1\cdot735$	$268\cdot0 - j182\cdot2$
2, 8, 20, 26	$1\cdot297 + j2\cdot630$	$150\cdot9 - j305\cdot9$	11, 17	$1\cdot212 + j3\cdot322$	$96\cdot9 - j265\cdot6$
3, 9, 21, 27	$1\cdot076 + j3\cdot102$	$99\cdot8 - j287\cdot7$	12, 18	$0\cdot847 + j3\cdot623$	$61\cdot2 - j261\cdot7$
4, 22	$2\cdot008 + j1\cdot269$	$355\cdot9 - j225\cdot0$	13	$2\cdot678 + j1\cdot830$	$254\cdot6 - j174\cdot0$
5, 23	$1\cdot357 + j2\cdot761$	$143\cdot4 - j291\cdot8$	14	$1\cdot245 + j3\cdot504$	$90\cdot0 - j253\cdot4$
6, 24	$1\cdot096 + j3\cdot257$	$92\cdot8 - j275\cdot8$	15	$0\cdot831 + j3\cdot798$	$55\cdot0 - j251\cdot3$

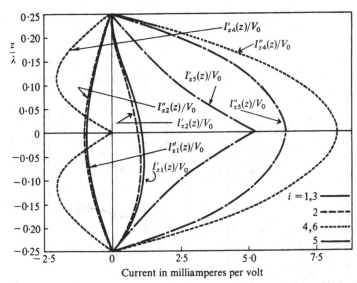

Fig. 7.19. Currents in the nine elements of a planar array. $I_{zi}(z_i) = I''_{zi}(z_i) + jI'_{zi}(z_i)$, $V_{01} = V_{02} = V_{03} = V_{07} = V_{08} = V_{09} = 0$, $V_{04} = V_{05} = V_{06} = V_0 = 1$. $a/\lambda = 0.007022$, $h/\lambda = 0.25$, $b/\lambda = 0.25$, $d/\lambda = 0.6$.

the rear elements if the forward direction along the positive x-axis is that of the maximum beam), receive the largest amount of power from the generators $(9.78V_0^2)$; the nine elements in the plane $x = 0$ the next largest amount $(5.78V_0^2)$; and the nine forward elements in the plane $x = b_x$ the smallest amount $(4.51V_0^2)$. However, the power is reasonably well divided among the elements. It is greatest in the middle elements 10, 13, 16 of the rear plane $(3.89V_0^2)$ where induced currents are relatively small; it is least in the middle elements 12, 15, 18 of the forward plane $(1.26V_0^2)$ where induced currents are relatively large.

The computed currents in 18 of the elements are shown graphically in Figs. 7.21a, b, c. The currents in elements 7, 8, 9, 16, 17, 18, 25, 26, 27 are obtained, respectively, from those in elements 1, 2, 3, 10, 11, 12, 19, 20, 21 with the substitution of $-z_i$ for z_i. Both the real and imaginary parts of the currents on differently situated but otherwise identical elements are seen to vary widely. Those in the outer tiers of elements with centres in the planes $z = \pm d$ exhibit a large asymmetry owing to the one-sidedness of the coupling.

If the full-wave elements in the array are replaced by half-wave elements $(h/\lambda = 0.25)$ with the same axial distance $(d-2h)/\lambda = 0.1$ between adjacent ends and all other conditions, including the

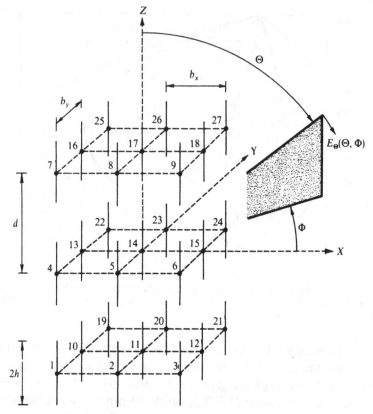

Fig. 7.20.   Three-dimensional array of 27 identical, equally-spaced elements.

driving voltages unchanged, the coefficients for the trigonometric components of the currents are computed to have the values listed in Table 7.8. The associated driving-point admittances and impedances are also given in Table 7.8. Note their very wide range.

As with the full-wave elements, the power in the nine elements in the rear plane $(x = -b_x)$ is greatest $(35 \cdot 56 \, V_0^2)$, in the nine elements in the middle plane next greatest $(4 \cdot 90 \, V_0^2)$, and in the nine elements in the forward plane least $(0 \cdot 94 \, V_0^2)$. The distribution of power is seen to be very uneven. Indeed, the currents induced in the central forward elements 12, 15, 18 are now so great that these act as negative resistances or generators. The assigned voltage at the terminals of these elements can be maintained only if loads are connected across their terminals instead of generators. This is also true of the central element 14 in the middle plane. In evaluating the powers in the elements in the three planes, the negative values were

Table 7.8. *Twenty-seven element three-dimensional endfire array*
Main beam in direction $\Phi = 0$

$a/\lambda = 0.007022$, $h/\lambda = 0.25$, $b_x/\lambda = b_y/\lambda = 0.25$, $d/\lambda = 0.6$

$V_{1+3n} = 1$, $V_{2+3n} = -j$, $V_{3+3n} = -1$, $n = 0, 1, 2, ... 8$

Coefficients of trigonometric functions in milliamperes per volt

i	A_i'	B_i'	D_i'	$Q_i†$	$R_i†$
1, 7, 19, 25	$-0.432 - j5.436$	$-9.874 - j12.025$	$62.110 + j20.189$	$-2.912 + j2.092$	$24.552 - j20.379$
2, 8, 20, 26	$-5.365 + j0.084$	$-11.235 + j1.644$	$12.666 - j11.274$	$2.203 + j0.779$	$-19.840 - j6.877$
3, 9, 21, 27	$0.033 + j5.467$	$0.526 + j12.608$	$-3.857 - j24.434$	$0.807 - j1.716$	$-7.485 + j16.453$
4, 22	$-0.411 - j5.509$	$-10.672 - j12.054$	$61.129 + j27.424$	$0 + j0$	$0 + j0$
5, 23	$-5.415 + j0.079$	$-11.015 + j1.880$	$17.355 - j11.122$	$0 + j0$	$0 + j0$
6, 24	$0.039 + j5.511$	$1.022 + j12.365$	$-5.463 - j28.184$	$0 + j0$	$0 + j0$
10, 16	$-0.373 - j5.669$	$-9.514 - j16.649$	$56.729 + j51.632$	$-2.990 + j0.788$	$26.280 - j10.367$
11, 17	$-5.461 - j0.004$	$-13.163 + j0.023$	$25.922 - j0.547$	$1.665 + j0.330$	$-15.317 - j3.610$
12, 18	$-0.024 + j5.556$	$-0.643 + j14.403$	$3.868 - j37.307$	$0.504 - j1.226$	$-5.271 + j12.429$
13	$-0.332 - j5.754$	$-10.184 - j17.271$	$53.761 + j60.828$	$0 + j0$	$0 + j0$
14	$-5.505 - j0.014$	$-13.073 - j0.071$	$30.191 + j1.477$	$0 + j0$	$0 + j0$
15	$-0.026 + j5.595$	$-0.359 + j14.166$	$3.450 - j39.822$	$0 + j0$	$0 + j0$

† Reverse signs for i = 7, 8, 9, 16, 17, 18, 25, 26, 27

Admittances in millimhos, impedances in ohms

i	$Y_{0i} = G_{0i} + jB_{0i}$	$Z_{0i} = R_{0i} + jX_{0i}$	i	$Y_{0i} = G_{0i} + jB_{0i}$	$Z_{0i} = R_{0i} + jX_{0i}$
1, 7, 19, 25	$8.749 - j0.675$	$113.6 + j8.8$	10, 16	$7.475 + j4.143$	$102.3 - j56.7$
2, 8, 20, 26	$1.742 - j2.160$	$226.2 + j280.5$	11, 17	$0.141 - j0.109$	$4432.9 + j3430.8$
3, 9, 21, 27	$0.637 + j0.015$	$1570.1 - j38.0$	12, 18	$-0.513 + j2.087$	$-111.1 - j451.9$
4, 22	$7.643 + j1.488$	$126.1 - j24.5$	13	$5.895 + j6.299$	$79.2 - j 84.6$
5, 23	$1.456 - j0.517$	$609.8 + j216.6$	14	$-0.367 + j1.275$	$-208.7 - j724.4$
6, 24	$0.617 + j1.400$	$263.5 - j598.0$	15	$-0.678 + j3.093$	$-67.6 - j308.5$

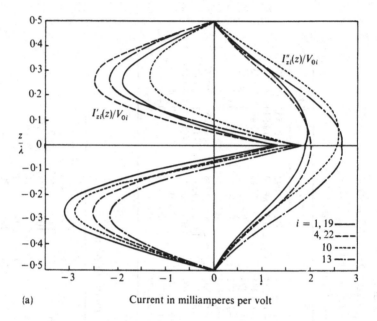

(a)   Current in milliamperes per volt

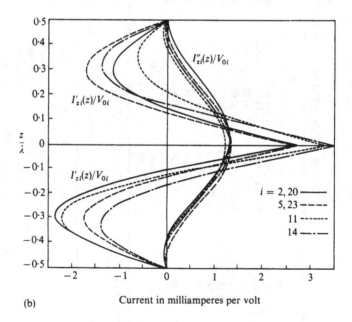

(b)   Current in milliamperes per volt

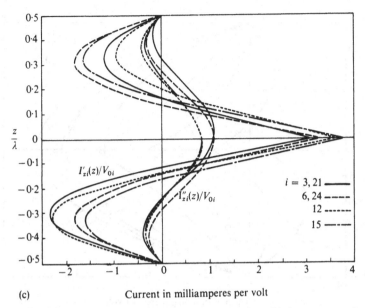

(c)   Current in milliamperes per volt

Fig. 7.21.   (a) Currents in elements Nos. 1, 4, 10 and 13 of the 27-element array shown in Fig. 7.20. $V_{1+3n} = 1$, $V_{2+3n} = -j$, $V_{3+3n} = -1$, $n = 0, 1, 2, \dots 8$; $a/\lambda = 0.007022$, $h/\lambda = 0.5$, $b_x/\lambda = b_y/\lambda = 0.25$, $d/\lambda = 1.1$. (b) Like (a) but for elements 2, 5, 11 and 14. (c) Like (a) but for elements 3, 6, 12 and 15.

subtracted since they represent power dissipated in a load, not radiated power. Note that the powers in elements 11 and 17 are not negative but very small. The entire admittance is very low, the input impedance correspondingly high. It might be supposed that these elements contribute negligibly to the radiation field. But this is not necessarily true. The fact that $I_{z11}(0)$ is near zero does not mean that $I_{z11}(z_{11})$ is everywhere equally small. It may be quite large.

The currents in the elements of the 27-element array of half-wave dipoles when the driving voltages are assigned to be $V_{1+3n} = 1$, $V_{2+3n} = -j$, $V_{3+3n} = -1$ with $n = 0, 1, \dots 8$ are shown in Fig. 7.22. Note that the currents on the elements in the rear plane (top figure) are greater than those in the middle plane (lower left) and still greater than those in the forward plane (lower right). Specifically, the current in element 11 is very small at $z = 0$, but quite comparable with the other currents out along the antenna. It is seen from Fig. 7.22 that even with half-wave elements the conventional assumption that the distributions of current along all elements are identical and cosinusoidal is not well satisfied. Since this assumption also implies that the phase of each current is the same along

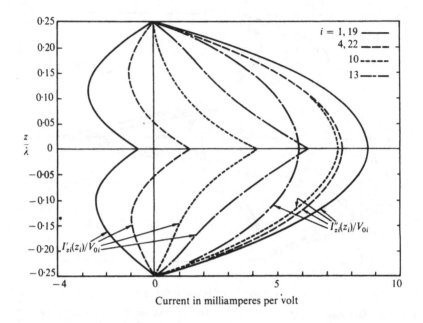

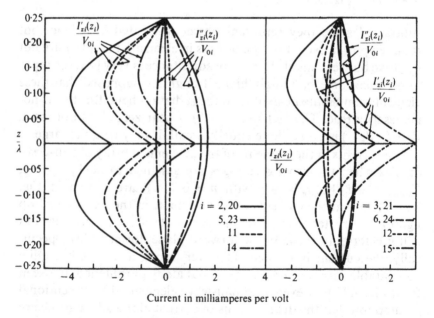

Fig. 7.22.   Like Figs. 7.21a, b, c but with $h/\lambda = 0{\cdot}25$ and $d/\lambda = 0{\cdot}6$.

the antenna as at the driving point, it is of interest to examine the relative phases referred to a common reference, viz. V_{01}. This is done in Fig. 7.23 where the phase angles of the currents along all elements are shown. For the elements in the rear plane where induced currents are not of major significance, the phase angle varies relatively little from $z = 0$ to $z = \pm h$, much as in an isolated antenna. On the other hand, when induced currents constitute the major parts of the currents in an element, the phase angle varies very widely—as much as $153°$ in the middle element 14. It is clear that when large currents are induced in some elements of an array, as in endfire arrangements which maintain a maximum field along the antennas, an assumed current with constant phase cannot be expected to represent even approximately the actual currents in an array.

Since with half-wave antennas the principal component of the current has its maximum value at $z = 0$, the progressive phases in the currents required for a specified field pattern can be approximated more closely when the maxima of the currents, i.e. the values $I_{zi}(0)$ are assigned instead of the voltages. Let the values $I_{1+3n}(0) = 2·5 \times 10^{-3}$, $I_{2+3n}(0) = -j2·5 \times 10^{-3}$, $I_{3+3n}(0) = -2·5 \times 10^{-3}$ amperes be specified for the same 27 element array. The corresponding coefficients for the currents as evaluated by computer are in Table 7.9 together with the required driving voltages and the associated admittances and impedances for the elements. Note that the voltages range from $V_{02} = -0·006 - j0·236$ to $V_{15} = -1·173 + j0·403$. The complete distributions of current are in Figs. 7.24a, b, c in the normalized form: $I_{zi}(z_i)/I_{zi}(0) = I_{zi}''(z_i)/I_{zi}(0) + jI_{zi}'(z_i)/I_{zi}(0)$.

With the driving-point currents specified, the power in each element is conveniently determined from $P_{0i} = \frac{1}{2}|I_{zi}(0)|^2 R_{0i}$. It is seen to be proportional to R_{0i} as given in Table 7.9. The distribution of power to the elements with the driving-point currents assigned is quite different from when the voltages are specified. Note that the nine elements in the rear plane ($x = -b_x$), which with voltages specified received the greatest power, now receive the least ($411·8 I_0^2$), the middle plane ($x = 0$) is again intermediate ($523·6 I_0^2$), and the elements in the forward plane ($x = b_x$), which with voltages assigned received the least power, now receive the greatest ($1484·8 I_0^2$). However, the division of power is not as extreme as before and there are no elements that have negative resistances and, therefore, feed power into a load instead of receiving

Table 7.9. *Twenty-seven-element three-dimensional endfire array*
Main beam in direction $\Phi = 0$

$$a/\lambda = 0\cdot007022, \quad h/\lambda = 0\cdot25, \quad b_x/\lambda = b_y/\lambda = 0\cdot25, \quad d/\lambda = 0\cdot6$$

$$I_{1+3n}(0) = 2\cdot5, \quad I_{2+3n}(0) = -j2\cdot5, \quad I_{3+3n}(0) = -2\cdot5 \text{ milliamperes}, \quad n = 0, 1, 2, \dots 8$$

Coefficients of trigonometric functions in milliamperes

i	A_i'	B_i'	D_i'	Q_i†	R_i†
1, 7, 19, 25	$-0\cdot981 - j1\cdot034$	$-4\cdot633 - j1\cdot870$	$21\cdot004 + j2\cdot854$	$-0\cdot096 + j0\cdot670$	$-0\cdot252 - j6\cdot390$
2, 8, 20, 26	$-1\cdot289 + j0\cdot157$	$-2\cdot825 - j2\cdot993$	$5\cdot244 - j18\cdot220$	$0\cdot702 + j0\cdot802$	$-6\cdot607 - j6\cdot543$
3, 9, 21, 27	$1\cdot088 + j3\cdot730$	$4\cdot704 + j8\cdot473$	$-20\cdot880 - j16\cdot196$	$-0\cdot066 - j1\cdot859$	$1\cdot803 + j17\cdot612$
4, 22	$-1\cdot615 - j1\cdot016$	$-6\cdot550 - j1\cdot345$	$25\cdot384 + j1\cdot124$	$0 + j0$	$0 + j0$
5, 23	$-1\cdot362 + j0\cdot257$	$-2\cdot582 + j3\cdot699$	$4\cdot165 - j0\cdot288$	$0 + j0$	$0 + j0$
6, 24	$2\cdot298 + j4\cdot128$	$8\cdot353 + j8\cdot561$	$-29\cdot207 - j15\cdot136$	$0 + j0$	$0 + j0$
10, 16	$-1\cdot524 - j1\cdot647$	$-5\cdot991 - j3\cdot109$	$23\cdot784 + j4\cdot991$	$0\cdot194 + j1\cdot044$	$-2\cdot972 - j9\cdot951$
11, 17	$-2\cdot062 + j0\cdot445$	$-4\cdot486 + j3\cdot812$	$8\cdot275 - j20\cdot031$	$1\cdot141 + j0\cdot698$	$-10\cdot697 - j5\cdot579$
12, 18	$0\cdot907 + j5\cdot463$	$4\cdot470 + j12\cdot196$	$-20\cdot701 - j22\cdot986$	$0\cdot010 - j2\cdot896$	$1\cdot103 + j26\cdot825$
13	$-2\cdot408 - j1\cdot738$	$-8\cdot521 - j2\cdot671$	$29\cdot404 + j3\cdot185$	$0 + j0$	$0 + j0$
14	$-2\cdot273 + j0\cdot712$	$-4\cdot393 + j4\cdot988$	$7\cdot235 - j23\cdot137$	$0 + j0$	$0 + j0$
15	$2\cdot341 + j0\cdot639$	$8\cdot894 + j13\cdot302$	$-30\cdot910 - j23\cdot585$	$0 + j0$	$0 + j0$

† Reverse signs for $i = 7, 8, 9, 16, 17, 18, 25, 26, 27$.

Admittances on millimhos, impedances in ohms, driving EMF's in volts

i	$Y_{0i} = G_{0i} + jB_{0i}$	$Z_{0i} = R_{0i} + jX_{0i}$	V_{0i}
1, 7, 19, 25	$7\cdot810 + j6\cdot470$	$75\cdot9 - j62\cdot9$	$0\cdot190 - j0\cdot157$
2, 8, 20, 26	$10\cdot580 + j0\cdot277$	$94\cdot4 - j2\cdot5$	$-0\cdot006 - j0\cdot236$
3, 9, 21, 27	$3\cdot430 + j0\cdot888$	$273\cdot2 - j70\cdot8$	$-0\cdot683 + j0\cdot177$
4, 22	$4\cdot299 + j6\cdot231$	$75\cdot0 - j108\cdot7$	$0\cdot186 - j0\cdot272$
5, 23	$9\cdot898 + j0\cdot910$	$100\cdot2 - j9\cdot2$	$-0\cdot023 - j0\cdot250$
6, 24	$2\cdot591 + j1\cdot353$	$303\cdot2 + j158\cdot3$	$-0\cdot758 + j0\cdot396$

i	$Y_{0i} = G_{0i} + jB_{0i}$	$Z_{0i} = R_{0i} + jX_{0i}$	V_{0i}
10, 16	$4\cdot804 + j4\cdot082$	$120\cdot9 - j102\cdot7$	$0\cdot302 - j0\cdot257$
11, 17	$6\cdot459 + j1\cdot007$	$151\cdot1 - j23\cdot6$	$-0\cdot059 - j0\cdot378$
12, 18	$2\cdot447 + j0\cdot351$	$400\cdot4 - j57\cdot5$	$-1\cdot001 + j0\cdot144$
13	$2\cdot896 + j3\cdot771$	$128\cdot1 - j166\cdot8$	$0\cdot320 - j0\cdot417$
14	$5\cdot622 + j1\cdot427$	$167\cdot1 - j42\cdot4$	$-0\cdot106 - j0\cdot418$
15	$1\cdot905 + j0\cdot654$	$469\cdot5 - j161\cdot2$	$-1\cdot173 + j0\cdot403$

power from a generator. A comparison of the relative powers in all of the elements is shown schematically in Fig. 7.23 in which boxes are located in the three-dimensional pattern of the array. The upper number in each box is the conductance G_{0i} when the conditions of Table 7.8 with voltages specified obtain; it is proportional to the power P_{0i} in each element. The lower number in each box is the resistance R_{0i} for the same array when the conditions of Table 7.9 are maintained with input currents assigned; it is proportional to the power P_{0i} in each element. The relative distribution of power is seen to be reversed.

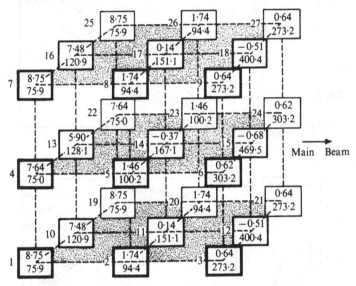

Fig. 7.23. Schematic diagram showing the relative powers supplied to the half-wave ($h/\lambda = 0.25$) elements in a 27-element endfire array. The upper number in each box is G_{0i} (in millimhos) which is proportional to power supplied when the V_{0i} are specified. The lower number is R_{0i} (in ohms) which is proportional to power when the driving-point currents $I_{zi}(0)$ are specified.

The distributions of current in Fig. 7.24a, b, c all have the same value at $z_i = 0$ and the components in phase with the input current are similarly distributed along the antenna in a rough sense. They range from a flattened cosine to a triangle. However, the quadrature currents are by no means negligible (they are presumed not to exist in conventional array theory). Indeed, they are of major significance in those elements which radiate most of the power. Note in particular the very large quadrature currents in all of the elements in the forward plane $x = b_x$ which are shown in Fig. 7.24c.

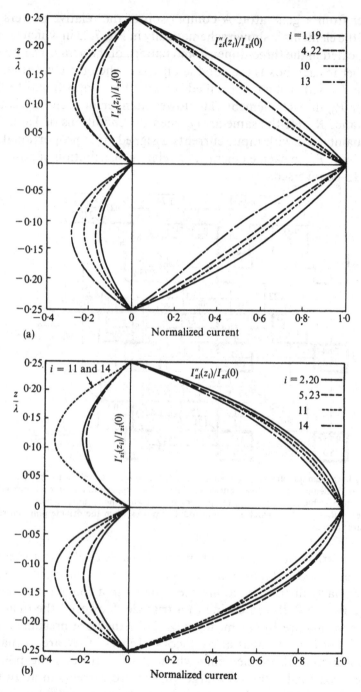

(a)

(b)

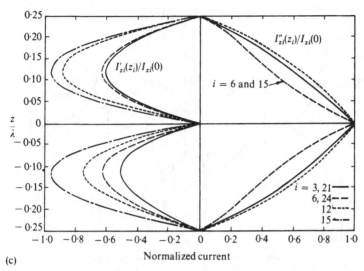

(c)

Fig. 7.24. (a) Currents in elements Nos. 1, 4, 10 and 13 of the 27-element array shown in Fig. 7.20. $I_{1+3n}(0) = 2.5$ mA, $I_{2+3n}(0) = -j2.5$ mA, $I_{3+3n}(0) = -2.5$ mA, with $n = 0, 1, 2, \ldots 8$. $a/\lambda = 0.007022$, $h/\lambda = 0.25$, $b_x/\lambda = b_y/\lambda = 0.25$, $d/\lambda = 0.6$. (b) Like Fig. 7.24a but for $i = 2$, 5, 11 and 14. (c) Like Fig. 7.24a but for $i = 3, 6, 12$ and 15.

(The currents in elements 9, 18 and 24 are like those in 3, 12 and 21 with $-z_i$ substituted for z_i.) These have distributions quite different from the conventionally assumed cosine curve. Evidently the phases are also as far from constant as those shown in Fig. 7.25 for the same array with assigned voltages.

The purpose of an array is to maintain a useful far field. The computed far-field patterns of the 27-element endfire array shown in Fig. 7.20 are in Fig. 7.26 for all the cases considered in this section, that is, for $h/\lambda = 0.5$ with voltages assigned, $h/\lambda = 0.25$ with voltages and currents assigned. The horizontal patterns in the equatorial plane $\Theta = 90°$ all have the principal maximum in the desired direction, $\Phi = 0$, $\Theta = 90°$. They also have a minor maximum in the backward direction, $\Phi = 180°$, $\Theta = 90°$. This is smallest with the array of half-wave elements with specified input currents, it is largest with the half-wave elements with voltages specified. The array of full-wave elements with voltages specified has a backward lobe of intermediate height. The vertical patterns for the array of half-wave elements are essentially the same when currents or voltages are specified. The former has a very slightly broader main beam and a correspondingly slightly lower minor lobe level. The array of full-wave elements has the narrowest main

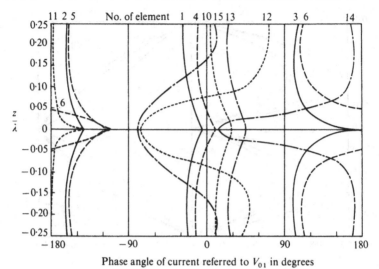

Fig. 7.25.   Phases of the currents in Fig. 7.22 all referred to V_{01}.

beam in the vertical pattern—the array is, of course, axially twice as long. On the other hand, its minor lobe level is correspondingly somewhat higher. Note that since very good approximations of actual currents on all of the elements are used, there are no nulls as would have been obtained with assumed sinusoidal currents with constant phase along each antenna. The details of the minor lobe structure derived from the five-term approximations of the several currents should have an accuracy comparable to that of the major lobe.

If all of the 27 elements are driven in phase, an approximately circular pattern with some undulations is obtained as would be expected; of interest is the fact that in this case, too, a number of the elements have negative driving-point conductances and resistances. This indicates that the induced currents in these elements predominate so that they act as generators and not as loads when connected to a transmission line. Elements with negative resistances are likely to occur in most arrays with large numbers of rather closely coupled elements.

7.9   Electrical beam scanning

The major lobe in the endfire patterns shown in Fig. 7.27 is in the direction $\Theta = 90°$, $\Phi = 0°$. This is readily switched electrically to the direction $\Theta = 90°$, $\Phi = 90°$ simply by interchanging the

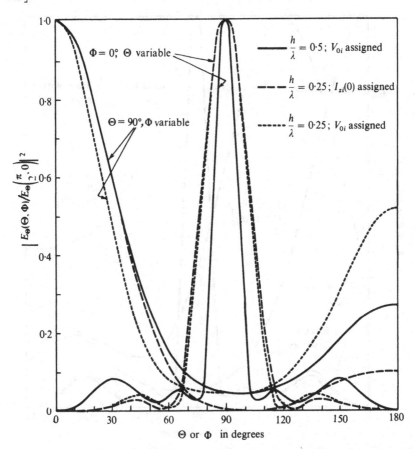

Fig. 7.26. Horizontal ($\Theta = 90°$) and vertical ($\Phi = 0°$) patterns of a three-dimensional endfire array of 27 elements; $a/\lambda = 0.007022$, $b_x/\lambda = b_y/\lambda = 0.25$, $(d-2h)/\lambda = 0.1$; $V_{1+3n} = 1$, $V_{2+3n} = -j$, $V_{3+3n} = -1$ or $I_{1+3n}(0) = 2.5$ mA, $I_{2+3n}(0) = -j2.5$ mA, $I_{3+3n}(0) = -2.5$ mA, with $n = 0, 1, 2, \ldots 8$.

phases of the voltages or currents in the broadside rows (parallel to the y-axis in Fig. 7.20) and the endfire rows (parallel to the x-axis). For example, the assigned voltages would be $V_{0i} = 1$, $1 \leqslant i \leqslant 9$; $V_{0i} = -j$, $10 \leqslant i \leqslant 18$; $V_{0i} = -1$, $19 \leqslant i \leqslant 27$ or, if the driving-point currents are assigned, $I_{zi}(0) = 2.5$ mA, $1 \leqslant i \leqslant 9$; $I_{zi}(0) = -j2.5$ mA, $10 \leqslant i \leqslant 18$; $I_{zi}(0) = -2.5$ mA, $19 \leqslant i \leqslant 27$.

More generally, the direction of the beam is specified by the far-field formula (7.57) in which the field factor of each individual antenna i is given by the square bracket, and the combination of these into a pattern for the array is determined by the phase factors $\exp(j\beta_0 \mathbf{r}_i \cdot \hat{\mathbf{R}}_0)$. The contribution to the pattern by each element is

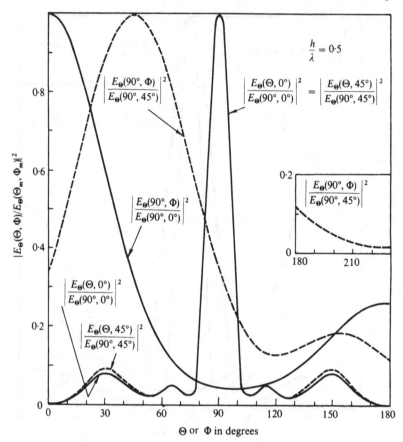

Fig. 7.27. Horizontal ($\Theta = 90°$) and vertical patterns of 27-element three-dimensional endfire array with beam in the directions $\Phi = 0°$ and $\Phi = 45°$. Driving voltages specified as in Table 7.10; $a/\lambda = 0.007022$, $h/\lambda = 0.5$, $b_x/\lambda = b_y/\lambda = 0.25$, $d/\lambda = 1.1$.

greatest when the amplitudes A_i', B_i', D_i', Q_i and R_i all include the common factor $\exp(-j\beta_0 \mathbf{r}_i \cdot \hat{\mathbf{R}}_0)$. When this is true the contributions from all the elements arrive in phase in the direction specified by particular values of Θ, Φ in $(\mathbf{r}_i \cdot \hat{\mathbf{R}}_0)$ as given in (7.58). This is, evidently, a necessary condition for a maximum in the field pattern. However, it is not a sufficient condition since the directional properties of the individual elements, as given by the square bracket in (7.57) for element i, are also involved. These may differ considerably from element to element owing to differences in the distributions of current so that no simple formula for the direction Θ_m, Φ_m of the main lobe on the field pattern can be written down. In the special case of maxima in the equatorial plane ($\Theta_m = 90°$)

the presence of the common phase factor

$$\exp(-j\beta_0 \mathbf{r}_i \cdot \hat{\mathbf{R}}_0) = \exp[-j\beta_0(X_i \cos \Phi_m + Y_i \sin \Phi_m)],$$

is sufficient to fix the main beam in the direction Φ_m. For example, when $\Phi_m = 0$, the factor is $\exp(-j\beta_0 X_i)$. This means that when voltages are assigned, these must have the relative phases $\exp(j\beta_0 b_x)$, 1, $\exp(-j\beta_0 b_x)$, respectively, for the elements in the planes $X_i = -b_x$, $X_i = 0$, and $X_i = b_r$. When $b_x = \lambda/4$ as in the arrays considered in this chapter, the phases are $\exp[j(\pi/2)] = j$, 1, and $\exp[-j(\pi/2)] = -j$; or, if j is removed as a common factor, the relative phases are given by 1, $-j$, -1, which are the values used in Table 7.7. When the beam is switched to $\Phi_m = 90°$, the phase factor is $\exp(-j\beta_0 Y_i)$ so that the voltages in the planes $Y_i = -b_y = -\lambda/4$, $Y_i = 0$, $Y_i = b_y = \lambda/4$ must have the relative phases $\exp[j(\pi/2)]$, 1, and $\exp[-j(\pi/2)]$. When driving-point currents instead of voltages are assigned, the coefficients apply to them unchanged.

If the direction of the maximum beam is to be $\Theta_m = 90°$, $\Phi_m = 45°$, the coefficients are given by

$$\exp(-j\beta_0 \mathbf{r}_i \cdot \hat{\mathbf{R}}_0) = \exp[-j\beta_0(X_i + Y_i)\sqrt{2}/2].$$

For the three-dimensional array of 27 elements shown in Fig. 7.20, the elements are located at $X_i = -b_x = -\lambda/4$, $X_i = 0$, $X_i = b_x = \lambda/4$ and $Y_i = -b_y = -\lambda/4$, $Y_i = 0$, $Y_i = b_y = \lambda/4$. Thus, the following relative phases must be assigned to the driving voltages (or currents if these are specified instead of the voltages):

$$X_i = Y_i = -\frac{\lambda}{4} \quad : \quad \exp(j\pi\sqrt{2}/2)$$

$$X_i = -\frac{\lambda}{4}, Y_i = 0; X_i = 0, Y_i = -\frac{\lambda}{4} \quad : \quad \exp(j\pi\sqrt{2}/4)$$

$$X_i = \frac{\lambda}{4}, Y_i = -\frac{\lambda}{4}; X_i = 0, Y_i = 0; X_i = -\frac{\lambda}{4}, Y_i = \frac{\lambda}{4} \quad : \quad 1$$

$$X_i = 0, Y_i = \frac{\lambda}{4}; X_i = \frac{\lambda}{4}, Y_i = 0 \quad : \quad \exp(-j\pi\sqrt{2}/4)$$

$$X_i = Y_i = \frac{\lambda}{4} \quad : \quad \exp(-j\pi\sqrt{2}/2).$$

Alternatively, if $\exp(j\pi\sqrt{2}/2)$ is removed as a common factor, the five-phase coefficients are, respectively, 1, $\exp\left(-j\pi\sqrt{2}/4\right)$, $\exp\left(-j\pi\sqrt{2}/2\right)$, $\exp\left(-j3\pi\sqrt{2}/4\right)$ and $\exp\left(-j\pi\sqrt{2}\right)$. With reference to Fig. 7.20, the required assigned voltages are listed in exponential form near the top of Table 7.10 and in complex numerical form later in the table. If these assigned voltages are used in the computer programme, the coefficients for the currents in the 27 elements of the same array analysed in Table 7.7, but with the beam rotated 45°, are as listed in Table 7.10. Note that the number of different currents is greater than when the main beam is in the direction $\Phi = 0$ or $\Phi = 90°$. The currents in the elements with centres in the plane $Z = d$ are, of course, the same as those in the plane $Z = -d$ with z_i replaced by $-z_i$. The associated admittances and impedances are also given together with the numerical values of the assigned voltages.

Since the voltage magnitudes are $|V_{0i}| = 1$, the relative powers to the elements are proportional to the driving-point conductances. In general, these are quite comparable in magnitude and range to those in Table 7.7 but the larger values are shifted to the new elements in the backward direction ($\Phi = 225°$), the smaller values to the new elements in the forward direction ($\Phi = 45°$).

The horizontal field pattern in the plane $\Theta = 90°$ and the vertical pattern in the plane $\Phi = 45°$ are shown in Fig. 7.27 together with the corresponding patterns from Fig. 7.26. It is seen that in the horizontal plane the main lobe has been rotated substantially unchanged through 45°, but that the minor lobe structure is somewhat different. The change in the vertical pattern is so small that it can be distinguished only near the peak of a minor lobe. Evidently, the rather narrow beam of a three-dimensional array of full-wave elements in collinear, broadside, and endfire combinations is readily rotated by appropriate changes in the phases of the driving voltages. A similar rotation of the corresponding array of half-wave elements is readily achieved with precisely the same changes in the phases of the assigned driving-point currents.

7.10   Problems with practical arrays

The theory developed in this and the preceding chapters provides a complete, practical tool for the quantitative determination of the properties of very general arrays when the active elements are driven by a concentrated EMF at their centres. In practice, antennas are driven from transmission lines that maintain the desired

Table 7.10. *Twenty-seven-element three-dimensional endfire array—Beam direction* $\Phi = 45°$

$$a/\lambda = 0·007022, \quad h/\lambda = 0·5, \quad b_x/\lambda = b_y/\lambda = 0·25, \quad d/\lambda = 1·1$$

$$V_{1+3n} = 1, \quad V_{2+3n} = V_{10+3n} = \exp(-j\pi\sqrt{2}/4), \quad V_{3+3n} = V_{11+3n} = V_{19+3n} = \exp(-j\pi\sqrt{2}/2),$$

$$V_{12+3n} = V_{20+3n} = \exp(-j3\pi\sqrt{2}/4), \quad V_{21+3n} = \exp(-j\pi\sqrt{2}), \quad n = 0, 1, 2$$

Coefficients of the trigonometric functions

i	A_i'	B_i'	D_i'	Q_i†	R_i†
1, 7	$0·096 - j3·641$	$0·147 - j0·558$	$1·204 + j1·871$	$0·184 + j0·572$	$0·334 + j0·528$
2, 8, 10, 16	$-3·219 - j1·657$	$-0·308 - j0·769$	$3·264 - j0·528$	$0·613 + j0·097$	$0·424 + j0·058$
3, 9, 19, 25	$-2·898 - j2·189$	$-0·974 - j0·365$	$2·792 - j2·055$	$0·300 - j0·442$	$0·104 + j0·025$
4	$0·096 - j3·631$	$0·216 - j0·427$	$1·142 + j1·641$	$0 + j0$	$0 + j0$
5, 13	$-3·210 - j1·652$	$-0·108 - j0·767$	$2·994 + j0·473$	$0 + j0$	$0 + j0$
6, 22	$-2·889 + j2·181$	$-0·878 - j0·552$	$2·632 - j0·181$	$0 + j0$	$0 + j0$
11, 17	$-2·878 + j2·160$	$-0·793 - j0·646$	$3·415 - j1·816$	$0·613 - j0·473$	$0·398 - j0·321$
12, 18, 20, 26	$0·701 + j3·532$	$-1·307 + j0·030$	$1·112 - j3·167$	$-0·141 - j0·697$	$-0·906 - j0·147$
14	$-2·867 + j2·154$	$-0·598 - j0·817$	$3·147 - j1·620$	$0 + j0$	$0 + j0$
15, 23	$0·702 + j3·519$	$-1·346 - j0·236$	$1·152 - j2·800$	$0 + j0$	$0 + j0$
21, 27	$3·500 + j0·888$	$-1·364 + j0·883$	$-0·930 - j2·181$	$-0·695 - j0·160$	$-0·207 + j0·471$
24	$3·491 + j0·879$	$-2·161 + j0·730$	$-0·596 - j1·941$	$0 + j0$	$0 + j0$

† Reverse signs for i = 7, 8, 9, 16, 17, 18, 25, 26, 27.

Admittances in millimhos, impedances in ohms

i	$Y_{0i} = G_{0i}+jB_{0i}$	$Z_{0i} = R_{0i}+jX_{0i}$	V_{0i}
1, 7	$1·498 + j0·755$	$532·3 - j268·2$	$1·0000 + j0·0000$
2, 8, 10, 16	$2·081 + j1·925$	$258·9 - j239·6$	$0·4440 + j0·8960$
3, 9, 19, 25	$1·704 + j2·358$	$201·3 - j278·6$	$-0·6057 - j0·7957$
4	$1·574 + j0·788$	$508·1 - j254·3$	$1·0000 + j0·0000$
5, 13	$2·186 + j2·019$	$246·9 - j228·0$	$0·4440 + j0·8960$
6, 22	$1·789 + j2·464$	$192·9 - j265·7$	$-0·6057 - j0·7957$

i	$Y_{0i} = G_{0i}+jB_{0i}$	$Z_{0i} = R_{0i}+jX_{0i}$	V_{0i}
11, 17	$1·366 + j3·339$	$105·0 - j256·6$	$-0·6057 - j0·7957$
12, 18, 20, 26	$0·886 + j3·335$	$74·4 - j280·1$	$-0·9819 + j0·1894$
14	$1·408 + j3·524$	$97·7 - j244·7$	$-0·6057 - j0·7957$
15, 23	$0·893 + j3·505$	$68·3 - j267·9$	$-0·9819 + j0·1894$
21, 27	$0·573 + j3·637$	$42·3 - j268·3$	$-0·2662 + j0·9639$
24	$0·550 + j3·779$	$37·4 - j257·8$	$-0·2662 + j0·9639$

voltage across the terminals of the antennas, but also introduce the complications that accompany transmission-line end-effects and the coupling between the antenna and the line. There is also the possibility of unbalanced currents on open-wire lines or on the outside surfaces of coaxial lines. These latter can be excited by asymmetrical conditions at the junctions of antennas and feeding lines, or by the intense near fields in an array whenever transmission lines are not in a neutral plane of these fields. Since such currents induced along transmission lines usually contribute significantly to the radiation field and can, therefore, constitute a non-negligible part of the load, both the circuit and field properties of an array can be modified greatly whenever they are excited. Important aspects of the problems relating to end-effects and coupling effects between antennas and transmission lines as well as techniques of measurement are considered in the next chapter. However, questions relating to the maintenance of the required voltages for antennas with positive conductances and loads for those with negative conductances in large arrays are not analysed since they involve the specific geometry of each array. A problem of this sort in which elements with both positive and negative resistances play important roles is treated in detail at the end of chapter 6 where the log-periodic array is analysed. This antenna includes not only radiating elements with specified geometrical properties but also a feeding line with definite electrical characteristics. Since it is in the neutral plane, the problems of unbalanced currents are avoided, but those relating to the transfer of power from the radiating elements to the line and vice versa constitute a major aspect of the analysis.

CHAPTER 8

TECHNIQUES AND THEORY OF MEASUREMENTS

The preceding chapters are devoted to the development of a theory to predict the characteristics of arrays of physically real dipoles and monopoles. This chapter is concerned with the experimental determination of these characteristics and with the correlation of measured and theoretical results. In practice, the mathematical intricacies of theoretical formulas can be largely avoided by means of a computer programme to which a user need only supply the parameters of a particular array to obtain radiation patterns, driving-point admittances or impedances, and other characteristics. When programmed in this manner, the computer becomes a simulator which can be substituted for the repeated testing and adjusting commonly required in designing an array. However, the value of such a simulator rests entirely on how well predictions agree with observation when the final model is assembled. Comparisons in the preceding chapters between measured and computed results indicate that the theory is capable of describing an actual experimental model with acceptable accuracy. However, in applying the theoretical results to different experimental systems, account must be taken of certain considerations and precautions if such agreement is to be obtained. It is these considerations which are of primary concern in the present chapter.

The required measuring techniques have been discussed in general in several books†; the purpose here is to examine difficulties and procedures which apply particularly to arrays of dipoles and monopoles. Although the discussions are not restricted to any particular range of frequencies, most of the procedures have been used in the 100–1200 MHz range.

8.1 Transmission lines with coupled loads

Owing to fundamental differences between theoretical and experimental models, theoretical and measured driving-point

† See [1] to [8].

impedances often do not agree if compared blindly. In the former, the radiating elements are in free space and coupled only to one another. In the latter, the radiating elements are coupled not only to one another but to transmission lines as well. Measurements are made on transmission lines that are not infinite but terminated, and equipped with probes that are rather crude approximations of the assumed ideal. Although these differences are commonly ignored, they can have important effects on the measured results.

Adequate theoretical expressions for predicting the principal characteristics of measured radiation patterns are relatively easy to obtain, for they are the result of an integration. This, in effect, yields a kind of average, so that a precise value of the current at any point on the dipole is not required, except possibly in determining the details of a minor lobe structure. In contrast, theoretical expressions for predicting the measured driving-point impedances are very difficult to obtain since they require precise values of the current near the driving point. When transmission lines are attached to antennas, they may have a negligible influence on the radiation patterns but produce important modifications in the boundary conditions and hence in the fields and currents near the driving point. The analysis of a complete system (consisting of the antennas together with their attached transmission lines) as a boundary-value problem has thus far proved to be too complicated to yield solutions, so that effects due to the presence of the transmission lines must be taken into account separately. An exception is the complete analysis of a coaxial line that drives a monopole through a hole in a highly conducting ground screen for which a formal solution is available [9].

In general, if the termination of a transmission line is to be described as a lumped circuit element and to be characterized in a useful manner as an impedance, it should ideally be independent of the circuit to which it is attached. However, such an ideal impedance can be defined only under very special conditions and these are frequently not adequately satisfied at microwave frequencies.† If they are violated, impedance measurements of a given physical load can be expected to yield different results when this is connected to physically or electrically different transmission lines.

Coupling between the transmission line and load usually extends throughout a small region near the line-load junction. This is called

† Chapter 2 of [6].

the terminal zone, and the coupling effects are called terminal-zone effects or end-effects.

Many properties of the terminal zones can be determined from the differential equations for the voltage and current along the transmission line. In the usual method of deriving these equations, the line is divided into identical infinitesimal sections, each section is represented by a lumped capacitance, an inductance, and a resistance, and Kirchhoff's laws of ordinary circuits are assumed to apply to each section. The results obtained by this method strictly apply only to infinite, unloaded lines. They can contain no information about terminal-zone effects or radiation which require a derivation based on a more complete theoretical model. Detailed steps of the more exact derivation are given in chapter 2 of [5] and [6]. Regardless of the particular transmission-line model and termination used in the derivation, the following generalized equations are obtained:

$$\frac{\partial^2 V(w)}{\partial w^2} - \gamma^2(w)V(w) = 0 \qquad (8.1a)$$

$$I_{1L}(w) = \frac{1}{z(w)}\frac{\partial V(w)}{\partial w}. \qquad (8.1b)$$

The distance, w, is measured along the transmission line from the line-load junction. $V(w)$ is the scalar potential difference or voltage between conductors of the transmission line at w, $I_{1L}(w)$ is the total current in one of the conductors at w, and the line is assumed to be perfectly balanced so that $I_{2L}(w) = -I_{1L}(w)$.

In the following definitions, $W(w)$ is the vector potential difference between the conductors at w, and a subscript p indicates the component of vector potential parallel to the transmission line. A subscript L denotes that part of a quantity which is determined only from currents and charges in the line, and a subscript T denotes that part which is determined only from currents and charges in the termination or load. The various quantities in (8.1a, b) are

$$\gamma(w) = \sqrt{z(w)y(w)a_1(w)\phi_1(w)}; \qquad \text{propagation 'constant' (8.2)}$$

$$a_1(w) = \frac{W_{pL}(w) + W_{pT}(w)}{W_{pL}(w)}; \qquad \text{coefficient of inductive (8.3)} \atop \text{coupling}$$

$$\phi_1(w) = \frac{V_L(w)}{V_L(w) + V_T(w)}; \qquad \text{coefficient of capacitive (8.4)} \atop \text{coupling}$$

$$z(w) = r(w) + j\omega l^e(w)$$

$$\doteq j\omega l^e(w)$$ impedance per unit   (8.5)

length

$$= j\omega[l_L^e(w) + l_T^e(w)]$$

$$y(w) = g(w) + j\omega c(w)$$ admittance per unit   (8.6)

$$\doteq j\omega c(w)$$ length

$$y^{-1}(w) = y_L^{-1}(w) + y_T^{-1}(w)$$

$$\beta^2 = \omega^2 \mu\varepsilon \tag{8.7}$$

$$Z_c(w) = \sqrt{\frac{z(w)}{y(w)}} \doteq \sqrt{\frac{l^e(w)}{c(w)}}$$ 'characteristic' impedance

(8.8)

μ and ε are the permeability and permittivity of the material in which the transmission-line conductors are embedded; $\mu = \mu_r\mu_0$, $\varepsilon = \varepsilon_r\varepsilon_0$, where μ_0, ε_0 are the values for free space.

Vector and scalar potentials are calculated from the Helmholtz integrals of currents and charges over all conductors. These integrals are defined in chapter 1. A detailed discussion of the transmission-line parameters is in [5] and [6], where it is shown that differences in the potential of equipotential rings located just outside the conductors of the transmission line at a distance w from the line-load junction are given approximately by

$$W_p(w) = W_{pL}(w) + W_{pT}(w) \doteq l^e(w)I_{1L}(w) \tag{8.9}$$

$$V(w) = V_L(w) + V_T(w) \doteq \frac{q_L(w)}{c(w)} \tag{8.10}$$

where $q_L(w)$ is the charge per unit length along one of the transmission-line conductors. Those parts due only to currents and charges in the line are

$$W_{pL}(w) \doteq l_L^e(w)I_{1L}(w) \tag{8.11}$$

$$V_L(w) \doteq \frac{q_L(w)}{c_L(w)}. \tag{8.12}$$

With these approximations, (8.3) and (8.4) become

$$a_1(w) \doteq l^e(w)/l_L^e(w) \tag{8.13}$$

$$\phi_1(w) \doteq c(w)/c_L(w). \tag{8.14}$$

When coupling between the line and its termination is expressed in terms of the vector potential, it is inductive; if there is no inductive coupling, $a_1(w) = 1$. Note that if all conductors of a termination are perpendicular to all conductors of the transmission line, there is no inductive coupling between them. When coupling between the line and its termination is expressed in terms of the scalar potential, it is capacitive; if there is no capacitive coupling, $\phi_1(w) = 1$.

Because the form of $\gamma(w)$ differs for each line and termination, (8.1a, b) have no general solution. For most specific lines and loads, they are too complicated to yield useful solutions although digital computer techniques [10] could be used to obtain numerical results. However, these would be of limited value since the computation would have to be repeated for each change in the geometrical arrangement of the line or load. An approximate but more general procedure, which is especially useful for experimental work, is developed in the following discussion.

A detailed examination of the parameters in (8.2)–(8.14) reveals that non-uniformities decrease rapidly with distance from the line-load junction. Along the transmission line, $a_1(w)$ and $\phi_1(w)$ usually differ negligibly from one and $z(w)$ and $y(w)$ are sensibly constant at distances from the line-load junction that exceed ten times the centre-to-centre spacing between the conductors of a two-wire line, or ten times the difference in radii between outer and inner conductors of a coaxial line. This is a rough measure of the extent of the terminal zone. For most transmission lines it is less than $0{\cdot}1\lambda$. At greater distances from the line-load junction, all parameters are constant and (8.1a, b) reduce to the usual linear form:

$$\frac{d^2 V(w)}{dw^2} - \gamma^2 V(w) = 0 \tag{8.15a}$$

$$I_{1L}(w) = \frac{1}{z_0}\frac{dV(w)}{dw} \tag{8.15b}$$

since

$$z(w) = z_0 = r + j\omega l^e \tag{8.16a}$$

$$y(w) = y_0 = g + j\omega c \tag{8.16b}$$

$$\gamma^2(w) = \gamma^2 = z_0 y_0, \tag{8.16c}$$

$$\gamma = \alpha + j\beta. \tag{8.16d}$$

Thus, except within a small terminal zone, conventional transmission-line theory applies and the usual measuring techniques are valid. Changes that occur in the line parameters over short distances near the line-load junction appear as lumped inductances and capacitances in series and parallel with the actual terminating impedance. When a load impedance is determined in the usual manner from measurements on the uniform part of the line, the quantity determined is always a combination of the actual load impedance with the inductances and capacitances caused by changes of the line parameters within the terminal zone. Approximate account can be taken of such changes if it is assumed that the uniform line parameters of an infinite unloaded line apply everywhere including the terminal zone, and the differences that occur within the terminal zone between the actual parameters and the assumed ones are represented by a balanced network of equivalent lumped series inductances and shunting capacitances, as shown in Figs. 8.1a or 8.1b. The lumped elements are defined as follows:

$$L_T = \int_0^d [l^e(w) - l_0^e]\, dw \tag{8.17a}$$

$$C_T = \int_0^d [c(w) - c_0]\, dw \tag{8.17b}$$

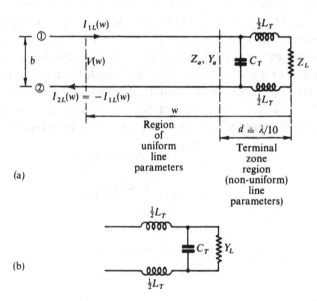

Fig. 8.1.   (a) Terminal-zone region and network. (b) Alternative representation for terminal-zone network.

where $l^e(w) = l_L^e(w) + l_T^e(w)$ is the true inductance per unit length, $c^{-1}(w) = c_L^{-1}(w) + c_T^{-1}(w)$ is the true reciprocal capacitance or elastance per unit length, and l_0^e and c_0^{-1} are the corresponding quantities for an infinite line. Everywhere along an infinite line or outside of the terminal zone of a terminated line, the ratio of the tangential component of vector potential difference to the current and the ratio of scalar potential difference to charge per unit length are constant and given by

$$W_p(w)/I_{1L}(w) = l_0^e \tag{8.18a}$$

$$V(w)/q_{1L}(w) = 1/c_0. \tag{8.18b}$$

With (8.13) and (8.14), the integrals for L_T and C_T are

$$L_T = \int_0^d [l_L^e(w)a_1(w) - l_0^e]\, dw \tag{8.19a}$$

$$C_T = \int_0^d [c_L(w)\phi_1(w) - c_0]\, dw. \tag{8.19b}$$

One advantage of the procedure of assuming line parameters to be uniform throughout the terminal zone and representing terminal-zone non-uniformities by a network of lumped elements is that useful approximate expressions for L_T and C_T can frequently be derived from considerations of the static and induction fields. Also, C_T and L_T can be obtained from measurements. If the load is characterized by its impedance and the terminal-zone network is like Fig. 8.1a, the actual load impedance Z_L, and the apparent load impedance Z_a (or the apparent load admittance Y_a), are related by

$$Y_a = \frac{1}{Z_a} = \frac{1}{Z_L + j\omega L_T} + j\omega C_T \tag{8.20}$$

$$Z_L = \frac{Z_a}{1 - j\omega C_T Z_a} - j\omega L_T = \frac{1}{Y_a - j\omega C_T} - j\omega L_T. \tag{8.21}$$

If the load is characterized by its admittance and the network of Fig. 8.1b is used,

$$Z_a = Y_a^{-1} = \frac{1}{Y_L + j\omega C_T} + j\omega L_T \tag{8.22}$$

$$Y_L = \frac{1}{Z_a - j\omega L_T} - j\omega C_T = \frac{Y_a}{1 - j\omega L_T Y_a} - j\omega C_T. \tag{8.23}$$

The terminology used here is that of the lower-frequency transmission lines. Problems involving waveguides and some of those involving coaxial lines are conveniently solved in terms of propagating and evanescent modes [11]. The latter decay rapidly with distance from a discontinuity and it is this distance which defines the extent of the terminal zone.

8.2   Equivalent lumped elements for terminal-zone networks

Whenever an antenna that is ideally approximated by an independent impedance Z_L is attached to a transmission line, the impedance that appears to be loading the transmission line is Z_a, not Z_L. The apparent impedance, Z_a, is a combination of Z_L and a terminal-zone network consisting of a series reactance $X_T = \omega L_T$, and a shunting susceptance, $B_T = \omega C_T$. This network takes account of changes in the parameters of the line as its end is approached and of coupling between the transmission line and the antenna.

L_T and C_T can be evaluated theoretically, or determined from measured values of Z_a and Z_L. The former is measured directly, the latter is obtained by repeating the measurement of Z_a as the distance between the conductors of the transmission line is decreased successively, and then extrapolating the results to a fictitious 'zero' spacing. The extrapolated value of Z_a is Z_L.

One common use of a terminal-zone network is to transform driving-point impedances which have been calculated from an established theory into those which can be measured on a particular transmission line. For this purpose, a single model of the desired antenna and its attached transmission line can be constructed, $Y_a = G_a + jB_a$ measured, and $Z_L = R_L + jX_L$ computed from the theory. Then, from (8.20),

$$\omega C_T = B_a \pm \sqrt{(G_a/R_L) - G_a^2} \qquad (8.24a)$$

$$\omega L_T = -X_L \pm \sqrt{(R_L/G_a) - R_L^2}. \qquad (8.24b)$$

For some models, Y_L may be more convenient to compute than Z_L. From (8.22),

$$\omega C_T = -B_L \pm \sqrt{(G_L/R_a) - G_L^2} \qquad (8.25a)$$

$$\omega L_T = X_a \pm \sqrt{(R_a/G_L) - R_a^2}. \qquad (8.25b)$$

The resulting values of ωL_T and ωC_T can then be used for all other elements of the same kind in the array, as long as the element spacing is not so small that the terminal zones are directly coupled to one another.

The sign to be used in (8.24) and (8.25) is usually the one which makes the magnitudes of ωL_T and ωC_T smallest; in any case, their correct values are the ones that satisfy the imaginary parts of (8.20) and (8.22). Equations (8.24) and (8.25) may involve small differences between quite large numbers so that high accuracy is required in Y_a or Z_a. Therefore account must be taken of adapters, bends, or connectors which are between the antenna and the point where the measurements are made.

Theoretical determinations of L_T and C_T can be based on (8.17) or (8.19). Expressions for the inductances and capacitances in the integrands of (8.17) are themselves integrals of the static or induction fields for the particular load and transmission line that is being analysed. An evaluation of these integrals is readily carried out by computer and numerical methods no more complicated than Simpson's rule. Approximate formulas that are applicable to dipoles as end-loads on two-wire lines and to monopoles as end-loads on coaxial lines are summarized in the following paragraphs.

Symmetrical dipole as a load on two-wire lines

This model is shown in Fig. 8.2. Approximate expressions for L_T and C_T are

$$L_T \doteq \frac{\mu}{2\pi}(b-a), \qquad -C_T/c_0 b \doteq 1{\cdot}5/\ln(b/a) \qquad (8.26)$$

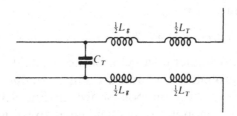

Fig. 8.2.　Network for terminal zone of dipole as end-load of two-wire line.

where a is the radius of the conductors and b the distance between the conductors of the transmission line. These expressions are accurate to within about 20 % for $b/a = 3$ and improve in accuracy as b/a increases. Expressions with higher accuracy have been derived.†

The inductance given by (8.26) accounts for non-uniformities near the end of the transmission line. Most theoretical models assume the antenna to extend from $z = 0$ to $z = \pm h$, whereas in some experimental models a section of the antenna may be missing between $z = \pm \frac{1}{2}b$. Account can be taken of this missing section by

† [6], p. 50.

subtracting from the measured input reactance the difference in zero-order input reactance between an antenna of length $(h - b/2)$ and one of length h [12]. That is,

$$X_g = \omega L_g = \frac{-\zeta_0 \psi}{2\pi}[\cot \beta_0(h - b/2) - \cot \beta_0 h] \qquad (8.27)$$

$$\psi = \begin{cases} \dfrac{|C_a(h, 0) \sin \beta_0 h - S_a(h, 0) \cos \beta_0 h|}{\sin \beta_0 h}, & \beta_0 h < \pi/2 \qquad (8.28a) \\[3mm] |C_a(h, h - \lambda/4) \sin \beta_0 h - S_a(h, h - \lambda/4) \cos \beta_0 h|, & \beta_0 h > \pi/2. \\ & \qquad\qquad\qquad (8.28b) \end{cases}$$

The integral functions $C_a(h, z)$ and $S_a(h, z)$ are defined in (1.56a, b); they are tabulated in the literature [13]. The total correction for end-effect is a shunt susceptance $B_T = \omega C_T$ with C_T given by (8.26) and a reactance $X_E/2$ in series with each conductor where

$$X_E = X_T + X_g = \omega(L_T + L_g) \qquad (8.29)$$

with L_T given by (8.26) and L_g by (8.27). The location of C_T in the network shown in Fig. 8.2 is arbitrary. It may be connected across the terminals of the dipole or across the line between L_g and L_T if more convenient.

Monopole over a ground plane fed by a coaxial line†

In this model the outer conductor of the coaxial line ends at the surface of the ground plane, and the inner conductor continues through it to form the monopole shown in Fig. 8.3a. When the coaxial line and monopole are perpendicular to the ground plane, currents on its surface are not coupled inductively to the antenna or the transmission line so that $L_T = 0$. Since the current on the inner conductor is continuous at the line-load junction, $L_g = 0$. Hence, the terminal network consists only of a shunting capacitance that is given in Fig. 8.3b. When considered on an admittance basis, this model is especially simple since the terminal-zone correction applies only to the susceptance

$$G_L + j(B_L + \omega C_T) = G_a + jB_a \qquad (8.30)$$

so that $G_a = G_L$ and B_a and B_L differ only by an additive constant for any particular combination of a coaxial transmission line and monopole antenna.

† [5], p. 430; also *Trans. I.R.E.*, **AP-3**, 66, April 1955.

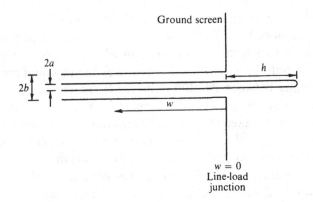

(a)

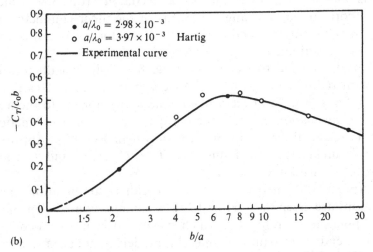

(b)

Fig. 8.3.   (a) Monopole over a ground plane. (b) Experimentally determined capacitive end correction for monopole over a ground screen driven by a coaxial line.

Change in conductor radius or spacing of two-wire lines† or coaxial lines‡

Applications frequently occur in which the conductors of the dipole or monopole must have diameters different from those of the attached feeding lines, or the base separation at the driving point must be different from the spacing between the conductors of the feeding lines. End-effects associated with these various changes have been analysed separately in the literature and the results can be applied directly, provided the conductors of the

---

† [5], p. 368 and p. 411.
‡ [5]. p. 377; [11]. p. 380; [14], p. 96.

dipole are extended to form a short section of two-wire line, or
the monopole and the associated hole in the ground plane are
extended to form a short section of coaxial line. This section of
line should be at least twice as long as the terminal zone to prevent
coupling between the different terminal regions.

Consider first a two-wire line (Fig. 8.4a) with constant spacing
between the axes of its conductors but with conductors of radius
a_l for $y \leqslant s$ and a_r for $y \geqslant s$. The current in each conductor is continuous and in only one direction near the junction at $y = s$;
therefore, $L_T = 0$ and the terminal-zone network consists only of
a shunting susceptance. The value of $B_T = \omega C_T$ for the two-wire
line is one-half of that obtained for a coaxial line (Fig. 8.4b) which
has a corresponding change in radius of the inner conductor, as
long as the ratios b/a_l and b/a_r are the same for the two-wire line
and the coaxial line.†

Consider next a two-wire line (Fig. 8.4c) with conductors of
constant radius a but with a distance b_l between their axes $y \leqslant s$
and b_r for $y \geqslant s$. Short sections of conductors normal to the axis
of the transmission line join the two parts at $y = s$. The junction
and its terminal-zone network are shown in Fig. 8.4c. Approxi-
mate formulas for calculating L_{Tl}, L_{Ty}, C_{Tl}, C_{Tr} and C_{Tc} are
straightforward but quite long.‡

Terminal-zone networks for many other combinations and
junctions are given in [5, 11 and 14]. A commonly used element is
the dipole shown in Fig. 8.5, which is centre-driven from a coaxial
line perpendicular to the antenna. This model has not been analysed
theoretically. Since it is unbalanced, currents are excited on the
outside surface of the coaxial line and this becomes a part of the
radiating source. The unbalanced condition can be avoided if a
shielded two-wire line is used instead of a coaxial line.

8.3   Voltages, currents, and impedances of uniform sections of lines

Whenever an array is driven from a single generator, the various
non-parasitic elements are connected to the generator through a
network of transmission lines that supply at the several terminals
currents and voltages which have the necessary amplitudes and
phases to produce the desired radiation pattern. In addition,

† Formulas and graphs for determining B_T for changes in the radius of both inner and outer
conductors of a coaxial line are in [11]. p. 368 and [14], p. 96.
‡ They are given respectively by (8), (13), (29), (38), and (39) of [5], p. 368 and p. 411.

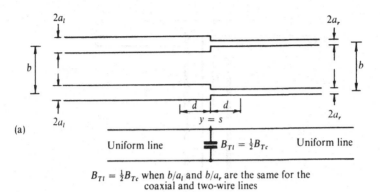

$B_{Tl} = \frac{1}{2}B_{Tc}$ when b/a_l and b/a_r are the same for the coaxial and two-wire lines

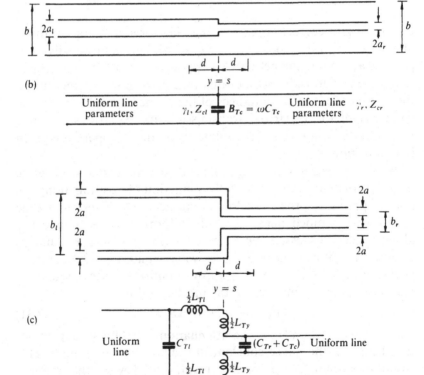

Fig. 8.4. Terminal-zone networks for changes in conductors of two-wire lines and coaxial lines. Within terminal zone of length $d \doteq \lambda/10$, line parameters are non-uniform. (a) Change in radius of two-wire line conductors; (b) change in radius of inner conductor of coaxial line; (c) change in spacing of two-wire line.

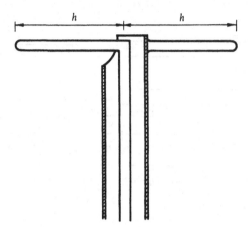

Fig. 8.5.   Dipole driven normal to the axis of a coaxial line.

they must give correct impedance matches for a maximum transfer of power. A detailed consideration of the design of power-dividers, phasing and matching networks is beyond the scope of this book.†
Experimental procedures for evaluating an array and for measuring the impedances and admittances of the elements are based on the solutions of the linearized transmission-line equations. A short review of relevant forms of the solution and their properties is given in this section.

Near line-load junctions and the ends of a transmission line the propagation 'constant' γ is usually a function of position along the line. A practical procedure for taking account of terminal-zone effects with lumped networks and uniform sections of line has already been discussed. Outside the terminal zones the line is essentially uniform and the simple wave equations with constant coefficients, (8.15a) and (8.15b), apply. Solutions of these equations may have many forms. One of the most useful is

$$V(w) = A e^{\gamma w} + B e^{-\gamma w} \tag{8.31}$$

where $\gamma = \alpha + j\beta$ and α is the attenuation constant in nepers per unit length, β the phase constant in radians per unit length. This solution is fitted to a particular line and load when the terminal conditions at the ends of the line are used to determine A and B. Note that these conditions must be specified within the terminal zones at the line-load junction or the line-generator junction. Hence, the apparent terminal impedance must be used in determining

† See [4] and [5].

A and *B*. If the line is terminated at $w = 0$ by an apparent impedance Z_a with current $I(0)$ and voltage drop $V(0)$, it follows that

$$V(0) = I(0)Z_a. \tag{8.32}$$

With (8.32), (8.31) and (8.15b) *A* and *B* are readily evaluated and the following expressions obtained:

$$V(w) = \frac{I(0)}{2}[(Z_a+Z_c)\,e^{\gamma w}+(Z_a-Z_c)\,e^{-\gamma w}] \tag{8.33a}$$

$$I(w) = \frac{I(0)}{2Z_c}[(Z_a+Z_c)\,e^{\gamma w}-(Z_a-Z_c)\,e^{-\gamma w}] \tag{8.33b}$$

where Z_c is the characteristic impedance of the transmission line.

These solutions suggest that an incident wave, travelling in the $-w$ direction from the generator at $w = s$ toward the load at $w = 0$, strikes the load and is partially or completely reflected back toward the generator. The incident and reflected parts are

$$V^+(w) = \frac{I(0)}{2}(Z_a+Z_c)\,e^{\gamma w}; \qquad \text{incident wave} \tag{8.34a}$$

$$V^-(w) = \frac{I(0)}{2}(Z_a-Z_c)\,e^{-\gamma w}; \qquad \text{reflected wave.} \tag{8.34b}$$

The relative amplitude and phase of the reflected wave are specified in terms of the apparent reflection coefficient of the load, a quantity defined by

$$\Gamma_a = \frac{V^-(0)}{V^+(0)} = \frac{I^-(0)}{I^+(0)} = \frac{Z_a-Z_c}{Z_a+Z_c} = \frac{Y_c - Y_a}{Y_c + Y_a} = |\Gamma_a|\,e^{j\psi_a}. \tag{8.35}$$

$Y_c = 1/Z_c$ is the characteristic admittance of the line. It follows that

$$V(w) = \frac{V(0)}{(1+\Gamma_a)}[e^{\gamma w}+\Gamma_a\,e^{-\gamma w}] = \frac{I(0)Z_c}{(1-\Gamma_a)}[e^{\gamma w}+\Gamma_a\,e^{-\gamma w}] \tag{8.36}$$

$$I(w) = \frac{I(0)}{(1-\Gamma_a)}[e^{\gamma w}-\Gamma_a\,e^{-\gamma w}] = \frac{V(0)Y_c}{(1+\Gamma_a)}[e^{\gamma w}-\Gamma_a\,e^{-\gamma w}]. \tag{8.37}$$

The superposition of the incident and reflected waves yields an interference pattern called a standing wave along the transmission line. When $Z_a = Z_c$, $\Gamma_a = 0$, $V^-(0) = 0$, the line is matched. For pure travelling waves outside the terminal zones the line appears to be infinite in length. When $|Z_a| \ll |Z_c|$, as when the load is a short circuit, $\Gamma_a \to -1$, and the incident wave is reflected with a 180° shift in phase. The voltage and current distributions are pure standing waves given by

$$V(w) = I(0)Z_c \sinh \gamma w, \qquad I(w) = I(0) \cosh \gamma w. \qquad (8.38)$$

When $Z_a \gg Z_c$, the entire incident wave is again reflected but with
no change in phase so that $\Gamma_a = 1$. The distributions of current and
voltage are given by (8.38) with the sinh and cosh interchanged.

The impedance looking toward the load at any point w is given
by

$$Z(w) = V(w)/I(w) = Z_c \left| \frac{1 + \Gamma_a e^{-2\gamma w}}{1 - \Gamma_a e^{-2\gamma w}} \right| \qquad (8.39)$$

and the admittance is $Y(w) = 1/Z(w)$.

Alternative expressions in terms of the hyperbolic functions are

$$V(w)/V(0) = \cosh \gamma w + (Y_a/Y_c) \sinh \gamma w \qquad (8.40a)$$

$$I(w)/I(0) = \cosh \gamma w + (Z_a/Z_c) \sinh \gamma w \qquad (8.40b)$$

$$V(w)/[I(0)Z_c] = \sinh \gamma w + (Z_a/Z_c) \cosh \gamma w \qquad (8.41a)$$

$$I(w)/[V(0)Y_c] = \sinh \gamma w + (Y_a/Y_c) \cosh \gamma w \qquad (8.41b)$$

$$Z(w)/Z_c = \frac{Z_a/Z_c + \tanh \gamma w}{1 + (Z_a/Z_c) \tanh \gamma w}. \qquad (8.42)$$

The preceding equations for current, voltage and impedance
express $V(w)$ and $I(w)$ in terms of $V(0)$ and $I(0)$ at the load. They do
not involve the actual driving voltage of the generator. A complete
solution is obtained by imposing boundary conditions at both
ends of the line† in terms of a generator with apparent internal
impedance Z_g and voltage V^e at $w = s$ or $y = s - w = 0$ and a
load with an apparent impedance Z_a at $w = 0$ or $y = s$. Specifically

$$y = 0: \qquad V_0 = V^e - I_0 Z_g$$

$$y = s: \qquad V_s = I_s Z_a.$$

The elimination of A and B in (8.31) yields

$$V(y) = \frac{V^e Z_c}{Z_c + Z_g} \frac{e^{-\gamma y} + \Gamma_a e^{-\gamma(2s-y)}}{1 - \Gamma_g \Gamma_a e^{-2\gamma s}} \qquad (8.43a)$$

$$I(y) = \frac{V^e}{Z_c + Z_g} \frac{e^{-\gamma y} - \Gamma_a e^{-\gamma(2s-y)}}{1 - \Gamma_g \Gamma_a e^{-2\gamma s}} \qquad (8.43b)$$

where Γ_g is the reflexion coefficient corresponding to Z_g.

The introduction of functions to describe separately the attenuation and the phase characteristics of the terminations makes it

† [5], p. 75.

possible to express currents and voltages in a completely hyperbolic form. These terminal functions are defined as follows:

$$\rho + j\phi = \coth^{-1}\frac{Z}{Z_c} = \tanh^{-1}\frac{Y}{Y_c} \tag{8.44a}$$

or
$$\rho + j\phi' = \coth^{-1}\frac{Y}{Y_c} = \tanh^{-1}\frac{Z}{Z_c}. \tag{8.44b}$$

The corresponding expressions for the currents and voltages are:

$$\frac{V(w)}{V^e} = \frac{\sinh(\rho_g + j\phi_g)\cosh[(\alpha w + \rho_a) + j(\beta w + \phi_a)]}{\sinh[(\alpha s + \rho_g + \rho_a) + j(\beta s + \phi_g + \phi_a)]} \tag{8.45a}$$

$$\frac{I(w)}{V^e} = \frac{\sinh(\rho_g + j\phi_g)\sinh[(\alpha w + \rho_a) + j(\beta w + \phi_a)]}{\sinh[(\alpha s + \rho_g + \rho_a) + j(\beta s + \phi_g + \phi_a)]}. \tag{8.45b}$$

The effects of a termination are now shown to be equivalent to those of a section of transmission line with a total loss specified by ρ and a total phase shift specified by ϕ. Note that the denominator of (8.45a, b) includes the total loss and total phase shift of the line plus its terminations at both ends. Impedance and admittance are given by

$$Z(w) = \coth[(\alpha w + \rho_a) + j(\beta w + \phi_a)] \tag{8.46a}$$
$$Y(w) = \tanh[(\alpha w + \rho_a) + j(\beta w + \phi_a)]. \tag{8.46b}$$

The terminal functions and the reflexion coefficient are related as follows:

$$\Gamma = |\Gamma|e^{j\psi} = e^{-2(\rho + j\phi)}$$
$$|\Gamma| = e^{-2\rho} = \frac{\coth\rho - 1}{\coth\rho + 1}, \qquad \psi = -2\phi \tag{8.47}$$

$$\rho = \tfrac{1}{2}\ln 1/|\Gamma| = \coth^{-1}\frac{1+|\Gamma|}{1-|\Gamma|}. \tag{8.48}$$

For most transmission lines that are useful as feeders for an array or for measuring sections, the line losses are very small and can often be neglected. Under these conditions

$$\gamma \doteq j\beta, \qquad Z_c \doteq R_c = \sqrt{l^e/c}$$

and (8.36) and (8.37) give

$$V(w) \doteq \frac{V(0)e^{j\beta w}}{1+\Gamma_a}[1+|\Gamma_a|e^{-j(2\beta w - \psi_a)}] \tag{8.49a}$$

$$I(w) \doteq \frac{I(0)e^{j\beta w}}{1-\Gamma_a}[1-|\Gamma_a|e^{-j(2\beta w - \psi_a)}]. \tag{8.49b}$$

These have the following maxima and minima:

$$|V_{\max}(w)| = \left|\frac{V(0)}{1+\Gamma_a}\right|[1+|\Gamma_a|], \qquad |I_{\min}(w)| = \left|\frac{I(0)}{1-\Gamma_a}\right|[1-|\Gamma_a|],$$

$$2\beta w - \psi_a = 0, 2\pi, \ldots = 2n\pi \qquad\qquad (8.50a)$$

$$|V_{\min}(w)| = \left|\frac{V(0)}{1+\Gamma_a}\right|[1-|\Gamma_a|], \qquad |I_{\max}(w)| = \left|\frac{I(0)}{1-\Gamma_a}\right|[1+|\Gamma_a|],$$

$$2\beta w - \psi_a = \pi, 3\pi, \ldots = (2n+1)\pi, n = 0, 1, 2 \ldots. \quad (8.50b)$$

The ratio of the maximum-to-minimum of either the current or the voltage on a lossless line is called the standing wave ratio and abbreviated SWR.

$$\text{SWR} = |V_{\max}(w)/V_{\min}(w)| = |I_{\max}(w)/I_{\min}(w)| = (1+|\Gamma_a|)/(1-|\Gamma_a|)$$

$$(8.51)$$

$$= \frac{|Z_a+Z_c|+|Z_a-Z_c|}{|Z_a+Z_c|-|Z_a-Z_c|} = \frac{|Y_c+Y_a|+|Y_c-Y_a|}{|Y_c+Y_a|-|Y_c-Y_a|} = \coth\rho. \quad (8.52)$$

The distributions of voltage and current are periodic and repeat every half wavelength; the adjacent maxima and minima of voltage or current are separated by a quarter wavelength; the current maxima occur at voltage minima and vice versa. The impedance and admittance looking toward the load also repeat every half wavelength. At maxima and minima of the current or voltage the impedance and admittance are real with the following values [from (8.39)]:

Voltage maxima or current minima

$$2\beta w - \psi_a = 0, 2\pi, 4\pi, \ldots$$

$$Z(w) = R_c\left[\frac{1+|\Gamma_a|}{1-|\Gamma_a|}\right] = R_c[\text{SWR}]; \qquad Y(w) = G_c\left[\frac{1-|\Gamma_a|}{1+|\Gamma_a|}\right]$$

$$= G_c/[\text{SWR}]. \quad (8.53a)$$

Voltage minima or current maxima

$$2\beta w - \psi_a = \pi, 3\pi, 5\pi, \ldots$$

$$Z(w) = R_c\left[\frac{1-|\Gamma_a|}{1+|\Gamma_a|}\right] = R_c/[\text{SWR}]; \qquad Y(w) = G_c\left[\frac{1+|\Gamma_a|}{1-|\Gamma_a|}\right]$$

$$= G_c[\text{SWR}]. \quad (8.53b)$$

The relative distributions of current and voltage for lossless lines are:

$$V(w)/V(0) = \cos\beta w + j(Y_a/G_c)\sin\beta w \qquad (8.54a)$$

$$I(w)/I(0) = \cos \beta w + j(Z_a/R_c) \sin \beta w \qquad (8.54b)$$

$$V(w)/[I(0)R_c] = (Z_a/R_c) \cos \beta w + j \sin \beta w \qquad (8.55a)$$

$$I(w)/[V(0)G_c] = (Y_a/G_c) \cos \beta w + j \sin \beta w. \qquad (8.55b)$$

The corresponding impedance is

$$Z(w) = R_c \left[\frac{(Z_a/R_c) + j \tan \beta w}{1 + j(Z_a/R_c) \tan \beta w} \right]. \qquad (8.56)$$

The admittance is given by an identical expression with Z_a and R_c replaced by Y_a and G_c.

The input power to a section of line is $P = VI^* = |V|^2/Z$ with the asterisk denoting the complex conjugate. At a voltage or current maximum (8.53a, b) give

$$P = |V_{max}|^2/R_c[SWR] = |I_{max}|^2 R_c/[SWR]. \qquad (8.57)$$

Equation (8.57) is sometimes of help in measuring the relative power in the branches of a feeding network.

Useful properties of a quarter-wave section of transmission line follow from (8.54) and (8.55). If $I(0)$ and $V(0)$ are the required driving-point current and voltage and, if $\beta w = \pi/2$,

$$V\left(\frac{\pi}{2}\right) = jR_c I(0), \qquad I\left(\frac{\pi}{2}\right) = jV(0)/R_c, \qquad Z\left(\frac{\pi}{2}\right) = R_c^2 Y_a.$$

8.4 Distribution-curve and resonance-curve measuring techniques

A number of techniques have been developed for measuring the apparent impedance or admittance of a load terminating a transmission line. When the load is a monopole or a dipole and the frequency ranges from a few hundred to a few thousand MHz, a choice is usually made between the distribution-curve or the resonance-curve methods. In the distribution-curve method, the total length of the transmission line and its excitation point remain fixed and a loosely-coupled probe is moved along the line to locate a current or voltage minimum and measure either the SWR or the curve width at twice-minimum power. In the resonance-curve or Chipman method a movable short circuit is used to tune the line with its terminations to resonance by adjusting its length. A small loop probe projecting from the short circuit is used to locate a resonance maximum and measure either the curve width at half-maximum power or the SWR.

There are variations of these methods which are sometimes convenient. Three fixed probes may be used with the distribution-

curve method† in place of a single movable one, or a pair of directional couplers may be used to sample the incident and reflected
wave directly.‡ An advantage of the resonance-curve method is
that it can be used with a receiving as well as with a transmitting
antenna. When the antenna is driven, the line can be excited
by a loosely coupled stub probe located on the transmission line
near the line-load junction. Also, in this method, the locations of
the generator and sampling probe can be interchanged.

Distribution-curve and resonance-curve methods are illustrated
schematically in Figs. 8.6 and 8.7. The quantities to be measured
are the position of a maximum or minimum and either a SWR or
a curve width. With reference to accuracy in the measured quantities, three significant figures in the SWR are usually sufficient.

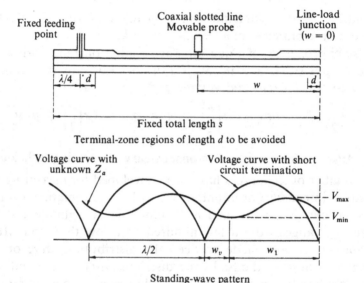

Fig. 8.6.   Distribution-curve method of measurement.

(For example, uncertainties of ± 0.02 in the SWR and $\pm 0.02\lambda$ in
the location of a current minimum of a thin quarter-wave monopole driven over a ground screen at a wavelength of about 44 cm
may introduce errors of up to 2% in measured values of G_a and up
to 5% in measured values of B_a [15].) This is readily achieved
except with very high or very low standing wave ratios. When the
actual ratio of maximum-to-minimum is measured, the detecting

† See chapter 4 of [5].
‡ See p. 235 of [2].

Coaxial line with sliding short-circuiting plunger

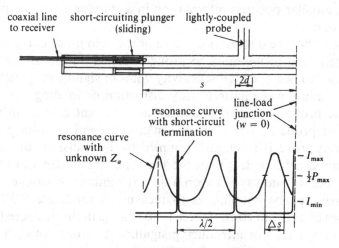

Resonance-curve pattern (current)

Fig. 8.7. Resonance-curve method of measurement.

and display system must be linear or obey a square law over a wide range, or a calibrated attenuator must be available. The curve width is generally more convenient to use if the ratios to be measured exceed ten with the distribution-curve method or five with the resonance-curve method. If Δw is the curve width at twice-minimum power for a distribution curve or a half-maximum power for a resonance curve, the SWR and apparent terminal attenuation function, ρ_a, are given by [5] in its eqs. (19) and (23).

$$\text{SWR} = \coth \pi \frac{\Delta w}{\lambda_g} \doteq \frac{\lambda_g}{\pi \Delta w} \doteq \frac{1}{\rho_a} \qquad (8.58)$$

where λ_g is the wavelength measured on the transmission line. The approximate equalities in (8.58) are true when $\Delta w/\lambda_g \ll 1$. When Δw is small or when the SWR is small, the distribution or resonance curve may require graphing on an enlarged scale in order to obtain sufficient accuracy. In determining the location of a distribution-curve minimum or a resonance-curve maximum, readings on each side of the minimum or maximum at several power levels should be averaged to increase the accuracy.

Much of the tedious work of graphing can be avoided with a recorder. The use of a miniature d.c. motor to drive the probe and a coupled miniature selsyn motor permits the direct acquisition of linear position information at a convenient scale to replace the usual angular position information in a standard antenna pattern recorder.

For either the distribution-curve or the resonance-curve method the detecting probe must be properly tuned and loosely coupled in order to provide high sensitivity without significantly distorting the standing wave pattern. Any distortion or loading introduced by the probe is most pronounced at a current maximum with a current probe and at a voltage maximum with a voltage probe. When a probe is too tightly coupled to the transmission line, the measured SWR is less than the true one and maxima are shifted from points midway between adjacent minima. A simple test for excessive probe coupling is to measure a moderate SWR with probes of different sizes. If there is no change in the measured SWR, the probes are not introducing significant errors. Another useful test is to measure the location of a maximum, the adjacent minima, and the curve width at half-maximum power with a short circuit as the termination. If the probe is introducing no errors, power variations about the maximum should be like $\cos^2 \beta w$ [from (8.54)] and the maximum should fall midway between the minima. This test is particularly severe, for the standing wave is very large and only the probe is absorbing power from the line.

A short circuit at or very near the line-load junction is commonly used as reference in determining electrical distances from the junction to a convenient distribution-curve minimum or resonance-curve maximum. If the feeding line is coaxial, this may simply be a conducting plug which makes good contacts with the inner and outer conductors. If the feeding line is an unshielded two-wire line, the short circuit may be provided by a conducting disk which makes good contact with both conductors and which has a radius of at least five times the centre-to-centre spacing of the conductors [16].

When standing wave measurements are made on open two-wire lines, the principal difficulty that is encountered is keeping the lines balanced so that $I_2(w) = -I_1(w)$. This requires that perfect symmetry be maintained everywhere in the vicinity of the lines. A small probe placed midway between the conductors and connected to a sensitive detector is usually necessary to monitor the condition of balance. When the lines are perfectly balanced, nothing is received on the monitor probe.

The resonance-curve method requires that the distances y and

w, which specify the location of the probes with respect to the ends of the line in Fig. 8.7, remain fixed and only the total length, s, be varied. One method already suggested is to use a probe attached to a sliding short circuit for sampling the current along the line.

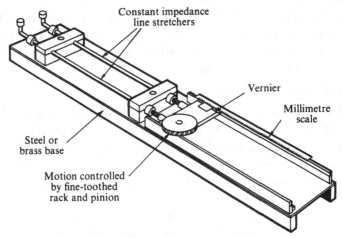

Fig. 8.8.   Trombone line stretcher or phase shifter for precision measurements.

However, a movable short circuit that maintains good electrical contact during its motion is difficult to construct. If the load being investigated has a small loss, the resonance-curve maximum is especially sensitive to erratic contacts. When there is sufficient room between inner and outer conductors, a 'non-contacting' short circuit can be used [17, 18]. An alternative method that is very satisfactory is to hold the short-circuit fixed and adjust the line length with a constant-impedance-trombone line as shown in Fig. 8.8. Additional characteristics of the individual components, the effects of probe errors, errors introduced by irregularities in the slotted line, and methods for correcting some of the errors are discussed in [1, 2, and 8].

The measured SWR and location for a minimum or maximum can be used to determine the apparent load impedance or admittance by means of a Smith chart, rectangular or other type of impedance chart, or by direct calculation from formulas. The graphical techniques are generally useful but they frequently introduce uncertainties as large as those involved in the measurement of SWR's and distances. In order to preserve all of the experimental precision it is usually advantageous to determine the impedances or admittances from formulas. In fact, if many measurements are to be converted, the use of a small digital computer is faster and cheaper since the required averaging can be done on the computer and it is then unnecessary to graph current or voltage distributions.

A useful property of the terminal functions defined in (8.44) is their comparatively slow variation with respect to $\beta_0 h$. Hence, graphs of these functions are useful for determining errors or irregularities in the measured data.

Distribution-curve method

Assume that a voltage probe is used to measure the SWR and the location of a voltage minimum. Let w_n be the distance from the line-load junction to the nth voltage minimum; in Fig. 8.6, $n = 1$. If transmission line losses are negligible over the section of line used in the measurements, the impedance at a voltage minimum is $R_c/[\text{SWR}]$. From (8.56),

$$\frac{1}{[\text{SWR}]} = \left[\frac{Z_a/R_c + j \tan \beta w_n}{1 + j(Z_a/R_c) \tan \beta w_n}\right].$$

When this equation is solved for $Z_a = R_a + jX_a$, the result is

$$Z_a = R_a + jX_a = \begin{cases} R_c\left[\dfrac{1 - j[\text{SWR}] \tan \beta w_n}{[\text{SWR}] - j \tan \beta w_n}\right] & \text{Voltage probe (8.59a)} \\[4mm] \dfrac{R_c}{[\text{SWR}]^2 + \tan^2 \beta w_n}\{[\text{SWR}](1 + \tan^2 \beta w_n) \\[3mm] \qquad + j(1 - [\text{SWR}]^2) \tan \beta w_n\}. \qquad\qquad (8.59\text{b}) \end{cases}$$

$\beta = 2\pi/\lambda$, n is the number of the minimum corresponding to w_n, λ is the wavelength along the transmission line.

If a current probe is used and w_n is the distance from the line-load junction to a current minimum,

$$Z_a = R_a + jX_a = \begin{cases} R_c\left[\dfrac{[\text{SWR}] - j \tan \beta w_n}{1 - j[\text{SWR}] \tan \beta w_n}\right] & \text{Current probe (8.59c)} \\[4mm] \dfrac{R_c}{1 + [\text{SWR}]^2 \tan^2 \beta w_n}\{[\text{SWR}](1 + \tan^2 \beta w_n) \\[3mm] \qquad + j([\text{SWR}]^2 - 1) \tan \beta w_n\}. \qquad\qquad (8.59\text{d}) \end{cases}$$

The reflexion coefficient is defined by (8.35) with magnitude and phase given by

$$|\Gamma_d| = \frac{[\text{SWR}] - 1}{[\text{SWR}] + 1} \qquad\qquad (8.60\text{a})$$

$$\psi_a = \begin{cases} 2[\beta w_n - (n + \tfrac{1}{2})\pi] & \text{Voltage probe} \quad\quad (8.60\text{b}) \\ 2[\beta w_n - n\pi] & \text{Current probe.} \quad\quad (8.60\text{c}) \end{cases}$$

For high SWR's the curve-width method may be used and the SWR calculated from (8.58). If the admittance is to be determined and a voltage probe is used, (8.59c) and (8.59d) apply with $G_a + jB_a$ substituted for $R_a + jX_a$ and with G_c substituted for R_c. If a current probe is used for admittance measurements, (8.59a) and (8.59b) apply with the indicated substitutions.

The terminal functions defined by (8.44) can be obtained from (8.47) and (8.48). The relations are

$$\rho_a = \coth^{-1} \frac{1 + |\Gamma_a|}{1 - |\Gamma_a|} = \coth^{-1} \text{SWR} = \tfrac{1}{2} \ln \frac{\text{SWR} + 1}{\text{SWR} - 1} \quad (8.61\text{a})$$

$$\doteq \pi \Delta w / \lambda, \quad \Delta w = \text{curve width} \quad\quad\quad (8.61\text{b})$$

$$\phi_a = -\psi_a / 2 = \begin{cases} (n + \tfrac{1}{2})\pi - \beta w_n & \text{Voltage probe} \quad (8.62\text{a}) \\ n\pi - \beta w_n & \text{Current probe.} \quad (8.62\text{b}) \end{cases}$$

Real and imaginary parts of impedances or admittances in terms of the terminal functions can be found from (8.44a, b). The results are

$$Z_a = R_a + jX_a = Z_c \left\{ \frac{\sinh 2\rho_a}{\cosh 2\rho_a - \cos 2\phi_a} - j \frac{\sin 2\phi_a}{\cosh 2\rho_a - \cos 2\phi_a} \right\}$$
$$(8.63\text{a})$$

$$Y_a = G_a + jB_a = Y_c \left\{ \frac{\sinh 2\rho_a}{\cosh 2\rho_a + \cos 2\phi_a} + j \frac{\sin 2\phi_a}{\cosh 2\rho_a + \cos 2\phi_a} \right\}.$$
$$(8.63\text{b})$$

Frequently the total distance w_n from the line-load junction to a convenient minimum is difficult to measure accurately. Since on a lossless line the impedance is repeated at intervals of $\lambda/2$ or $\beta w = \pi$ radians, it is necessary only to determine the location of a minimum with respect to an integral number of half wavelengths from the junction. If the load being investigated is removed and a short circuit is placed at the junction, a voltage null will appear at the junction and along the line at each half wavelength from the junction. Let w_v be the distance from the nth voltage null with the short circuit as a load to the nearest voltage minimum with the antenna as a load. Distances toward the generator are positive, those toward the load are negative. Then,

$$\beta w_n = n\pi \pm \beta w_v \quad\quad\quad (8.64)$$

and
$$\tan(n\pi \pm \beta w_v) = \frac{\tan n\pi \pm \tan \beta w_v}{1 \pm \tan n\pi \tan \beta w_v} = \pm \tan \beta w_v$$

so that w_v, the shift in the location of a minimum when the unknown impedance is substituted for the short circuit, can be used directly in (8.59a) and (8.59b) if due regard is given to the sign. Similar results hold for a current minimum and (8.59c) and (8.59d). In terms of the minimum shift, the phase of the terminal functions and of the reflexion coefficient is

$$\phi_a = -\psi_a/2 = \begin{cases} \pi/2 \mp \beta w_v & \text{Voltage probe} & (8.65a) \\ \mp \beta w_v & \text{Current probe.} & (8.65b) \end{cases}$$

The various distances are illustrated in Fig. 8.6.

Resonance-curve method

Let the solution of (8.45) be written for a balanced generator located at an arbitrary point $y = y_g$ along the transmission line instead of at $y = 0$. Also let the hyperbolic functions be separated into their real and imaginary parts.† Then the magnitudes of currents and voltages along the line are:

$$|I| = \frac{V^e}{R_c}\ \frac{\sqrt{\sinh^2(\alpha y_g+\rho_g)+\sin^2(\beta y_g+\phi_g)}\ \times\sqrt{\sinh^2(\alpha w+\rho_a)+\sin^2(\beta w+\phi_a)}}{\sqrt{\sinh^2(\alpha s+\rho_g+\rho_a)+\sin^2(\beta s+\phi_g+\phi_a)}}$$

$$|V| = V^e\ \frac{\sqrt{\sinh^2(\alpha y_g+\rho_g)+\sin^2(\beta y_g+\phi_g)}\ \times\sqrt{\sinh^2(\alpha w+\rho_a)+\cos^2(\beta w+\phi_a)}}{\sqrt{\sinh^2(\alpha s+\rho_g+\rho_a)+\sin^2(\beta s+\phi_g+\phi_a)}}$$

$$y_g \leqslant y \leqslant s$$

$$|I| = \frac{V^e}{R_c}\ \frac{\sqrt{\sinh^2(\alpha w_g+\rho_a)+\sin^2(\beta w_g+\phi_a)}\ \times\sqrt{\sinh^2(\alpha y+\rho_g)+\sin^2(\beta y+\phi_g)}}{\sqrt{\sinh^2(\alpha s+\rho_g+\rho_a)+\sin^2(\beta s+\phi_g+\phi_a)}}$$

$$|V| = V^e\ \frac{\sqrt{\sinh^2(\alpha w_g+\rho_a)+\sin^2(\beta w_g+\phi_a)}\ \times\sqrt{\sinh^2(\alpha y+\rho_g)+\cos^2(\beta y+\phi_g)}}{\sqrt{\sinh^2(\alpha s+\rho_g+\rho_a)+\sin^2(\beta s+\phi_g+\phi_a)}}$$

$$0 \leqslant y \leqslant y_g$$

where $w_g = s - y_g$ is the distance from the load to the generator.

† See chapter 4 of [5].

The first two equations give current and voltage distributions between the load at $y = s$ and the generator as illustrated in Fig. 8.7. The last two equations give the distributions between the generator and the end of the line at $y = 0$. If the distance y_g or w_g between the appropriate ends of the line and the generator, and the distance w or y between the appropriate ends of the line and the probe are held constant, while the total length s is changed, the current and voltage vary in the same manner,

$$I(w) \sim V(w) \sim [\sinh^2 (\alpha s + \rho_g + \rho_a) + \sin^2 (\beta s + \phi_g + \phi_a)]^{-1/2}. \quad (8.66)$$

The line is resonant when (8.66) has its maximum value. Maxima and minima of (8.66) are defined by

$$(\beta s + \phi_g + \phi_a) = \begin{cases} n\pi - \frac{1}{2}\sin^{-1}\left[\dfrac{\alpha}{\beta}\sinh 2(\alpha s + \rho_g + \rho_a)\right] & \text{Maxima} \\[2ex] (n+\frac{1}{2})\pi + \frac{1}{2}\sin^{-1}\left[\dfrac{\alpha}{\beta}\sinh 2(\alpha s + \rho_g + \rho_a)\right] & \text{Minima.} \end{cases}$$

For lossless lines, $\alpha/\beta = 0$, and

$$(\beta s_{\max} + \phi_g + \phi_a) = n\pi \qquad \text{Maxima} \qquad (8.67a)$$

$$(\beta s_{\min} + \phi_g + \phi_a) = (n+\frac{1}{2})\pi \qquad \text{Minima.} \qquad (8.67b)$$

With a short circuit at $w = 0$, $\phi_a = \pi/2$; this value is commonly used as a reference. Let s_s be a convenient resonant length with a short circuit as the termination at $w = 0$ and let s_1 be the corresponding resonant length with the unknown impedance at $w = 0$. When (8.67a) is written successively for both loads and the one is subtracted from the other, the result is

$$\phi_a = \pi/2 + \beta(s_s - s_1). \qquad (8.68)$$

For lines with very small losses, $\alpha s \ll 1$, the ratio of maximum-to-minimum in (8.66) is

$$\text{SWR} \doteq \coth (\rho_g + \rho_a). \qquad (8.69)$$

This equation illustrates an important additional requirement in the resonance-curve method that does not occur in the distribution-curve method. In the distribution-curve method, only the parameters that characterize the generator are involved in the measurements; in the resonance-curve method, ρ_g must be known or the generator must be lightly coupled so that $\rho_g \ll \rho_a$ for the loads under investigation. When these conditions are satisfied, ρ_a is given by (8.69), and impedances or admittances can be computed directly from (8.63a) or (8.63b). The magnitude of the reflexion coefficient can be calculated from (8.60a) and its phase, ψ_a, from

$$\psi_a = -2\phi_a = \beta(s_1 - s_s) - \pi. \tag{8.70}$$

With the resonance-curve method, it is frequently more convenient to measure the curve width than the SWR, and sometimes it is simpler to measure the curve widths at power levels other than 1/2. For low-loss transmission lines and symmetrical resonance curves, (8.58) is simply

$$\rho_a \doteq \frac{\pi}{\sqrt{p^2 - 1}} \frac{\Delta s}{\lambda}$$

where Δs is the width of the resonance curve at a level $1/p$ of the maximum.

8.5   The measurement of self- and mutual impedance or admittance

At the driving points of the several elements in an array, currents and voltages are related by the usual coupled circuit equations. Let V_k be the driving voltage across the terminals of element k in an array of N elements; let $I_k(0)$ be the current in the same terminals. Then, if a Kirchhoff equation is written for each element, the following set is obtained:

$$V_1 = I_1(0)Z_{11} + I_2(0)Z_{12} + ... I_k(0)Z_{1k} + ... I_p(0)Z_{1p} + ... I_N(0)Z_{1N}$$
$$\vdots$$
$$V_k = I_1(0)Z_{k1} + I_2(0)Z_{k2} + ... I_k(0)Z_{kk} + ... I_p(0)Z_{kp} + ... I_N(0)Z_{kN} \tag{8.71}$$
$$\vdots$$
$$V_N = I_1(0)Z_{N1} + I_2(0)Z_{N2} + ... I_k(0)Z_{Nk} + ... I_p(0)Z_{Np} + ... I_N(0)Z_{NN}.$$

The coefficient $Z_{kp}, p \neq k$, is the mutual impedance between element k and element p. As long as the array is in an isotropic medium such as air, $Z_{kp} = Z_{pk}$. Z_{kk} is the self-impedance of element k. The input or driving-point impedance of element k is

$$Z_{kin} = \frac{V_k}{I_k(0)} = \frac{I_1(0)}{I_k(0)}Z_{k1} + ... Z_{kk} + ... \frac{I_p(0)}{I_k(0)}Z_{kp} + ... \frac{I_N(0)}{I_k(0)}Z_{kN}. \tag{8.72}$$

If the elements are fed by transmission lines, Z_{kin} is the apparent load impedance of the transmission line. The driving terminals of an antenna coincide with the line-load junction between it and its feeding transmission line. The self-impedance of an element is the input impedance at the terminals of that element when the driving-point currents of all other elements in the array are zero—

that is, when all other elements are open-circuited at their driving points. The mutual impedance between element k and element p is the open-circuit voltage at the driving point of element p per unit current at the driving terminals of element k, with the driving points of all elements but k open-circuited.

The relations that involve the admittances are the duals of (8.71). They are

$$
\left.
\begin{aligned}
I_1(0) &= V_1 Y_{11} + V_2 Y_{12} + \dots V_k Y_{1k} + \dots V_p Y_{1p} + \dots V_N Y_{1N} \\
&\vdots \\
I_k(0) &= V_1 Y_{k1} + V_2 Y_{k2} + \dots V_k Y_{kk} + \dots V_p Y_{kp} + \dots V_N Y_{kN} \\
&\vdots \\
I_N(0) &= V_1 Y_{N1} + V_2 Y_{N2} + \dots V_k Y_{Nk} + \dots V_p Y_{Np} + \dots V_N Y_{NN}
\end{aligned}
\right\}
\quad (8.73)
$$

$$
Y_{kin} = \frac{I_k(0)}{V_k} = \frac{V_1}{V_k} Y_{k1} + \frac{V_2}{V_k} Y_{k2} + \dots Y_{kk} + \dots \frac{V_p}{V_k} Y_{kp} + \dots \frac{V_N}{V_k} Y_{kN}. \quad (8.74)
$$

The self-admittance of an element is its input admittance when the driving-point voltages of all other elements in the array are zero—that is, when all other elements are short-circuited at their driving points. The mutual admittance between element k and element p is the short-circuit current at the driving point of element p per unit voltage at the terminals of element k, when the terminals of all elements but k are short-circuited.

Self- and mutual impedances or admittances depend upon the geometrical configuration of each element, the relative orientation and location of the elements in the array, and the total number of elements. Once the self- and mutual impedances or admittances have been determined for an array, they can be used in equations like (8.72) and (8.74) to predict the driving-point impedances or admittances for any set of driving voltages or currents that may be applied to the array.

In principle, there is no difficulty in determining self- and mutual impedances. If known sets of currents or voltages are maintained at the terminals of the several elements and a sufficient number of input impedances or admittances are measured, (8.72) or (8.74) can be inverted and the self- and mutual impedances evaluated. There are, however, two practical difficulties. The first is that the only set of excitation coefficients useful in measuring self- and mutual impedances is that which can be adjusted with high accuracy

independently of the driving-point impedances. The second
difficulty is that so many quantities must be determined. In a linear
array of N identical, equally spaced elements there are $N/2$ different
self-impedances and $N^2/4$ different mutual impedances if N is
even; $(N+1)/2$ different self-impedances and $(N^2-1)/4$ different
mutual impedances if N is odd. For $N = 100$ there are 2550
different quantities to be determined. Fortunately, many of the
mutual impedances are sufficiently small so that they can be
neglected and many of the self-impedances are alike within tolerable
limits. The measuring procedure should provide a rapid indication
of such possible simplifications as well as relatively simple steps for
determining the significant self- and mutual impedances.

In the case of dipole or monopole arrays, the laborious measurement of self- and mutual quantities may be avoided by applying the
two- or three-term theory as discussed in the preceding chapters.
The theory has been shown to give good agreement with measured
results in applications to circular and Yagi arrays, and this agreement will generally hold for other arrays.

An important especially simple array consists of identical
elements uniformly spaced about the circumference of a circle.
Since all elements have the same self-impedance, symmetry reduces
the total number of unknowns to $N/2$ if N is even or $(N+1)/2$ if N
is odd. Self- and mutual admittances can be measured as follows:
Let element k be driven, all others short-circuited at their driving
points. Measure the apparent input admittance of element k. From
(8.74)

$$Y_{kin1} = Y_{kk}. \tag{8.75}$$

Next, let elements k and p be driven with the voltages $V_p = -V_k$;
let the terminals of all other elements be short-circuited. Again
measure the apparent input admittance of element k. Then

$$Y_{kin2} = Y_{kk} - Y_{kp}, \qquad Y_{kp} = Y_{kin1} - Y_{kin2}. \tag{8.76}$$

Thus, for the circular array, all mutual admittances can be found
by successively driving element k and one other element with equal
voltages in opposite phases, and each time measuring the input
admittance of only element k. The driving voltages $V_k = -V_p$
produce a null in the electromagnetic field along the perpendicular
bisector of a chord joining elements k and p. This property can be
used to obtain the correct voltages by locating a probe at the centre
of the array and adjusting the phase and amplitude of element p or

k until no signal is observed in the probe. A 30–40 db range of receiving sensitivity provides an accurate adjustment of the voltages. The short circuits in the elements other than p and k can be placed at the driving terminals or at the ends of lossless sections of transmission lines which are electrically a half wavelength long. This length is critical and must take account of phase shifts in connectors, terminal zones, etc. For monopoles driven over a ground plane by coaxial lines in the manner shown in Fig. 8.3a, the short circuits can consist of very thin plugs in the ends of the coaxial lines. End-effects are quite simple for this model since the terminal-zone network consists of a shunting capacitance. If the measured apparent input admittances given in (8.75) and (8.76) are Y_{ka1} and Y_{ka2}, then

$$Y_{kk} = Y_{ka1} - j\omega C_T \qquad (8.77a)$$
$$Y_{pp} = Y_{ka1} - Y_{ka2}. \qquad (8.77b)$$

Note that the end-effect contributes only to the self-susceptance.

The self- and mutual impedances of a circular array can be measured with analogous procedures. To determine the self-impedance, drive one element, open-circuit all others so that all the driving-point currents are zero, and measure the input impedance of the driven element. To determine the mutual admittances, drive the elements successively in pairs with a receiving probe at the centre of the circle to aid in setting $I_p(0) = -I_k(0)$, and measure the input impedance of a driven element. Equations (8.75) and (8.76) apply with the corresponding impedances substituted for admittances. On an impedance basis, the experimental model of monopoles driven by coaxial lines offers little simplification in terminal-zone effects, and the actual load or input impedances are obtained from the measured apparent ones with (8.21).

In a linear, planar or more general array, the self-admittances or impedances can be measured by the method used for circular arrays, i.e. by driving the element of interest, loading the other elements with short circuits for admittances or open circuits for impedances, and measuring the input admittance or impedance of the driven element. Difficulties arise in measuring mutual admittances by the method of driving the elements in pairs so that $I_p(0) = -I_k(0)$ or $V_p = -V_k$, since, in general, there is no simple null in the field at which a receiving probe can be located to aid in the adjustment of the current or voltage. For example, in a

7-element curtain array of identical equispaced elements, there are 4 self- and 8 mutual impedances which must be determined. Among the 8 mutual impedances only Z_{17}, Z_{26} and Z_{35} correspond to pairs of elements symmetrically placed about the centre of the array. For these elements, a null at the centre ensures that each pair is driven by voltages with equal amplitudes and opposite phases.

The open circuit-short circuit method is a traditional procedure for measuring self- and mutual impedances [3, 6]. The self-impedances are measured as already discussed by driving the element of interest and open circuiting the driving points of all other elements. If element k is the driven element, (8.72) becomes

$$Z_{kin1} = V_k/I_k(0) = Z_{kk}. \tag{8.78}$$

To determine a given mutual impedance a short circuit is substituted for the open circuit in the appropriate element, and the input impedance of the driven element is again measured. If element k is driven and element p is short-circuited, the applicable pair of equations is

$$V_k = I_k(0)Z_{kk} + I_p(0)Z_{kp} \tag{8.79a}$$

$$0 = I_k(0)Z_{kp} + I_p(0)Z_{pp}. \tag{8.79b}$$

From (8.79a)

$$Z_{kin2} = \frac{V_k}{I_k(0)} = Z_{kk} + \frac{I_p(0)}{I_k(0)}Z_{kp}. \tag{8.79c}$$

The use of (8.79b) to eliminate the current ratio in (8.79c), the subsequent solution for Z_{kp}, and the expression of Z_{kk} and Z_{pp} in terms of their measured values, Z_{kin1} and Z_{pin1}, yields

$$Z_{kp} = \pm\sqrt{Z_{pin1}(Z_{kin1} - Z_{kin2})}. \tag{8.80}$$

Mutual admittances may be determined in the same manner by an interchange of open and short circuits and the substitution of the appropriate admittances in (8.80). A satisfactory method of providing the required open circuits is to short-circuit the feeding line at an electrical quarter wavelength from the line-load junction.

An alternative procedure† for determining mutual impedances is based on the measurement of both the relative amplitude and phase of the driving-point currents or voltages. Let element k in an array be driven and all other elements be open-circuited. The complete

† See [6], p. 349.

set (8.71) then becomes

$$V_1 = I_k(0)Z_{1k}$$
$$\vdots$$
$$V_k = I_k(0)Z_{kk}$$
$$\vdots \qquad\qquad\qquad (8.81)$$
$$V_p = I_k(0)Z_{pk}$$
$$\vdots$$
$$V_N = I_k(0)Z_{Nk}.$$

It follows that

$$Z_{kin} = \frac{V_k}{I_k(0)} = Z_{kk} \qquad\qquad (8.82a)$$

$$\frac{V_p}{V_k} = \frac{Z_{pk}}{Z_{kk}}. \qquad\qquad (8.82b)$$

The relative amplitudes of the voltages immediately indicate which mutual impedances are large enough to be important, and the relative phases need be measured only for these.

Let the open circuits in the elements other than k be provided by identical sections of transmission line that are terminated in short circuits. It follows from (8.56) that the transmission line may be either $\lambda/4$ or $3\lambda/4$ in length, and (8.54) suggests that the apparent voltages across the loads can be measured by a probe placed at a distance $w = \lambda/2$ from the line-load junction. Thus, if a section of transmission line is assembled with a short circuit at $w = 3\lambda/4$ and a probe at $w = \lambda/2$, the apparent driving-point voltages can be measured by interchanging this measuring section with the other loads. When the short circuit is removed, the measuring section can be incorporated in the line feeding the driven element and used to measure V_k. Note that the probe must be loosely coupled to the line and the electrical distances carefully adjusted. A coaxial measuring section for use with this procedure is shown in Fig. 8.9.

For the measurement of admittances, one element is driven while the others are short-circuited. From (8.73) it is seen that

$$Y_{kin} = I_k(0)/V_k = Y_{kk} \qquad\qquad (8.83a)$$

$$I_p(0)/I_k(0) = Y_{pk}/Y_{kk}. \qquad\qquad (8.83b)$$

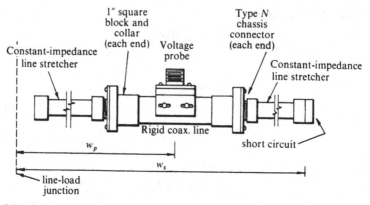

Fig. 8.9.   Coaxial measuring section. For apparent open-circuited load voltages: $w_p = \lambda/2$, $w_s = 3\lambda/4$. For apparent open-circuit load currents: $w_p = \lambda/4$, $w_s = \lambda/2$.

If a current probe is used in the measuring section, it must be placed at $w = \lambda/2$, with the short circuit at $w = \lambda$. However, from (8.55a) with $\beta w = \pi/2$, it is seen that

$$V(\lambda/4) = jI(0)R_c \qquad (8.84)$$

so that a voltage probe may be used at $w = \lambda/4$ and the short circuit placed at $w = \lambda/2$.

8.6   Theory and properties of probes

Successful techniques for sampling fields, currents and charges must be based on the responses of physically real probes, not ideal infinitesimal electric and magnetic doublets. Electric-field or charge probes are usually one-dimensional, short thin dipoles or monopoles that have a simple behaviour without serious errors. The usual magnetic-field or current probes, on the other hand, are small loops that have complicated behaviour because they are two-dimensional and can be excited in more than one mode. For electrically small loops only the first two modes are important. Because of the manner in which current is distributed around the loop, they are called the circulating or transmission-line mode and the dipole mode, respectively. In the transmission-line mode, there is a continuous current circulating around the loop; currents on opposite sides are equal but in opposite directions in space. In the dipole mode, currents on opposite sides are equal but in opposite directions around the loop, hence in the same direction in space; there is no net circulating current and the probe resembles a small folded dipole. As is shown below, currents in the transmission-line mode are related to the amplitude of the magnetic field at the centre of the loop; currents in the dipole mode are related to the amplitude of the electric field at the centre of the loop. Generally, currents in both modes can maintain a potential difference across a load. Hence, when the objective is to measure magnetic fields, the presence of dipole-mode currents in the loop may introduce an error that must be eliminated or corrected.

Charge or electric-field probes

To examine the properties of a small charge or electric field probe† consider the short thin centre-loaded dipole in a linearly polarized field of E^i volts per metre shown in Fig. 8.10a. In the figure E^i and E_p^i are in the plane wave front perpendicular to the propagation vector $\mathbf{k}$; E_p^i and $\mathbf{k}$ are in the plane containing the axis of the antenna. The equivalent circuit, Fig. 8.10b, consists of a Thévenin generator of voltage $V_g(Z_L = \infty)$ in a series combination with the load impedance Z_L and the input impedance of the antenna Z_0. $V_g(Z_L = \infty)$ is the open-circuit voltage at the terminals.

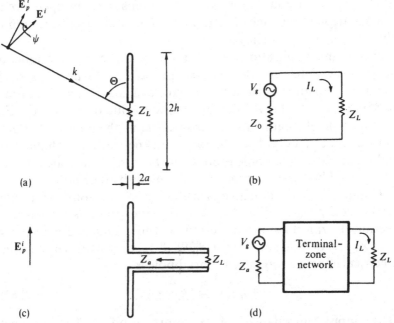

Fig. 8.10. Centre-loaded receiving dipole for electric field probe. (*a*) idealized with no feeding lines; (*b*) idealized equivalent circuit; (*c*) actual with feeding lines; (*d*) actual equivalent circuit.

† See [6], [19] and [20]. In particular [6], p. 184 and p. 475.

It is given by

$$V_g(Z_L = \infty) = -2h_e(\Theta)E^i \cos \psi \qquad (8.85)$$

where $h_e(\Theta)$ is the complex effective length of a short dipole and $E^i \cos \psi = E_p^i$ is the projection of E^i onto the plane containing the axis of the antenna and the direction of advance of the incident plane wave through the centre of the antenna. The load current is

$$I_L = \frac{V_g(Z_L = \infty)}{Z_0 + Z_L} = \frac{-2h_e(\Theta)E^i \cos \psi}{Z_0 + Z_L} = S_c E_p^i \qquad (8.86)$$

where

$$S_c = \frac{-2h_e(\Theta)}{Z_0 + Z_L}$$

is a sensitivity constant. As indicated in (8.86), the load current is proportional to the average tangential electric field along the dipole. Directions of the field can be determined by rotating the probe until I_L is maximum. If the incident electric field is elliptically polarized, it can be resolved into two linearly polarized components along the major and minor axes of the ellipse and an open-circuit voltage defined for each. The total current is the algebraic sum of the currents due to each generator.

For many applications a short monopole over a conducting surface is an effective electric field probe. Such a probe is easily made by extending the inner conductor of a coaxial line. Equations (8.85)–(8.86) still apply if the appropriate value of Z_0 is used. For a monopole of length h over a ground plane, the input admittance is twice that of a dipole of the same thickness and length $2h$. With either dipole or monopole probes, most errors of measurement are introduced because the probe is too long or bent or both.

Computed and measured sensitivities S_c of some monopole probes are given in Fig. 8.11. Usually, the probe is loaded by a section of transmission line terminated in a matched detector so that Z_L can be calculated from (8.39), (8.42) or (8.46a). The complex electrical effective length of a short dipole is

$$\beta_0 h_e(\Theta) = \tfrac{1}{2}\beta_0 h \sin \Theta. \qquad (8.87a)$$

The input impedance of a short dipole, $\beta_0 h \leqslant 1$, with $\Omega = 2 \ln (2h/a) = 10$ is [6]

$$Z_0 = 18.3\beta_0^2 h^2 (1 + 0.086\beta_0^2 h^2) - j(396.0/\beta_0 h)(1 - 0.383\beta_0^2 h^2). \qquad (8.87b)$$

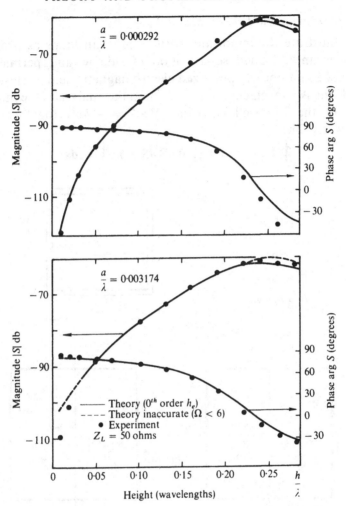

Fig. 8.11.   Relative sensitivity $S = S_c$ of monopole probes (Whiteside).

When $\beta_0 h \leqslant 0.5$ and $a \ll h$, the reactance is quite accurately given by

$$X_0 \doteq \frac{-60(\Omega - 3.39)}{\beta_0 h} \qquad (8.87\text{c})$$

the resistance by

$$R_0 \doteq 20\beta_0^2 h^2(1 + 0.133\beta_0^2 h^2). \qquad (8.87\text{d})$$

If terminal-zone effects are significant, account must be taken of them in determining the apparent resistance and reactance.

Current or magnetic-field probes

To illustrate the important features of small loops as probes,† consider an unloaded square loop of side w, and perimeter l immersed in a linearly polarized electromagnetic field as shown in Fig. 8.12a. A convenient starting point in the analysis is the integral form of the Maxwell equation, $\nabla \times \mathbf{E} = -j\omega\mathbf{B}$ together with $\mathbf{B} = \nabla \times \mathbf{A}$. That is,

$$\oint_s \mathbf{E} \cdot \mathbf{ds} = j\omega \iint_S \hat{\mathbf{n}} \cdot \mathbf{B} \, dS = j\omega \oint \mathbf{A} \cdot \mathbf{ds} \qquad (8.88)$$

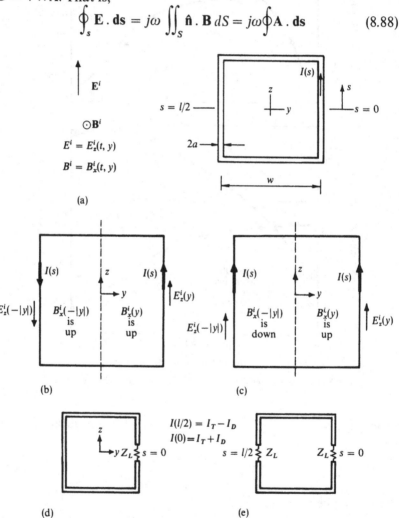

(a)

(b)   (c)

(d)   (e)

Fig. 8.12.   Loop probe in linearly polarized field. (a) Coordinates; (b) even magnetic field; (c) odd magnetic field; (d) singly-loaded loop; (e) doubly-loaded loop.

† [19] p. 270, [20], [21], [22] and [23].

where s is measured along the contour of the loop, S is the plane area bounded by the contour, $\hat{n}$ is a unit normal to this area in the right-hand screw sense with respect to integration around s, $\mathbf{E}$ is the complex amplitude of the total electric field and $\mathbf{B}$ of the total magnetic field at any point on the surface S of the loop. The radius of the wire, a, is assumed small, so that a quasi-one-dimensional analysis is adequate. The analysis is no more complicated for a rectangular than for a square loop, but the latter can be shown to have the optimum shape for minimizing averaging errors in a general incident field. $\mathbf{E}$ on the surface of the wire can be related to the total axial current by $\mathbf{E} \cdot \mathbf{ds} = z^i I \, ds$ where z^i is the internal impedance per unit length of the wire. In general, it is convenient to treat the total field in two parts, the incident field and the reradiated field maintained by the currents induced in the loop, i.e. $\mathbf{E} = \mathbf{E}^i + \mathbf{E}^r$. With these relations and (1.8a), (8.88) becomes

$$-j\omega \iint_S \hat{n} \cdot \mathbf{B}^i \, dS = \oint_s z^i I(s) \, ds + \frac{j\omega\mu_0}{4\pi} \oint_{s'} \oint_s I(s') \frac{e^{-j\beta_0 R}}{R} \, \mathbf{ds}' \cdot \mathbf{ds} \quad (8.89)$$

where $\beta_0 = 2\pi/\lambda$ and R is the distance from the element $\mathbf{ds}'$ at s' along the axis of the wire to the element $\mathbf{ds}$ on its surface. At this point, the assumption is usually made that the loop is sufficiently small to replace $\mathbf{B}^i$ by its value at the centre of the loop and $I(s)$ by a constant I. Actually a more careful treatment is often required.

Suppose that $\mathbf{B}^i$ can be resolved into an even and an odd part with respect to an axis through the centre of the loop. For example, if the loop lies in the yz plane and $\mathbf{B}^i$ is a function of y only and is directed parallel to the x-axis as in Fig. 8.12, the even and odd parts of $\mathbf{B}^i$ with respect to y are

$$\mathbf{B}^i = \mathbf{B}_T^i + \mathbf{B}_D^i \quad (8.90)$$

$$\text{even:} \quad \mathbf{B}_T^i(y) = \tfrac{1}{2}[\mathbf{B}^i(y) + \mathbf{B}^i(-y)] \quad (8.91)$$

$$\text{odd:} \quad \mathbf{B}_D^i(y) = \tfrac{1}{2}[\mathbf{B}^i(y) - \mathbf{B}^i(-y)] \quad (8.92)$$

with the following symmetry conditions:

$$\mathbf{B}_T^i(-y) = \mathbf{B}_T^i(y); \quad \mathbf{B}_D^i(-y) = -\mathbf{B}_D^i(y). \quad (8.93)$$

The subscripts T and D denote the transmission-line and dipole modes of the induced currents.

The electric field is related to the magnetic field by the Maxwell equation

$$\mathbf{E}^i = \frac{j}{\omega\mu_0\varepsilon_0} \nabla \times \mathbf{B}^i. \quad (8.94)$$

Since $\mathbf{E}^i$ is obtained from the first spatial derivative of $\mathbf{B}^i$, $\mathbf{E}^i$ is odd when $\mathbf{B}^i$ is even, and vice versa. That is,

$$\mathbf{E}^i_T(-y) = -\mathbf{E}^i_T(y); \qquad \mathbf{E}^i_D(-y) = \mathbf{E}^i_D(y). \qquad (8.95)$$

Note that $\mathbf{B}^i_T \doteq \hat{\mathbf{x}} B^i_0$ over the area bounded by the loop when $\beta_0 w \ll 1$; the current, $I_T(s)$, is, then, essentially constant around the loop as indicated in Fig. 8.12b. With the symmetry conditions (8.91) and (8.94), (8.89) becomes

$$-j\omega B^i_0 S = I_T \left\{ \oint_s z^i \, ds + \frac{j\omega\mu}{4\pi} \oint_s \oint_s \frac{e^{-jkR}}{R} \, \mathbf{ds}' . \, \mathbf{ds} \right\} = I_T Z_0. \qquad (8.96)$$

The quantity in braces in (8.96) is the impedance $Z_0 = Y_0^{-1}$ of the loop with constant current.† Therefore

$$I_T = I_T(0) = -j\omega B^i_0 Y_0 S = \lambda S_B (c B^i_0) \qquad (8.97)$$

where the sensitivity constant S_B for the unloaded loop has been introduced. It is defined by

$$S_B = -jkS Y_0 / \lambda \qquad (8.98)$$

and depends only on the geometry of the probe. The magnetic field is conveniently multiplied by $c = 3 \times 10^8 \text{m/sec}$ to give it the same dimensions as $\mathbf{E}^i$. Note that the current I_T is directly proportional to and, hence, a measure of the incident magnetic field $B^i_0 = B^i_T(0)$ at the centre of the loop.

$\mathbf{B}^i_D(y)$ is odd in y and therefore zero at the centre of the loop; it makes no contribution to the surface integral in (8.89). The associated electric field $\mathbf{E}^i_D$ in (8.95) has a non-zero value at the centre of the loop and is approximately constant over the space occupied by the small loop. It maintains equal and codirectional currents in the two sides of the loop which are parallel to the z-axis and hence parallel to $\mathbf{E}^i_D$ as shown in Fig. 8.12c. These dipole-mode currents are zero at $s = \pm l/4$ ($y = 0$, $z = \pm w/2$). In so far as they are concerned, the loop could be cut at these points and treated as an array of two bent receiving antennas. The current at the centre of each side is

$$I_D(0) = h_{eD} Y_D E^i_0 = \lambda S_E E^i_0 \qquad (8.99)$$

where h_{eD} is the effective length of each half of the array for the dipole mode, Y_D is the input admittance at the centre of each antenna when both are driven with equal and codirectional currents, and $E^i_0 = E^i_0(0)$ is the electric field at the centre of the loop.

† See chapter 6 of [24].

The electric sensitivity constant S_E for the unloaded loop has been introduced in (8.99). It is defined as follows:

$$S_E = h_{eD} Y_D / \lambda. \tag{8.100}$$

Note that $I_D(0)$ is proportional to and, therefore, a measure of the incident electric field E_0^i at the centre of the loop.

Transmission-line and dipole-mode currents have the following symmetries with respect to the anti-clockwise direction:

$$I_T\left(s + \frac{l}{2}\right) = I_T(s); \qquad I_D\left(s + \frac{l}{2}\right) = -I_D(s). \tag{8.101}$$

Hence, $I_T(s)$ corresponds to the zero-sequence current $I^{(0)}(s)$, and $I_D(s)$ to the first-sequence current $I^{(1)}(s)$. Higher-order sequence currents are assumed to be negligible in a sufficiently small loop. The total currents at points s and $s + l/2$ in the loop are

$$I(s) = I_T(s) + I_D(s) \tag{8.102}$$

$$I\left(s + \frac{l}{2}\right) = I_T(s) - I_D(s). \tag{8.103}$$

A very important fact is now evident. The magnetic field that contributes to the surface integral is related only to loop currents in the transmission-line (zero-sequence) mode and not to those in the dipole (first-sequence) mode. Conversely, if the magnetic field at the centre of the loop is to be determined from the current induced in the loop, measurements must involve currents in the transmission-line mode only.

When a small loop is used as a probe, the quantity of primary concern is the load current. Let a load Z_L be located at $s = 0$, as shown in Fig. 8.12d. The loaded loop can be analysed by replacing the load by a Thévenin generator of voltage $V = -I_L(0)Z_L$, where $I_L(0)$ is the current in the load. This generator maintains a current $VY(s) = -I(0)Z_L Y(s)$ where $Y(s)$ is the input admittance when the loop is driven at the point s. The total current at $s = 0$ is then

$$I_L(0) = I_T(0) + I_D(0) - I_L(0)Z_L Y(0) \tag{8.104}$$

With (8.97) and (8.99), it follows that

$$I_L(0) = \lambda S_B^{(1)} c B_0^i + \lambda S_E^{(1)} E_0^i \tag{8.105}$$

where the sensitivity constants for the singly-loaded loop are defined as follows:

$$S_B^{(1)} = \frac{Y_L}{Y_L + Y(0)} S_B \tag{8.106}$$

$$S_E^{(1)} = \frac{Y_L}{Y_L + Y(0)} S_E.$$ (8.107)

The importance of the location of the load with respect to the incident fields is now evident. When the load is located at $s = l/2$ instead of $s = 0$ (a change that is equivalent to rotating the loop through $180°$ about the x-axis), the input admittance of the loop is still the same but the current is given by (8.103) so that

$$I_L\left(\frac{\lambda}{2}\right) = \lambda S_B^{(1)} c B_0^i - \lambda S_E^{(1)} E_0^i.$$ (8.108)

This equation and (8.105) are useful for determining the relative importance of I_D. If the load current remains constant when the probe is rotated $180°$ about the x-axis, I_D is negligible and $I_L = I_T$. If the load current does not remain constant, readings of amplitude and phase may be taken in both positions. Then

$$I_T = \lambda S_B^{(1)} c B_0^i = \tfrac{1}{2}\left[I_L(0) + I_L\left(\frac{l}{2}\right)\right]$$ (8.109a)

$$I_D = \lambda S_E^{(1)} E_0^i = \tfrac{1}{2}\left[I_L(0) - I_L\left(\frac{l}{2}\right)\right].$$ (8.109b)

Instead of rotating the probe and taking two readings of amplitude and phase, one may use two loads with a hybrid junction to evaluate the sum and the difference.

In the simple example of a linearly polarized electric field indicated in Fig. 8.12, I_D is easily eliminated by a simple rotation of the loop until the side containing the load is perpendicular to the electric field. That is, the load is located at $s = \pm l/4$ in Fig. 8.12. However, when the electric field is elliptically polarized this expedient is unavailable and a doubly-loaded probe is probably the simplest solution in spite of the increased constructional difficulties.

The analysis of a doubly-loaded loop with identical loads Z_L at $s = 0$ and $s = l/2$, as shown in Fig. 8.12e, parallels that of the singly-loaded loop. The two load currents are

$$I_{L1} = I(0) = I_T(0) + I_D(0) - \left[I(0) Y(0) + I\left(\frac{l}{2}\right) Y\left(\frac{l}{2}\right)\right] / Y_L$$ (8.110a)

$$I_{L2} = I\left(\frac{l}{2}\right) = I_T(0) - I_D(0) - \left[I(0) Y\left(\frac{l}{2}\right) + I\left(\frac{l}{2}\right) Y(0)\right] / Y_L.$$ (8.110b)

The driving-point admittances $Y(0)$ and $Y(l/2)$ may be resolved into the zero- and first-sequence admittances $Y^{(0)}$ and $Y^{(1)}$. These can be introduced as follows:

$$Y(0) = Y^{(0)} + Y^{(1)}; \qquad Y\left(\frac{l}{2}\right) = Y^{(0)} - Y^{(1)}.$$

Let
$$I_\Sigma = I_{L1} + I_{L2} = \frac{2Y_L}{Y_L + 2Y^{(0)}} I_T(0) = \lambda S_B^{(2)} c B_0^i \qquad (8.111a)$$

$$I_\Delta = I_{L1} - I_{L2} = \frac{2Y_L}{Y_L + 2Y^{(1)}} I_D(0) = \lambda S_E^{(2)} E_0^i. \qquad (8.111b)$$

The sensitivity constants for the doubly-loaded probe are defined as follows:

$$S_B^{(2)} = \frac{2Y_L}{Y_L + 2Y^{(0)}} S_B \qquad (8.111c)$$

$$S_E^{(2)} = \frac{2Y_L}{Y_L + 2Y^{(1)}} S_E. \qquad (8.111d)$$

In actual practice, the hybrid junctions used to perform the summing and differencing operations will have good but not infinite isolation, so that the actual measurable currents are

$$I_B = I_\Sigma + \gamma I_\Delta = \lambda S_B^{(2)} c B_0^i + \gamma \lambda S_E^{(2)} E_0^i \qquad (8.112a)$$

$$I_E = I_\Delta + \gamma' I_\Sigma = \lambda S_E^{(2)} E_0^i + \gamma' \lambda S_B^{(2)} c B_0^i \qquad (8.112b)$$

where γ and γ' are the coefficients of cross-coupling between the adding and subtracting circuits. It is assumed that they are small.

In the measurement of the magnetic field (especially near the end of a dipole antenna where the polarization of the electric field is highly elliptical) it is particularly important that those parts of the current in the load that are excited in the dipole mode, viz. I_D, be negligible. To provide a measure of the ability of a probe and loading system to discriminate against such currents, a system error ratio $\varepsilon^{(n)}$ can be defined as the ratio of the output current due to unit parallel electric field ($E_0^i = 1$ volt/metre) to the output current due to unit normal magnetic field ($c B_0^i = 1$ volt/metre),

$$\varepsilon^{(1)} = S_E^{(1)}/S_B^{(1)} \qquad (8.113a)$$

$$\varepsilon^{(2)} = S_E^{(2)}/S_B^{(2)} \qquad (8.113b)$$

where the superscript indicates the number of loads in the probe. Note that (8.113b) applies to the combination of the probe and its summing and differencing circuits. The actual ratio of the two currents depends on the ratio of the fields $E_0^i/c B_0^i$ and generally equals $\varepsilon^{(n)}$ only in a plane-wave field. For a system to be capable of measuring the magnetic field with an error of no more than 10%, it is necessary that $\varepsilon^{(n)} \leqslant -20$ dB, where $\varepsilon^{(n)}$ in dB $= 20 \log_{10} \varepsilon^{(n)}$.

So far, the discussion has been concerned with square loops, although circular loops are often more desirable. Actually, a comparable analysis of circular loops follows precisely the steps outlined for the square loop including the definition of sensitivity constants, error ratios, etc; differences between the two shapes arise in the theoretical expressions for evaluating the sensitivity constants.

The final step in the practical analysis of loops as probes is to obtain expressions for the sensitivity constants. Consider first the square loop. Y_0, required in the definition of S_B in (8.98), may be found from (8.96) by an expansion of the exponential in a power series. The result for a loop of side w and wire radius a is [23]

$$Y_0 = \frac{-j\pi}{\zeta_0 \beta_0 w (\Omega - 4 \cdot 32 + 0 \cdot 37 \beta_0^2 w^2)} \tag{8.114}$$

where $\zeta_0 = \sqrt{\mu_0/\varepsilon_0} \doteq 120\pi$ ohms and $\Omega = 2 \ln (4w/a)$. Hence

$$S_B = \frac{-\pi w}{\lambda \zeta_0 (\Omega - 4 \cdot 32 + 14 \cdot 6 w^2/\lambda^2)}. \tag{8.115}$$

The unloaded electric sensitivity S_E is defined by (8.100) in terms of h_{eD} and Y_D. The effective length for the dipole mode is found by cutting the loop at $s = \pm l/4$, treating the two halves as a transmitting array, and applying the Rayleigh–Carson reciprocal theorem.† The result is

$$h_{eD} = \frac{2}{I_0} \int_0^{l/4} I(s)\,ds \tag{8.116}$$

where $I(s)$ is the transmitting current when the array is driven with codirectional currents, and I_0 is its value at the driving point. To zero order, this current is

$$I(z) \approx \frac{j 2\pi V}{\zeta(\Omega - 3 \cdot 17)} \frac{\sin \beta_0 (w - |z|)}{\cos \beta_0 w}. \tag{8.117}$$

With (8.117), (8.116) becomes

$$h_{eD} = \frac{\cos \frac{1}{2}\beta_0 w - \cos \beta_0 w}{\beta_0 \sin \frac{1}{2}\beta_0 w}. \tag{8.118}$$

† [6], p. 568.

The input admittance is†

$$Y_D = \frac{j2\pi \tan \beta_0 w}{\zeta(\Omega - 3 \cdot 17)}. \tag{8.119}$$

It follows that for the square loop,

$$S_E = \frac{j}{\zeta_0(\Omega - 3 \cdot 17)} \frac{\tan \beta_0 w(\cos \frac{1}{2}\beta_0 w - \cos \beta_0 w)}{\sin \frac{1}{2}\beta_0 w}. \tag{8.120}$$

In order to calculate the sensitivity constants S_E, S_B for loaded loops, note that $Y^{(1)} = Y_D/2$ and $Y^{(0)} \approx Y_0$ so that the values from (8.119) and (8.114) can be directly substituted into the following equations for $S_B^{(2)}$ and $S_E^{(2)}$: (8.106), (8.107), (8.111c) and (8.111d).

For a circular loop of diameter w and wire radius a,

$$Y_0 = \frac{-j4}{\zeta_0 \beta_0 w(\Omega - 3 \cdot 52 + 0 \cdot 33\beta_0^2 w^2)} \tag{8.121}$$

and

$$S_B = \frac{-\pi w}{\lambda \zeta_0(\Omega - 3 \cdot 52 + 13 \cdot 0 w^2/\lambda^2)} \tag{8.122}$$

where $\Omega = 2 \ln (\pi w/a)$.

The unloaded electric sensitivity is found from the response of the loop to an electric field that is uniform in the plane of the loop and pointing in the z-direction. The result is‡

$$I(\phi) = E_{z0}^i \frac{w}{j\zeta_0 a_1} \cos \phi \tag{8.123a}$$

where ϕ is the angular coordinate measured from the y-axis and a_1 is an expansion parameter calculated by Storer [26]. For $w \leqslant 0 \cdot 1\lambda$,

$$a_1 \doteq -(\Omega - 3 \cdot 52)(1 - \beta_0^2 w^2/4)/\pi\beta_0 w \tag{8.123b}$$

and

$$S_E = \frac{j2\pi^2 w^2/\lambda^2}{\zeta_0(\Omega - 3 \cdot 52)(1 - 9 \cdot 8w^2/\lambda^2)} \tag{8.123c}$$

for the circular loop. Zero and first-phase-sequence admittances are again needed for evaluating the sensitivity constant for the loaded loop. They can be found from Storer's results [26]. The expressions are complicated but subject to the conditions $w \leqslant 0 \cdot 03\lambda$ and $Y_L > 10Y^{(1)}$, $Y^{(0)}$ and $Y^{(1)}$ are

$$Y^{(0)} \doteq -j2\lambda[\pi w \zeta_0(\Omega - 3 \cdot 52)]^{-1} \tag{8.124a}$$

$$Y^{(1)} \doteq j4\pi w[\lambda \zeta_0(\Omega - 3 \cdot 52)]^{-1}. \tag{8.124b}$$

Generally (8.114)–(8.124b) provide quite accurate results for loop

---

† [6], p. 568.

‡ [25], chapter 10.

diameters or sides $w \leqslant 0.03\lambda$ and serve as a useful guide for $w \leqslant 0.1\lambda$. When $w \leqslant 0.03\lambda$, most of the expressions can be simplified. Note that $Y^{(0)} \gg Y^{(1)}$. They are summarized in Table 8.1.

Table 8.1. *Probe characteristics of electrically small loops*

Square loop, side w. $w \leqslant 0.03\lambda$

$$S_B^{(1)} = \frac{Y_L}{Y_L + Y^{(0)}} \frac{-\pi w}{\lambda \zeta_0 (\Omega - 4.32)} \qquad S_E^{(1)} = \frac{Y_L}{Y_L + Y^{(0)}} \frac{j3\pi^2 w^2}{\lambda^2 \zeta_0 (\Omega - 3.17)}$$

$$\varepsilon^{(1)} = -j3\pi \frac{w}{\lambda} \frac{\Omega - 4.32}{\Omega - 3.17}$$

$$Y^{(0)} = -j\lambda [2w\zeta_0 (\Omega - 4.32)]^{-1} \qquad Y^{(1)} = j2\pi^2 w [\lambda \zeta_0 (\Omega - 3.17)]^{-1}$$

$$\Omega = 2\ln(4w/a)$$

Circular loop, diameter w. $w \leqslant 0.03\lambda$

$$S_B^{(1)} = \frac{Y_L}{Y_L + Y^{(0)}} \frac{-\pi w}{\lambda \zeta_0 (\Omega - 3.52)} \qquad S_E^{(1)} = \frac{Y_L}{Y_L + Y^{(0)}} \frac{j2\pi^2 w^2}{\lambda^2 \zeta_0 (\Omega - 3.52)}$$

$$\varepsilon^{(1)} = -j2\pi \frac{w}{\lambda}$$

$$Y^{(0)} = -j2\lambda [\pi w \zeta_0 (\Omega - 3.52)]^{-1} \qquad Y^{(1)} = j4\pi w [\lambda \zeta_0 (\Omega - 3.52)]^{-1}$$

$$\Omega = 2\ln(\pi w/a)$$

Square and circular loops. $w \leqslant 0.03\lambda$

$$S_B^{(2)} = 2\frac{Y_L + Y^{(0)}}{Y_L + 2Y^{(0)}} S_B^{(1)} \qquad S_E^{(2)} = 2\frac{Y_L + Y^{(0)}}{Y_L + 2Y^{(1)}} S_E^{(1)}$$

$$\varepsilon^{(2)} = \frac{Y_L + 2Y^{(0)}}{Y_L + 2Y^{(1)}} \varepsilon^{(1)}$$

The simplified relations of Table 8.1 reveal that the error ratio $\varepsilon^{(1)}$ of singly-loaded probes is independent of the load and approximately a linear function of the length of the side of square probes and of the diameter for circular probes. The magnetic fields measured with a circular loop will have an error less than 10% provided $w \leqslant 0.016\lambda$. At 600 MHz this corresponds to $w \sim 0.5$ cm; at 3000 MHz to $w \sim 1$ mm. It is obviously advantageous to make such measurements at frequencies below 1000 MHz.

Sensitivities and error ratios as functions of loop size are shown in Fig. 8.13 for typical square loops, in Fig. 8.14 for circular loops. The sensitivities are in dB referred to 1 mho, the error ratios are in dB referred to 1, and magnitudes are absolute, phases are relative. Important characteristics of the graphs are the relatively slow increase in sensitivity as w/λ increases beyond about 0.03 or 0.04, and the minimum of $\varepsilon^{(2)}$ at about $w/\lambda = 0.04$, indicating that this size may be a good compromise for probes. From the curves of

Figs. 8.13c and 8.14c, a singly-loaded loop with dimensions as large as $w/\lambda = 0.1$ is seen to respond nearly as well to the electric field in the dipole mode as to the normal magnetic field in the circulating mode.

A comparison of the theoretical and experimental results in Figs. 8.13–8.14 shows good agreement and suggests that the theoretical results are adequate guides for the design of probes. Graphs of the limiting loop sizes and wire thicknesses required to keep error ratios below given limits are in Fig. 8.15 for singly-loaded loops, in Fig. 8.16 for doubly-loaded loops. In Fig. 8.16, γ, the cross-coupling coefficient between the adding and subtracting circuits, is assumed to be 1; in an actual system it will be at least -20 dB, reducing the indicated error ratio by this amount. Since the effects of changes in wire thickness and load resistance are similar for circular and square loops, the curves of Figs. 8.15 and 8.16 provide a useful qualitative guide for the former.

8.7 Construction and use of field probes

In many applications probes are used either in a free-standing arrangement or in conjunction with an image plane.† In the former, shown in Fig. 8.17, the probe is supported at the end of a long rigid tube which contains the feeder lines to the receiving equipment. The supporting tube is attached by a movable carriage to a track on a pivoting arm that permits the accurate placement and orientation of the probe. A loop is usually mounted with its plane perpendicular to the axis of the supporting tube. The free-standing arrangement is versatile and useful for measuring near-zone fields and surface currents on three-dimensional models. A principal disadvantage is that the supporting tube is always present in the field. Its disturbing effects can be reduced with quarter-wave sleeves and absorbing material. An alternative procedure incorporates a rectifying crystal directly in the probe and makes use of resistive wire to measure the d.c. voltage.

The image-plane arrangement, Fig. 8.18, is well-suited for measuring the surface currents on symmetrical models that can themselves be mounted on an image plane [28]. It has the advantage that all cables and supports are contained within the metal walls of the object under investigation or behind the image plane.

† See [6], p. 127 and [27].

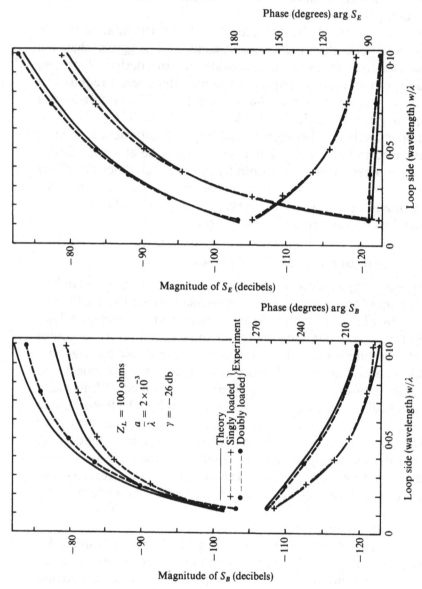

Fig. 8.13b.   Typical electric sensitivity of square loops.

Fig. 8.13a.   Typical magnetic sensitivity of square loops.

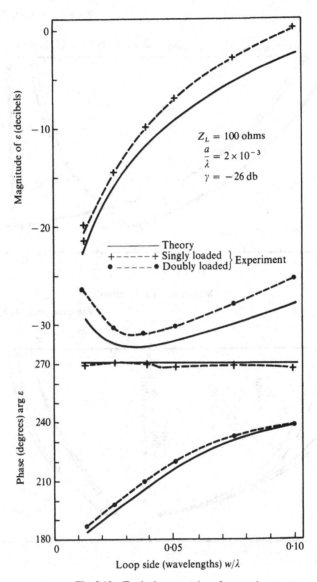

Fig. 8.13c. Typical error ratios of square loops.

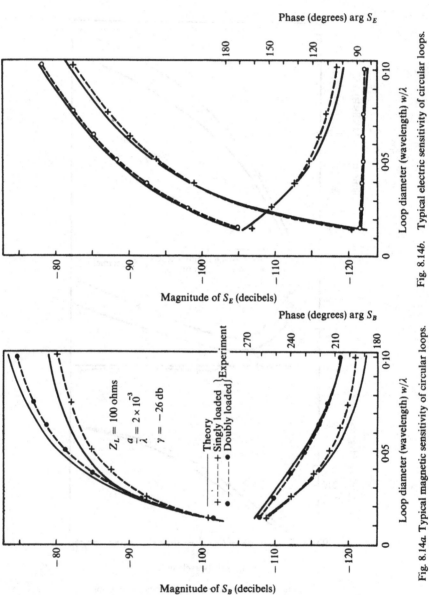

Fig. 8.14b. Typical electric sensitivity of circular loops.

Fig. 8.14a. Typical magnetic sensitivity of circular loops.

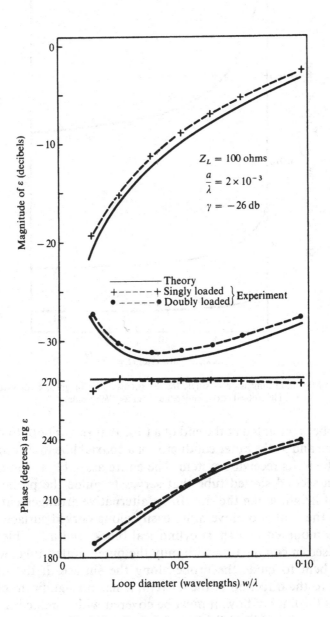

Fig. 8.14c.   Typical error ratios of circular loops.

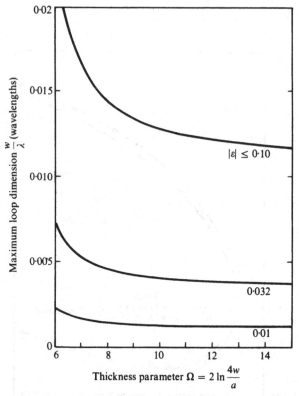

Fig. 8.15. Maximum dimension of singly-loaded square loop for given error ratio $|\varepsilon^{(1)}|$. Theoretical curves independent of Z_L (Whiteside).

The probe is mounted at the end of a tube that serves both to move the probe and as the outer conductor of a coaxial line that connects the probe to its receiving system. The entire assembly is contained within a second slotted tube that serves to guide the probe at a constant height along the slot. In an alternative arrangement that permits the probe to move more easily along curved surfaces, the probe is mounted in a short cylindrical block, and a flexible feed line is used in conjunction with only the outer slotted tube, which can be bent to guide the probe along the surface. If the slot is parallel to the direction of the current, it has no significant effect. If it cuts the lines of flow, it must be covered with conducting tape except in the immediate vicinity of the probe.

Examples of probes that may be used in a free-standing arrangement are shown in Fig. 8.19. A balanced charge probe consists of an electrically short dipole formed by 90° bends in the conductors of

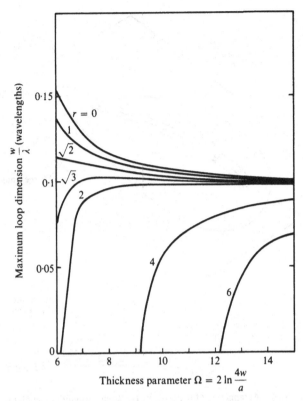

Fig. 8.16. Maximum dimension of doubly-loaded square loop for given error ratio $|\varepsilon^{(2)}| \leqslant 1$. Theoretical curves, load impedance $Z_L = 60\,r$ ohms (Whiteside).

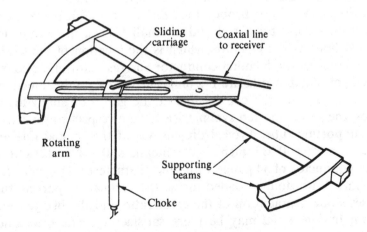

Fig. 8.17. Arrangement for free-space probe measurements.

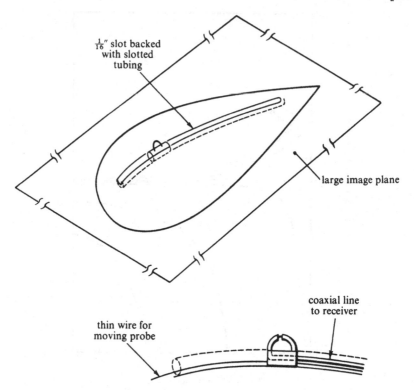

$\frac{1}{16}''$ slot backed
with slotted
tubing

large image plane

coaxial line
to receiver

thin wire for
moving probe

Fig. 8.18.   Arrangement for probe measurements on an image plane.

a two-wire transmission line, or by bends in the inner conductors
of a shielded-pair line or in a pair of adjacent miniature coaxial
lines (Fig. 8.19a). Loop probes (Figs. 8.19b and 8.19c) are made one-
half from a solid brass rod and one-half from a miniature rigid
coaxial line. At the junction, which is the location of the load, a
small gap is left in the outer conductor and insulator of the coaxial
line. Typical gap widths are 1–2 mm. The load Z_L is the impedance
seen looking into the coaxial line at the gap. For singly-loaded
loops, the gap is located symmetrically with respect to the axis of
the supporting tube. Typical dimensions of the coaxial line are:
outer diameter 0·032 inch, wall thickness 0·004 inch, and an
inner conductor of 34 gauge wire. The characteristic impedance is
50 ohms. With doubly-loaded loops, the vertical supporting tube
causes some degradation of the error ratio and the bridged loop
shown in Fig. 8.19d may be more satisfactory. The mechanical
construction must preserve a high degree of symmetry.

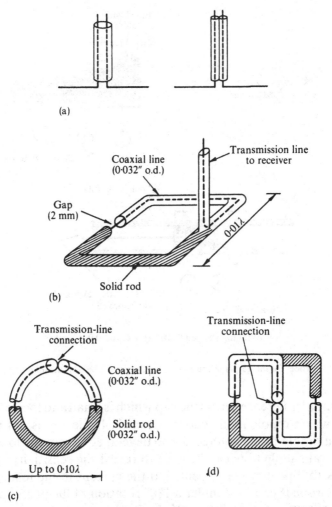

Fig. 8.19. Probes for free-space system. (a) Balanced charge probes; (b) singly-loaded square loop; (c) doubly-loaded circular loop; (d) bridged loop.

Probes for use with the image-plane technique (Fig. 8.20) have the same basic construction as those just described, except that the charge probes are usually monopoles and the current probes are half-loops. The half-loops may be either singly- or doubly-loaded.

The probes just described are shielded loops. Open-wire loops could be used instead of shielded loops,† but they are less convenient in the elimination of dipole-mode currents. With shielded loops the currents induced on the outside of the shield maintain

† See [5], p. 209 and [4], p. 231.

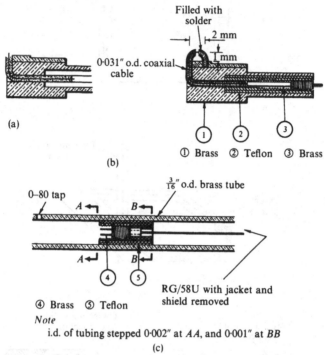

Fig. 8.20.   Probes for image-plane equipment. (a) Charge probe; (b) current probe; (c) probe socket.

a potential difference across the gap which is small and located at the point where dipole-mode currents vanish. If the gap is not at this point a dipole-mode voltage is also developed across the load.

Several simple tests can be used to reveal the sensitivity of loop probes to dipole-mode currents.[†] If the current in the load circuit of the probe is constant under a 180° rotation of the probe, dipole-mode currents are negligible. If the load current is not constant, readings must be taken with the probe in each position and averaged. With the image-plane technique, the probe cannot be rotated but can be tested in a short-circuited coaxial line. Let the probe current be measured with the probe successively at $w = w_0 = \lambda/4$ and $w = w_1 = \lambda/2$. Then

$$\varepsilon_{\text{db}}^{(1)} = 20 \log |I_L(w_0)/I_L(w_1)|.$$

When doubly-loaded probes are used with summing and differencing circuits, the output currents must be balanced because neither the probe nor the attached lines and loads are perfectly

[†][20], chapter VII, p. 5.

symmetrical. For this purpose, variable attenuators and phase shifters (or line stretchers) are used in the feed lines. The differencing circuit may be balanced by placing the probe so that it is symmetrically excited and adjusting the attenuators and phase shifters until the difference current is constant under 180° rotations of the probe. An alternative procedure is to measure individually the difference currents $I_{\Delta 1}$ and $I_{\Delta 2}$ due to the probe loads 1 and 2 when the probe is placed so that it is symmetrically excited. The probe is rotated 180° and the new currents $I'_{\Delta 1}$ and $I'_{\Delta 2}$ are measured, and the attenuators and phase shifters adjusted until

$$I_{\Delta 2}/I_{\Delta 1} = I'_{\Delta 1}/I'_{\Delta 2}. \tag{8.125}$$

Similar procedures can be used to balance the summing circuits. If a single hybrid junction is used to provide the outputs for both the sum and difference, the balancing adjustment cannot be optimized for both arms simultaneously, but a satisfactory compromise can usually be found.

The measurement of surface distributions of current and charge on good conductors actually involves the measurement of magnetic and electric fields near the surface. Most of the current in a good conductor at high frequencies is concentrated within a very small distance of the surface, d_s, called the skin depth and given by†

$$d_s = (2/\omega\mu\sigma)^{1/2}. \tag{8.126}$$

Such a thin layer of current is well approximated by the surface density $\mathbf{K}$ on a perfect conductor and is related to the total magnetic field at the surface by the boundary condition

$$\hat{n} \times \mathbf{B} = \hat{t}B_t = -\mathbf{K}\mu_0. \tag{8.127}$$

Similarly, the surface charge η is related to the total electric field by

$$\hat{n} \cdot \mathbf{E} = E_n = -\eta/\varepsilon_0 \tag{8.128}$$

where $\hat{n}$ is an outward unit normal from the surface. On thin cylinders, $\mathbf{K}$ and η have no angular variation around the cylinder, so that the total axial current and charge per unit length are $\mathbf{I}(z) = 2\pi a\mathbf{K}(z)$ and $q(z) = 2\pi a\eta(z)$. Except near the ends or edges of conductors distributions of B_t and E_n are often unchanged at very small fractions of a wavelength from the surface, so that probes placed sufficiently near the surface and moved parallel to it sample fields which are proportional to $\mathbf{K}$ and η. In the image-plane method, the effective centre of the probe is usually quite

† [20], chapter VII, p. 5.

near the surface; in the free-standing method, it is at least a probe radius away. At distances from a surface that are less than a few probe diameters, a probe is tightly coupled to its image so that its distance from the surface must be kept constant. Since the coupling between a probe and an ideal image does not exist near edges and corners, meaningful measurements cannot be made.

The most difficult fields to measure are linearly polarized magnetic fields associated with elliptically polarized electric fields. Figures 8.21–8.23 show the effects of size and orientation of probes in such fields and illustrate the effective use of singly- and doubly-loaded loops to make measurements. The near-zone elliptically polarized electric field of a quarter-wave monopole over an image plane is shown in confocal coordinates in Fig. 8.21a.† In Fig. 8.22a are graphs of measurements made along the coordinate $k_e = 2$ with a singly-loaded square loop with $w/\lambda = 0.013$, oriented in the four positions indicated in Fig. 8.21b. Owing to its small size, this loop was relatively insensitive to dipole-mode fields and its orientation with respect to the electric field was not critical. In contrast, for a loop with $w/\lambda = 0.1$, the orientation is seen in Fig. 8.23a to be very important. When the probe is oriented in the positions marked 'In' and 'Out' the dipole-mode current excited by the component E_ε maintains a voltage across the load, that excited by E_ρ does not. In the positions marked 'Right' and 'Left', the dipole-mode currents due to E_ρ maintain a voltage across the load, those due to E_ε do not. Since E_ε is nearly proportional to B_Φ, whereas E_ρ is not, no significant error is introduced in a relative measurement with the probe in the 'In' and 'Out' positions, a very large error in the 'Right' and 'Left' positions. The doubly-loaded loop with its summing and differencing circuits is seen to provide accurate results regardless of orientation even for sizes as large as $w/\lambda = 0.1$.

8.8   Equipment for measuring amplitude and phase

Measurements of the relative amplitudes of fields, currents, and charges are straightforward when the sampling probes are correctly used. The requirements are a signal source with sufficient power and stability in frequency and amplitude during a measurement, and a receiving system that is accurate and obeys a square law or is linear over the relevant range of power levels. For transmission-line measurements, a signal source must be isolated from

† See chapter 5 in [6].

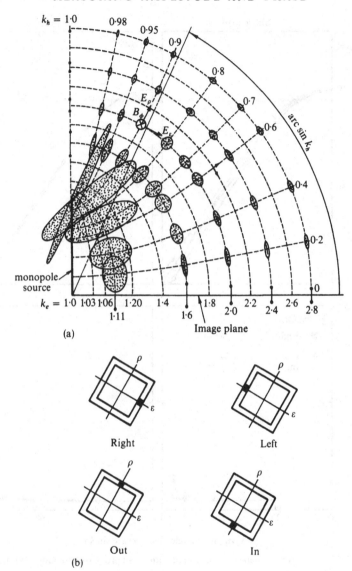

(a)

(b)

Fig. 8.21.   (a) Elliptically polarized electric field near quarter-wave monopole over an image
plane; (b) probe orientations for measurements of Figs. 8.22 and 8.23.

the line by at least 10 dB in order that its output be independent of
changes in the apparent load impedances. The isolation can be
provided by attenuators, resistive cables, isolators, or loosely
coupled probes. Bolometers and barretters obey a square law at
powers below their saturation and burn-out points. Crystals such as

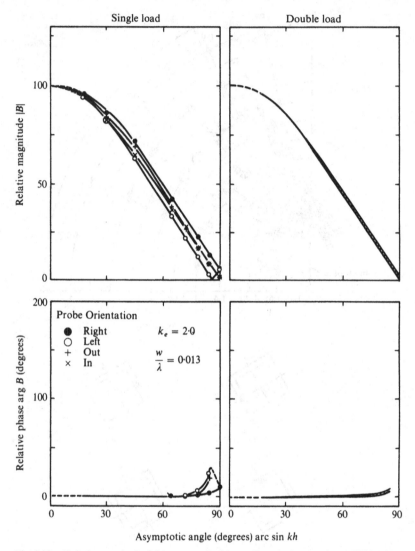

Fig. 8.22.   Relative magnetic field measured with small probe (square loop) (Whiteside).

the 1N21 or 1N23 are more sensitive than bolometers by about
15–20 dB but may have response laws with exponents either greater
or less than 2 and several crystals may have to be tested before a
satisfactory one is found. The superheterodyne receiving system is
most sensitive but is susceptible to interference from stray leakage
signals. This can be reduced by shielding the receiver with a metal
screening, covering coaxial connectors and other possible sources

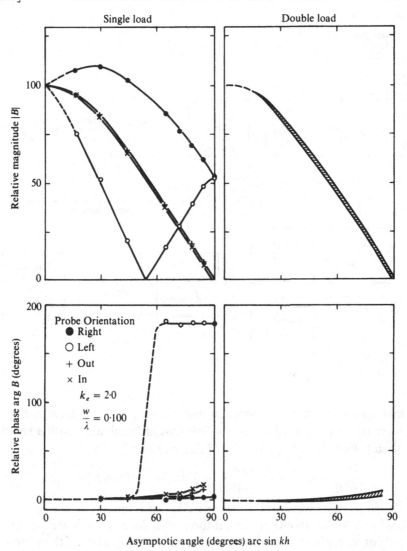

Fig. 8.23. Relative magnetic field measured with large probe (square loop) (Whiteside).

of spurious radiation with aluminium foil, and by a careful arrangement of the parts of the circuit. The components required for amplitude measurements with a superheterodyne or amplitude-modulated system are indicated in the block diagrams of Figs. 8.24a and 8.24b.

The complete receiving system (consisting of a probe, detector, amplifiers, and a display meter or recorder) can be calibrated by

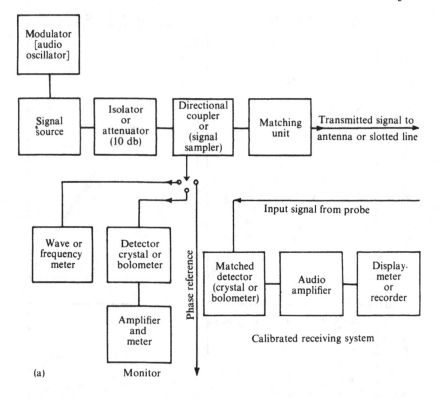

(a)

measuring the distance between half-power points of the standing-wave pattern on a transmission line that is terminated with a short circuit. For $Z_a = 0$, (8.54) and (8.55) give

$$\frac{I(w)}{I(0)} = \cos \beta w; \quad \frac{V(w)}{[I(0)R_c]} = j \sin \beta w = j \cos \beta \left(w + \frac{\lambda}{4}\right).$$

Power is proportional to I^2 or V^2 and hence to $\cos^2 \beta w$ or $\cos^2 \beta(w + \lambda/4)$. In either case, if the power level as indicated on the output meter is P and the exponent of the response law of the system is n,

$$P = P_{max}\cos^n \left(\frac{2\pi d}{\lambda}\right)$$

where d is distance measured from a maximum. Let Δw be the total distance between points at which $P/P_{max} = 1/2$; then, the response law is

$$n = \frac{\log \frac{1}{2}}{\log \cos \left(\frac{\pi \Delta w}{\lambda}\right)}. \tag{8.129}$$

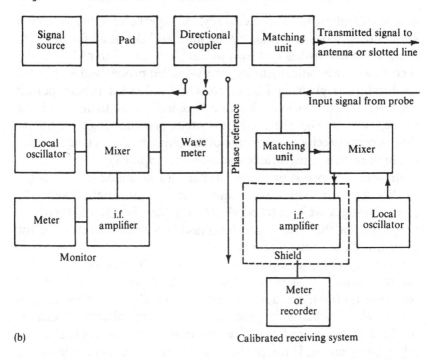

(b)   Calibrated receiving system

Fig. 8.24.   (a) Amplitude-modulated system for amplitude measurements; (b) superheterodyne system for amplitude measurements.

Techniques for measuring phase are commonly based on the interference of two signals from the same transmitter. The principal signal is fed to the system under investigation, sampled by a probe, and returned to the receiver. The reference signal, obtained usually from a directional coupler, is fed through an adjustable attenuator and a precision phase shifter and then combined with the signal from the probe. If the signal from the probe is $e_1(t) = U \cos \omega t$ and the reference signal is $e_2(t) = R \cos(\omega t + \phi_d)$, where ϕ_d is the difference in phase or electrical path length between the signals, it follows that

$$e(t) = e_1(t) + e_2(t) = U \cos \omega t + R \cos(\omega t + \phi_d) = [U - R] \cos \omega t,$$

$$\phi_d = (2n+1)\pi. \tag{8.130}$$

A minimum is observed when ϕ_d is an odd integral multiple of π and a null when, in addition, the amplitudes are equal. In practice, the difference signal is displayed on a meter and, for each probe position, the phase shifter is adjusted until a minimum is indicated by the meter. As in determining transmission-line minima, readings may be taken at equal power levels on each side of a minimum and averaged to increase accuracy. A superheterodyne receiving system that is based on these principles and designed for use with doubly-loaded probes is shown in the block diagram of Fig. 8.25a. A comparable arrangement for singly-loaded probes and amplitude modulation is given in Fig. 8.25b. Doubly-loaded probes permit accurate measurements with one mechanical adjustment of the probe, and the superheterodyne system provides high sensitivity that permits precise adjustments of the phase but, as is evident from the figures, at a considerable additional complexity.

An effective precision phase shifter can be made from a high-quality constant-impedance line stretcher or alternatively the power along a well-matched slotted line can be sampled with an r.f. probe. In the latter case the phase is a linear function of the position of the probe along the line.

Coaxial hybrid junctions are convenient devices for combining the two signals. In a perfect coaxial hybrid junction, a signal fed into one terminal divides equally between the two opposite terminals with 0° change in phase at the far terminal and 90° change in phase at the nearest opposite terminal. No signal appears at the adjacent terminal. If the probe signal and the reference signal are fed into adjacent terminals and the electrical path length of the reference signal is properly adjusted, the two signals will be in phase and add at one of the two opposite terminals, while at the other they will be in opposite phase and subtract. The loads at all four terminals of the hybrid junction must be carefully matched.

When carefully used, the procedure just described is capable of yielding accurate results, but it has the serious inconvenience that the depth of the null in the difference signal depends on the relative amplitudes of the input signals. Since deep nulls are required for high precision, frequent readjustment of an attenuator in the phase reference line is necessary. Unfortunately, attenuators either introduce shifts in phase or have very high insertion losses. Figure 8.26 shows a system that avoids this difficulty and produces a deep null independently of the relative amplitude of the two input signals.†

In the arrangement of Fig. 8.26, let the signal from the probe be $e_1(t) = U(1 + m \cos \omega_m t) \cos \omega t$, that from the phase-reference line $e_2(t) = R(1 + m \cos \omega_m t) \cos (\omega t + \phi_d)$; ϕ_d is their phase difference,

† [5], p. 233 and [29].

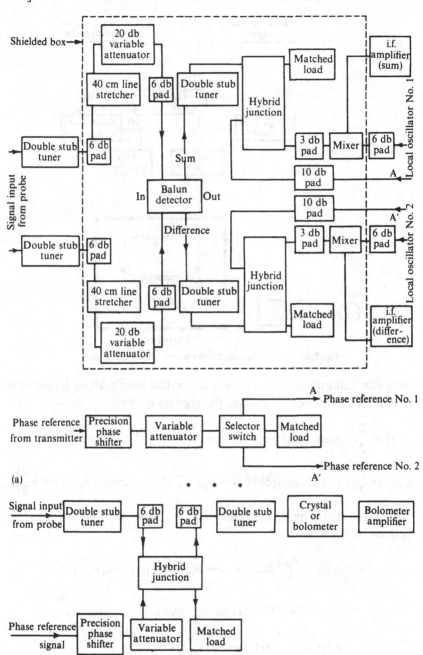

Fig. 8.25. (a) Phase-sensitive superheterodyne receiving equipment for use with doubly-loaded probes (Whiteside); (b) equipment for phase measurements with amplitude-modulated systems.

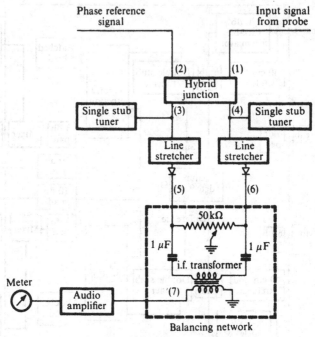

Fig. 8.26. Amplitude-insensitive phase detector (Burton).

m is the modulation factor, and ω_m is the modulation frequency. The hybrid junction combines the signals to give

$$e_3(t) = \frac{R}{2}(1 + m\cos\omega_m t)\cos\left(\omega t + \phi_d + \frac{\pi}{2}\right) + \frac{U}{2}(1 + m\cos\omega_m t)\cos\omega t$$

$$e_4(t) = \frac{R}{2}(1 + m\cos\omega_m t)\cos(\omega t + \phi_d) + \frac{U}{2}(1 + m\cos\omega_m t)\cos\left(\omega t + \frac{\pi}{2}\right).$$

After detection by the square-law crystals, the signals can be represented as

$$e_5(t) = \frac{a_2'}{4}\left[R(1 + m\cos\omega_m t)\cos\left(\omega t + \phi_d + \frac{\pi}{2}\right)\right.$$
$$\left. + U(1 + m\cos\omega_m t)\cos\omega t\right]^2$$

$$e_6(t) = \frac{a_2}{4}\left[R(1 + m\cos\omega_m t)\cos(\omega t + \phi_d)\right.$$
$$\left. + U(1 + m\cos\omega_m t)\cos\left(\omega t + \frac{\pi}{2}\right)\right]^2.$$

The coefficients a_2 and a_2' can be equated by the balancing potentiometer. With balanced currents, the output of the i.f. transformer circuit is

$$e_7(t) = a_2 R U_m \sin \phi_d \cos \omega_m t. \qquad (8.131)$$

When ϕ_d is an integral multiple of π, $e_7(t) = 0$ for all values of R and U that differ from zero.

The principal errors with this procedure come from improper balancing of the coefficients a_2 and a_2', mismatches looking into the crystals, and a standing wave on the phase-reference line. With either system it is important that the phase-reference line have no standing wave. A SWR of 1·1 to 1·5 can introduce phase errors of 2·5° to 12°.

More sophisticated techniques for measuring phase are available [30]. The procedures outlined above are relatively simple and capable of precise results; they are convenient for general laboratory use. Additional details of the individual components are discussed in [1] and [2].

8.9 Measurement of radiation patterns

The radiation pattern of an antenna is one of the most important and frequently measured parameters. In the usual procedure the test antenna is mounted at a distance $R = 2D^2/\lambda$ from a transmitting antenna and the received power is recorded as a function of the angular intervals of interest. D is the larger of the maximum dimensions of the test antenna and the transmitting antenna. R is chosen to satisfy the requirements for far-field conditions. Its minimum value implies that the incident field closely approximates a plane wave in the volume occupied by the test antenna during the measurement. Also $D/R \ll 1$. The quantity actually measured is the square of the magnitude of the generalized effective length of the receiving antenna.† As discussed in section 6.10 the reciprocal theorem shows that this is the same as the radiation pattern of the antenna when used for transmitting. Occasionally, measurements of the magnitude and phase of the far-zone fields are required. The equipment for either measurement is substantially that described in section 8.8 with the probe replaced by a receiving antenna.

The gain of an antenna in a given direction may be defined as the ratio of the magnitude of the Poynting vector $|S| = |E^2|/2\zeta_0$

† [6], p. 690.

maintained in that direction by the test antenna to the magnitude of S at the same location due to a fictitious isotropic radiator when both have the same input power. It may be measured by comparing the power received on the test antenna to the power received on a reference antenna when both are in the same transmitted field. The reference antenna is one for which the gain relative to an isotropic radiator can be computed. Wave-guide horns and dipoles near resonance are useful for this purpose. The gain of a sufficiently thin half-wave dipole compared to an isotropic radiator is 1·64 or 2·15 dB.

When the radiated fields of antennas or arrays are not very directive, there are two potential sources of error in measuring their radiation patterns and gain. The first arises from the coupling of the array to its mount, the second is due to extraneous reflexions of the transmitted field, notably from the ground. Interference from ground reflexions can be reduced by arranging the transmitting antenna with a null of its pattern along the ground and by mounting it on a sufficiently high tower. It can be eliminated by an image-plane form of testing site in which the antennas are mounted close to a level conducting earth. For grazing incidence, the reflexion coefficient of the forward reflexion is close to -1 and the vertical variation of the field at the test antenna can be represented by

$$E(h_r) = E_0 \sin\left(\frac{2\pi h_r h_t}{\lambda R}\right) \tag{8.132}$$

where h_r and h_t are the heights of the test antenna and the trans-mitting antenna, respectively, and R is the distance between them. The test antenna is located at the lowest maximum given by (8.132). This type of testing facility is particularly convenient for use on a flat roof.†

A complete image-plane technique [33] is useful for investigating all of the properties—radiated fields, gain, current distributions, and apparent driving-point impedances or admittances—of small arrays of monopoles. In this arrangement, the test array is mounted on a large circular disk that is located several wavelengths from the nearest edges of a large rectangular ground screen; an auxiliary antenna is mounted near one corner. A monopole with corner reflector makes a satisfactory auxiliary antenna that may be used

† Additional discussions on the measurement of radiation patterns and gain may be found in [3], [31] and [32], chapters 15 and 16.

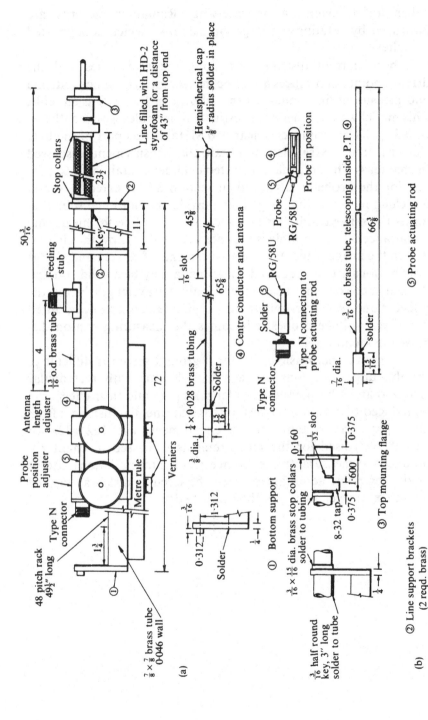

Fig. 8.27. Combined monopole antenna and measuring line: (a) assembled line; (b) parts detail; dimensions are in inches.

either for transmitting or receiving. Radiation patterns are measured by rotating the large circular disk which is supported on rollers.

If both current distributions along the monopoles and the driving-point admittances are to be measured, a combined antenna and measuring line similar to that shown in Fig. 8.27 is suitable. This model was designed to investigate monopoles with lengths up to $h/\lambda = 3/4$ at frequencies near 650 MHz. It was part of the five-element circular array used to obtain the measurements discussed in chapter 4. The antenna is a slotted cylinder containing a coaxial line for the probe as discussed in section 8.7. It extends several wavelengths below the image plane inside an additional tube that forms the outer conductor of a second slotted line for measuring admittances. The assembly is mounted vertically below the circular disk in the image plane. When measurements of current and charge distributions are not required, an antenna may be a solid rod and the outer coaxial line need extend only a wavelength below the image plane. When terminated in a suitable connector and fitted with fixed voltage probes it becomes the measuring section discussed in section 8.5.

A minimum permissible size for a ground screen is difficult to specify. Edge reflexions are less troublesome with rectangular shapes than with square or circular ones [34], and their effects can be reduced by locating the test antenna at unequal distances from all edges. The rectangular screen (dimensions: $16\lambda \times 32\lambda$ or 24×38 feet) used for the measurements discussed in chapter 4 produced no measurable interference in either the radiation patterns or driving-point admittances when the circular disk was located about six wavelengths from the nearest edge.

APPENDIX I

Tables of Ψ_{dR}, $T^{(m)}$ or $T'^{(m)}$
and
Self- and mutual admittances for single elements and circular arrays

Notation:
Self- and mutual admittances are written in the form

$$Y_{1(m+1)} = G_{1(m+1)} + jB_{1(m+1)}$$

with the self-admittance given by the row corresponding to $m = 0$, the first mutual Y_{12} by the row corresponding to $m = 1$, etc. A factor of 10^{-3} has been suppressed in the admittances; hence, tabulated values are in millimhos. The characteristic impedance of free space, ζ_0, was taken to be $\zeta_0 = 376 \cdot 730$ ohms.

Table 1. Ψ_{dR}, $T(h)$ or $T'\left(\frac{\lambda}{4}\right)$ and admittances for isolated antenna

$(N = 1,\ a/\lambda = 7.022 \times 10^{-3})$

$\beta_0 h$	$\frac{h}{\lambda}$	Ψ_{dR}	$T(h)$ Real	or $T'\left(\frac{\lambda}{4}\right)$ Imag.	G_0	B_0	$B_0 + 0.72$
1·200	0·1910	5·31670	0·27602	−0·74791	4·12999	9·59504	10·315
1·350	0·2150	5·69058	−0·37945	−1·16933	12·28333	9·12585	9·846
1·432	0·2280	5·88844	−0·98304	−0·87529	15·51296	2·93684	3·657
1·501	0·2390	6·05385	−1·13164	−0·36538	13·56922	−2·22898	−1·509
1·570	0·2500	6·21771	2·65166	3·79157	10·17040	−4·43037	−3·710
1·652	0·2630	6·37511	−0·78390	0·20461	7·09591	−4·77136	−4·051
1·796	0·2860	6·54009	−0·51411	0·30471	4·24183	−3·92426	−3·204
1·997	0·3180	6·61380	−0·32658	0·30590	2·63298	−2·72756	−2·008
2·355	0·3750	6·44947	−0·19988	0·26239	1·63816	−1·33809	−0·618
2·751	0·4380	6·05835	−0·16459	0·21601	1·23747	−0·18730	0·533
3·141	0·5000	5·73687	−0·17204	0·17559	1·02096	1·00032	1·720
3·398	0·5410	5·66947	−0·19117	0·15329	0·91729	1·91899	2·639
3·649	0·5810	5·74850	−0·22145	0·14008	0·87181	2·99706	3·717
3·800	0·6050	5·86161	−0·24791	0·14155	0·91248	3·80520	4·525
3·926	0·6250	5·98717	−0·27769	0·15443	1·03854	4·65317	5·373

Table 2. Ψ_{dR}, $T'^{(m)}$ and admittances of circular array† of N elements

$$(a/\lambda = 7 \cdot 022 \times 10^{-3})$$
$$h/\lambda = 0 \cdot 25, \quad \beta_0 h = \pi/2, \quad \Omega = 8 \cdot 54$$

Sequence m	Ψ_{dR}	Re $T'^{(m)}$	Im $T'^{(m)}$	Sequence admittance		Self- and mutual admittances		$\dfrac{d}{\lambda}$	N
				$G^{(m)}$	$B^{(m)}$	$G_{1(m+1)}$	$B_{1(m+1)}$		
0	6·21771	1·08745	2·72968	7·32201	−0·23458	6·29694	−7·25537	0·1875	2
1	6·21771	6·32221	1·96537	5·27186	−14·27616	1·02507	7·02079	0·1875	2
0	6·21771	0·89928	3·12073	8·37095	0·27016	6·92620	−5·08924	0·2500	2
1	6·21771	4·89530	2·04351	5·48146	−10·44863	1·44475	5·35939	0·2500	2
0	6·21771	0·74401	3·60819	9·67851	0·68667	7·71191	−3·72010	0·3125	2
1	6·21771	4·02973	2·14188	5·74531	−8·12687	1·96660	4·40677	0·3125	2
0	6·21771	0·68981	4·23444	11·35834	0·83205	8·70338	−2·84976	0·3750	2
1	6·21771	3·43500	2·25488	6·04843	−6·53158	2·65496	3·68181	0·3750	2
0	6·21771	0·88064	5·00919	13·43651	0·32017	9·91292	−2·50992	0·4375	2
1	6·21771	2·99078	2·38197	6·38933	−5·34001	3·52359	2·83009	0·4375	2
0	6·21771	1·53690	5·78029	15·50489	−1·44016	11·13910	−2·91835	0·5000	2
1	6·21771	2·63905	2·52512	6·77331	−4·39654	4·36579	1·47819	0·5000	2
0	6·21771	2·69527	6·08303	16·31694	−4·54733	11·76374	−4·08367	0·5625	2
1	6·21771	2·34955	2·68812	7·21053	−3·62001	4·55321	−0·46366	0·5625	2
0	6·21771	3·78922	5·55813	14·90897	−7·48172	11·31236	−5·22543	0·6250	2
1	6·21771	2·10691	2·87646	7·71574	−2·96914	3·59661	−2·25629	0·6250	2
0	6·21771	4·22845	4·62546	12·40720	−8·65991	10·35762	−5·54486	0·6875	2
1	6·21771	1·90584	3·09727	8·30803	−2·42981	2·04958	−3·11505	0·6875	2
0	6·21771	4·12243	3·84925	10·32511	−8·37552	9·66718	−5·19435	0·7500	2
1	6·21771	1·75053	3·35869	9·00926	−2·01319	0·65793	−3·18117	0·7500	2
0	6·21771	0·65962	1·97164	5·28868	−0·91303	5·27747	−9·21309	0·1875	3

† Note that $T'^{(m)} = T'^{(m)}\left(\dfrac{\lambda}{4}\right)$.

Table 2 (continued)

Sequence				Sequence admittance		Self- and mutual admittances			
m	Ψ_{dR}	$\mathrm{Re}\,T'^{(m)}$	$\mathrm{Im}\,T'^{(m)}$	$G^{(m)}$	$B^{(m)}$	$G_{1(m+1)}$	$B_{1(m+1)}$	$\dfrac{d}{\lambda}$	N
1	6·21771	6·32221	1·96537	5·27186	−14·27616	0·00561	5·06306	0·1875	3
0	6·21771	0·33433	2·29021	6·14320	1·78558	5·70204	−6·37056	0·2500	3
1	6·21771	4·89530	2·04351	5·48146	−10·44863	0·22058	4·07807	0·2500	3
0	6·21771	−0·08089	2·69241	7·22203	2·89935	6·23755	−4·45146	0·3125	3
1	6·21771	4·02973	2·14188	5·74531	−8·12687	0·49224	3·67541	0·3125	3
0	6·21771	−0·62328	3·27474	8·78406	4·35423	6·96030	−2·90297	0·3750	3
1	6·21771	3·43500	2·25488	6·04843	−6·53158	0·91188	3·62860	0·3750	3
0	6·21771	−1·33434	4·27609	11·47006	6·26158	8·08291	−1·47282	0·4375	3
1	6·21771	2·99078	2·38197	6·38933	−5·34001	1·69358	3·86720	0·4375	3
0	6·21771	−2·06303	6·38546	17·12817	8·21617	10·22493	−0·19230	0·5000	3
1	6·21771	2·63905	2·52512	6·77331	−4·39654	3·45162	4·20424	0·5000	3
0	6·21771	−0·52523	10·97641	29·44281	4·09124	14·62129	−1·04959	0·5625	3
1	6·21771	2·34955	2·68812	7·21053	−3·62001	7·41076	2·57042	0·5625	3
0	6·21771	7·51808	10·33528	27·72306	−17·48391	14·38485	−7·80740	0·6250	3
1	6·21771	2·10691	2·87646	7·71574	−2·96914	6·66911	−4·83826	0·6250	3
0	6·21771	7·73778	4·56984	12·25800	−18·07322	9·62469	−7·64428	0·6875	3
1	6·21771	1·90584	3·09727	8·30803	−2·42981	1·31666	−5·21447	0·6875	3
0	6·21771	5·86472	2·79150	7·48783	−13·04899	8·50211	−5·69179	0·7500	3
1	6·21771	1·75053	3·35869	9·00926	−2·01319	−0·50714	−3·67860	0·7500	3
0	6·21771	0·40910	1·61526	4·33274	1·58500	4·23392	−9·94442	0·1875	4
1	6·21771	4·64950	2·06586	5·54141	−9·78933	0·70316	5·84226		
2	6·21771	9·12118	0·56670	1·52011	−21·78403	−1·30749	−0·15509		
0	6·21771	−0·01844	1·89994	5·09635	2·73183	4·80273	−6·56582	0·2500	4
1	6·21771	3·61790	2·21455	5·94025	−7·02217	0·71557	4·42065		
2	6·21771	6·57371	0·83287	2·23406	−14·95077	−1·13752	0·45635		

Table 2 (continued)

Sequence				Sequence admittance		Self- and mutual admittances		$\dfrac{d}{\lambda}$	N
m	Ψ_{dR}	Re $T'^{(m)}$	Im $T'^{(m)}$	$G^{(m)}$	$B^{(m)}$	$G_{1(m+1)}$	$B_{1(m+1)}$		
0	6·21771	−0·63097	2·25790	6·05652	4·37485	5·44649	−4·28694	0·3125	4
1	6·21771	2·96320	2·39158	6·41511	−5·26603	0·78932	3·84135		
2	6·21771	5·09732	1·08085	2·89924	−10·99055	−0·96862	0·97909		
0	6·21771	−1·59127	2·81668	7·55539	6·95074	6·25385	−2·36381	0·3750	4
1	6·21771	2·49201	2·60144	6·97803	−4·00212	1·01286	3·83811		
2	6·21771	4·13220	1·30629	3·50394	−8·40172	−0·72418	1·63832		
0	6·21771	−3·34395	4·05396	10·87422	11·65209	7·56365	−0·23979	0·4375	4
1	6·21771	2·12940	2·85620	7·66138	−3·02948	1·70415	4·55110		
2	6·21771	3·44273	1·51270	4·05763	−6·55232	−0·09773	2·78968		
0	6·21771	−6·93658	9·19894	24·67498	21·28884	11·57104	2·89748	0·5000	4
1	6·21771	1·85172	3·17452	8·51526	−2·28462	5·02408	6·60463		
2	6·21771	2·91237	1·70695	4·57867	−5·12969	3·05578	5·18210		
0	6·21771	12·59562	19·02475	51·03015	−31·10376	18·82977	−9·67092	0·5625	4
1	6·21771	1·67463	3·57861	9·59917	−1·80960	11·48554	−6·78576		
2	6·21771	2·47658	1·89731	5·08929	−3·96072	9·23060	−7·86132		
0	6·21771	9·49153	3·80603	10·20918	−22·77744	9·42201	−7·31992	0·6250	4
1	6·21771	1·66405	4·07540	10·93173	−1·78124	1·14845	−4·95942		
2	6·21771	2·09596	2·09344	5·61538	−2·93977	−1·50973	−5·53868		
0	6·21771	6·13225	2·43284	6·52578	−13·76660	9·33672	−5·19927	0·6875	4
1	6·21771	1·93857	4·59131	12·31559	−2·51759	0·08397	−2·94282		
2	6·21771	1·74386	2·30763	6·18991	−1·99530	−2·97887	−2·68168		
0	6·21771	5·74651	2·36377	6·34051	−12·73189	9·43757	−5·02606	0·7000	4
1	6·21771	2·03672	4·67782	12·54766	−2·78086	0·00652	−2·73032		
2	6·21771	1·67501	2·35405	6·31443	−1·81063	−3·11010	−2·24520		
0	6·21771	4·63831	2·26944	6·08749	−9·75930	9·77373	−4·78684	0·7500	4
1	6·21771	2·54993	4·87384	13·07345	−4·15749	−0·19326	−2·17155		
2	6·21771	1·40005	2·55764	6·86054	−1·07309	−3·29972	−0·62935		

Table 2 (continued)

Sequence				Sequence admittance		Self- and mutual admittances			
m	Ψ_{dR}	Re $T'^{(m)}$	Im $T'^{(m)}$	$G^{(m)}$	$B^{(m)}$	$G_{1(m+1)}$	$B_{1(m+1)}$	$\frac{d}{\lambda}$	N
0	6·21771	0·21514	1·40228	3·76143	2·10529	3·54866	−10·26034	0·1875	5
1	6·21771	3·38329	1·96748	5·27752	−6·39288	0·85015	6·20351		
2	6·21771	8·57189	0·63877	1·71342	−20·31061	−0·74377	−0·02069		
0	6·21771	−0·33422	1·66115	4·45583	3·57888	4·20839	−6·53350	0·2500	5
1	6·21771	2·57552	2·16720	5·81324	−4·22614	0·80724	4·69057		
2	6·21771	6·18088	0·92448	2·47981	−13·89704	−0·68352	0·36562		
0	6·21771	−1·21851	1·99594	5·35384	5·95086	4·92644	−3·95320	0·3125	5
1	6·21771	1·99727	2·40790	6·45887	−2·67505	0·83996	4·15493		
2	6·21771	4·79641	1·18564	3·18031	−10·18338	−0·62626	0·79710		
0	6·21771	−2·92127	2·65004	7·10840	10·51830	5·86009	−1·54167	0·3750	5
1	6·21771	1·50745	2·71553	7·28405	−1·36117	1·08846	4·44406		
2	6·21771	3·89003	1·42112	3·81196	−7·75215	−0·46430	1·58593		
0	6·21771	−4·15869	3·23605	8·68029	13·83752	6·43347	−0·37358	0·4000	5
1	6·21771	1·32071	2·86853	7·69445	−0·86027	1·37683	4·92397		
2	6·21771	3·60681	1·50951	4·04908	−6·99243	−0·25343	2·18158		
0	6·21771	−7·54050	5·70219	15·29540	22·90879	8·19364	2·13276	0·4375	5
1	6·21771	1·04301	3·14776	8·44346	−0·11536	2·68116	6·51145		
2	6·21771	3·23949	1·63771	4·39293	−6·00714	0·86972	3·87657		
0	6·21771	−9·81216	8·71608	23·37976	29·00223	10·01690	3·62199	0·45313	5
1	6·21771	0·92770	3·28782	8·81917	0·19394	4·29908	7·64959		
2	6·21771	3·10265	1·69000	4·53321	−5·64009	2·38234	5·04053		
0	6·21771	−3·91456	26·67988	71·56534	13·18266	20·12878	0·97151	0·48438	5
1	6·21771	0·70033	3·62681	9·72844	0·80383	13·95875	4·34305		
2	6·21771	2·85149	1·79350	4·81084	−4·96639	11·75953	1·76253		
0	6·21771	11·71804	22·64512	60·74263	−28·74975	18·24252	−7·17352	0·5000	5
1	6·21771	0·59141	3·83461	10·28586	1·09600	11·81836	−4·10811		
2	6·21771	2·73538	1·84506	4·94912	−4·65492	9·43170	−6·68000		

Table 2 (continued)

Sequence m	Ψ_{dR}	Re $T'^{(m)}$	Im $T'^{(m)}$	Sequence admittance $G^{(m)}$	Sequence admittance $B^{(m)}$	Self- and mutual admittances $G_{1(m+1)}$	Self- and mutual admittances $B_{1(m+1)}$	$\dfrac{d}{\lambda}$	N
0	6·21771	13·94954	12·57851	33·74024	−34·73548	13·15613	−8·14286	0·51563	5
1	6·21771	0·48991	4·07569	10·93251	1·36826	6·45297	−5·36780		
2	6·21771	2·62455	1·89671	5·08769	−4·35766	3·83909	−7·92851		
0	6·21771	8·42858	4·54611	12·19435	−19·92622	10·09068	−4·65915	0·5625	5
1	6·21771	0·31035	5·07731	13·61924	1·84991	2·33913	−2·61274		
2	6·21771	2·31774	2·05426	5·51030	−3·53468	−1·28730	−5·02080		
0	6·21771	5·44546	3·30165	8·85624	−11·92436	11·93591	−3·68524	0·6250	5
1	6·21771	1·26107	7·19498	19·29960	−0·70028	2·17891	−1·64603		
2	6·21771	1·95089	2·27860	6·11205	−2·55063	−3·71874	−2·47353		
0	6·21771	4·22835	3·18268	8·53713	−8·65964	12·14514	−6·63446	0·6875	5
1	6·21771	4·95667	7·19221	19·29217	−10·61325	1·89086	−2·51209		
2	6·21771	1·61255	2·53586	6·80212	−1·64308	−3·69486	1·49950		
0	6·21771	3·96685	3·20896	8·60762	−7·95818	10·98986	−7·20681	0·7100	5
1	6·21771	5·73836	5·99691	16·08594	−12·71004	1·41714	−2·73297		
2	6·21771	1·49504	2·64128	7·08489	−1·32788	−2·60826	2·35729		
0	6·21771	3·64504	3·29276	8·83240	−7·09499	9·21245	−6·73060	0·7500	5
1	6·21771	5·65993	4·08688	10·96254	−12·49968	0·64516	−2·71185		
2	6·21771	1·29053	2·85284	7·65238	−0·77932	−0·83519	2·52965		
0	14·83942	0·76543	1·60211	1·80064	0·26364	1·12981	−44·14067	0·0625	8
1	8·04324	5·28009	1·57354	3·26285	−8·87507	0·87168	24·88961		
2	3·73120	9·62409	0·15395	0·68816	−38·54906	0·01520	−2·53078		
3	4·39219	22·19073	−0·04765	−0·18092	−80·46642	−0·34588	−0·42171		
4	2·56902	16·03502	−0·04660	−0·30252	−97·60802	−0·41117	0·53006		
0	8·48656	0·21045	1·19329	2·34512	1·55166	1·77822	−18·48555	0·1250	8
1	7·82203	2·78655	1·76586	3·76519	−3·80931	0·93309	10·62944		
2	6·21771	7·19363	0·75322	2·02043	−16·61362	−0·21774	−0·81075		
3	4·61340	10·31149	0·04917	0·17777	−33·66265	−0·33526	0·07470		
4	3·94887	10·77021	−0·01094	−0·04619	−41·26493	−0·19327	0·25043		

Table 2 (continued)

Sequence				Sequence admittance		Self- and mutual admittances			
m	Ψ_{dR}	Re $T^{\cdot(m)}$	Im $T^{\cdot(m)}$	$G^{(m)}$	$B^{(m)}$	$G_{1(m+1)}$	$B_{1(m+1)}$	$\dfrac{d}{\lambda}$	N
0	6·21771	−0·30497	1·04952	2·81521	3·50041	2·43903	−10·55638	0·1875	8
1	6·21771	1·51389	1·60947	4·31719	−1·37844	0·97192	6·99193		
2	6·21771	4·52861	1·20601	3·23497	−9·46505	−0·44869	−0·40912		
3	6·21771	8·56067	0·28472	0·76372	−20·28053	−0·28443	0·30903		
4	6·21771	10·58235	0·02433	0·06525	−25·70343	−0·10143	0·27310		
0	6·21771	−1·61604	1·26374	3·38982	7·01720	3·18540	−6·05024	0·2500	8
1	6·21771	0·86391	1·85441	4·97421	0·36504	0·97095	5·48956		
2	6·21771	3·22073	1·58968	4·26410	−5·95683	−0·59988	0·18818		
3	6·21771	5·99746	0·61091	1·63868	−13·40505	−0·20834	0·62110		
4	6·21771	7·49628	0·12651	0·33936	−17·42544	−0·12105	0·46977		
0	6·21771	−3·07274	1·59436	4·27665	10·92459	3·66117	−4·11294	0·28125	8
1	6·21771	0·51587	1·99731	5·35752	1·29862	1·03474	5·39567		
2	6·21771	2·76714	1·76689	4·73947	−4·74013	−0·57855	0·72395		
3	6·21771	5·16270	0·79124	2·12240	−11·16592	−0·10905	0·98879		
4	6·21771	6·44788	0·21396	0·57391	−14·61324	−0·07879	0·82071		
0	6·21771	−6·26287	3·48135	9·33828	19·48172	4·68497	−1·79227	0·3125	8
1	6·21771	0·11521	2·16787	5·81502	2·37333	1·62530	6·06000		
2	6·21771	2·38182	1·94322	5·21243	−3·70656	−0·02668	1·81987		
3	6·21771	4·50764	0·97178	2·60668	−9·40879	0·49098	1·89439		
4	6·21771	5·59887	0·32552	0·87318	−12·33586	0·47411	1·72546		
0	6·21771	−3·84144	14·56160	39·05962	12·98655	8·83417	−1·45440	0·34375	8
1	6·21771	−0·37948	2·39001	6·41089	3·70027	5·31836	4·99896		
2	6·21771	2·03733	2·12667	5·70452	−2·78250	3·60935	1·01128		
3	6·21771	3·98168	1·14798	3·07932	−7·99797	4·14048	0·86301		
4	6·21771	4·90003	0·45641	1·22425	−10·46134	4·08906	0·69445		

Table 2 (continued)

Sequence				Sequence admittance		Self- and mutual admittances			
m	Ψ_{dR}	Re $T'^{(m)}$	Im $T'^{(m)}$	$G^{(m)}$	$B^{(m)}$	$G_{1(m+1)}$	$B_{1(m+1)}$	$\dfrac{d}{\lambda}$	N
0	6·21771	4·74469	8·23353	22·08539	−10·04466	7·22937	−3·18958	0·3750	8
1	6·21771	−1·03911	2·71812	7·29101	5·46964	3·22335	2·03307		
2	6·21771	1·71403	2·32753	6·24331	−1·91530	1·40099	−1·88938		
3	6·21771	3·54993	1·31820	3·53589	−6·83985	1·89571	−2·31900		
4	6·21771	4·31831	0·59990	1·60915	−8·90094	1·81592	−2·50447		
0	6·21771	3·60511	5·20813	13·97015	−6·98788	6·92141	−1·54358	0·40625	8
1	6·21771	−1·99948	3·29915	8·84954	8·04573	2·35624	2·53477		
2	6·21771	1·39645	2·56053	6·86828	−1·06342	0·28046	−1·55617		
3	6·21771	3·18782	1·48290	3·97768	−5·86853	0·63578	−2·38466		
4	6·21771	3·82897	0·74937	2·01009	−7·58834	0·50779	−2·63218		
0	6·21771	2·61065	4·38490	11·76194	−4·32035	7·89420	0·37131	0·4375	8
1	6·21771	−3·51453	4·63664	12·43719	12·10963	2·58768	3·29992		
2	6·21771	1·07151	2·84911	7·64236	−0·19182	−0·13872	−1·30088		
3	6·21771	2·87791	1·64397	4·40975	−5·03724	−0·25045	−2·76242		
4	6·21771	3·41217	0·89960	2·41305	−6·47035	−0·52927	−3·16489		
0	6·21771	0·97143	2·43161	6·52248	0·07664	1·43500	−12·49961	0·1700	20
1	6·21771	−0·58939	0·96375	2·58513	4·26334	1·04357	8·55105		
2	6·21771	0·64311	1·12504	3·01778	0·95730	0·30332	−0·63750		
3	6·21771	1·72937	1·05916	2·84107	−1·95645	−0·02529	−0·20626		
4	6·21771	3·18240	0·70144	1·88152	−5·85400	0·06437	−0·05895		
5	6·21771	5·16224	0·26348	0·70676	−11·16468	0·19925	−0·12010		
6	6·21771	7·38030	0·04658	0·12494	−17·11434	0·22424	−0·22323		
7	6·21771	9·39761	−0·00104	−0·00279	−22·52551	0·20842	−0·27115		
8	6·21771	10·99347	−0·00840	−0·02253	−26·80619	0·20638	−0·28878		
9	6·21771	12·02388	−0·01052	−0·02823	−29·57014	0·21217	−0·30248		
10	6·21771	12·38084	−0·01120	−0·03004	−30·52763	0·21460	−0·30893		

Table 2 (continued)

Sequence				Sequence admittance		Self- and mutual admittances		$\dfrac{d}{\lambda}$	N
m	Ψ_{dR}	Re $T'^{(m)}$	Im $T'^{(m)}$	$G^{(m)}$	$B^{(m)}$	$G_{1(m+1)}$	$B_{1(m+1)}$		
0	6·21771	−1·02714	1·71496	4·60017	5·43753	2·85913	−5·81890	0·2500	20
1	6·21771	1·00569	3·01550	8·08869	−0·01526	1·73291	5·81862		
2	6·21771	−1·67840	2·02008	5·41861	7·18446	0·20562	−0·80449		
3	6·21771	0·34365	1·59237	4·27133	1·76059	−0·01856	−0·36063		
4	6·21771	1·51934	1·42374	3·81899	−1·39307	−0·09549	−0·41945		
5	6·21771	2·71476	1·03755	2·78309	−4·59963	−0·06486	−0·39799		
6	6·21771	4·11322	0·53003	1·42174	−8·35082	−0·20028	−0·11090		
7	6·21771	5·62312	0·16006	0·42935	−12·40092	−0·23897	0·22478		
8	6·21771	6·89325	0·02510	0·06733	−15·80788	−0·25558	0·52470		
9	6·21771	7·69158	−0·00100	−0·00270	−17·94931	−0·25782	0·74258		
10	6·21771	7·96098	−0·00395	−0·01059	−18·67192	−0·25483	0·82200		
0	6·21771	1·05103	3·85805	10·34872	−0·13687	6·02583	−1·62248	0·3400	20
1	6·21771	−2·34603	3·18249	8·53661	8·97530	2·94009	4·10157		
2	6·21771	0·65817	3·07359	8·24451	0·91691	−0·50792	−0·95743		
3	6·21771	−0·87100	7·04184	18·88884	5·01871	−1·01616	−0·19390		
4	6·21771	−0·43591	2·47257	6·63236	3·85164	−0·83124	0·06052		
5	6·21771	1·11157	1·93507	5·19058	−0·29926	0·06403	0·32155		
6	6·21771	2·22382	1·45216	3·89524	−3·28274	1·02681	0·06635		
7	6·21771	3·25400	0·88492	2·37368	−6·04608	1·15323	−0·41290		
8	6·21771	4·24964	0·37783	1·01349	−8·71674	0·41960	−0·78701		
9	6·21771	5·02288	0·10007	0·26843	−10·79085	−0·57409	−0·95892		
10	6·21771	5·31204	0·02995	0·08034	−11·56648	−1·02580	−0·99404		

Table 3. Ψ_{dR}, $T^{(m)}$ and admittances of circular array† of N elements

$$(a/\lambda = 7.022 \times 10^{-3})$$
$$h/\lambda = 3/8, \quad \beta_0 h = 3\pi/4, \quad \Omega = 9.34$$

Sequence m	Ψ_{dR}	Re $T^{(m)}$	Im $T^{(m)}$	Sequence admittance		Self- and mutual admittances		$\dfrac{d}{\lambda}$	N
				$G^{(m)}$	$B^{(m)}$	$G_{1(m+1)}$	$B_{1(m+1)}$		
0	6·44947	−0·34647	0·34841	2·17514	−0·42294	1·32680	−1·12131	0·1875	2
1	6·44947	−0·12274	0·07664	0·47845	−1·81968	0·84834	0·69837		
0	6·44947	−0·30504	0·39203	2·44750	−0·68161	1·55035	−1·05823	0·2500	2
1	6·44947	−0·18438	0·10463	0·65319	−1·43485	0·89716	0·37662		
0	6·44947	−0·24661	0·42127	2·63003	−1·04640	1·72902	−1·11269	0·3125	2
1	6·44947	−0·22537	0·13263	0·82801	−1·17899	0·90101	0·06630		
0	6·44947	−0·17613	0·42372	2·64534	−1·48639	1·82429	−1·24458	0·3750	2
1	6·44947	−0·25359	0·16070	1·00324	−1·00277	0·82105	−0·24181		
0	6·44947	−0·11086	0·39101	2·44112	−1·89389	1·81048	−1·38879	0·4375	2
1	6·44947	−0·27267	0·18898	1·17984	−0·88369	0·63064	−0·50510		
0	6·44947	−0·07291	0·33159	2·07016	−2·13080	1·71430	−1·47116	0·5000	2
1	6·44947	−0·28423	0·21759	1·35844	−0·81153	0·35586	−0·65964		
0	6·44947	−0·07068	0·26849	1·67624	−2·14471	1·60735	−1·46392	0·5625	2
1	6·44947	−0·28878	0·24643	1·53846	−0·78313	0·06889	−0·68079		
0	6·44947	−0·09430	0·22053	1·37682	−1·99728	1·54683	−1·39861	0·6250	2
1	6·44947	−0·28608	0·27500	1·71683	−0·79995	−0·17001	−0·59867		
0	6·44947	−0·12872	0·19294	1·20452	−1·78236	1·54531	−1·32429	0·6875	2
1	6·44947	−0·27546	0·30211	1·88610	−0·86622	−0·34079	−0·45807		
0	6·44947	−0·16365	0·18248	1·13923	−1·56431	1·58566	−1·27515	0·7500	2
1	6·44947	−0·25628	0·32549	2·03209	−0·98599	−0·44643	−0·28916		
0	6·44947	−0·47138	0·35091	2·19075	0·35693	1·04922	−1·09414	0·1875	3
1	6·44947	−0·12274	0·07664	0·47845	−1·81968	0·57076	−0·72553		

† Note that $T^{(m)} = T^{(m)}\left(\dfrac{3\lambda}{8}\right)$.

Table 3 (continued)

Sequence				Sequence admittance		Self- and mutual admittances		$\dfrac{d}{\lambda}$	N
m	Ψ_{dR}	Re $T^{(m)}$	Im $T^{(m)}$	$G^{(m)}$	$B^{(m)}$	$G_{1(m+1)}$	$B_{1(m+1)}$		
0	6·44947	−0·44657	0·43938	2·74308	0·20198	1·34982	−0·88924	0·2500	3
1	6·44947	−0·18438	0·10463	0·65319	−1·43485	0·69663	0·54561		
0	6·44947	−0·38446	0·54228	3·38553	−0·18574	1·68051	−0·84791	0·3125	3
1	6·44947	−0·22537	0·13263	0·82801	−1·17899	0·85251	0·33108		
0	6·44947	−0·24901	0·63644	3·97336	−1·03141	1·99328	−1·01232	0·3750	3
1	6·44947	−0·25359	0·16070	1·00324	−1·00277	0·99004	−0·00954		
0	6·44947	−0·02896	−0·63299	3·95182	−2·40516	2·10384	−1·39084	0·4375	3
1	6·44947	−0·27267	0·18898	1·17984	−0·88369	0·92399	−0·50716		
0	6·44947	0·13984	0·45236	2·82412	−3·45905	1·84700	−1·69403	0·5000	3
1	6·44947	−0·28423	0·21759	1·35844	−0·81153	0·48856	−0·88251		
0	6·44947	−0·12450	0·23956	1·49557	−3·36327	1·52417	−1·64317	0·5625	3
1	6·44947	−0·28878	0·24643	1·53846	−0·78313	−0·01430	−0·86005		
0	6·44947	0·02068	0·12744	0·79560	−2·71512	1·40975	−1·43834	0·6250	3
1	6·44947	−0·28608	0·27500	1·71683	−0·79995	−0·30708	−0·63839		
0	6·44947	−0·07859	0·09350	0·58373	−2·09536	1·45198	−1·27594	0·6875	3
1	6·44947	−0·27546	0·30211	1·88610	−0·86622	−0·43412	−0·40971		
0	6·44947	−0·15342	0·09624	0·60083	−1·62816	1·55500	−1·20004	0·7500	3
1	6·44947	−0·25628	0·32549	2·03209	−0·98599	−0·47709	−0·21406		
0	6·44947	−0·55329	0·35643	2·22523	0·86829	0·92702	−1·05015	0·1875	4
1	6·44947	−0·19581	0·11142	0·69561	−1·36349	0·53340	0·80255		
2	6·44947	−0·03909	0·01468	0·09163	−2·34192	0·23141	0·31334		
0	6·44947	−0·54954	0·48678	3·03902	0·84487	1·28146	−0·76256	0·2500	4
1	6·44947	−0·24507	0·15105	0·94301	−1·05600	0·70955	−0·65699		
2	6·44947	−0·12860	0·03216	0·20081	−1·78309	0·33845	0·29345		
0	6·44947	−0·48519	0·68375	4·26873	0·44314	1·75057	−0·67078	0·3125	4
1	6·44947	−0·27373	0·19101	1·19247	−0·87708	0·98003	0·45381		
2	6·44947	−0·19443	0·05584	0·34860	−1·37211	0·55810	0·20630		

Table 3 (continued)

Sequence				Sequence admittance		Self- and mutual admittances		$\dfrac{d}{\lambda}$	N
m	Ψ_{dR}	$\operatorname{Re} T^{(m)}$	$\operatorname{Im} T^{(m)}$	$G^{(m)}$	$B^{(m)}$	$G_{1(m+1)}$	$B_{1(m+1)}$		
0	6·44947	−0·19693	0·91255	5·69715	−1·35650	2·27841	−0·99994	0·3750	4
1	6·44947	−0·28731	0·23157	1·44574	−0·79228	1·29303	−0·07445		
2	6·44947	−0·24463	0·08409	0·52501	−1·05872	0·83267	−0·20767		
0	6·44947	0·32272	0·70498	4·40129	−4·60077	2·13046	−1·75241	0·4375	4
1	6·44947	−0·28670	0·27216	1·69915	−0·79610	0·91976	−0·94602		
2	6·44947	−0·28340	0·11569	0·72225	−0·81668	0·43131	−0·95631		
0	6·44947	0·29755	0·22178	1·38458	−4·44362	1·54804	−1·71731	0·5000	4
1	6·44947	−0·27039	0·30999	1·93530	−0·89794	0·11189	−0·95347		
2	6·44947	−0·31334	0·15009	0·93700	−0·62975	−0·38725	−0·81938		
0	6·44947	0·07700	0·06927	0·43245	−3·06671	1·45636	−1·44191	0·5625	4
1	6·44947	−0·23702	0·33817	2·11123	−1·10623	−0·18452	−0·64456		
2	6·44947	−0·33597	0·18749	1·17052	−0·48847	−0·65487	−0·33568		
0	6·44947	−0·06893	0·05805	0·36244	−2·15564	1·52608	−1·33380	0·6250	4
1	6·44947	−0·19080	0·34549	2·15694	−1·39480	−0·26640	−0·44142		
2	6·44947	−0·35175	0·22874	1·42802	−0·38996	−0·63085	0·06100		
0	6·44947	−0·15616	0·08162	0·50959	−1·61108	1·56757	−1·32067	0·6875	4
1	6·44947	−0·14737	0·32377	2·02131	−1·66597	−0·30213	−0·31785		
2	6·44947	−0·35980	0·27520	1·71810	−0·33968	−0·45373	0·34529		
0	6·44947	−0·16904	0·08762	0·54702	−1·53067	1·56935	−1·31981	0·7000	4
1	6·44947	−0·14096	0·31630	1·97472	−1·70593	−0·30848	−0·29848		
2	6·44947	−0·36028	0·28527	1·78094	−0·33673	−0·40537	0·38611		
0	6·44947	−0·21007	0·11293	0·70504	−1·27449	1·56698	−1·30361	0·7500	4
1	6·44947	−0·12720	0·28125	1·75587	−1·79184	−0·33652	−0·22955		
2	6·44947	−0·35715	0·32854	2·05113	−0·35628	−0·18889	0·48822		
0	6·44947	−0·61970	0·36751	2·29438	1·28290	0·86350	−0·99648	0·1875	5
1	6·44947	−0·27081	0·14413	0·89982	−0·89527	0·53394	0·86995		
2	6·44947	−0·05584	0·01790	0·11174	−2·23737	0·18150	0·26974		

Table 3 (continued)

Sequence				Sequence admittance		Self- and mutual admittances		$\frac{d}{\lambda}$	N
m	Ψ_{dR}	Re $T^{(m)}$	Im $T^{(m)}$	$G^{(m)}$	$B^{(m)}$	$G_{1(m+1)}$	$B_{1(m+1)}$		
0	6·44947	−0·64486	0·55417	3·45972	1·43995	1·28095	−0·64140	0·2500	5
1	6·44947	−0·31193	0·19746	1·23277	−0·63857	0·76674	0·75430		
2	6·44947	−0·14433	0·03840	0·23975	−1·68489	0·32265	0·28637		
0	6·44947	−0·54344	0·93294	5·82443	0·80675	1·96639	−0·55634	0·3125	5
1	6·44947	−0·33241	0·25558	1·59564	−0·51070	1·23005	0·51358		
2	6·44947	−0·20862	0·06537	0·40812	−1·28353	0·69897	0·16796		
0	6·44947	−0·32595	1·11700	6·97358	−4·62090	2·43923	−1·52249	0·3750	5
1	6·44947	−0·33217	0·32142	2·00669	−0·51224	1·44711	−0·66922		
2	6·44947	−0·25667	0·09684	0·60459	−0·98354	0·82007	−0·87999		
0	6·44947	−0·57666	0·70317	4·39000	−6·18612	2·02805	−1·81601	0·4000	5
1	6·44947	−0·32429	0·35016	2·18607	−0·56143	0·92523	−1·02005		
2	6·44947	−0·27237	0·11037	0·68904	−0·88555	0·25574	−1·16500		
0	6·44947	−0·42187	0·23114	1·44306	−5·21980	1·60258	−1·63129	0·4375	5
1	6·44947	−0·30048	0·39465	2·46383	−0·71005	0·32744	−0·88634		
2	6·44947	−0·29276	0·13152	0·82111	−0·75828	−0·40720	−0·90791		
0	6·44947	−0·31785	0·14604	0·91174	−4·57038	1·56449	−1·52095	0·45313	5
1	6·44947	−0·28518	0·41283	2·57737	−0·80556	0·21681	−0·78337		
2	6·44947	−0·30023	0·14063	0·87798	−0·71161	−0·54318	−0·74135		
0	6·44947	−0·14751	0·07588	0·47375	−3·50693	1·60478	−1·38079	0·48438	5
1	6·44947	−0·24284	0·44530	2·78003	−1·06989	0·11638	−0·63020		
2	6·44947	−0·31352	0·15938	0·99503	−0·62863	−0·68189	−0·43286		
0	6·44947	−0·08327	0·06638	0·41440	−3·10585	1·64758	−1·35482	0·5000	5
1	6·44947	−0·21527	0·45754	2·85645	−1·24202	0·09445	−0·58309		
2	6·44947	−0·31937	0·16903	1·05530	−0·59209	−0·71104	−0·29243		

Table 3 (continued)

| Sequence | | | | Sequence admittance | | Self- and mutual admittances | | | |
m	Ψ_{dR}	Re $T^{(m)}$	Im $T^{(m)}$	$G^{(m)}$	$B^{(m)}$	$G_{1(m+1)}$	$B_{1(m+1)}$	$\dfrac{d}{\lambda}$	N
0	6·44947	0·03044	0·06516	0·40683	-2·77603	1·69002	-1·35474	0·51563	5
1	6·44947	-0·18355	0·46529	2·90484	-1·44006	0·07902	-0·55239		
2	6·44947	-0·32471	0·17888	1·11679	-0·55877	-0·72062	-0·15826		
0	6·44947	-0·07925	0·08411	0·52513	-2·09120	1·74275	-1·45800	0·5625	5
1	6·44947	-0·07441	0·44607	2·78486	-2·12141	0·02551	-0·52578		
2	6·44947	-0·33765	0·20974	1·30944	-0·47798	-0·63432	0·20918		
0	6·44947	-0·15781	0·12414	0·77502	-1·60074	1·56994	-1·56058	0·6250	5
1	6·44947	0·01552	0·31211	1·94857	-2·68287	-0·11828	-0·51643		
2	6·44947	-0·34722	0·25448	1·58877	-0·41822	-0·27918	0·49635		
0	6·44947	-0·19469	0·16288	1·01690	-1·37050	1·40669	-1·40661	0·6875	5
1	6·44947	-0·02918	0·17778	1·10991	-2·40384	-0·27375	-0·43291		
2	6·44947	-0·34575	0·30407	1·89837	-0·42743	0·07886	0·45097		
0	6·44947	-0·20126	0·17513	1·09336	-1·32949	1·40201	-1·32741	0·7100	5
1	6·44947	-0·06173	0·15075	0·94115	-2·20057	-0·31778	-0·39124		
2	6·44947	-0·34162	0·32311	2·01720	-0·45322	0·16345	0·39020		
0	6·44947	-0·20643	0·19347	1·20785	-1·29723	1·45463	-1·20877	0·7500	5
1	6·44947	-0·12027	0·12776	0·79763	-1·83511	-0·38310	-0·31211		
2	6·44947	-0·32801	0·35800	2·23501	-0·53821	0·25971	0·26788		
0	8·34383	-0·02515	0·40052	1·93278	-1·87751	0·51502	-1·13221	0·1500	20
1	8·25111	-0·83304	0·41564	2·02828	2·04386	0·43561	0·84404		
2	7·98204	-0·53929	0·24478	1·23478	0·63094	0·25643	0·19158		

Table 3 (continued)

Sequence m	Ψ_{dR}	Re $T^{(m)}$	Im $T^{(m)}$	Sequence admittance		Self- and mutual admittances		$\dfrac{d}{\lambda}$	N
				$G^{(m)}$	$B^{(m)}$	$G_{1(m+1)}$	$B_{1(m+1)}$		
3	7·56295	−0·37900	0·12488	0·66487	−0·18747	0·09682	−0·09386		
4	7·03486	−0·24489	0·03860	0·22095	−0·96914	0·01625	−0·14187		
5	6·44947	−0·15155	0·00584	0·03649	−1·63983	−0·00505	−0·16493		
6	5·86408	−0·11432	0·00036	0·00248	−2·05918	−0·00889	−0·19133		
7	5·33600	−0·11294	−0·00012	−0·00089	−2·27337	−0·01514	−0·21710		
8	4·91690	−0·12607	−0·00015	−0·00122	−2·35962	−0·02320	−0·23417		
9	4·64783	−0·13954	−0·00015	−0·00131	−2·37952	−0·02871	−0·24255		
10	4·55512	−0·14497	−0·00015	−0·00133	−2·37994	−0·03047	−0·24494		
0	6·44947	0·35836	0·88997	5·55618	−4·82325	1·53249	−0·96196	0·2800	20
1	6·44947	−0·44839	0·45566	2·84473	0·21335	1·00305	0·09476		
2	6·44947	0·09217	0·37118	2·31731	−3·16144	0·17140	−0·59347		
3	6·44947	−0·62997	0·63638	3·97298	1·34698	−0·10521	−0·31997		
4	6·44947	−0·50165	0·29433	1·83756	0·54586	0·01334	−0·02415		
5	6·44947	−0·40107	0·15795	0·98612	−0·08206	0·18889	0·13000		
6	6·44947	−0·31037	0·06941	0·43331	−0·64832	0·30382	0·11621		
7	6·44947	−0·22595	0·02062	0·12870	−1·17535	0·29188	−0·07330		
8	6·44947	−0·16422	0·00380	0·02375	−1·56073	0·16145	−0·34748		
9	6·44947	−0·13077	0·00038	0·00240	−1·76954	0·01048	−0·57893		
10	6·44947	−0·12053	−0·00003	−0·00016	−1·83351	−0·05450	−0·66864		

Table 4. Ψ_{dR}, $T^{(m)}$ and admittances of circular array† of N elements

$$(a/\lambda = 7{\cdot}022 \times 10^{-3})$$

$$h/\lambda = 0{\cdot}5, \quad \beta_0 h = \pi, \quad \Omega = 9{\cdot}92$$

Sequence m	Ψ_{dR}	$\mathrm{Re}\,T^{(m)}$	$\mathrm{Im}\,T^{(m)}$	Sequence admittance		Self- and mutual admittances		$\dfrac{d}{\lambda}$	N
				$G^{(m)}$	$B^{(m)}$	$G_{1(m+1)}$	$B_{1(m+1)}$		
0	5·73687	−0·24563	0·26072	1·51595	1·42818	0·90483	1·15817	0·1875	2
1	5·73687	−0·15275	0·05051	0·29370	0·88815	0·61112	0·27001	0·2500	2
0	5·73687	−0·21332	0·28152	1·63686	1·24030	1·02739	1·15671	0·3125	2
1	5·73687	−0·18456	0·07187	0·41791	1·07312	0·60948	0·08359	0·3750	2
0	5·73687	−0·17004	0·28854	1·67767	0·98868	1·11071	1·09623	0·4375	2
1	5·73687	−0·20704	0·09352	0·54375	1·20379	0·56696	−0·10755	0·5000	2
0	5·73687	−0·12443	0·27467	1·59702	0·72347	1·13360	1·00855	0·5625	2
1	5·73687	−0·22249	0·11526	0·67017	1·29364	0·46342	−0·28509	0·6250	2
0	5·73687	−0·09046	0·23989	1·39480	0·52596	1·09581	0·93779	0·6875	2
1	5·73687	−0·23212	0·13704	0·79681	1·34962	0·29900	−0·41183	0·7500	2
0	5·73687	−0·07897	0·19532	1·13564	0·45915	1·02932	0·91689	0·1875	3
1	5·73687	−0·23642	0·15874	0·92300	1·37462	0·10632	−0·45774		
0	5·73687	−0·08921	0·15582	0·90602	0·51869	0·97645	0·94372		
1	5·73687	−0·23541	0·18005	1·04688	1·36875	−0·07043	−0·42503		
0	5·73687	−0·11205	0·12974	0·75438	0·65153	0·95933	0·99088		
1	5·73687	−0·22878	0·20024	1·16429	1·33024	−0·20495	−0·33936		
0	5·73687	−0·13840	0·11752	0·68333	0·80471	0·97538	1·03084		
1	5·73687	−0·21618	0·21798	1·26743	1·25697	−0·29205	−0·22613		
0	5·73687	−0·16285	0·11602	0·67459	0·94690	1·00918	1·04805		
1	5·73687	−0·19765	0·23111	1·34377	1·14920	−0·33459	−0·10115		
0	5·73687	−0·32803	0·29439	1·71168	1·90727	0·76636	1·22786		
1	5·73687	−0·15275	0·05051	0·29370	0·88815	0·47266	0·33971		

† Note that $T^{(m)} = T^{(m)}\left(\dfrac{\lambda}{2}\right)$.

Table 4 (continued)

Sequence m	Ψ_{dR}	$\mathrm{Re}\,T^{(m)}$	$\mathrm{Im}\,T^{(m)}$	Sequence admittance $G^{(m)}$	$B^{(m)}$	Self- and mutual admittances $G_{1(m+1)}$	$B_{1(m+1)}$	$\dfrac{d}{\lambda}$	N
0	5·73687	−0·29727	0·35169	2·04486	1·72846	0·96023	1·29157	0·2500	3
1	5·73687	−0·18456	0·07187	0·41791	1·07312	0·54232	0·21845		
0	5·73687	−0·22826	0·40826	2·37378	1·32720	1·15376	1·24493	0·3125	3
1	5·73687	−0·20704	0·09352	0·54375	1·20379	0·61001	0·04114		
0	5·73687	−0·10932	0·42957	2·49770	0·63563	1·27935	1·07430	0·3750	3
1	5·73687	−0·22249	0·11526	0·67017	1·29364	0·60918	−0·21934		
0	5·73687	0·02165	0·35891	2·08686	−0·12586	1·22683	0·85779	0·4375	3
1	5·73687	−0·23212	0·13704	0·79681	1·34962	1·43002	−0·49183		
0	5·73687	0·06652	0·21555	1·25327	−0·38679	1·03309	0·78749	0·5000	3
1	5·73687	−0·23642	0·15874	0·92300	1·37462	0·11009	−0·58714		
0	5·73687	0·01613	0·10388	0·60400	−0·09378	0·89925	0·88124	0·5625	3
1	5·73687	−0·23541	0·18005	1·04688	1·36875	−0·14763	−0·48751		
0	5·73687	−0·05914	0·05632	0·32744	0·34386	0·88534	1·00145	0·6250	3
1	5·73687	−0·22878	0·20024	1·16429	1·33024	−0·27895	−0·32879		
0	5·73687	−0·12272	0·04855	0·28230	0·71352	0·93905	1·07582	0·6875	3
1	5·73687	−0·21618	0·21798	1·26743	1·25697	−0·32838	−0·18115		
0	5·73687	−0·16955	0·05915	0·34394	0·98581	1·01050	1·09474	0·7500	3
1	5·73687	−0·19765	0·23111	1·34377	1·14920	−0·33328	−0·05446		
0	5·73687	−0·38958	0·32298	1·87795	2·26516	0·70749	1·28318	0·1875	4
1	5·73687	−0·19076	0·07711	0·44835	1·10916	0·45566	0·40398		
2	5·73687	−0·11166	0·00951	0·05532	0·64925	0·25914	0·17403		
0	5·73687	−0·36068	0·42208	2·45413	2·09712	0·95895	1·38535	0·2500	4
1	5·73687	−0·21789	0·10779	0·62675	1·26688	0·58149	0·29665		
2	5·73687	−0·15660	0·02205	0·12818	0·91051	0·33220	0·11847		
0	5·73687	−0·24528	0·54869	3·19030	1·42614	1·25828	1·31478	0·3125	4
1	5·73687	−0·23260	0·13859	0·80581	1·35240	0·73978	0·07449		
2	5·73687	−0·19403	0·03976	0·23119	1·12817	0·45247	−0·03762		

Table 4 (continued)

Sequence				Sequence admittance		Self- and mutual admittances			
m	Ψ_{dR}	Re $T^{(m)}$	Im $T^{(m)}$	$G^{(m)}$	$B^{(m)}$	$G_{1(m+1)}$	$B_{1(m+1)}$	$\dfrac{d}{\lambda}$	N
0	5·73687	0·03603	0·57029	3·31589	−0·20949	1·41026	0·96170	0·3750	4
1	5·73687	−0·23660	0·16917	0·98359	1·37571	0·73948	−0·37859		
2	5·73687	−0·22442	0·06156	0·35796	1·30489	0·42667	−0·41400		
0	5·73687	0·21558	0·28252	1·64266	−1·25348	1·11283	0·71545	0·4375	4
1	5·73687	−0·22971	0·19830	1·15297	1·33565	0·28499	−0·67437		
2	5·73687	−0·24835	0·08646	0·50271	1·44399	−0·04014	−0·62020		
0	5·73687	0·09549	0·06553	0·38103	−0·55524	0·90838	0·86177	0·5000	4
1	5·73687	−0·21098	0·22272	1·29499	1·22672	−0·07037	−0·52603		
2	5·73687	−0·26639	0·11394	0·66250	1·54890	−0·38661	−0·36495		
0	5·73687	−0·03885	0·01820	0·10581	0·22588	0·92235	0·98819	0·5625	4
1	5·73687	−0·18107	0·23615	1·37308	1·05281	−0·18291	−0·34885		
2	5·73687	−0·27884	0·14403	0·83744	1·62126	−0·45073	−0·06462		
0	5·73687	−0·12339	0·02822	0·16409	−0·71746	0·96987	1·02010	0·6250	4
1	5·73687	−0·14652	0·23091	1·34259	0·85195	−0·21653	−0·23539		
2	5·73687	−0·28533	0·17718	1·03020	1·65902	−0·37272	0·16814		
0	5·73687	−0·17437	0·05243	0·30486	1·01388	0·98500	1·01961	0·6875	4
1	5·73687	−0·12132	0·20555	1·19513	0·70539	−0·23500	−0·15998		
2	5·73687	−0·28443	0·21410	1·24487	1·65379	−0·21013	0·31422		
0	5·73687	−0·18188	0·05774	0·33572	1·05751	0·98483	1·02087	0·7000	4
1	5·73687	−0·11866	0·19889	1·15643	0·68991	−0·23876	−0·14716		
2	5·73687	−0·28312	0·22199	1·29075	1·64615	−0·17160	0·33096		
0	5·73687	−0·20546	0·07919	0·46045	1·19464	0·98512	1·03626	0·7500	4
1	5·73687	−0·11730	0·17165	0·99803	0·68205	−0·25588	−0·09792		
2	5·73687	−0·27283	0·25522	1·48397	1·58631	−0·01291	0·35421		
0	5·73687	−0·44337	0·35405	2·05857	2·57796	0·68373	1·33527	0·1875	5
1	5·73687	−0·23258	0·10524	0·61191	1·35233	0·46530	0·45724		
2	5·73687	−0·11985	0·01172	0·06813	0·69687	0·22212	0·16411		

Table 4 (continued)

Sequence m	Ψ_{dR}	Re $T^{(m)}$	Im $T^{(m)}$	Sequence admittance $G^{(m)}$	$B^{(m)}$	Self- and mutual admittances $G_{1(m+1)}$	$B_{1(m+1)}$	$\dfrac{d}{\lambda}$	N
0	5·73687	−0·41387	0·51736	3·00811	2·40640	1·00601	1·46290	0·2500	5
1	5·73687	−0·25661	0·14733	0·85663	1·49204	0·65756	0·35440		
2	5·73687	−0·16545	0·02654	0·15434	0·96200	0·34349	0·11736		
0	5·73687	−0·14964	0·75703	4·40167	0·87007	1·43714	1·26009	0·3125	5
1	5·73687	−0·26445	0·19253	1·11942	1·53760	0·93049	−0·01700		
2	5·73687	−0·20253	0·04688	0·27258	1·17759	0·55178	−0·17801		
0	5·73687	0·35041	0·42661	2·48049	−2·03744	1·22362	0·72145	0·3750	5
1	5·73687	−0·25376	0·24152	1·40429	1·47545	0·53554	−0·66098		
2	5·73687	−0·23165	0·07129	0·41452	1·34690	0·09290	−0·71847		
0	5·73687	0·30621	0·20277	1·17897	−1·78041	1·03469	0·76885	0·4000	5
1	5·73687	−0·24246	0·26157	1·52085	1·40974	0·26962	−0·63572		
2	5·73687	−0·24123	0·08193	0·47638	1·40260	−0·19748	−0·63891		
0	5·73687	0·14471	0·04886	0·28409	−0·84143	0·96034	0·92211	0·4375	5
1	5·73687	−0·21531	0·28980	1·68502	1·25190	0·07942	−0·49057		
2	5·73687	−0·25353	0·09868	0·57378	1·47410	−0·41754	−0·39120		
0	5·73687	0·08545	0·02679	0·15579	−0·49686	0·97479	0·96515	0·45313	5
1	5·73687	−0·19977	0·29980	1·74317	1·16155	0·04731	−0·44113		
2	5·73687	−0·25794	0·10593	0·61591	1·49975	−0·45681	−0·28988		
0	5·73687	−0·00645	0·01543	0·08970	0·03748	1·02753	0·99888	0·48438	5
1	5·73687	−0·16073	0·31321	1·82113	0·93455	0·01560	−0·37660		
2	5·73687	−0·26553	0·12088	0·70284	1·54390	−0·48451	−0·10409		

Table 4 (continued)

Sequence				Sequence admittance		Self- and mutual admittances		$\frac{d}{\lambda}$	N
m	Ψ_{dR}	Re $T^{(m)}$	Im $T^{(m)}$	$G^{(m)}$	$B^{(m)}$	$G_{1(m+1)}$	$B_{1(m+1)}$		
0	5·73687	−0·04103	0·01824	0·10608	0·23859	1·05256	0·99281	0·5000	5
1	5·73687	−0·13765	0·31486	1·83072	0·80033	0·00556	−0·35896		
2	5·73687	−0·26871	0·12859	0·74764	1·56240	−0·47880	−0·01815		
0	5·73687	−0·06970	0·02385	0·13868	0·40525	1·07040	0·97532	0·51563	5
1	5·73687	−0·11303	0·31186	1·81328	0·65720	−0·00487	−0·34852		
2	5·73687	−0·27148	0·13645	0·79337	1·57848	−0·46099	−0·06349		
0	5·73687	−0·12986	0·04795	0·27882	0·75507	1·05588	0·89736	0·5625	5
1	5·73687	−0·04373	0·26898	1·56397	0·25424	−0·05392	−0·33909		
2	5·73687	−0·27718	0·16104	0·93633	1·61162	−0·33461	0·26795		
0	5·73687	−0·17222	0·08301	0·48267	1·00134	0·92501	0·89242	0·6250	5
1	5·73687	−0·01972	0·15994	0·92995	0·11463	−0·15783	−0·30843		
2	5·73687	−0·27789	0·19628	1·14124	1·61574	−0·06334	−0·36289		
0	5·73687	−0·18928	0·11386	0·66204	1·10052	0·87378	1·02151	0·6875	5
1	5·73687	−0·07632	0·08466	0·49228	0·44377	−0·24722	−0·22979		
2	5·73687	−0·26826	0·23410	1·36115	1·55976	0·14135	−0·26930		
0	5·73687	−0·19114	0·12315	0·71602	1·11135	0·89069	1·06776	0·7100	5
1	5·73687	−0·10210	0·07336	0·42653	0·59366	−0·27077	−0·19626		
2	5·73687	−0·26143	0·24804	1·44218	1·52007	0·18344	−0·21805		
0	5·73687	−0·19034	0·13638	0·79297	1·10669	0·95087	1·12366	0·7500	5
1	5·73687	−0·14431	0·06840	0·39769	0·83910	−0·30452	−0·13340		
2	5·73687	−0·24366	0·27226	1·58301	1·41671	0·22557	0·12492		

APPENDIX II

Summary of the two-term theory for applications

To facilitate practical applications of the theory apart from its detailed development and verification, principal points of the two-term theory, together with the steps required in its utilization, are summarized in this appendix. Numerical results for arrays containing only a few elements can be computed by hand with the aid of the tables in appendix I or those in the literature [1]. Calculations for larger arrays or for element parameters not included in the tables generally require a computer. For these applications, the theory can be conveniently packaged as a computer programme to which the user need only supply input data cards specifying the parameters of the elements and array to obtain a numerical evaluation of any properties of the array [2]. In this form, the theory can be used without considering either intricate mathematics or complicated programming steps once the initial programming is completed. In a reasonably general programme, the parameters that can be specified as input data are the number N of antennas in the array, their radius a, length $2h$, spacing d, and the relative driving voltages V_i or currents $I_i(0)$.

The theory applies to arrays of thin, identical, parallel, non-staggered, centre-driven, highly conducting dipoles which are uniformly spaced around a circle. If the admittances and currents are multiplied by two, they also apply to arrays of vertical monopoles over a large highly-conducting ground plane. The normalized radiation patterns apply to either dipoles or monopoles, but for the latter the ground plane must extend beyond the radius at which the field is evaluated and measured. Dipole lengths should not exceed $1\cdot2\lambda$ and the distance between adjacent elements should be at least $\lambda/8$; the complete three-term theory is required for high accuracy when $\beta_0 d_{ik} < 1$. To satisfy completely the mathematical conditions of the theory, the element thickness should satisfy $a \leqslant 0\cdot01\lambda$. However, useful results can be obtained for elements that are two or three times as thick. Comparisons of theoretical and measured results show that predicted radiation patterns are usually well within $\pm 1\,\mathrm{dB}$ of the measured ones, and that calculated self- and mutual admittances are generally within 5%–10% of the measured values except near resonant lengths or spacings. Near resonance, theoretical values are shifted toward larger values of h/λ and d/λ by approximately one-half of the element radius. The comparison of self- and mutual admittances assumes that proper account has been taken of end-effects. The theoretical driving-point admittances apply only to dipoles with no attached transmission lines, and the presence of these lines must be taken into account as discussed in sections 2.8, 4.3, and 8.2.

The central point of the two-term theory is that the current along each dipole can be accurately represented in the form $I_k(z) = s_k \sin \beta_0(h-|z|) + c_k(\cos \beta_0 z - \cos \beta_0 h)$. Driving-point admittances are then given by $Y_k = I_k(0)/V_k = s_k \sin \beta_0 h + c_k(1 - \cos \beta_0 h)$ and the far-zone fields by eqs. (4.23). The principal computing problem is to evaluate the s_k and c_k coefficients which are functions of all of the parameters of the array. Because the s_k and c_k coefficients corresponding to different elements are not simply related, the field expressions cannot be summed in a closed form as with the usual one-term sinusoidal theory. However, in the single-term sinusoidal theory the current is always represented by the sine term alone and this is accurate only for very thin dipoles with lengths very near $2h = \lambda/2$. With the two-term theory, general expressions for the currents become indeterminate at $2h = \lambda/2$ and correct expressions are found by taking limits as $\beta_0 h \to \pi/2(2h \to \lambda/2)$. The special expressions applicable to $\beta_0 h = \pi/2$ are indicated in the text (e.g. (2.35) on p. 56).

For a general array, the evaluation of s_k and c_k or their equivalents requires the solution of a system of coupled integral equations as in chapters 5 and 6. However, the symmetry of a circular array of identical equispaced elements permits a reduction of the system of equations to a single integral equation. This reduction is accomplished by the method of symmetrical components in which the elements are all excited with equal amplitudes but with uniformly progressive phase so that the total phase change around the circle is $2\pi m$ radians where m is an integer ranging from zero to $N-1$. Each value of m gives one of the N phase sequences which is designated by a superscript m. The resulting functions are called sequence functions and they form a set that is characteristic of the elements and their spacing in the array. The associated properties of the elements such as the driving-point admittances for arbitrary driving conditions can be obtained from the sequence functions.

In the engineering design of arrays, the quantities of primary concern are radiation patterns and driving-point admittances or impedances. Regardless of whether the objective is to calculate only radiation patterns and driving-point admittances or to perform a complete analysis of the array, the same basic sequence of steps is required in a programme of calculation that utilizes the two-term theory. These steps are as follows:
1. Evaluation of the C_{bi}, S_{bi}, and E_{bi} integrals (right part, eqs. (4.13)).
2. Calculation of the sequence sums (left part, eqs. (4.13)).
3. Calculation of required differences of the sequence sums (eqs. (4.12)).
4. Calculation of Ψ_{dR}—from eq. (4.16).
5. Calculation of the remaining Ψ functions—from eqs. (4.7)–(4.11).
6. Calculation of $T^{(m)}$ or $T'^{(m)}$—from eqs. (4.5).
7. Calculation of $s^{(m)}$ and $c^{(m)}$ or $s'^{(m)}$ and $c'^{(m)}$—from eqs. (4.15).
   Results obtainable at this point:
   a. Sequence currents $I^{(m)}(z)$—from eqs. (4.16).
   b. Sequence admittances $Y^{(m)}$—from eqs. (4.17).

c. Self- and mutual admittances Y_{1k}—from eqs. (4.19).
d. Self- and mutual impedances Z_{1k}—from eqs. (4.26) and (4.27).
e. Driving-point admittances Y_k—from eq. (4.24b).
f. Driving-point impedances Z_k—from reciprocals of driving-point admittances.

8. Calculation of sequence voltage $V^{(m)}$—from eqs. (4.20a).
   Note: If driving-point currents are specified, sequence currents $I^{(m)}(0)$ are calculated from eq. (4.20b) and sequence voltages from eq. (4.17).
9. Calculation of s_k and c_k—from eqs. (4.22).

Final results obtainable:

g. Element currents $I_k(z)$—from eqs. (4.21).
h. Driving-point admittances—from eqs. (4.24a).
i. Far-zone fields—from eqs. (4.23).
j. Normalized power radiation pattern—from eq. (4.28).

Steps 1–3 will be discussed in detail. The driving-point admittances can be calculated either as in steps e or h. Also, driving-point impedances are simply the reciprocals of corresponding driving-point admittances. If the driving-point voltages or currents are to be specified according to some rule so as to provide a particular distribution such as a Tchebyscheff, provision can be made for a subroutine to calculate them as part of step 8.

The special case of $\beta_0 h = \pi/2(h = \lambda/4)$ is most easily incorporated in a general programme by providing a subroutine to carry out parts of the calculation for this length. If the programme is designed to examine the input data and use the appropriate subroutine whenever $h = \lambda/4$ is specified, the range of h/λ for which the programme can be used is continuous except for values of h/λ that differ from $\frac{1}{4}$ by less than about 0.001λ. At these values some of the functions may be sufficiently close to the indeterminate form $0/0$ to produce computer overflow.

The integrals of eqs. (4.13) must be evaluated numerically. Although more efficient methods may be found, Simpson's Rule is adequate. Let the integrand $f(z)$ of an integral I be evaluated at an odd number of points which are uniformly spaced at multiples of the interval $\Delta z'$. Then, according to Simpson's Rule,

$$I = \int_r^s f(z')\,dz'$$
$$\doteq \frac{\Delta z'}{3}\{f(r)+f(s)+4[f(r+\Delta z')+f(r+3\Delta z')+ \ldots f(s-\Delta z')]$$
$$+2[f(r+2\Delta z')+f(r+4\Delta z')+ \ldots f(s-2\Delta z')]\} \tag{1}$$

with an approximate error of integration ε given by

$$|\varepsilon| \leqslant \left|\frac{I(\Delta z')-I(2\Delta z')}{10}\right|. \tag{2}$$

As a convenient starting point in their evaluation, the integrals may be written in the following form:

$$E(h, z) = \text{Re } E(h, z) + j \text{ Im } E(h, z)$$

$$= \int_0^{h/\lambda} \left[\left(\frac{\cos 2\pi \bar{R}_1}{\bar{R}_1} + \frac{\cos 2\pi \bar{R}_2}{\bar{R}_2} \right) - j \left(\frac{\sin 2\pi \bar{R}_1}{\bar{R}_1} + \frac{\sin 2\pi \bar{R}_2}{\bar{R}_2} \right) \right] dz_1' \quad (3a)$$

$$C(h, z) = \text{Re } C(h, z) + j \text{ Im } C(h, z)$$

$$= \int_0^{h/\lambda} \cos 2\pi \, z_1' \left[\left(\frac{\cos 2\pi \bar{R}_1}{\bar{R}_1} + \frac{\cos 2\pi \bar{R}_2}{\bar{R}_2} \right) \right.$$

$$\left. - j \left(\frac{\sin 2\pi \bar{R}_1}{\bar{R}_1} + \frac{\sin 2\pi \bar{R}_2}{\bar{R}_2} \right) \right] dz_1' \quad (3b)$$

$$S(h, z) = \int_0^{h/\lambda} \sin 2\pi z_1' \left[\left(\frac{\cos 2\pi \bar{R}_1}{\bar{R}_1} + \frac{\cos 2\pi \bar{R}_2}{\bar{R}_2} \right) \right.$$

$$\left. - j \left(\frac{\sin 2\pi \bar{R}_1}{\bar{R}_1} + \frac{\sin 2\pi \bar{R}_2}{\bar{R}_2} \right) \right] dz_1' \quad (3c)$$

$$\bar{R}_1 = \sqrt{(z/\lambda - z_1')^2 + (b_i/\lambda)^2}, \qquad \bar{R}_2 = \sqrt{(z/\lambda + z_1')^2 + (b_i/\lambda)^2}. \quad (3d)$$

One set of integrals must be evaluated for each element of the array and the sum of these values in subsequent steps takes account of the intercoupling of the elements. The distance from antenna 1 to the ith antenna is b_i given by

$$b_i/\lambda = \frac{(d/\lambda) \sin (i-1)\pi/N}{\sin \pi/N}. \quad (4)$$

Thus, for $i = 2$, b_2/λ corresponds to the distance between adjacent antennas and equals d/λ.

When $i = 1$, the integrals are to be evaluated at the dipole surface and the value to be used for b_i is $b_1/\lambda = a/\lambda$, the dipole radius. Very small integration intervals $\Delta z'$ are required for these integrals because the element is thin, $a/\lambda \ll 1$, and some of the integrands rise rapidly to a sharp maximum of λ/a at $z_1' = z/\lambda$. These integrals with $b_1 = a$ correspond to a single isolated dipole; they are functions only of h/λ and a/λ and remain constant as the other array parameters are changed. Their evaluation comprises one of the longest parts of the calculation and considerable computer time can be saved by evaluating them once and supplying their values as input data for subsequent analyses of different arrays that use dipoles of the same h/λ and a/λ.

As a guide in choosing the interval size $\Delta z'$, or the number of points n_p at which the integrands are to be evaluated, the error ε of eq. (2) is less than 10^{-5} for any of the 'b' integrals when $d/\lambda > 1/8$ with the following choice of n_p: $h = \lambda/4$, $n_p = 17$; $h = 3\lambda/8$, $n_p = 25$; $h = \lambda/2$, $n_p = 33$. For the 'a' integrals, the following choice of n_p produces an error of 10^{-5} or less in the integrals with $a = 0{\cdot}007\lambda$: $h = \lambda/4$, $n_p = 193$; $h = \lambda/2$, $n_p = 385$. For $a = 0{\cdot}002\lambda$ and $h = \lambda/4$, $n_p = 1537$ points are necessary to ensure that the error remains less than 10^{-5} for all of the integrals.

Each integral is a function of the three parameters h/λ, z/λ, and b_i/λ. The distances b_i/λ are obtained from eq. (4) with d/λ supplied as part of the input data; the element length h/λ and radius a/λ must also be specified with the input data. For each h/λ and b_i/λ, the two-term theory requires an evaluation of the integrals at the values of z in the following list:

$h > \lambda/4$: 'b' integrals $(i > 1)$ 'a' integrals $(i = 1)$

     $C(h, h)$, $C(h, 0)$ $C(h, h)$, $C(h, 0)$ Re $C(h, h - \lambda/4)$

     $S(h, h)$, $S(h, 0)$ $S(h, h)$, Im $S(h, 0)$, Re $S(h, h - \lambda/4)$

     $E(h, h)$, $E(h, 0)$ $E(h, h)$, $E(h, 0)$

$h < \lambda/4$: Required 'b' integrals are the same but in 'a' integrals, Re $C(h, 0)$ and Re $S(h, 0)$ are required instead of Re $C(h, h - \lambda/4)$ and Re $S(h, h - \lambda/4)$.

$h = \lambda/4$: 'b' integrals $(i > 1)$ 'a' integrals $(i = 1)$

     $C(h, h)$ $C(h, h)$, Re $C(h, 0)$

     $S(h, h)$ $S(h, h)$, Re $S(h, 0)$

     $E(h, h)$ $E(h, h)$

After the required integrals have been evaluated and stored in an ordered array, the sums to be used in calculating Ψ functions may be computed. Four different kinds of sums are used in eqs. (4.6)–(4.11) and each is designated by a subscript as follows: A subscript Σ means the sum is to be extended over all values of i; a subscript $\Sigma 1$ means that only the integral corresponding to $i = 1$ with $b_i/\lambda = b_1/\lambda = a/\lambda$ is to be used; a subscript $\Sigma 2$ means that the sum excludes $i = 1$ but includes all other values of i; and a subscript d indicates that a difference of two sums is to be used as in eq. (4.12).

The general arrangement of a programme that might be used to package the two-term theory for applications is shown schematically in Fig. 1 (pp. 419–21).

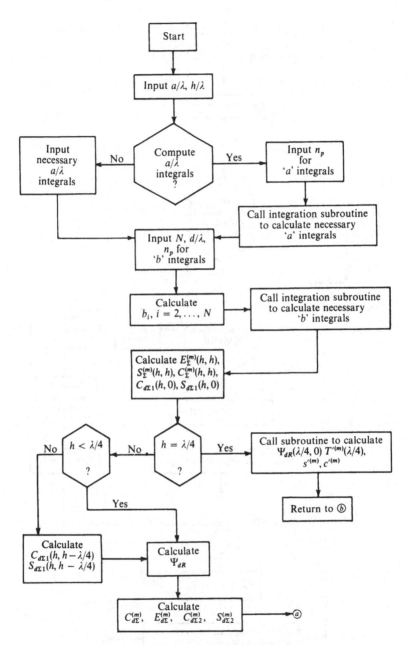

Fig. 1. Flow chart for applications of two-term theory.

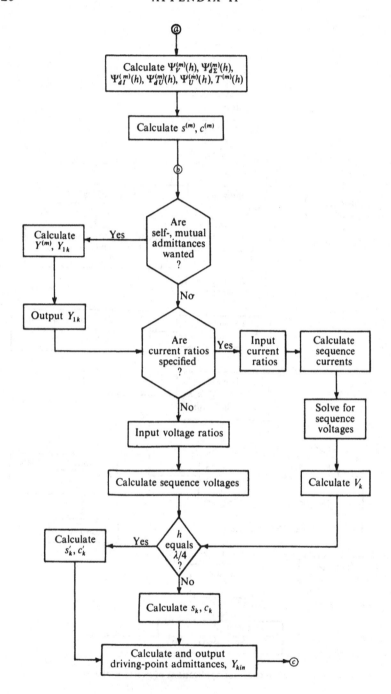

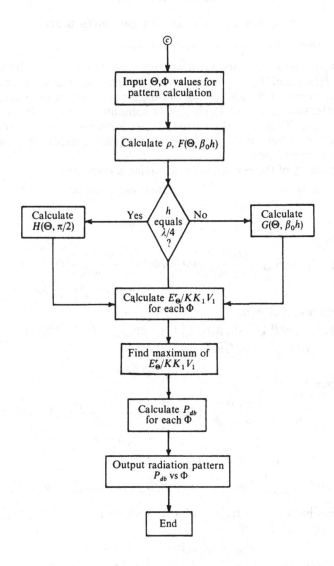

APPENDIX III

Summary of formulas for the curtain array

This appendix contains (1) a summary of formulas for the curtain array, (2) tables of Ψ_{dR}, Φ_u and Φ_v for single elements as functions of Ω and h/λ and tables of Φ_u and Φ_v for off-diagonal elements as functions of $(k-i)(b/\lambda)$ and h/λ (Table 1), (3) sample computer printouts for two cases are included. They give the currents, impedances and field patterns for curtain arrays both for the case of driving-point currents specified and driving voltages specified.

A summary of the curtain array formulas is given below.

I. Driving-point current, voltages, admittances and impedances:
  1. Base currents specified.

$$\{I_z(0)\} \quad \text{specified.}$$

  Then

$$\{V_0\} = -j60\Psi_{dR}\frac{\cos\beta_0 h}{1-\cos\beta_0 h}[\Phi_u]\left[[\Phi_v]+\frac{\sin\beta_0 h}{1-\cos\beta_0 h}[\Phi_u]\right]^{-1}\{I_z(0)\}$$

$$\beta_0 h \neq n\pi/2, n \text{ odd} \qquad \text{(A3-1)}$$

  and near $\beta_0 h = n\pi/2$,

$$\{V_0\} = j60\Psi_{dR}[-\sin\beta_0 h[\Phi_u']+(1-\cos\beta_0 h)[\Phi_v']]^{-1}[\Phi_u']\{V_0\}. \qquad \text{(A3-2)}$$

  2. Base voltages specified

$$\{V_0\} \quad \text{specified}$$

  Then

$$\{I_z(0)\} = \frac{j}{60\Psi_{dR}}\frac{1-\cos\beta_0 h}{\cos\beta_0 h}[\Phi_u]^{-1}\left[[\Phi_v]+\frac{\sin\beta_0 h}{1-\cos\beta_0 h}[\Phi_u]\right]\{V_0\}$$

$$\beta_0 h \neq n\pi/2, n \text{ odd} \qquad \text{(A3-3)}$$

  and near $\beta_0 h = n\pi/2$, n odd

$$\{I_z(0)\} = \frac{-j}{60\Psi_{dR}}[\Phi_u']^{-1}[-\sin\beta_0 h[\Phi_u']+(1-\cos\beta_0 h)[\Phi_v']]\{V_0\}. \qquad \text{(A3-4)}$$

  The individual driving-point impedances Z_{0i} and admittances Y_{0i}, $i = 1, 2, 3, \ldots N$ are

$$Z_{0i} = V_{0i}/I_{zi}(0), \qquad Y_{0i} = I_{zi}(0)/V_{0i}. \qquad \text{(A3-5)}$$

II. Distributions of current on the individual elements:

  1.　$$I_{zi}(z) = \frac{j2\pi V_{0i}}{\zeta_0\Psi_{dR}\cos\beta_0 h}[\sin\beta_0(h-|z|) + T(\cos\beta_0 z - \cos\beta_0 h)],$$

$$\beta_0 h \neq \frac{n\pi}{2}, n \text{ odd} \qquad \text{(A3-6)}$$

where
$$T = -\frac{j(Y_{0i}/A)+\sin\beta_0 h}{1-\cos\beta_0 h}$$

and
$$A = \frac{2\pi}{\zeta_0 \Psi_{dR}\cos\beta_0 h}.$$

2. $I_z(z) = \dfrac{-j2\pi}{\zeta_0 \Psi_{dR}\cos\beta_0 h}[\sin\beta_0 z - \sin\beta_0 h + T'(\cos\beta_0 z - \cos\beta_0 h)].$

$$\beta_0 h \text{ near } \frac{n\pi}{2}, n \text{ odd} \qquad \text{(A3-7)}$$

$$T' = \frac{j(Y_{0i}/A)+\sin\beta_0 h}{1-\cos\beta_0 h}.$$

III. Field patterns
   1. Array of N elements, $\beta_0 h \neq n\pi/2$, n odd

$$E^r_\Theta(\Theta, \Phi) = \frac{-V_{01}}{\Psi_{dR}}\frac{e^{-j\beta_0 R_0}}{R_0 \cos\beta_0 h}\sum_{i=1}^{N} C_i e^{-j\beta_0 b[(N-2i+1)/2]\sin\Theta\cos\Phi} \qquad \text{(A3-8)}$$

where
$$C_i = \xi_i\left[F_m(\Theta, \beta_0 h) - \frac{[j(Y_{0i}/A)+\sin\beta_0 h]}{1-\cos\beta_0 h}G_m(\Theta, \beta_0 h)\right]. \qquad \xi_i = \frac{V_{0i}}{V_{01}}. \qquad \text{(A3-9)}$$

   2. Array of N elements, $\beta_0 h$ near $n\pi/2$

$$E^r_\Theta(\Theta, \Phi) = \frac{V_{01}}{\Psi_{dR}}\frac{e^{-j\beta_0 R_0}}{R_0}\sum_{i=1}^{N} C'_i e^{-j\beta_0 b[(N-2i+1)/2]\sin\Theta\cos\Phi} \qquad \text{(A3-10)}$$

where
$$c'_i = \xi_i\left[H_m(\Theta, \beta_0 h) + \frac{[j(Y_{0i}/A)+\sin\beta_0 h]}{1-\cos\beta_0 h}G_m(\Theta, \beta_0 h)\right]. \qquad \text{(A3-11)}$$

Numerical values for the functions Ψ_{dR}, Φ_u and Φ_v needed to solve the matrix equations for the driving-point impedances and currents of curtain arrays are given in Table 1. Note that the off-diagonal matrix elements are dependent only on the spacing between the two elements. The integration programme used to compute these functions is an improved Romberg method. The programme is arranged so as to include the correct contributions from the major singularities.

A sample calculation for the case of a three-element endfire array with quarter-wavelength spacing and half-wavelength elements follows. The following headings appear in the printout:

N   Number of elements
BETAH $\beta_0 h$, electric half-length of the antenna
BETAB $\beta_0 b$, electrical distance between two adjacent elements
IZ(0) $I_{zi}(0)$, driving-point currents
V0 V_{0i}, base voltages
ADMTC Y_{0i}, admittances

IMPDC	Z_{0i}, impedances
PSI	$\xi_i = V_{0i}/V_{01}$, ratios of base voltages
C	C_i, source strength coefficients for radiation patterns
IZ(Z)	$I_{zi}(z)$, element currents where $z = 0, 1/4\beta, 1/2\beta, 3/4\beta, \ldots 3/\beta$, π/β where $i = 1, 2, 3, \ldots N$

In the sample printout, values are read from left to right under each heading. For example, driving voltages are read as follows:

$$\text{V0}$$

$$\text{Re}\,(V_{01})\,\text{Im}\,(V_{01}) \qquad \text{Re}\,(V_{02})\,\text{Im}\,(V_{02})$$
$$\text{Re}\,(V_{03})\ldots$$
$$\vdots$$

IZ(Z), or $I_{zi}(z)$, is arranged so that all 14 values for one particular value of i are in one group. The groups are then read vertically down each of the two columns on the page. For example:

$$\text{IZ(Z)} = I_{zi}(z)$$
$$\text{Re}\,(I_{z1}(0))\,\text{Im}\,(I_{z1}(0))\,\text{Re}\,(I_{z1}(1/4\beta))\,\text{Im}\,(I_{z1}(1/4\beta))$$
$$\vdots$$

$$\text{Re}\,(I_{z1}(3/\beta))\,\text{Im}\,(I_{z1}(3/\beta))\,\text{Re}\,(I_{z1}(\pi/\beta))\,\text{Im}\,(I_{z1}(\pi/\beta))$$
$$\text{Re}\,(I_{z2}(0))\ldots$$
$$\vdots$$

$$\text{Re}\,(I_{z3}(0))\ldots$$
$$\vdots$$

Note that only terms for values of z equal to or less than h are used. For example, when $h = \pi/2\beta$, the last term used is $3/2\beta$. If $h = \pi/\beta$ all terms are used.

NOTE

The off-diagonal values of Φ_u and Φ_v on pages 431–49 depend only on the distance b_{ik}. The left-hand column, i.e. $(k-i)b/\lambda$, is the distance between elements k and i. The two columns under each Φ_{kiu} and Φ_{kiv} are the real and imaginary parts.

Sample output from FAP programme to compute currents, impedances and field patterns for curtain arrays with driving-point currents specified

N = 3   PSIDR = 5·83400   BETAH = 3·14159   BETAB = 1·57080   FM = 2·00000   GM = 3·14159

IZ(O)	1·00000000	0·		0·	-1·00000000			
	-1·00000000	0·						
PHIV	0·67210	-1·66050	-0·68130	-0·85700	-0·62960	0·40710	-0·68130	-0·85700
	0·67210	-1·66050	-0·68130	-0·85700	-0·62960	0·40710	-0·68130	-0·85700
	0·67210	-1·66050						
PHIU	2·89590	-6·68980	1·02170	1·52470	1·10300	-0·66350	1·02170	1·52470
	2·89590	-6·68980	1·02170	1·52470	1·10300	-0·66350	1·02170	1·52470
	2·89590	-6·68980						
VO	611·59042358	-591·04013824		-590·09282684	-160·23533821			
	-61·54054213	435·34058380						
IMPDNC	611·59042358	-591·04013824		160·23533821	-590·09282684			
	61·54054213	-435·34058380						
ADMTNC	0·00084547	0·00081706		0·00042856	0·00157827			
	0·00031835	0·00225204						
PSI	1·00000000	-0·		-0·36798476	-0·61761774			
	-0·40773164	0·31778590						

Sample output from FAP programme to compute currents, impedances and field patterns for curtain arrays with driving-point currents specified—contd.

IZ(Z)YOI

0·00084547	0·00081706	0·00083233	0·00009757
−0·00079372	−0·00060257	0·00073204	−0·00123987
−0·00065114	−0·00177466	0·00055603	−0·00217371
−0·00045264	−0·00241222	0·00034738	−0·00247534
−0·00024682	−0·00235917	0·00015718	−0·00207090
−0·00008406	−0·00162847	0·00003199	−0·00105940
−0·00000422	−0·00039905	−0·	−0·00000000
−0·00081706	−0·00084547	0·00036784	−0·00057224
−0·00007885	−0·00028972	−0·00049524	−0·00001546
−0·00085544	0·00023346	−0·00113706	0·00044160
−0·00132257	0·00059599	−0·00140045	0·00068704
−0·00136586	0·00070909	−0·00122094	0·00066077
−0·00097471	0·00054508	−0·00064248	0·00036922
−0·00024490	0·00014412	−0·00000000	−0·00000000
−0·00084547	−0·00081706	−0·00060772	−0·00051618
−0·00035847	−0·00020861	−0·00011321	0·00008653
−0·00011279	0·00035089	0·00030550	0·00056803
−0·00045294	0·00072446	0·00054593	0·00081044
−0·00057869	0·00082064	0·00054919	0·00075440
−0·00045925	0·00061586	0·00031449	0·00041363
−0·00012388	0·00016028	−0·00000000	−0·00000000

Sample output from FAP programme to compute currents, impedances and field patterns for curtain arrays with driving-point currents specified—contd.

C

0·00153787 −0·00126944	−0·00461196 0·00253471	−0·00361550	−0·00157552

E

0	0·01162040	2	0·01162091	4	0·01162239	6	0·01162473
8	0·01162772	10	0·01163108	12	0·01163444	14	0·01163737
16	0·01163937	18	0·01163985	20	0·01163815	22	0·01163358
24	0·01162538	26	0·01161272	28	0·01159474	30	0·01157056
32	0·01153924	34	0·01149984	36	0·01145143	38	0·01139304
40	0·01132374	42	0·01124261	44	0·01114879	46	0·01104144
48	0·01091982	50	0·01078323	52	0·01063106	54	0·01046284
56	0·01027817	58	0·01007678	60	0·00985856	62	0·00962351
64	0·00937182	66	0·00910378	68	0·00881989	70	0·00852080
72	0·00820733	74	0·00788044	76	0·00754129	78	0·00719119
80	0·00683161	82	0·00644618	84	0·00609070	86	0·00571312
88	0·00533357	90	0·00495435	92	0·00457800	94	0·00420730
96	0·00384534	98	0·00349566	100	0·00316235	102	0·00285022
104	0·00256502	106	0·00231355	108	0·00210349	110	0·00194265
112	0·00183734	114	0·00179002	116	0·00179765	118	0·00185201
120	0·00194210	122	0·00205668	124	0·00218605	126	0·00232258
128	0·00246059	130	0·00259605	132	0·00272615	134	0·00284899
136	0·00296339	138	0·00306866	140	0·00316450	142	0·00325087
144	0·00332797	146	0·00339614	148	0·00345585	150	0·00350764
152	0·00355213	154	0·00358995	156	0·00362176	158	0·00364819
160	0·00366990	162	0·00368751	164	0·00370158	166	0·00371265
168	0·00372122	170	0·00372771	172	0·00373250	174	0·00373590
176	0·00373816	178	0·00373944	180	0·00373985	182	

Sample output from FAP programme to compute currents, impedances and field patterns for curtain arrays with driving-point currents specified—contd.

E1

Index	Value	Index	Value	Index	Value	Index	Value
0	0·00492699	2	0·00492784	4	0·00493038	6	0·00493454
8	0·00494024	10	0·00494736	12	0·00495573	14	0·00496517
16	0·00497545	18	0·00498629	20	0·00499744	22	0·00500855
24	0·00501928	26	0·00502925	28	0·00503807	30	0·00504531
32	0·00505054	34	0·00505330	36	0·00505313	38	0·00504956
40	0·00504210	42	0·00503029	44	0·00501366	46	0·00499177
48	0·00496417	50	0·00493045	52	0·00489025	54	0·00484321
56	0·00478903	58	0·00472746	60	0·00465829	62	0·00458138
64	0·00449665	66	0·00440407	68	0·00430369	70	0·00419565
72	0·00408011	74	0·00395735	76	0·00382772	78	0·00369162
80	0·00354953	82	0·00340200	84	0·00324966	86	0·00309318
88	0·00293329	90	0·00277083	92	0·00260662	94	0·00244159
96	0·00227673	98	0·00211307	100	0·00195174	102	0·00179394
104	0·00164104	106	0·00149450	108	0·00135605	110	0·00122765
112	0·00111156	114	0·00101037	116	0·00092682	118	0·00086350
120	0·00082221	122	0·00080330	124	0·00080527	126	0·00082500
128	0·00085846	130	0·00090162	132	0·00095089	134	0·00100342
136	0·00105704	138	0·00111014	140	0·00116163	142	0·00121072
144	0·00125691	146	0·00129991	148	0·00133955	150	0·00137579
152	0·00140866	154	0·00143824	156	0·00146467	158	0·00148810
160	0·00150872	162	0·00152669	164	0·00154220	166	0·00155541
168	0·00156651	170	0·00157563	172	0·00158291	174	0·00158844
176	0·00159233	178	0·00159464	180	0·00159541	182	

Sample output from FAP programme to compute currents, impedances and field patterns for curtain arrays with base voltages specified

	N = 3	PSIDR = 5·83400		BETAH = 3·14159	BETAB = 1·57080	FM = 2·00000		GM = 3·14159
IZ(0)	0·00097904	0·00070133		0·00133570	-0·00100300			
	-0·00080815	-0·00202375						
PHIV	0·67210	-1·66050	-0·68130	-0·85700	-0·62960	0·40710	-0·68130	-0·85700
	0·67210	-1·66050	-0·68130	-0·85700	-0·62960	0·40710	-0·68130	-0·85700
	0·67210	-1·66050						
PHIU	-6·68980	2·89590	1·02170	1·52470	1·10300	-0·66350	1·02170	1·52470
	-6·68980	2·89590	1·02170	1·52470	1·10300	-0·66350	1·02170	1·52470
	-6·68980	2·89590						
VO	1·00000000	0·		0·	-1·00000000			
	-1·00000000	0·						
IMPDNC	675·01622772	-483·54592133		359·48430634	-478·72311783			
	170·18375969	-426·17085266						
ADMTNC	0·00097904	0·00070133		0·00100300	0·00133570			
	0·00080815	0·00202375						
PSI	1·00000000	0·		-0·	-1·00000000			
	-1·00000000	0·						

Sample output from FAP programme to compute currents, impedances and field patterns for curtain arrays with base voltages specifed—contd.

IZ(Z)

0·00097904	0·00070133	0·00096382	−0·00001635
0·00091912	−0·00071122	0·00084770	−0·00134006
0·00075401	−0·00186379	0·00064388	−0·00224983
0·00052415	−0·00247418	0·00040226	−0·00252289
0·00028580	−0·00239295	0·00018201	−0·00209241
0·00009734	−0·00163998	0·00003705	−0·00106378
0·00000490	−0·00039963	−0·	0·00000000
0·00133570	−0·00100300	0·000s0815	−0·00098742
−0·00011568	−0·00094161	−0·00079080	−0·00086845
−0·00137523	−0·00077247	−0·00183263	−0·00065964
−0·00213456	−0·00053698	−0·00226225	−0·00041211
−0·00220776	−0·00029280	−0·00197448	−0·00018647
−0·00157691	−0·00009973	−0·00103977	−0·00003796
−0·00039646	−0·00000501	0·00000000	−0·
−0·00080815	−0·00202375	−0·00079559	−0·00128551
−0·00075868	−0·00053025	−0·00069973	0·00019506
−0·00062239	0·00084533	−0·00053149	0·00138012
−0·00043266	0·00176620	−0·00033205	0·00197954
−0·00023592	0·00200690	−0·00015024	0·00184656
−0·00008035	0·00159850	−0·00003058	0·00101373
−0·00000404	0·00039302	0·	−0·00000000

Sample output from FAP programme to compute currents, impedances and field patterns for curtain arrays with base voltages specified—contd.

C

| 0·00132806 | −0·00443017 | −0·00224539 | 0·00077446 |
| 0·00048764 | 0·00104618 | | |

E

0	0·00772204	2	0·00772032	4	0·00771517	6	0·00770652
8	0·00769426	10	0·00767826	12	0·00765837	14	0·00763436
16	0·00760600	18	0·00757303	20	0·00753516	22	0·00749207
24	0·00744344	26	0·00738894	28	0·00732822	30	0·00726093
32	0·00718673	34	0·00710532	36	0·00701638	38	0·00691964
40	0·00681487	42	0·00670187	44	0·00658051	46	0·00645070
48	0·00631244	50	0·00616582	52	0·00601098	54	0·00584818
56	0·00567778	58	0·00550026	60	0·00531621	62	0·00512636
64	0·00493156	66	0·00473284	68	0·00453135	70	0·00432844
72	0·00412563	74	0·00392462	76	0·00372729	78	0·00353575
80	0·00335229	82	0·00317939	84	0·00301960	86	0·00287558
88	0·00274985	90	0·00264467	92	0·00256182	94	0·00250234
96	0·00246640	98	0·00245317	100	0·00246091	102	0·00248718
104	0·00252901	106	0·00258327	108	0·00264687	110	0·00271690
112	0·00279079	114	0·00286629	116	0·00294158	118	0·00301513
120	0·00308578	122	0·00315262	124	0·00321502	126	0·00327255
128	0·00332496	130	0·00337216	132	0·00341418	134	0·00345115
136	0·00348332	138	0·00351093	140	0·00353434	142	0·00355388
144	0·00356995	146	0·00358290	148	0·00359313	150	0·00360101
152	0·00360689	154	0·00361107	156	0·00361389	158	0·00361560
160	0·00361645	162	0·00361665	164	0·00361641	166	0·00361586
168	0·00361513	170	0·00361437	172	0·00361362	174	0·00361297
176	0·00361248	178	0·00361217	180	0·00361206	182	

Sample output from FAP programme to compute currents, impedances and field patterns for curtain arrays with base voltages specified—contd.

El

0	2·99999997	2	2·99999905	4	2·99998534	6	2·99992591
8	2·99976629	10	2·99943051	12	2·99882182	14	2·99782327
16	2·99629834	18	2·99409226	20	2·99103278	22	2·98693159
24	2·98158598	26	2·97478038	28	2·96628857	30	2·95587531
32	2·94329914	34	2·92831475	36	2·91067564	38	2·89013717
40	2·86645940	42	2·83941054	44	2·80877000	46	2·77433190
48	2·73590815	50	2·69333220	52	2·64646170	54	2·59518206
56	2·53940907	58	2·47909155	60	2·41421363	62	2·34479681
64	2·27090105	66	2·19262639	68	2·11011279	70	2·02354059
72	1·93312947	74	1·83913770	76	1·74185981	78	1·64162478
80	1·53879254	82	1·43375127	84	1·32691267	86	1·21870857
88	1·10958548	90	1·0000045	92	0·89041533	94	0·78129232
96	0·67308814	98	0·56624962	100	0·46120823	102	0·35837606
104	0·25814093	106	0·16086308	108	0·06687122	110	0·02353985
112	0·11011214	114	0·19262574	116	0·27090044	118	0·34479616
120	0·41421304	122	0·47909095	124	0·53940848	126	0·59518150
128	0·64646117	130	0·69333168	132	0·73590770	134	0·77433146
136	0·80876961	138	0·83941018	140	0·86645908	142	0·89013688
144	0·91067540	146	0·92831454	148	0·94329897	150	0·95587515
152	0·96628844	154	0·97478033	156	0·98158590	158	0·98693153
160	0·99103274	162	0·99409228	164	0·99629838	166	0·99782326
168	0·99882185	170	0·99943053	172	0·99976631	174	0·99992595
176	0·99998536	178	0·99999908	180	0·99999999	182	

Table of elements of Φ_u and Φ_v matrices for $h/\lambda = 0.125$
Values for single element (Φ_{kku}, Φ_{kkv} and $\Psi_{kkdR} = \Psi_{dR}$)

Ω	Ψ_{dR}	Φ_{kku}		Φ_{kkv}	
7.00	4.05418	2.64345	0.19772	0.78608	−0.38223
7.50	4.51933	2.97163	0.19776	0.79628	−0.38230
8.00	4.99202	3.30540	0.19778	0.80428	−0.38234
8.50	5.47065	3.64353	0.19779	0.81054	−0.38237
9.00	5.95394	3.98508	0.19780	0.81544	−0.38238
9.50	6.44089	4.32928	0.19781	0.81926	−0.38239
10.00	6.93071	4.67555	0.19781	0.82224	−0.38240
10.50	7.42276	5.02343	0.19781	0.82457	−0.38240
11.00	7.91657	5.37257	0.19781	0.82638	−0.38240
11.50	8.41174	5.72268	0.19781	0.82780	−0.38240
12.00	8.90797	6.07356	0.19781	0.82890	−0.38240
12.50	9.40504	6.42503	0.19781	0.82976	−0.38240
13.00	9.90275	6.77696	0.19781	0.83043	−0.38241
15.00	11.89766	8.18757	0.19781	0.83192	−0.38241

Off diagonal values of Φ_u and Φ_v (see note on page 424)

$(k-i)(b/\lambda)$	Φ_{kiu}		Φ_{kiv}	
0.250	0.09815	0.11183	−0.19198	−0.21626
0.500	0.08361	−0.03104	−0.16178	0.05987
0.750	−0.01455	−0.06033	0.02796	0.11660
1.000	−0.04655	0.00834	0.08992	−0.01602
1.250	0.00539	0.03774	−0.01034	−0.07287
1.500	0.03167	−0.00376	−0.06116	0.00722
1.750	−0.00277	−0.02727	0.00532	0.05265
2.000	−0.02393	0.00213	0.04619	−0.00408
2.250	0.00168	0.02131	−0.00323	−0.04114
2.500	0.01921	−0.00136	−0.03708	0.00262
2.750	−0.00113	−0.01748	0.00216	0.03374
3.000	−0.01603	0.00095	0.03095	−0.00182
3.250	0.00081	0.01481	−0.00155	−0.02859
3.500	0.01376	−0.00070	−0.02656	0.00134
3.750	−0.00061	−0.01285	0.00117	0.02480
4.000	−0.01205	0.00053	0.02325	−0.00102
4.250	0.00047	0.01134	−0.00091	−0.02189
4.500	0.01071	−0.00042	−0.02068	0.00081
4.750	−0.00038	−0.01015	0.00073	0.01960
5.000	−0.00965	0.00034	0.01862	−0.00066
5.250	0.00031	0.00919	−0.00059	−0.01773
5.500	0.00877	−0.00028	−0.01693	0.00054
5.750	−0.00026	−0.00839	0.00050	0.01620
6.000	−0.00804	0.00024	0.01552	−0.00046
6.250	0.00022	0.00772	−0.00042	−0.01490
6.500	0.00742	−0.00020	−0.01433	0.00039
6.750	−0.00019	−0.00715	0.00036	0.01380
7.000	−0.00690	0.00017	0.01331	−0.00033
7.250	0.00016	0.00666	−0.00031	−0.01285
7.500	0.00644	−0.00015	−0.01242	0.00029
7.750	−0.00014	−0.00623	0.00027	0.01202
8.000	−0.00603	0.00013	0.01165	−0.00026
8.250	0.00013	0.00585	−0.00024	−0.01129
8.500	0.00568	−0.00012	−0.01096	0.00023
8.750	−0.00011	−0.00552	0.00021	0.01065
9.000	−0.00536	0.00011	0.01035	−0.00020

Table of elements of Φ_u and Φ_v matrices for $h/\lambda = 0.125$
Values for single element (Φ_{kku}, Φ_{kkv} and $\Psi_{kkdR} = \Psi_{dR}$)—contd.

$(k-i)(b/\lambda)$	Φ_{kiu}		Φ_{kiv}	
9·250	0·00010	0·00522	−0·00019	−0·01007
9·500	0 00508	−0·00009	−0·00981	0·00018
9·750	−0·00009	−0·00495	0·00017	0·00956
10·000	−0·00483	0·00009	0·00932	−0·00016
10·250	0·00008	0·00471	−0·00016	−0·00909
10·500	0·00460	−0·00008	−0·00888	0·00015
10·750	−0·00007	−0·00449	0·00014	0·00867
11·000	−0·00439	0·00007	0·00847	−0·00014
11·250	0·00007	0·00429	−0·00013	−0·00828
11·500	0·00420	−0·00006	−0·00810	0·00012
11·750	−0·00006	−0·00411	0·00012	0·00793
12·000	−0·00402	0·00006	0·00777	−0·00011
12·250	0·00006	0·00394	−0·00011	−0·00761
12·500	0·00386	−0·00005	−0·00746	0·00011
12·750	−0·00005	−0·00379	0·00010	0·00731
13·000	−0·00371	0·00005	0·00717	−0·00010
13·250	0·00005	0·00364	−0·00009	−0·00703
13·500	0·00358	−0·00005	−0·00690	0·00009
13·750	−0·00005	−0·00351	0·00009	0·00678
14·000	−0·00345	0·00004	0·00666	−0·00008
14·250	0·00004	0·00339	−0·00008	−0·00654
14·500	0·00333	−0·00004	−0·00643	0·00008
14·750	−0·00004	−0·00327	0·00008	0·00632
15·000	−0·00322	0·00004	0·00621	−0·00007
15·250	0·00004	0·00317	−0·00007	−0·00611
15·500	0·00312	−0·00004	−0·00601	0·00007
15·750	−0·00003	−0·00307	0·00007	0·00592
16·000	−0·00302	0·00003	0·00583	−0·00006
16·250	0·00003	0·00297	−0·00006	−0·00574
16·500	0·00293	−0·00003	−0·00565	0·00006
16·750	−0·00003	−0·00288	0·00006	0·00557
17·000	−0·00284	0·00003	0·00548	−0·00006
17·250	0·00003	0·00280	−0·00006	−0·00540
17·500	0·00276	−0·00003	−0·00533	0·00005
17·750	−0·00003	−0·00272	0·00005	0·00525
18·000	−0·00268	0·00003	0·00518	−0·00005
18·250	0·00003	0·00265	−0·00005	−0·00511
18·500	0·00261	−0·00003	−0·00504	0·00005
18·750	−0·00002	−0·00258	0·00005	0·00497
19·000	−0·00254	0·00002	0·00491	−0·00005
19·250	0·00002	0·00251	−0·00004	−0·00484
19·500	0·00248	−0·00002	−0·00478	0·00004
19·750	−0·00002	−0·00245	0·00004	0·00472
20·000	−0·00241	0·00002	0·00466	−0·00004
20·250	0·00002	0·00239	−0·00004	−0·00460
20·500	0·00236	−0·00002	−0·00455	0·00004
20·750	−0·00002	−0·00233	0·00004	0·00449
21·000	−0·00230	0·00002	0·00444	−0·00004
21·250	0·00002	0·00227	−0·00004	−0·00439
21·500	0·00225	−0·00002	−0·00434	0·00004
21·750	−0·00002	−0·00222	0·00003	0·00429
22·000	−0·00220	0·00002	0·00424	−0·00003
22·250	0·00002	0·00217	−0·00003	−0·00419
22·500	0·00215	−0·00002	−0·00414	0·00003
22·750	−0·00002	−0·00212	0·00003	0·00410

Table of elements of Φ_u and Φ_v matrices for $h/\lambda = 0.125$
Values for single element (Φ_{kku}, Φ_{kkv} and $\Psi_{kkdR} = \Psi_{dp}$)—contd.

$(k-i)(b/\lambda)$	Φ_{kiu}		Φ_{kiv}	
23·000	−0·00210	0·00002	0·00405	−0·00003
23·250	0·00002	0·00208	−0·00003	−0·00401
23·500	0·00206	−0·00002	−0·00397	0·00003
23·750	−0·00002	−0·00203	0·00003	0·00393
24·000	−0·00201	0·00001	0·00388	−0·00003
24·250	0·00001	0·00199	−0·00003	−0·00384
24·500	0·00197	−0·00001	−0·00381	0·00003
24·750	−0·00001	−0·00195	0·00003	0·00377
25·000	−0·00193	0·00001	0·00373	−0·00003
25·250	0·00001	0·00191	−0·00003	−0·00369
25·500	0·00189	−0·00001	−0·00366	0·00003
25·750	−0·00001	−0·00188	0·00002	0·00362
26·000	−0·00186	0·00001	0·00359	−0·00002
26·250	0·00001	0·00184	−0·00002	−0·00355
26·500	0·00182	−0·00001	−0·00352	0·00002
26·750	−0·00001	−0·00181	0·00002	0·00349
27·000	−0·00179	0·00001	0·00345	−0·00002
27·250	0·00001	0·00177	−0·00002	−0·00342
27·500	0·00176	−0·00001	−0·00339	0·00002
27·750	−0·00001	−0·00174	0·00002	0·00336
28·000	−0·00173	0·00001	0·00333	−0·00002
28·250	0·00001	0·00171	−0·00002	−0·00330
28·500	0·00169	−0·00001	−0·00327	0·00002
28·750	−0·00001	−0·00168	0·00002	0·00324
29·000	−0·00167	0·00001	0·00321	−0·00002
29·250	0·00001	0·00165	−0·00002	−0·00319
29·500	0·00164	−0·00001	−0·00316	0·00002
29·750	−0·00001	−0·00162	0·00002	0·00313
30·000	−0·00161	0·00001	0·00311	−0·00002
30·250	0·00001	0·00160	−0·00002	−0·00308
30·500	0·00158	−0·00001	−0·00306	0·00002
30·750	−0·00001	−0·00157	0·00002	0·00303
31·000	−0·00156	0·00001	0·00301	−0·00002
31·250	0·00001	0·00155	−0·00002	−0·00298
31·500	0·00153	−0·00001	−0·00296	0·00002
31·750	−0·00001	−0·00152	0·00002	0·00294
32·000	−0·00151	0·00001	0·00291	−0·00002
32·250	0·00001	0·00150	−0·00002	−0·00289
32·500	0·00149	−0·00001	−0·00287	0·00002
32·750	−0·00001	−0·00147	0·00002	0·00285
33·000	−0·00146	0·00001	0·00283	−0·00002
33·250	0·00001	0·00145	−0·00001	−0·00280
33·500	0·00144	−0·00001	−0·00278	0·00001
33·750	−0·00001	−0·00143	0·00001	0·00276
34·000	−0·00142	0·00001	0·00274	−0·00001
34·250	0·00001	0·00141	−0·00001	−0·00272
34·500	0·00140	−0·00001	−0·00270	0·00001
34·750	−0·00001	−0·00139	0·00001	0·00268
35·000	−0·00138	0·00001	0·00266	−0·00001
35·250	0·00001	0·00137	−0·00001	−0·00264
35·500	0·00136	−0·00001	−0·00263	0·00001
35·750	−0·00001	−0·00135	0·00001	0·00261
36·000	−0·00134	0·00001	0·00259	−0·00001
36·250	0·00001	0·00133	−0·00001	−0·00257
36·500	0·00132	−0·00001	−0·00255	0·00001

Table of elements of Φ_u and Φ_v matrices for $h/\lambda = 0.125$
Values for single element (Φ_{kku}, Φ_{kkv} and $\Psi_{kkdR} = \Psi_{dR}$)—contd.

$(k-i)(b/\lambda)$	Φ_{kiu}		Φ_{kiv}	
36·750	−0·00001	−0·00131	0·00001	0·00254
37·000	−0·00131	0·00001	0·00252	−0·00001
37·250	0·00001	0·00130	−0·00001	−0·00250
37·500	0·00129	−0·00001	−0·00249	0·00001
37·750	−0·00001	−0·00128	0·00001	0·00247
38·000	−0·00127	0·00001	0·00245	−0·00001
38·250	0·00001	0·00126	−0·00001	−0·00244
38·500	0·00125	−0·00001	−0·00242	0·00001
38·750	−0·00001	−0·00125	0·00001	0·00241
39·000	−0·00124	0·00001	0·00239	−0·00001
39·250	0·00001	0·00123	−0·00001	−0·00238
39·500	0·00122	−0·00001	−0·00236	0·00001
39·750	−0·00001	−0·00122	0·00001	0·00235
40·000	−0·00121	0·00001	0·00233	−0·00001
40·250	0·00001	0·00120	−0·00001	−0·00232
40·500	0·00119	−0·00001	−0·00230	0·00001
40·750	−0·00001	−0·00119	0·00001	0·00229
41·000	−0·00118	0·00001	0·00227	−0·00001
41·250	0·00001	0·00117	−0·00001	−0·00226
41·500	0·00116	−0·00000	−0·00225	0·00001
41·750	−0·00000	−0·00116	0·00001	0·00223
42·000	−0·00115	0·00000	0·00222	−0·00001
42·250	0·00000	0·00114	−0·00001	−0·00221
42·500	0·00114	−0·00000	−0·00219	0·00001
42·750	−0·00000	−0·00113	0·00001	0·00218
43·000	−0·00112	0·00000	0·00217	−0·00001
43·250	0·00000	0·00112	−0·00001	−0·00216
43·500	0·00111	−0·00000	−0·00214	0·00001
43·750	−0·00000	−0·00110	0·00001	0·00213
44·000	−0·00110	0·00000	0·00212	−0·00001
44·250	0·00000	0·00109	−0·00001	−0·00211
44·500	0·00109	−0·00000	−0·00210	0·00001
44·750	−0·00000	−0·00108	0·00001	0·00208
45·000	−0·00107	0·00000	0·00207	−0·00001
45·250	0·00000	0·00107	−0·00001	−0·00206
45·500	0·00106	−0·00000	−0·00205	0·00001
45·750	−0·00000	−0·00106	0·00001	0·00204
46·000	−0·00105	0·00000	0·00203	−0·00001
46·250	0·00000	0·00104	−0·00001	−0·00202
46·500	0·00104	−0·00000	−0·00200	0·00001
46·750	−0·00000	−0·00103	0·00001	0·00199
47·000	−0·00103	0·00000	0·00198	−0·00001
47·250	0·00000	0·00102	−0·00001	−0·00197
47·500	0·00102	−0·00000	−0·00196	0·00001
47·750	−0·00000	−0·00101	0·00001	0·00195
48·000	−0·00101	0·00000	0·00194	−0·00001
48·250	0·00000	0·00100	−0·00001	−0·00193
48·500	0·00100	−0·00000	−0·00192	0·00001
48·750	−0·00000	−0·00099	0·00001	0·00191
49·000	−0·00099	0·00000	0·00190	−0·00001
49·250	0·00000	0·00098	−0·00001	−0·00189
49·500	0·00098	−0·00000	−0·00188	0·00001
49·750	−0·00000	−0·00097	0·00001	0·00187
50·000	−0·00097	0·00000	0·00186	−0·00001

Table of elements of Φ_u and Φ_v matrices for $h/\lambda = 0.250$
Values for single element (Φ_{kku}, Φ_{kkv} and $\Psi_{kkdR} = \Psi_{dR}$)

Ω	Ψ_{dR}	Φ_{kku}		Φ_{kkv}	
7·00	4·73675	0·61497	−1·21658	4·73675	−0·00000
7·50	5·21607	0·63565	−1·21746	5·21607	−0·00000
8·00	5·69991	0·65181	−1·21800	5·69991	−0·00000
8·50	6·18729	0·66443	−1·21832	6·18729	−0·00000
9·00	6·67744	0·67427	−1·21852	6·67744	−0·00000
9·50	7·16976	0·68196	−1·21864	7·16976	−0·00000
10·00	7·66377	0·68794	−1·21871	7·66377	−0·00000
10·50	8·15911	0·69261	−1·21876	8·15911	−0·00000
11·00	8·65547	0·69625	−1·21878	8·65547	−0·00000
11·50	9·15263	0·69908	−1·21880	9·15263	−0·00000
12·00	9·65042	0·70129	−1·21881	9·65042	−0·00000
12·50	10·14870	0·70301	−1·21882	10·14870	−0·00000
13·00	10·64736	0·70435	−1·21882	10·64736	−0·00000
15·00	12·64438	0·70734	−1·21883	12·64438	−0·00000

Off diagonal values of Φ_u and Φ_v (see note on page 424)

$(k-i)(b/\lambda)$	Φ_{kiu}		Φ_{kiv}	
0·250	−0·47248	−0·67976	0·00000	−0·00000
0·500	−0·49881	0·20887	−0·00000	−0·00000
0·750	0·11054	0·37495	−0·00000	0·00000
1·000	0·29570	−0·06686	0·00000	−0·00000
1·250	−0·04437	−0·24260	0·00000	−0·00000
1·500	−0·20507	0·03146	−0·00000	0·00000
1·750	0·02341	0·17734	−0·00000	0·00000
2·000	0·15607	−0·01807	0·00000	−0·00000
2·250	−0·01436	−0·13929	0·00000	−0·00000
2·500	−0·12573	0·01168	−0·00000	−0·00000
2·750	0·00968	0·11455	−0·00000	0·00000
3·000	0·10517	−0·00816	0·00000	−0·00000
3·250	−0·00696	−0·09721	0·00000	−0·00000
3·500	−0·09036	0·00601	−0·00000	−0·00000
3·750	0·00524	0·08440	−0·00000	0·00000
4·000	0·07918	−0·00461	0·00000	−0·00000
4·250	−0·00409	−0·07457	−0·00000	−0·00000
4·500	−0·07046	0·00365	−0·00000	−0·00000
4·750	0·00328	0·06678	−0·00000	0·00000
5·000	0·06346	−0·00296	0·00000	0·00000
5·250	−0·00269	−0·06046	0·00000	−0·00000
5·500	−0·05772	0·00245	−0·00000	−0·00000
5·750	0·00224	0·05522	0·00000	0·00000
6·000	0·05293	−0·00206	0·00000	0·00000
6·250	−0·00190	−0·05083	−0·00000	−0·00000
6·500	−0·04888	0·00175	−0·00000	−0·00000
6·750	0·00163	0·04707	−0·00000	0·00000
7·000	0·04540	−0·00151	0·00000	0·00000
7·250	−0·00141	−0·04384	0·00000	−0·00000
7·500	−0·04238	0·00132	−0·00000	−0·00000
7·750	0·00124	0·04102	−0·00000	0·00000
8·000	0·03974	−0·00116	0·00000	−0·00000
8·250	−0·00109	−0·03854	−0·00000	−0·00000
8·500	−0·03741	0·00103	−0·00000	0·00000
8·750	0·00097	0·03634	−0·00000	0·00000
9·000	0·03533	−0·00092	0·00000	0·00000

APPENDIX III

Table of elements of Φ_u and Φ_v matrices for $h/\lambda = 0.250$
Values for single element (Φ_{kku}, Φ_{kkv} and $\Psi_{kkdR} = \Psi_{dR}$)—contd.

$(k-i)(b/\lambda)$	Φ_{kiu}		Φ_{kiv}	
9·250	−0·00087	−0·03438	0·00000	−0·00000
9·500	−0·03348	0·00082	−0·00000	−0·00000
9·750	0·00078	0·03262	−0·00000	0·00000
10·000	0·03181	−0·00074	0·00000	0·00000
10·250	−0·00071	−0·03103	0·00000	−0·00000
10·500	−0·03029	0·00067	−0·00000	−0·00000
10·750	0·00064	0·02959	−0·00000	0·00000
11·000	0·02892	−0·00061	0·00000	0·00000
11·250	−0·00059	−0·02828	0·00000	−0·00000
11·500	−0·02766	0·00056	−0·00000	0·00000
11·750	0·00054	0·02707	0·00000	0·00000
12·000	0·02651	−0·00052	0·00000	−0·00000
12·250	−0·00049	−0·02597	−0·00000	−0·00000
12·500	−0·02545	0·00048	−0·00000	−0·00000
12·750	0·00046	0·02495	−0·00000	0·00000
13·000	0·02447	−0·00044	0·00000	0·00000
13·250	−0·00042	−0·02401	0·00000	−0·00000
13·500	−0·02357	0·00041	−0·00000	−0·00000
13·750	0·00039	0·02314	−0·00000	0·00000
14·000	0·02273	−0·00038	0·00000	0·00000
14·250	−0·00037	−0·02233	0·00000	−0·00000
14·500	−0·02194	0·00035	−0·00000	−0·00000
14·750	0·00034	0·02157	−0·00000	0·00000
15·000	0·02121	−0·00033	0·00000	0·00000
15·250	−0·00032	−0·02087	0·00000	−0·00000
15·500	−0·02053	0·00031	−0·00000	−0·00000
15·750	0·00030	0·02020	−0·00000	0·00000
16·000	0·01989	−0·00029	0·00000	−0·00000
16·250	−0·00028	−0·01958	−0·00000	−0·00000
16·500	−0·01929	0·00027	−0·00000	0·00000
16·750	0·00026	0·01900	0·00000	0·00000
17·000	0·01872	−0·00026	0·00000	−0·00000
17·250	−0·00025	−0·01845	−0·00000	−0·00000
17·500	−0·01818	0·00024	−0·00000	−0·00000
17·750	0·00024	0·01793	−0·00000	0·00000
18·000	0·01768	−0·00023	0·00000	0·00000
18·250	−0·00022	−0·01744	0·00000	−0·00000
18·500	−0·01720	0·00022	−0·00000	−0·00000
18·750	0·00021	0·01697	−0·00000	0·00000
19·000	0·01675	−0·00021	0·00000	0·00000
19·250	−0·00020	−0·01653	0·00000	−0·00000
19·500	−0·01632	0·00020	−0·00000	−0·00000
19·750	0·00019	0·01611	−0·00000	0·00000
20·000	0·01591	−0·00019	0·00000	0·00000
20·250	−0·00018	−0·01572	0·00000	−0·00000
20·500	−0·01552	0·00018	−0·00000	−0·00000
20·750	0·00017	0·01534	−0·00000	0·00000
21·000	0·01515	−0·00017	0·00000	0·00000
21·250	−0·00016	−0·01498	0·00000	−0·00000
21·500	−0·01480	0·00016	−0·00000	−0·00000
21·750	0·00016	0·01463	−0·00000	0·00000
22·000	0·01447	−0·00015	0·00000	0·00000
22·250	−0·00015	−0·01430	0·00000	−0·00000
22.500	−0·01414	0·00015	−0·00000	−0·00000

Table of elements of Φ_u and Φ_v matrices for $h/\lambda = 0.250$
Values for single element (Φ_{kku}, Φ_{kkv} and $\Psi_{kkdR} = \Psi_{dR}$)—contd.

$(k-i)(b/\lambda)$	Φ_{kiu}		Φ_{kiv}	
22·750	0·00014	0·01399	0·00000	0·00000
23·000	0·01384	−0·00014	0·00000	−0·00000
23·250	−0·00014	−0·01369	−0·00000	−0·00000
23·500	−0·01354	0·00013	−0·00000	0·00000
23·750	0·00013	0·01340	0·00000	0·00000
24·000	0·01326	−0·00013	0·00000	−0·00000
24·250	−0·00013	−0·01312	−0·00000	−0·00000
24·500	−0·01299	0·00012	−0·00000	0·00000
24·750	0·00012	0·01286	0·00000	0·00000
25·000	0·01273	−0·00012	0·00000	0·00000
25·250	−0·00012	−0·01260	0·00000	−0·00000
25·500	−0·01248	0·00011	−0·00000	−0·00000
25·750	0·00011	0·01236	−0·00000	0·00000
26·000	0·01224	−0·00011	0·00000	0·00000
26·250	−0·00011	−0·01212	0·00000	−0·00000
26·500	−0·01201	0·00011	−0·00000	−0·00000
26·750	0·00010	0·01190	−0·00000	0·00000
27·000	0·01179	−0·00010	0·00000	0·00000
27·250	−0·00010	−0·01168	0·00000	−0·00000
27·500	−0·01157	0·00010	−0·00000	−0·00000
27·750	0·00010	0·01147	−0·00000	0·00000
28·000	0·01137	−0·00009	0·00000	0·00000
28·250	−0·00009	−0·01127	0·00000	−0·00000
28·500	−0·01117	0·00009	−0·00000	−0·00000
28·750	0·00009	0·01107	−0·00000	0·00000
29·000	0·01098	−0·00009	0·00000	0·00000
29·250	−0·00009	−0·01088	0·00000	−0·00000
29·500	−0·01079	0·00009	−0·00000	−0·00000
29·750	0·00008	0·01070	−0·00000	0·00000
30·000	0·01061	−0·00008	0·00000	0·00000
30·250	−0·00008	−0·01052	0·00000	−0·00000
30·500	−0·01044	0·00008	−0·00000	−0·00000
30·750	0·00008	0·01035	−0·00000	0·00000
31·000	0·01027	−0·00008	0·00000	0·00000
31·250	−0·00008	−0·01019	0·00000	−0·00000
31·500	−0·01010	0·00007	−0·00000	−0·00000
31·750	0·00007	0·01002	−0·00000	0·00000
32·000	0·00995	−0·00007	0·00000	−0·00000
32·250	−0·00007	−0·00987	−0·00000	−0·00000
32·500	−0·00979	0·00007	−0·00000	0·00000
32·750	0·00007	0·00972	0·00000	0·00000
33·000	0·00965	−0·00007	0·00000	−0·00000
33·250	−0·00007	−0·00957	−0·00000	−0·00000
33·500	−0·00950	0·00007	−0·00000	0·00000
33·750	0·00007	0·00943	0·00000	0·00000
34·000	0·00936	−0·00006	0·00000	−0·00000
34·250	−0·00006	−0·00929	−0·00000	−0·00000
34·500	−0·00923	0·00006	−0·00000	0·00000
34·750	0·00006	0·00916	0·00000	0·00000
35·000	0·00909	−0·00006	0·00000	0·00000
35·250	−0·00006	−0·00903	0·00000	−0·00000
35·500	−0·00897	0·00006	−0·00000	−0·00000
35·750	0·00006	0·00890	−0·00000	0·00000
36·000	0·00884	−0·00006	0·00000	0·00000
36·250	−0·00006	−0·00878	0·00000	−0·00000

Table of elements of Φ_u and Φ_v matrices for $h/\lambda = 0.250$
Values for single element (Φ_{kku}, Φ_{kkv} and $\Psi_{kkdR} = \Psi_{dR}$)—contd.

$(k-i)(b/\lambda)$	Φ_{kiu}		Φ_{kiv}	
36·500	−0·00872	0·00006	−0·00000	−0·00000
36·750	0·00006	0·00866	−0·00000	0·00000
37·000	0·00860	−0·00005	0·00000	0·00000
37·250	−0·00005	−0·00854	0·00000	−0·00000
37·500	−0·00849	0·00005	−0·00000	−0·00000
37·750	0·00005	0·00843	−0·00000	0·00000
38·000	0·00838	−0·00005	0·00000	0·00000
38·250	−0·00005	−0·00832	0·00000	−0·00000
38·500	−0·00827	0·00005	−0·00000	−0·00000
38·750	0·00005	0·00821	−0·00000	0·00000
39·000	0·00816	−0·00005	0·00000	0·00000
39·250	−0·00005	−0·00811	0·00000	−0·00000
39·500	−0·00806	0·00005	−0·00000	−0·00000
39·750	0·00005	0·00801	−0·00000	0·00000
40·000	0·00796	−0·00005	0·00000	0·00000
40·250	−0·00005	−0·00791	0·00000	−0·00000
40·500	−0·00786	0·00005	−0·00000	−0·00000
40·750	0·00004	0·00781	−0·00000	0·00000
41·000	0·00776	−0·00004	0·00000	0·00000
41·250	−0·00004	−0·00772	0·00000	−0·00000
41·500	−0·00767	0·00004	−0·00000	−0·00000
41·750	0·00004	0·00762	−0·00000	0·00000
42·000	0·00758	−0·00004	0·00000	0·00000
42·250	−0·00004	−0·00753	0·00000	−0·00000
42·500	−0·00749	0·00004	−0·00000	−0·00000
42·750	0·00004	0·00745	−0·00000	0·00000
43·000	0·00740	−0·00004	0·00000	0·00000
43·250	−0·00004	−0·00736	0·00000	−0·00000
43·500	−0·00732	0·00004	−0·00000	−0·00000
43·750	0·00004	0·00728	−0·00000	0·00000
44·000	0·00723	−0·00004	0·00000	0·00000
44·250	−0·00004	−0·00719	0·00000	−0·00000
44·500	−0·00715	0·00004	−0·00000	−0·00000
44·750	0·00004	0·00711	−0·00000	0·00000
45·000	0·00707	−0·00004	0·00000	0·00000
45·250	−0·00004	−0·00703	0·00000	−0·00000
45·500	−0·00700	0·00004	−0·00000	0·00000
45·750	0·00004	0·00696	0·00000	0·00000
46·000	0·00692	−0·00004	0·00000	−0·00000
46·250	−0·00003	−0·00688	−0·00000	−0·00000
46·500	−0·00685	0·00003	−0·00000	0·00000
46·750	0·00003	0·00681	0·00000	0·00000
47·000	0·00677	−0·00003	0·00000	−0·00000
47·250	−0·00003	−0·00674	−0·00000	−0·00000
47·500	−0·00670	0·00003	−0·00000	0·00000
47·750	0·00003	0·00667	0·00000	0·00000
48·000	0·00663	−0·00003	0·00000	−0·00000
48·250	−0·00003	−0·00660	−0·00000	−0·00000
48·500	−0·00656	0·00003	−0·00000	0·00000
48·750	0·00003	0·00653	0·00000	0·00000
49·000	0·00650	−0·00003	0·00000	−0·00000
49·250	−0·00003	−0·00646	−0·00000	−0·00000
49·500	−0·00643	0·00003	−0·00000	−0·00000
49·750	0·00003	0·00640	−0·00000	0·00000
50·000	0·00637	−0·00003	0·00000	0·00000

Table of elements of Φ_u and Φ_v matrices for $h/\lambda = 0.375$
Values for single element (Φ_{kku}, Φ_{kkv} and $\Psi_{kkdR} = \Psi_{dR}$)

Ω	Ψ_{dR}	Φ_{kku}		Φ_{kkv}	
7·00	4·19919	−3·23123	2·57558	0·21311	−1·80626
7·50	4·67051	−3·60298	2·57993	0·24365	−1·80933
8·00	5·14769	−3·97013	2·58257	0·26761	−1·81119
8·50	5·62962	−4·33386	2·58417	0·28638	−1·81232
9·00	6·11538	−4·69506	2·58514	0·30105	−1·81301
9·50	6·60417	−5·05439	2·58573	0·31251	−1·81343
10·00	7·09538	−5·41231	2·58609	0·32146	−1·81368
10·50	7·58849	−5·76918	2·58631	0·32844	−1·81383
11·00	8·08311	−6·12526	2·58644	0·33388	−1·81393
11·50	8·57890	−6·48075	2·58652	0·33812	−1·81398
12·00	9·07561	−6·83578	2·58657	0·34143	−1·81402
12·50	9·57305	−7·19047	2·58660	0·34401	−1·81404
13·00	10·07105	−7·54490	2·58661	0·34602	−1·81405
15·00	12·06658	−8·96103	2·58664	0·35049	−1·81407

Off diagonal values of Φ_u and Φ_v (see note on page 424)

$(k-i)(b/\lambda)$	Φ_{kiu}		Φ_{kiv}	
0·250	0·91495	1·41028	−0·64760	−0·98392
0·500	1·02366	−0·50347	−0·71398	0·36254
0·750	−0·30409	−0·80374	0·22402	0·56460
1·000	−0·65434	0·19794	0·46271	−0·14798
1·250	0·13707	0·54781	−0·10339	−0·38918
1·500	0·46915	−0·09974	−0·33434	0·07566
1·750	−0·07549	−0·40923	0·05748	0·29226
2·000	−0·36234	0·05896	0·25917	−0·04501
2·250	0·04724	0·32477	−0·03613	−0·23255
2·500	0·29408	−0·03866	−0·21074	0·02960
2·750	−0·03220	−0·26857	0·02468	0·19258
3·000	−0·24705	0·02721	0·17724	−0·02088
3·250	0·02329	0·22868	−0·01788	−0·16412
3·500	0·21281	−0·02016	−0·15278	0·01548
3·750	−0·01761	−0·19898	0·01353	0·14288
4·000	−0·18682	0·01552	0·13418	−0·01193
4·250	0·01377	0·17604	−0·01059	−0·12646
4·500	0·16643	−0·01231	−0·11957	0·00946
4·750	−0·01106	−0·15781	0·00851	0·11339
5·000	−0·15003	0·01000	0·10782	−0·00769
5·250	0·00908	0·14298	−0·00698	−0·10276
5·500	0·13656	−0·00828	−0·09815	0·00637
5·750	−0·00758	−0·13068	0·00583	0·09393
6·000	−0·12529	0·00697	0·09006	−0·00536
6·250	0·00642	0·12033	−0·00494	−0·08650
6·500	0·11574	−0·00594	−0·08320	0·00457
6·750	−0·00551	−0·11148	0·00424	0·08015
7·000	−0·10753	0·00513	0·07731	−0·00395
7·250	0·00478	0·10385	−0·00368	−0·07467
7·500	0·10041	−0·00447	−0·07219	0·00344
7·750	−0·00419	−0·09719	0·00322	0·06988
8·000	−0·09417	0·00393	0·06771	−0·00303
8·250	0·00370	0·09133	−0·00285	−0·06567
8·500	0·08866	−0·00348	−0·06375	0·00268
8·750	−0·00329	−0·08613	0·00253	0·06194
9·000	−0·08375	0·00311	0·06023	−0·00239
9·250	0·00294	0·08150	−0·00227	−0·05861

Table of elements of Φ_u and Φ_v matrices for $h/\lambda = 0.375$
Values for single element (Φ_{kku}, Φ_{kkv} and $\Psi_{kkdR} = \Psi_{dR}$)—contd.

$(k-i)(b/\lambda)$	Φ_{kiu}		Φ_{kiv}	
9·500	0·07936	−0·00279	−0·05707	0·00215
9·750	−0·00265	−0·07733	0·00204	0·05561
10·000	−0·07541	0·00252	0·05423	−0·00194
10·250	0·00240	0·07358	−0·00185	−0·05291
10·500	0·07183	−0·00229	−0·05166	0·00176
10·750	−0·00218	−0·07016	0·00168	0·05046
11·000	−0·06857	0·00208	0·04932	−0·00161
11·250	0·00199	0·06705	−0·00153	−0·04822
11·500	0·06560	−0·00191	−0·04718	0·00147
11·750	−0·00183	−0·06421	0·00141	0·04618
12·000	−0·06287	0·00175	0·04522	−0·00135
12·250	0·00168	0·06159	−0·00130	−0·04430
12·500	0·06036	−0·00161	−0·04342	0·00124
12·750	−0·00155	−0·05918	0·00120	0·04257
13·000	−0·05805	0·00149	0·04175	−0·00115
13·250	0·00144	0·05695	−0·00111	−0·04096
13·500	0·05590	−0·00138	−0·04021	0·00107
13·750	−0·00134	−0·05489	0·00103	0·03948
14·000	−0·05391	0·00129	0·03877	−0·00099
14·250	0·00124	0·05297	−0·00096	−0·03809
14·500	0·05205	−0·00120	−0·03744	0·00092
14·750	−0·00116	−0·05117	0·00089	0·03681
15·000	−0·05032	0·00112	0·03619	−0·00086
15·250	0·00109	0·04950	−0·00084	−0·03560
15·500	0·04870	−0·00105	−0·03503	0·00081
15·750	−0·00102	−0·04793	0·00078	0·03447
16·000	−0·04718	0·00099	0·03394	−0·00076
16·250	0·00096	0·04646	−0·00074	−0·03341
16·500	0·04575	−0·00093	−0·03291	0·00071
16·750	−0·00090	−0·04507	0·00069	0·03242
17·000	−0·04441	0·00087	0·03194	−0·00067
17·250	0·00085	0·04377	−0·00065	−0·03148
17·500	0·04314	−0·00082	−0·03103	0·00064
17·750	−0·00080	−0·04253	0·00062	0·03059
18·000	−0·04194	0·00078	0·03017	−0·00060
18·250	0·00076	0·04137	−0·00058	−0·02976
18·500	0·04081	−0·00074	−0·02936	0·00057
18·750	−0·00072	−0·04027	0·00055	0·02896
19·000	−0·03974	0·00070	0·02858	−0·00054
19·250	0·00068	0·03922	−0·00053	−0·02821
19·500	0·03872	−0·00066	−0·02785	0·00051
19·750	−0·00065	−0·03823	0·00050	0·02750
20·000	−0·03775	0·00063	0·02716	−0·00049
20·250	0·00062	0·03729	−0·00047	−0·02682
20·500	0·03683	−0·00060	−0·02649	0·00046
20·750	−0·00059	−0·03639	0·00045	0·02618
21·000	−0·03596	0·00057	0·02586	−0·00044
21·250	0·00056	0·03553	−0·00043	−0·02556
21·500	0·03512	−0·00055	−0·02526	0·00042
21·750	−0·00053	−0·03472	0·00041	0·02497
22·000	−0·03432	0·00052	0·02469	−0·00040
22·250	0·00051	0·03394	−0·00039	−0·02441
22·500	0·03356	−0·00050	−0·02414	0·00038
22·750	−0·00049	−0·03319	0·00038	0·02388
23·000	−0·03283	0·00048	0·02362	−0·00037

Table of elements of Φ_u and Φ_v matrices for $h/\lambda = 0.375$
Values for single element (Φ_{kku}, Φ_{kkv} and $\Psi_{kkdR} = \Psi_{dR}$)—contd.

$(k-i)(b/\lambda)$	Φ_{kiu}		Φ_{kiv}	
23·250	0·00047	0·03248	−0·00036	−0·02336
23·500	0·03123	−0·00046	−0·02311	0·00035
23·750	−0·00045	−0·03180	0·00035	0·02287
24·000	−0·03147	0·00044	0·02263	−0·00034
24·250	0·00043	0·03114	−0·00033	−0·02240
24·500	0·03082	−0·00042	−0·02217	0·00032
24·750	−0·00041	−0·03051	0·00032	0·02195
25·000	−0·03021	0·00040	0·02173	−0·00031
25·250	0·00040	0·02991	−0·00031	−0·02151
25·500	0·02962	−0·00039	−0·02130	0·00030
25·750	−0·00038	−0·02933	0·00029	0·02110
26·000	−0·02905	0·00037	0·02089	−0·00029
26·250	0·00037	0·02877	−0·00028	−0·02069
26·500	0·02850	−0·00036	−0·02050	0·00028
26·750	−0·00035	−0·02823	0·00027	0·02031
27·000	−0·02797	0·00035	0·02012	−0·00027
27·250	0·00034	0·02771	−0·00026	−0·01994
27·500	0·02746	−0·00033	−0·01975	0·00026
27·750	−0·00033	−0·02722	0·00025	0·01958
28·000	−0·02697	0·00032	0·01940	−0·00025
28·250	0·00032	0·02673	−0·00024	−0·01923
28·500	0·02650	−0·00031	−0·01906	0·00024
28·750	−0·00031	−0·02627	0·00024	0·01890
29·000	−0·02604	0·00030	0·01873	−0·00023
29·250	0·00030	0·02582	−0·00023	−0·01857
29·500	0·02560	−0·00029	−0·01842	0·00022
29·750	−0·00029	−0·02539	0·00022	0·01826
30·000	−0·02518	0·00028	0·01811	−0·00022
30·250	0·00028	0·02497	−0·00021	−0·01796
30·500	0·02476	−0·00027	−0·01781	0·00021
30·750	−0·00027	−0·02456	0·00021	0·01767
31·000	−0·02436	0·00026	0·01753	−0·00020
31·250	0·00026	0·02417	−0·00020	−0·01738
31·500	0·02398	−0·00025	−0·01725	0·00020
31·750	−0·00025	−0·02379	0·00019	0·01711
32·000	−0·02360	0·00025	0·01698	−0·00019
32·250	0·00024	0·02342	−0·00019	−0·01685
32·500	0·02324	−0·00024	−0·01672	0·00018
32·750	−0·00024	−0·02306	0·00018	0·01659
33·000	−0·02289	0·00023	0·01646	−0·00018
33·250	0·00023	0·02272	−0·00018	−0·01634
33·500	0·02255	−0·00023	−0·01622	0·00017
33·750	−0·00022	−0·02238	0·00017	0·01610
34·000	−0·02221	0·00022	0·01598	−0·00017
34·250	0·00022	0·02205	−0·00017	−0·01586
34·500	0·02189	−0·00021	−0·01575	0·00016
34·750	−0·00021	−0·02174	0·00016	0·01563
35·000	−0·02158	0·00021	0·01552	−0·00016
35·250	0·00020	0·02143	−0·00016	−0·01541
35·500	0·02128	−0·00020	−0·01530	0·00015
35·750	−0·00020	−0·02113	0·00015	0·01520
36·000	−0·02098	0·00019	0·01509	−0·00015
36·250	0·00019	0·02084	−0·00015	−0·01499
36·500	0·02069	−0·00019	−0·01489	0·00015

Table of elements of Φ_u and Φ_v matrices for $h/\lambda = 0.375$
Values for single element (Φ_{kku}, Φ_{kkv} and $\Psi_{kkdR} = \Psi_{dR}$)—contd.

$(k-i)(b/\lambda)$	Φ_{kiu}		Φ_{kiv}	
36·750	−0·00019	−0·02055	0·00014	0·01478
37·000	−0·02041	0·00018	0·01468	−0·00014
37·250	0·00018	0·02028	−0·00014	−0·01459
37·500	0·02014	−0·00018	−0·01449	0·00014
37·750	−0·00018	−0·02001	0·00014	0·01439
38·000	−0·01988	0·00018	0·01430	−0·00013
38·250	0·00017	0·01975	−0·00013	−0·01420
38·500	0·01962	−0·00017	−0·01411	0·00013
38·750	−0·00017	−0·01949	0·00013	0·01402
39·000	−0·01937	0·00017	0·01393	−0·00013
39·250	0·00016	0·01924	−0·00013	−0·01384
39·500	0·01912	−0·00016	−0·01375	0·00012
39·750	−0·00016	−0·01900	0·00012	0·01367
40·000	−0·01888	0·00016	0·01358	−0·00012
40·250	0·00016	0·01877	−0·00012	−0·01350
40·500	0·01865	−0·00015	−0·01342	0·00012
40·750	−0·00015	−0·01854	0·00012	0·01333
41·000	−0·01842	0·00015	0·01325	−0·00012
41·250	0·00015	0·01831	−0·00011	−0·01317
41·500	0·01820	−0·00015	−0·01309	0·00011
41·750	−0·00014	−0·01809	0·00011	0·01301
42·000	−0·01798	0·00014	0·01294	−0·00011
42·250	0·00014	0·01788	−0·00011	−0·01286
42·500	0·01777	−0·00014	−0·01278	0·00011
42·750	−0·00014	−0·01767	0·00011	0·01271
43·000	−0·01757	0·00014	0·01264	−0·00011
43·250	0·00014	0·01746	−0·00010	−0·01256
43·500	0·01736	−0·00013	−0·01249	0·00010
43·750	−0·00013	−0·01726	0·00010	0·01242
44·000	−0·01717	0·00013	0·01235	−0·00010
44·250	0·00013	0·01707	−0·00010	−0·01228
44·500	0·01697	−0·00013	−0·01221	0·00010
44·750	−0·00013	−0·01688	0·00010	0·01214
45·000	−0·01679	0·00012	0·01207	−0·00010
45·250	0·00012	0·01669	−0·00010	−0·01201
45·500	0·01660	−0·00012	−0·01194	0·00009
45·750	−0·00012	−0·01651	0·00009	0·01188
46·000	−0·01642	0·00012	0·01181	−0·00009
46·250	0·00012	0·01633	−0·00009	−0·01175
46·500	0·01624	−0·00012	−0·01168	0·00009
46·750	−0·00012	−0·01616	0·00009	0·01162
47·000	−0·01607	0·00011	0·01156	−0·00009
47·250	0·00011	0·01599	−0·00009	−0·01150
47·500	0·01590	−0·00011	−0·01144	0·00009
47·750	−0·00011	−0·01582	0·00009	0·01138
48·000	−0·01574	0·00011	0·01132	−0·00008
48·250	0·00011	0·01565	−0·00008	−0·01126
48·500	0·01557	−0·00011	−0·01120	0·00008
48·750	−0·00011	−0·01549	0·00008	0·01115
49·000	−0·01542	0·00011	0·01109	−0·00008
49·250	0·00010	0·01534	−0·00008	−0·01103
49·500	0·01526	−0·00010	−0·01098	0·00008
49·750	−0·00010	−0·01518	0·00008	0·01092
50·000	−0·01511	0·00010	0·01087	−0·00008

Table of elements of Φ_u and Φ_v matrices for $h/\lambda = 0.50$
Values for single element (Φ_{kku}, Φ_{kkv} and $\Psi_{kkdR} = \Psi_{dR}$)

Ω	Ψ_{dR}	Φ_{kku}		Φ_{kkv}	
7.00	2.96068	− 3.65469	2.87405	0.48701	− 1.64559
7.50	3.42057	− 4.16756	2.88308	0.52709	− 1.65089
8.00	3.88899	− 4.67618	2.88857	0.55866	− 1.65410
8.50	4.36420	− 5.18191	2.89191	0.58344	− 1.65605
9.00	4.84477	− 5.68568	2.89393	0.60287	− 1.65724
9.50	5.32957	− 6.18815	2.89516	0.61807	− 1.65796
10.00	5.81769	− 6.68975	2.89590	0.62995	− 1.65839
10.50	6.30841	− 7.19079	2.89635	0.63923	− 1.65866
11.00	6.80117	− 7.69146	2.89662	0.64647	− 1.65882
11.50	7.29552	− 8.19189	2.89679	0.65212	− 1.65891
12.00	7.79112	− 8.69217	2.89689	0.65653	− 1.65897
12.50	8.28768	− 9.19235	2.89695	0.65996	− 1.65901
13.00	8.78501	− 9.69246	2.89699	0.66264	− 1.65903
15.00	10.77905	− 11.69262	2.89704	0.66860	− 1.65906

Off diagonal values of Φ_u and Φ_v (see note on page 424)

$(k-i)(b/\lambda)$	Φ_{kiu}		Φ_{kiv}	
0.250	1.02173	1.52473	− 0.68124	− 0.85698
0.500	1.10298	− 0.66347	− 0.62956	0.40704
0.750	− 0.45462	− 0.90710	0.29109	0.51517
1.000	− 0.76996	0.32113	0.44407	− 0.21742
1.250	0.23434	0.66433	− 0.16567	− 0.39044
1.500	0.58100	− 0.17648	− 0.34702	0.12866
1.750	− 0.13673	− 0.51431	0.10187	0.31107
2.000	− 0.46021	0.10856	0.28103	− 0.08216
2.250	0.08803	0.41571	− 0.06739	− 0.25573
2.500	0.37862	− 0.07267	− 0.23423	0.05612
2.750	− 0.06093	− 0.34732	0.04737	0.21582
3.000	− 0.32062	0.05177	0.19993	− 0.04046
3.250	0.04450	0.29760	− 0.03492	− 0.18610
3.500	0.27757	− 0.03864	− 0.17397	0.03042
3.750	− 0.03385	− 0.26000	0.02673	0.16327
4.000	− 0.24448	0.02989	0.15376	− 0.02365
4.250	0.02659	0.23067	− 0.02107	− 0.14527
4.500	0.21832	− 0.02379	− 0.13764	0.01889
4.750	− 0.02141	− 0.20720	0.01702	0.13076
5.000	− 0.19714	0.01937	0.12451	− 0.01542
5.250	0.01761	0.18800	− 0.01403	− 0.11882
5.500	0.17966	− 0.01607	− 0.11362	0.01282
5.750	− 0.01473	− 0.17202	0.01175	0.10885
6.000	− 0.16500	0.01354	0.10446	− 0.01082
6.250	0.01250	0.15852	− 0.00999	− 0.10040
6.500	0.15253	− 0.01157	− 0.09664	0.00925
6.750	− 0.01074	− 0.14697	0.00859	0.09315
7.000	− 0.14180	0.00999	0.08990	− 0.00800
7.250	0.00932	0.13698	− 0.00746	− 0.08687
7.500	0.13247	− 0.00872	− 0.08403	0.00698
7.750	− 0.00817	− 0.12825	0.00654	0.08137
8.000	− 0.12429	0.00767	0.07888	− 0.00615
8.250	0.00722	0.12056	− 0.00578	− 0.07653
8.500	0.11706	− 0.00680	− 0.07431	0.00545
8.750	− 0.00642	− 0.11374	0.00515	0.07222
9.000	− 0.11061	0.00607	0.07024	− 0.00487

Table of elements of Φ_u and Φ_v matrices for $h/\lambda = 0.50$
Values for single element (Φ_{kku}, Φ_{kkv} and $\Psi_{kkdR} = \Psi_{dR}$)—contd.

$(k-i)(b/\lambda)$	Φ_{kiu}		Φ_{kiv}	
9·250	0·00575	0·10765	−0·00461	−0·06837
9·500	0·10484	−0·00545	−0·06659	0·00438
9·750	−0·00518	−0·10217	0·00416	0·06491
10·000	−0·09964	0·00493	0·06330	−0·00395
10·250	0·00469	0·09722	−0·00376	−0·06177
10·500	0·09492	−0·00447	−0·06032	0·00359
10·750	−0·00427	−0·09273	0·00342	0·05893
11·000	−0·09063	0·00407	0·05760	−0·00327
11·250	0·00390	0·08863	−0·00313	−0·05633
11·500	0·08672	−0·00373	−0·05512	0·00300
11·750	−0·00357	−0·08488	0·00287	0·05396
12·000	−0·08312	0·00343	0·05284	−0·00275
12·250	0·00329	0·08143	−0·00264	−0·05177
12·500	0·07981	−0·00316	−0·05074	0·00254
12·750	−0·00304	−0·07826	0·00244	0·04976
13·000	−0·07676	0·00292	0·04881	−0·00235
13·250	0·00281	0·07531	−0·00226	−0·04789
13·500	0·07393	−0·00271	−0·04701	0·00218
13·750	−0·00261	−0·07259	0·00210	0·04616
14·000	−0·07130	0·00252	0·04534	−0·00203
14·250	0·00243	0·07005	−0·00196	−0·04455
14·500	0·06885	−0·00235	−0·04379	0·00189
14·750	−0·00227	−0·06768	0·00183	0·04305
15·000	−0·06656	0·00220	0·04233	−0·00177
15·250	0·00212	0·06547	−0·00171	−0·04164
15·500	0·06442	−0·00206	−0·04097	0·00165
15·750	−0·00199	−0·06340	0·00160	0·04033
16·000	−0·06241	0·00193	0·03970	−0·00155
16·250	0·00187	0·06145	−0·00150	−0·03909
16·500	0·06052	−0·00182	−0·03850	0·00146
16·750	−0·00175	−0·05962	0·00142	0·03793
17·000	−0·05875	0·00171	0·03737	−0·00138
17·250	0·00166	0·05790	−0·00134	−0·03683
17·500	0·05707	−0·00161	−0·03631	0·00130
17·750	−0·00157	−0·05627	0·00126	0·03580
18·000	−0·05549	0·00153	0·03531	−0·00123
18·250	0·00148	0·05473	−0·00119	−0·03482
18·500	0·05400	−0·00144	−0·03435	0·00116
18·750	−0·00141	−0·05328	0·00113	0·03390
19·000	−0·05258	0·00137	0·03345	−0·00110
19·250	0·00133	0·05190	−0·00107	−0·03302
19·500	0·05123	−0·00130	−0·03260	0·00105
19·750	−0·00127	−0·05059	0·00102	0·03219
20·000	−0·04995	0·00124	0·03179	−0·00099
20·250	0·00121	0·04934	−0·00097	−0·03139
20·500	0·04874	−0·00118	−0·03101	0·00095
20·750	−0·00115	−0·04815	0·00092	0·03064
21·000	−0·04758	0·00112	0·03028	−0·00090
21·250	0·00110	0·04702	−0·00088	−0·02992
21·500	0·04647	−0·00107	−0·02957	0·00086
21·750	−0·00105	−0·04594	0·00084	0·02923
22·000	−0·04542	0·00102	0·02890	−0·00082
22·250	0·00100	0·04491	−0·00080	−0·02858
22·500	0·04441	−0·00098	−0·02826	0·00079

Table of elements of Φ_u and Φ_v matrices for $h/\lambda = 0.50$
Values for single element (Φ_{kku}, Φ_{kkv} and $\Psi_{kkdR} = \Psi_{dR}$)—contd.

$(k-i)(b/\lambda)$	Φ_{kiu}		Φ_{kiv}	
22·750	−0·00096	−0·04392	0·00077	0·02795
23·000	−0·04345	0·00094	0·02765	−0·00075
23·250	0·00092	0·04298	−0·00074	−0·02735
23·500	0·04252	−0·00090	−0·02706	0·00072
23·750	−0·00088	−0·04208	0·00071	0·02678
24·000	−0·04164	0·00086	0·02650	−0·00069
24·250	0·00084	0·04121	−0·00068	−0·02623
24·500	0·04079	−0·00082	−0·02596	0·00066
24·750	−0·00081	−0·04038	0·00065	0·02570
25·000	−0·03998	0·00079	0·02544	−0·00064
25·250	0·00078	0·03958	−0·00062	−0·02519
25·500	0·03919	−0·00076	−0·02494	0·00061
25·750	−0·00075	−0·03881	0·00060	0·02470
26·000	−0·03844	0·00073	0·02446	−0·00059
26·250	0·00072	0·03807	−0·00058	−0·02423
26·500	0·03772	−0·00070	−0·02400	0·00057
26·750	−0·00069	−0·03736	0·00056	0·02378
27·000	−0·03702	0·00068	0·02356	−0·00055
27·250	0·00067	0·03668	−0·00054	−0·02334
27·500	0·03635	−0·00065	−0·02313	0·00053
27·750	−0·00064	−0·03602	0·00052	0·02292
28·000	−0·03570	0·00063	0·02272	−0·00051
28·250	0·00062	0·03538	−0·00050	−0·02252
28·500	0·03507	−0·00061	−0·02232	0·00049
28·750	−0·00060	−0·03477	0·00048	0·02213
29·000	−0·03447	0·00059	0·02194	−0·00047
29·250	0·00058	0·03417	−0·00047	−0·02175
29·500	0·03388	−0·00057	−0·02157	0·00046
29·750	−0·00056	−0·03360	0·00045	0·02139
30·000	−0·03332	0·00055	0·02121	−0·00044
30·250	0·00054	0·03304	−0·00044	−0·02103
30·500	0·03277	−0·00053	−0·02086	0·00043
30·750	−0·00052	−0·03251	0·00042	0·02069
31·000	−0·03225	0·00052	0·02052	−0·00041
31·250	0·00051	0·03199	−0·00041	−0·02036
31·500	0·03173	−0·00050	−0·02020	0·00040
31·750	−0·00049	−0·03148	0·00040	0·02004
32·000	−0·03124	0·00048	0·01988	−0·00039
32·250	0·00048	0·03100	−0·00038	−0·01973
32·500	0·03076	−0·00047	−0·01958	0·00038
32·750	−0·00046	−0·03052	0·00037	0·01943
33·000	−0·03029	0·00045	0·01928	−0·00037
33·250	0·00045	0·03007	−0·00036	−0·01914
33·500	0·02984	−0·00044	−0·01899	0·00035
33·750	−0·00043	−0·02962	0·00035	0·01885
34·000	−0·02940	0·00043	0·01871	−0·00034
34·250	0·00042	0·02919	−0·00034	−0·01858
34·500	0·02898	−0·00042	−0·01844	0·00033
34·750	−0·00041	−0·02877	0·00033	0·01831
35·000	−0·02856	0·00040	0·01818	−0·00033
35·250	0·00040	0·02836	−0·00032	−0·01805
35·500	0·02816	−0·00039	−0·01792	0·00032
35·750	−0·00039	−0·02796	0·00031	0·01780
36·000	−0·02777	0·00038	0·01768	−0·00031
36·250	0·00038	0·02758	−0·00030	−0·01755
36·500	0·02739	−0·00037	−0·01743	0·00030

448 APPENDIX III

Table of elements of Φ_u and Φ_v matrices for $h/\lambda = 0.50$
Values for single element (Φ_{kku}, Φ_{kkv} and $\Psi_{kkdR} = \Psi_{dR}$)—contd.

$(k-i)(b/\lambda)$	Φ_{kiu}		Φ_{kiv}	
36·750	−0·00037	−0·02720	0·00029	0·01732
37·000	−0·02702	0·00036	0·01720	−0·00029
37·250	0·00036	0·02684	−0·00029	−0·01708
37·500	0·02666	−0·00035	−0·01697	0·00028
37·750	−0·00035	−0·02648	0·00028	0·01686
38·000	−0·02631	0·00034	0·01675	−0·00028
38·250	0·00034	0·02614	−0·00027	−0·01664
38·500	0·02597	−0·00033	−0·01653	0·00027
38·750	−0·00033	−0·02580	0·00027	0·01642
39·000	−0·02563	0·00033	0·01632	−0·00026
39·250	0·00032	0·02547	−0·00026	−0·01621
39·500	0·02531	−0·00032	−0·01611	0·00026
39·750	−0·00031	−0·02515	0·00025	0·01601
40·000	−0·02499	0·00031	0·01591	−0·00025
40·250	0·00031	0·02484	−0·00025	−0·01581
40·500	0·02469	−0·00030	−0·01571	0·00024
40·750	−0·00030	−0·02453	0·00024	0·01562
41·000	−0·02438	0·00029	0·01552	−0·00024
41·250	0·00029	0·02424	−0·00023	−0·01543
41·500	0·02409	−0·00029	−0·01534	0·00023
41·750	−0·00028	−0·02395	0·00023	0·01524
42·000	−0·02380	0·00028	0·01515	−0·00023
42·250	0·00028	0·02366	−0·00022	−0·01506
42·500	0·02352	−0·00027	−0·01497	0·00022
42·750	−0·00027	−0·02339	0·00022	0·01489
43·000	−0·02325	0·00027	0·01480	−0·00022
43·250	0·00026	0·02312	−0·00021	−0·01472
43·500	0·02298	−0·00026	−0·01463	0·00021
43·750	−0·00026	−0·02285	0·00021	0·01455
44·000	−0·02272	0·00026	0·01446	−0·00021
44·250	0·00025	0·02259	−0·00020	−0·01438
44·500	0·02247	−0·00025	−0·01430	0·00020
44·750	−0·00025	−0·02234	0·00020	0·01422
45·000	−0·02222	0·00024	0·01414	−0·00020
45·250	0·00024	0·02210	−0·00019	−0·01407
45·500	0·02197	−0·00024	−0·01399	0·00019
45·750	−0·00024	−0·02185	0·00019	0·01391
46·000	−0·02174	0·00023	0·01384	−0·00019
46·250	0·00023	0·02162	−0·00019	−0·01376
46·500	0·02150	−0·00023	−0·01369	0·00018
46·750	−0·00023	−0·02139	0·00018	0·01361
47·000	−0·02127	0·00022	0·01354	−0·00018
47·250	0·00022	0·02116	−0·00018	−0·01347
47·500	0·02105	−0·00022	−0·01340	0·00018
47·750	−0·00022	−0·02094	0·00017	0·01333
48·000	−0·02083	0·00021	0·01326	−0·00017
48·250	0·00021	0·02072	−0·00017	−0·01319
48·500	0·02062	−0·00021	−0·01312	0·00017
48·750	−0·00021	−0·02051	0·00017	0·01306
49·000	−0·02041	0·00021	0·01299	−0·00017
49·250	0·00020	0·02030	−0·00016	−0·01292
49·500	0·02020	−0·00020	−0·01286	0·00016
49·750	−0·00020	−0·02010	0·00016	0·01279
50·000	−0·02000	0·00020	0·01273	−0·00016

Table of elements of Φ_u and Φ_v matrices for $h/\lambda = 0.625$
Values for single element (Φ_{kku}, Φ_{kkv} and $\Psi_{kkdR} = \Psi_{dR}$)

Ω	Ψ_{dR}	Φ_{kku}		Φ_{kkv}	
7·00	2·81599	−2·06379	1·28068	0·83173	−1·02685
7·50	3·26383	−2·40814	1·28795	0·88239	−1·03206
8·00	3·72310	−2·75320	1·29236	0·92221	−1·03523
8·50	4·19134	−3·09919	1·29505	0·95343	−1·03716
9·00	4·66659	−3·44620	1·29668	0·97785	−1·03833
9·50	5·14730	−3·79420	1·29766	0·99694	−1·03904
10·00	5·63227	−4·14312	1·29826	1·01185	−1·03947
10·50	6·12057	−4·49286	1·29863	1·02348	−1·03973
11·00	6·61145	−4·84330	1·29885	1·03255	−1·03989
11·50	7·10435	−5·19435	1·29898	1·03963	−1·03998
12·00	7·59882	−5·54589	1·29906	1·04514	−1·04004
12·50	8·09452	−5·89783	1·29911	1·04944	−1·04008
13·00	8·59116	−6·25011	1·29914	1·05278	−1·04010
15·00	10·58370	−7·66140	1·29918	1·06024	−1·04013

Off diagonal values of Φ_u and Φ_v (see note on page 424)

$(k-i)(b/\lambda)$	Φ_{kiu}		Φ_{kiv}	
0·250	0·45164	0·59842	−0·56525	−0·53538
0·500	0·42851	−0·45552	−0·40583	0·24351
0·750	−0·38272	−0·42507	0·17624	0·28315
1·000	−0·41223	0·29489	0·23084	−0·16312
1·250	0·22288	0·38447	−0·15177	−0·21024
1·500	0·35176	−0·16988	−0·19944	0·13674
1·750	−0·13186	−0·31989	0·12048	0·19073
2·000	−0·29107	0·10445	0·18209	−0·10500
2·250	0·08436	0·26579	−0·09121	−0·17329
2·500	0·24383	−0·06935	−0·16453	0·07932
2·750	−0·05790	−0·22480	0·06922	0·15603
3·000	−0·20826	0·04901	0·14794	−0·06068
3·250	0·04198	0·19382	−0·05348	−0·14033
3·500	0·18113	−0·03635	−0·13324	0·04738
3·750	−0·03176	−0·16993	0·04221	0·12666
4·000	−0·15997	0·02798	0·12058	−0·03778
4·250	0·02484	0·15108	−0·03399	−0·11496
4·500	0·14310	−0·02219	−0·10977	0·03071
4·750	−0·01994	−0·13589	0·02787	0·10497
5·000	−0·12936	0·01801	0·10053	−0·02539
5·250	0·01635	0·12342	−0·02322	−0·09641
5·500	0·11799	−0·01491	−0·09259	0·02131
5·750	−0·01365	−0·11300	0·01962	0·08904
6·000	−0·10842	0·01255	0·08574	−0·01812
6·250	0·01157	0·10419	−0·01678	−0·08265
6·500	0·10027	−0·01070	−0·07977	0·01558
6·750	−0·00993	−0·09663	0·01450	0·07707
7·000	−0·09325	0·00923	0·07454	−0·01353
7·250	0·00861	0·09009	−0·01266	−0·07217
7·500	0·08714	−0·00805	−0·06993	0·01186
7·750	−0·00754	−0·08437	0·01114	0·06783
8·000	−0·08177	0·00707	0·06584	−0·01047
8·250	0·00665	0·07933	−0·00987	−0·06396
8·500	0·07702	−0·00627	−0·06218	0·00932
8·750	−0·00592	−0·07485	0·00881	0·06050

Table of elements of Φ_u and Φ_v matrices for $h/\lambda = 0.625$
Values for single element (Φ_{kku}, Φ_{kkv} and $\Psi_{kkdR} = \Psi_{dR}$)—contd.

$(k-i)(b/\lambda)$	Φ_{kiu}		Φ_{kiv}	
9·000	−0·07279	0·00559	0·05890	−0·00834
9·250	0·00530	0·07085	−0·00791	−0·05739
9·500	0·06900	−0·00502	−0·05594	0·00751
9·750	−0·00477	−0·06725	0·00713	0·05457
10·000	−0·06558	0·00453	0·05326	−0·00679
10·250	0·00431	0·06400	−0·00647	−0·05201
10·500	0·06249	−0·00411	−0·05082	0·00617
10·750	−0·00392	−0·06105	0·00589	0·04968
11·000	−0·05967	0·00375	0·04859	−0·00563
11·250	0·00358	0·05835	−0·00539	−0·04754
11·500	0·05709	−0·00343	−0·04654	0·00516
11·750	−0·00328	−0·05589	0·00495	0·04558
12·000	−0·05473	0·00315	0·04465	−0·00475
12·250	0·00302	0·05362	−0·00456	−0·04377
12·500	0·05255	−0·00290	−0·04291	0·00438
12·750	−0·00279	−0·05153	0·00421	0·04209
13·000	−0·05054	0·00268	0·04130	−0·00406
13·250	0·00258	0·04959	−0·00391	−0·04054
13·500	0·04868	−0·00249	−0·03981	0·00376
13·750	−0·00240	−0·04780	0·00363	0·03910
14·000	−0·04695	0·00231	0·03842	−0·00350
14·250	0·00223	0·04613	−0·00338	−0·03776
14·500	0·04534	−0·00216	−0·03712	0·00327
14·750	−0·00209	−0·04457	0·00316	0·03650
15·000	−0·04383	0·00202	0·03590	−0·00306
15·250	0·00195	0·04312	−0·00296	−0·03532
15·500	0·04242	−0·00189	−0·03476	0·00286
15·750	−0·00183	−0·04175	0·00277	0·03422
16·000	−0·04110	0·00177	0·03370	−0·00269
16·250	0·00172	0·04047	−0·00261	−0·03318
16·500	0·03986	−0·00167	−0·03269	0·00253
16·750	−0·00162	−0·03927	0·00246	0·03221
17·000	−0·03869	0·00157	0·03174	−0·00238
17·250	0·00153	0·03813	−0·00232	−0·03129
17·500	0·03759	−0·00148	−0·03085	0·00225
17·750	−0·00144	−0·03706	0·00219	0·03042
18·000	−0·03655	0·00140	0·03000	−0·00213
18·250	0·00136	0·03605	−0·00207	−0·02959
18·500	0·03556	−0·00133	−0·02920	0·00202
18·750	−0·00129	−0·03509	0·00196	0·02881
19·000	−0·03463	0·00126	0·02844	−0·00191
19·250	0·00122	0·03418	−0·00186	−0·02807
19·500	0·03375	−0·00119	−0·02772	0·00182
19·750	−0·00116	−0·03332	0·00177	0·02737
20·000	−0·03290	0·00113	0·02703	−0·00173
20·250	0·00111	0·03250	−0·00168	−0·02670
20·500	0·03210	−0·00108	−0·02638	0·00164
20·750	−0·00105	−0·03172	0·00160	0·02606
21·000	−0·03134	0·00103	0·02576	−0·00157
21·250	0·00101	0·03097	−0·00153	−0·02546
21·500	0·03061	−0·00098	−0·02516	0·00150
21·750	−0·00096	−0·03026	0·00146	0·02488
22·000	−0·02992	0·00094	0·02460	−0·00143
22·250	0·00092	0·02958	−0·00140	−0·02432
22·500	0·02925	−0·00090	−0·02405	0·00137

Table of elements of Φ_u and Φ_v matrices for $h/\lambda = 0.625$
Values for single element (Φ_{kku}, Φ_{kkv} and $\Psi_{kkdR} = \Psi_{dR}$)—contd.

$(k-i)(b/\lambda)$	Φ_{kiu}		Φ_{kiv}	
22·750	−0·00088	−0·02893	0·00134	0·02379
23·000	−0·02862	0·00086	0·02354	−0·00131
23·250	0·00084	0·02831	−0·00128	−0·02328
23·500	0·02801	−0·00082	−0·02304	0·00125
23·750	−0·00080	−0·02772	0·00123	0·02280
24·000	−0·02743	0·00079	0·02256	−0·00120
24·250	0·00077	0·02715	−0·00118	−0·02233
24·500	0·02687	−0·00076	−0·02210	0·00115
24·750	−0·00074	−0·02660	0·00113	0·02188
25·000	−0·02633	0·00073	0·02167	−0·00111
25·250	0·00071	0·02607	−0·00109	−0·02145
25·500	0·02582	−0·00070	−0·02124	0·00106
25·750	−0·00068	−0·02557	0·00104	0·02104
26·000	−0·02532	0·00067	0·02084	−0·00102
26·250	0·00066	0·02508	−0·00100	−0·02064
26·500	0·02484	−0·00065	−0·02045	0·00099
26·750	−0·00063	−0·02461	0·00097	0·02026
27·000	−0·02439	0·00062	0·02007	−0·00095
27·250	0·00061	0·02416	−0·00093	−0·01989
27·500	0·02394	−0·00060	−0·01971	0·00092
27·750	−0·00059	−0·02373	0·00090	0·01953
28·000	−0·02352	0·00058	0·01936	−0·00088
28·250	0·00057	0·02331	−0·00087	−0·01919
28·500	0·02310	−0·00056	−0·01902	0·00085
28·750	−0·00055	−0·02290	0·00084	0·01885
29·000	−0·02271	0·00054	0·01869	−0·00082
29·250	0·00053	0·02251	−0·00081	−0·01853
29·500	0·02232	−0·00052	−0·01838	0·00080
29·750	−0·00051	−0·02213	0·00078	0·01822
30·000	−0·02195	0·00050	0·01807	−0·00077
30·250	0·00050	0·02177	−0·00076	−0·01792
30·500	0·02159	−0·00049	−0·01778	0·00074
30·750	−0·00048	−0·02141	0·00073	0·01763
31·000	−0·02124	0·00047	0·01749	−0·00072
31·250	0·00046	0·02107	−0·00071	−0·01735
31·500	0·02091	−0·00046	−0·01722	0·00070
31·750	−0·00045	−0·02074	0·00069	0·01708
32·000	−0·02058	0·00044	0·01695	−0·00068
32·250	0·00044	0·02042	−0·00067	−0·01682
32·500	0·02026	−0·00043	−0·01669	0·00066
32·750	−0·00042	−0·02011	0·00065	0·01656
33·000	−0·01996	0·00042	0·01644	−0·00064
33·250	0·00041	0·01981	−0·00063	−0·01631
33·500	0·01966	−0·00040	−0·01619	0·00062
33·750	−0·00040	−0·01951	0·00061	0·01607
34·000	−0·01937	0·00039	0·01595	−0·00060
34·250	0·00039	0·01923	−0·00059	−0·01584
34·500	0·01909	−0·00038	−0·01572	0·00058
34·750	−0·00038	−0·01895	0·00057	0·01561
35·000	−0·01882	0·00037	0·01550	−0·00057
35·250	0·00037	0·01868	−0·00056	−0·01539
35·500	0·01855	−0·00036	−0·01528	0·00055
35·750	−0·00036	−0·01842	0·00054	0·01518
36·000	−0·01829	0·00035	0·01507	−0·00054
36·250	0·00035	0·01817	−0·00053	−0·01497

Table of elements of Φ_u and Φ_v matrices for $h/\lambda = 0.625$
Values for single element (Φ_{kku}, Φ_{kkv} and $\Psi_{kkdR} = \Psi_{dR}$)—contd.

$(k-i)(b/\lambda)$	Φ_{kiu}		Φ_{kiv}	
36·500	0·01804	−0·00034	−0·01486	0·00052
36·750	−0·00034	−0·01792	0·00051	0·01476
37·000	−0·01780	0·00033	0·01466	−0·00051
37·250	0·00033	0·01768	−0·00050	−0·01457
37·500	0·01756	−0·00032	−0·01447	0·00049
37·750	−0·00032	−0·01745	0·00049	0·01437
38·000	−0·01733	0·00031	0·01428	−0·00048
38·250	0·00031	0·01722	−0·00047	−0·01419
38·500	0·01711	−0·00031	−0·01409	0·00047
38·750	−0·00030	−0·01700	0·00046	0·01400
39·000	−0·01689	0·00030	0·01391	−0·00046
39·250	0·00029	0·01678	−0·00045	−0·01383
39·500	0·01667	−0·00029	−0·01374	0·00044
39·750	−0·00029	−0·01657	0·00044	0·01365
40·000	−0·01647	0·00028	0·01357	−0·00043
40·250	0·00028	0·01636	−0·00043	−0·01348
40·500	0·01626	−0·00028	−0·01340	0·00042
40·750	−0·00027	−0·01616	0·00042	0·01332
41·000	−0·01606	0·00027	0·01324	−0·00041
41·250	0·00027	0·01597	−0·00041	−0·01316
41·500	0·01587	−0·00026	−0·01308	0·00040
41·750	−0·00026	−0·01578	0·00040	0·01300
42·000	−0·01568	0·00026	0·01292	−0·00039
42·250	0·00025	0·01559	−0·00039	−0·01285
42·500	0·01550	−0·00025	−0·01277	0·00038
42·750	−0·00025	−0·01541	0·00038	0·01270
43·000	−0·01532	0·00025	0·01262	−0·00038
43·250	0·00024	0·01523	−0·00037	−0·01255
43·500	0·01514	−0·00024	−0·01248	0·00037
43·750	−0·00024	−0·01505	0·00036	0·01241
44·000	−0·01497	0·00023	0·01234	−0·00036
44·250	0·00023	0·01488	−0·00035	−0·01227
44·500	0·01480	−0·00023	−0·01220	0·00035
44·750	−0·00023	−0·01472	0·00035	0·01213
45·000	−0·01464	0·00022	0·01206	−0·00034
45·250	0·00022	0·01456	−0·00034	−0·01200
45·500	0·01448	−0·00022	−0·01193	0·00034
45·750	−0·00022	−0·01440	0·00033	0·01187
46·000	−0·01432	0·00021	0·01180	−0·00033
46·250	0·00021	0·01424	−0·00032	−0·01174
46·500	0·01416	−0·00021	−0·01167	0·00032
46·750	−0·00021	−0·01409	0·00032	0·01161
47·000	−0·01401	0·00021	0·01155	−0·00031
47·250	0·00020	0·01394	−0·00031	−0·01149
47·500	0·01387	−0·00020	−0·01143	0·00031
47·750	−0·00020	−0·01379	0·00030	0·01137
48·000	−0·01372	0·00020	0·01131	−0·00030
48·250	0·00019	0·01365	−0·00030	−0·01125
48·500	0·01358	−0·00019	−0·01119	0·00030
48·750	−0·00019	−0·01351	0·00029	0·01114
49·000	−0·01344	0·00019	0·01108	−0·00029
49·250	0·00019	0·01337	−0·00029	−0·01102
49·500	0·01331	−0·00019	−0·01097	0·00028
49·750	−0·00018	−0·01324	0·00028	0·01091
50·000	−0·01317	0·00018	0·01086	−0·00028

APPENDIX IV

Tables of admittance and impedance curtain arrays

This table is abridged from the report 'Tables for Curtain Arrays' by Ronold W. P. King, Barbara H. Sandler and Sheldon S. Sandler, Cruft Laboratory Scientific Report No. 4 (Series 3), Harvard University, May 1964.

The calculations for the individual elements are given for $\Omega = 2 \ln 2h/a = 8{\cdot}6138$ for $\beta_0 h = \pi/4$ and for $\Omega = 10$ for all other electrical lengths. The admittances are given in millimhos and the impedances in ohms. The vertical listings begin with the first element at the top. The unilateral endfire patterns are prescribed to point in the direction away from the first element toward the last element.

Broadside array (driving-point currents specified)

$$\beta_0 h = 0.78539$$

Admittance $\beta_0 b = 1.57080$	Impedance	Admittance $\beta_0 b = 3.14159$	Impedance
	$N = 1$		$N = 1$
$0.122 + j3.250$	$11.564 - j307.244$	$0.122 + j3.250$	$11.564 - j307.244$
	$N = 4$		$N = 4$
$0.123 + j3.135$	$12.532 - j318.521$	$0.100 + j3.204$	$9.690 - j311.824$
$0.223 + j3.047$	$23.894 - j326.465$	$0.078 + j3.170$	$7.786 - j315.273$
$0.223 + j3.047$	$23.894 - j326.465$	$0.078 + j3.170$	$7.786 - j315.273$
$0.123 + j3.135$	$12.532 - j318.521$	$0.100 + j3.204$	$9.690 - j311.824$
	$N = 10$		$N = 10$
$0.152 + j3.154$	$15.277 - j316.276$	$0.101 + j3.210$	$9.786 - j311.261$
$0.202 + j3.084$	$21.113 - j322.911$	$0.076 + j3.161$	$7.615 - j316.127$
$0.170 + j3.020$	$18.620 - j330.103$	$0.083 + j3.185$	$8.207 - j313.805$
$0.151 + j3.030$	$16.446 - j329.235$	$0.080 + j3.172$	$7.963 - j315.015$
$0.166 + j3.078$	$17.493 - j323.927$	$0.081 + j3.178$	$8.058 - j314.481$
$0.166 + j3.078$	$17.493 - j323.927$	$0.081 + j3.178$	$8.058 - j314.481$
$0.151 + j3.030$	$16.446 - j329.235$	$0.080 + j3.172$	$7.963 - j315.015$
$0.170 + j3.020$	$18.620 - j330.103$	$0.083 + j3.185$	$8.207 - j313.805$
$0.202 + j3.084$	$21.113 - j322.911$	$0.076 + j3.161$	$7.615 - j316.127$
$0.152 + j3.154$	$15.277 - j316.276$	$0.101 + j3.210$	$9.786 - j311.261$
	$N = 20$		$N = 20$
$0.140 + j3.146$	$14.158 - j317.252$	$0.101 + j3.211$	$9.804 - j311.079$
$0.210 + j3.071$	$22.183 - j324.119$	$0.076 + j3.159$	$7.599 - j316.331$
$0.184 + j3.029$	$19.973 - j328.914$	$0.084 + j3.187$	$8.233 - j313.549$
$0.141 + j3.045$	$15.134 - j327.710$	$0.080 + j3.169$	$7.934 - j315.316$
$0.148 + j3.066$	$15.754 - j325.408$	$0.082 + j3.182$	$8.106 - j314.096$
$0.180 + j3.057$	$19.217 - j325.963$	$0.081 + j3.173$	$7.998 - j314.970$
$0.177 + j3.046$	$18.957 - j327.166$	$0.082 + j3.179$	$8.071 - j314.345$
$0.150 + j3.053$	$16.056 - j326.795$	$0.081 + j3.175$	$8.024 - j314.780$
$0.151 + j3.058$	$16.158 - j326.263$	$0.081 + j3.178$	$8.052 - j314.506$
$0.176 + j3.052$	$18.842 - j326.573$	$0.081 + j3.176$	$8.039 - j314.639$
$0.176 + j3.052$	$18.842 - j326.573$	$0.081 + j3.176$	$8.039 - j314.639$
$0.151 + j3.058$	$16.158 - j326.263$	$0.081 + j3.178$	$8.052 - j314.506$
$0.150 + j3.053$	$16.056 - j326.795$	$0.081 + j3.175$	$8.024 - j314.780$
$0.177 + j3.046$	$18.957 - j327.166$	$0.082 + j3.179$	$8.071 - j314.345$
$0.180 + j3.057$	$19.217 - j325.963$	$0.081 + j3.173$	$7.998 - j314.970$
$0.148 + j3.066$	$15.754 - j325.408$	$0.082 + j3.182$	$8.106 - j314.096$
$0.141 + j3.045$	$15.134 - j327.710$	$0.080 + j3.169$	$7.934 - j315.316$
$0.184 + j3.029$	$19.973 - j328.914$	$0.084 + j3.187$	$8.233 - j313.549$
$0.210 + j3.071$	$22.183 - j324.119$	$0.076 + j3.159$	$7.599 - j316.331$
$0.140 + j3.146$	$14.158 - j317.252$	$0.101 + j3.211$	$9.804 - j311.079$

Broadside array (base voltages specified)

$$\beta_0 h = 0.78539$$

Admittance $\beta_0 b = 1.57080$	Impedance	Admittance $\beta_0 b = 3.14159$	Impedance
N = 1		*N* = 1	
0·122 + *j*3·250	11·564 − *j*307·244	0·122 + *j*3·250	11·564 − *j*307·244
N = 4		*N* = 4	
0·118 + *j*3·137	11·937 − *j*318·359	0·100 + *j*3·204	9·724 − *j*311·807
0·229 + *j*3·045	24·512 − *j*326·608	0·078 + *j*3·170	7·753 − *j*315·290
0·229 + *j*3·045	24·512 − *j*326·608	0·078 + *j*3·170	7·753 − *j*315·290
0·118 + *j*3·137	11·937 − *j*318·359	0·100 + *j*3·204	9·724 − *j*311·807
N = 10		*N* = 10	
0·153 + *j*3·157	15·298 − *j*316·064	0·102 + *j*3·210	9·848 − *j*311·209
0·204 + *j*3·087	21·338 − *j*322·502	0·075 + *j*3·161	7·517 − *j*316·212
0·168 + *j*3·014	18·457 − *j*330·739	0·084 + *j*3.185	8·263 − *j*313·749
0·148 + *j*3·026	16·125 − *j*329·718	0·080 + *j*3·172	7·931 − *j*315·051
0·169 + *j*3·083	17·702 − *j*323·419	0·082 + *j*3·178	8·068 − *j*314·470
0·169 + *j*3·083	17·702 − *j*323·419	0·082 + *j*3·178	8·068 − *j*314·470
0·148 + *j*3·026	16·125 − *j*329·718	0·080 + *j*3·172	7·931 − *j*315·051
0·168 + *j*3·014	18·457 − *j*330·739	0·084 + *j*3·185	8·263 − *j*313·749
0·204 + *j*3·087	21·338 − *j*322·502	0·075 + *j*3·161	7·517 − *j*316·212
0·153 + *j*3·157	15·298 − *j*316·064	0·102 + *j*3·210	9·848 − *j*311·209
N = 20		*N* = 20	
0·137 + *j*3·147	13·778 − *j*317·153	0·102 + *j*3·211	9·844 − *j*311·084
0·215 + *j*3·071	22·654 − *j*324·092	0·074 + *j*3·158	7·449 − *j*316·530
0·187 + *j*3·025	20·346 − *j*329·273	0·084 + *j*3·187	8·278 − *j*313·535
0·136 + *j*3·046	14·591 − *j*327·645	0·079 + *j*3·168	7·842 − *j*315·477
0·146 + *j*3·069	15·423 − *j*325·119	0·082 + *j*3·181	8·116 − *j*314·110
0·184 + *j*3·056	19·673 − *j*326·009	0·080 + *j*3·172	7·932 − *j*315·103
0·180 + *j*3·044	19·327 − *j*327·375	0·082 + *j*3·179	8·064 − *j*314·379
0·146 + *j*3·054	15·636 − *j*326·695	0·080 + *j*3·174	7·966 − *j*314·901
0·148 + *j*3·059	15·791 − *j*326·090	0·081 + *j*3·178	8·062 − *j*314·490
0·180 + *j*3·050	19·235 − *j*326·688	0·081 + *j*3·176	8·036 − *j*314·645
0·180 + *j*3·050	19·235 − *j*326·688	0·081 + *j*3·176	8·036 − *j*314·645
0·148 + *j*3·059	15·791 − *j*326·090	0·081 + *j*3·178	8·062 − *j*314·490
0·146 + *j*3·054	15·636 − *j*326·695	0·080 + *j*3·174	7·966 − *j*314·901
0·180 + *j*3·044	19·327 − *j*327·375	0·082 + *j*3·179	8·064 − *j*314·379
0·184 + *j*3·056	19·673 − *j*326·009	0·080 + *j*3·172	7·932 − *j*315·103
0·146 + *j*3·069	15·423 − *j*325·119	0·082 + *j*3·181	8·116 − *j*314·110
0·136 + *j*3·046	14·591 − *j*327·645	0·079 + *j*3·168	7·842 − *j*315·477
0·187 + *j*3·025	20·346 − *j*329·273	0·084 + *j*3·187	8·278 − *j*313·535
0·215 + *j*3·071	22·654 − *j*324·092	0·074 + *j*3·158	7·449 − *j*316·530
0·137 + *j*3·147	13·778 − *j*317·153	0·102 + *j*3·211	9·844 − *j*311·084

Endfire array (driving-point current specified)

$$\beta_0 h = 0.78539$$

Admittance $\beta_0 b = 1.57080$	Impedance	Admittance $\beta_0 b = 3.14159$	Impedance
N = 1		*N = 1*	
$0.122 + j3.250$	$11.564 - j307.244$	$0.122 + j3.250$	$11.564 - j307.244$
N = 4		*N = 4*	
$0.059 + j3.193$	$5.778 - j313.123$	$0.168 + j3.367$	$14.796 - j296.304$
$0.151 + j3.309$	$13.761 - j301.623$	$0.195 + j3.404$	$16.735 - j292.822$
$0.150 + j3.308$	$13.659 - j301.691$	$0.195 + j3.404$	$16.735 - j292.822$
$0.257 + j3.427$	$21.769 - j290.196$	$0.168 + j3.367$	$14.796 - j296.304$
N = 10		*N = 10*	
$0.063 + j3.203$	$6.120 - j312.060$	$0.183 + j3.442$	$15.431 - j289.678$
$0.142 + j3.292$	$13.121 - j303.166$	$0.217 + j3.498$	$17.657 - j284.743$
$0.169 + j3.333$	$15.135 - j299.277$	$0.228 + j3.526$	$18.267 - j282.387$
$0.192 + j3.399$	$16.544 - j293.286$	$0.233 + j3.541$	$18.520 - j281.159$
$0.196 + j3.409$	$16.847 - j292.350$	$0.235 + j3.548$	$18.619 - j280.616$
$0.215 + j3.469$	$17.814 - j287.133$	$0.235 + j3.548$	$18.619 - j280.616$
$0.209 + j3.457$	$17.416 - j288.192$	$0.233 + j3.541$	$18.520 - j281.159$
$0.236 + j3.528$	$18.875 - j282.157$	$0.228 + j3.526$	$18.267 - j282.387$
$0.203 + j3.482$	$16.706 - j286.198$	$0.217 + j3.498$	$17.657 - j284.743$
$0.309 + j3.580$	$23.931 - j277.234$	$0.183 + j3.442$	$15.431 - j289.678$
N = 20		*N = 20*	
$0.064 + j3.207$	$6.188 - j311.699$	$0.193 + j3.500$	$15.723 - j284.807$
$0.141 + j3.288$	$13.044 - j303.581$	$0.229 + j3.564$	$17.992 - j279.463$
$0.170 + j3.338$	$15.223 - j298.777$	$0.243 + j3.599$	$18.651 - j276.631$
$0.189 + j3.392$	$16.416 - j293.886$	$0.250 + j3.621$	$18.964 - j274.831$
$0.199 + j3.418$	$17.012 - j291.597$	$0.255 + j3.637$	$19.145 - j273.590$
$0.211 + j3.458$	$17.564 - j288.089$	$0.258 + j3.649$	$19.258 - j272.700$
$0.215 + j3.472$	$17.802 - j286.910$	$0.260 + j3.657$	$19.335 - j272.063$
$0.224 + j3.507$	$18.174 - j283.963$	$0.261 + j3.663$	$19.385 - j271.619$
$0.226 + j3.513$	$18.255 - j283.463$	$0.262 + j3.667$	$19.415 - j271.339$
$0.235 + j3.547$	$18.571 - j280.729$	$0.263 + j3.668$	$19.428 - j271.203$
$0.234 + j3.546$	$18.548 - j280.770$	$0.263 + j3.668$	$19.428 - j271.203$
$0.243 + j3.580$	$18.869 - j278.028$	$0.262 + j3.667$	$19.415 - j271.339$
$0.240 + j3.573$	$18.740 - j278.613$	$0.261 + j3.663$	$19.385 - j271.619$
$0.251 + j3.611$	$19.130 - j275.641$	$0.260 + j3.657$	$19.335 - j272.063$
$0.245 + j3.595$	$18.836 - j276.916$	$0.258 + j3.649$	$19.258 - j272.700$
$0.259 + j3.640$	$19.423 - j273.359$	$0.255 + j3.637$	$19.145 - j273.590$
$0.245 + j3.610$	$18.749 - j275.767$	$0.250 + j3.621$	$18.964 - j274.831$
$0.272 + j3.673$	$20.021 - j270.784$	$0.243 + j3.599$	$18.651 - j276.631$
$0.232 + j3.611$	$17.682 - j275.761$	$0.229 + j3.564$	$17.992 - j279.463$
$0.344 + j3.706$	$24.834 - j267.513$	$0.193 + j3.500$	$15.723 - j284.807$

Endfire array (*base voltages specified*)

$$\beta_0 h = 0.78539$$

Admittance $\beta_0 b = 1.57080$	Impedance	Admittance $\beta_0 b = 3.14159$	Impedance
N = 1		**N = 1**	
$0.122 + j3.250$	$11.564 - j307.244$	$0.122 + j3.250$	$11.564 - j307.244$
N = 4		**N = 4**	
$0.053 + j3.191$	$5.215 - j313.290$	$0.169 + j3.367$	$14.886 - j296.227$
$0.149 + j3.309$	$13.563 - j301.579$	$0.193 + j3.403$	$16.646 - j292.899$
$0.142 + j3.306$	$12.988 - j301.907$	$0.193 + j3.403$	$16.646 - j292.899$
$0.245 + j3.423$	$20.806 - j290.646$	$0.169 + j3.367$	$14.886 - j296.226$
N = 10		**N = 10**	
$0.059 + j3.202$	$5.753 - j312.223$	$0.187 + j3.446$	$15.675 - j289.334$
$0.140 + j3.294$	$12.839 - j303.024$	$0.217 + j3.499$	$17.654 - j284.681$
$0.161 + j3.329$	$14.500 - j299.689$	$0.227 + j3.526$	$18.207 - j282.464$
$0.187 + j3.396$	$16.167 - j293.555$	$0.232 + j3.540$	$18.439 - j281.305$
$0.189 + j3.403$	$16.230 - j292.965$	$0.234 + j3.546$	$18.531 - j280.791$
$0.210 + j3.464$	$17.472 - j287.643$	$0.234 + j3.546$	$18.531 - j280.791$
$0.201 + j3.450$	$16.837 - j288.855$	$0.232 + j3.540$	$18.439 - j281.305$
$0.230 + j3.520$	$18.527 - j282.918$	$0.227 + j3.526$	$18.207 - j282.464$
$0.198 + j3.477$	$16.294 - j286.701$	$0.217 + j3.499$	$17.654 - j284.681$
$0.293 + j3.571$	$22.792 - j278.154$	$0.187 + j3.446$	$15.675 - j289.334$
N = 20		**N = 20**	
$0.061 + j3.206$	$5.895 - j311.808$	$0.199 + j3.508$	$16.092 - j284.146$
$0.138 + j3.289$	$12.707 - j303.472$	$0.231 + j3.567$	$18.068 - j279.167$
$0.163 + j3.335$	$14.649 - j299.156$	$0.243 + j3.600$	$18.651 - j276.528$
$0.184 + j3.389$	$15.985 - j294.162$	$0.249 + j3.621$	$18.933 - j274.843$
$0.192 + j3.411$	$16.458 - j292.213$	$0.254 + j3.636$	$19.096 - j273.675$
$0.205 + j3.453$	$17.166 - j288.556$	$0.257 + j3.647$	$19.200 - j272.836$
$0.208 + j3.464$	$17.287 - j287.644$	$0.259 + j3.655$	$19.270 - j272.234$
$0.219 + j3.501$	$17.813 - j284.541$	$0.260 + j3.660$	$19.316 - j271.815$
$0.219 + j3.504$	$17.772 - j284.267$	$0.261 + j3.664$	$19.343 - j271.550$
$0.229 + j3.539$	$18.242 - j281.386$	$0.261 + j3.666$	$19.356 - j271.421$
$0.227 + j3.536$	$18.087 - j281.618$	$0.261 + j3.666$	$19.356 - j271.421$
$0.238 + j3.572$	$18.568 - j278.749$	$0.261 + j3.664$	$19.343 - j271.550$
$0.233 + j3.563$	$18.293 - j279.483$	$0.260 + j3.660$	$19.316 - j271.815$
$0.246 + j3.601$	$18.853 - j276.425$	$0.259 + j3.655$	$19.270 - j272.234$
$0.237 + j3.584$	$18.398 - j277.783$	$0.257 + j3.647$	$19.200 - j272.836$
$0.254 + j3.629$	$19.162 - j274.232$	$0.254 + j3.636$	$19.096 - j273.675$
$0.239 + j3.600$	$18.329 - j276.574$	$0.249 + j3.621$	$18.933 - j274.843$
$0.265 + j3.659$	$19.715 - j271.845$	$0.243 + j3.600$	$18.651 - j276.528$
$0.228 + j3.605$	$17.447 - j276.327$	$0.231 + j3.567$	$18.068 - j279.167$
$0.324 + j3.693$	$23.582 - j268.702$	$0.199 + j3.508$	$16.092 - j284.146$

Broadside array (driving-point currents specified)

$$\beta_0 h = 1.57079$$

Admittance $\beta_0 b = 1.57080$	Impedance	Admittance $\beta_0 b = 3.14159$	Impedance
$N = 1$		$N = 1$	
$10.449 - j3.889$	$84.059 + j31.286$	$10.499 - j3.889$	$84.059 + j31.286$
$N = 4$		$N = 4$	
$12.709 + j3.576$	$72.913 - j20.518$	$14.944 - j0.824$	$66.715 + j\ 3.678$
$5.807 + j3.319$	$129.803 - j74.198$	$18.493 + j1.493$	$53.723 - j\ 4.336$
$5.807 + j3.319$	$129.803 - j74.198$	$18.493 + j1.493$	$53.723 - j\ 4.336$
$12.709 + j3.576$	$72.913 - j20.518$	$14.944 - j0.824$	$66.715 + j\ 3.678$
$N = 10$		$N = 10$	
$10.113 + j1.428$	$96.946 - j13.689$	$14.720 - j0.850$	$67.711 + j\ 3.909$
$7.202 + j2.746$	$121.234 - j46.219$	$18.832 + j1.491$	$52.771 - j\ 4.177$
$5.923 + j5.391$	$92.343 - j84.042$	$17.703 + j1.442$	$56.116 - j\ 4.572$
$6.572 + j6.018$	$82.765 - j75.783$	$18.161 + j1.434$	$54.721 - j\ 4.321$
$8.176 + j3.532$	$103.068 - j44.525$	$17.978 + j1.440$	$55.268 - j\ 4.428$
$8.176 + j3.532$	$103.068 - j44.525$	$17.978 + j1.440$	$55.268 - j\ 4.428$
$6.572 + j6.018$	$82.765 - j75.783$	$18.161 + j1.434$	$54.721 - j\ 4.321$
$5.923 + j5.391$	$92.343 - j84.042$	$17.703 + j1.442$	$56.116 - j\ 4.572$
$7.202 + j2.746$	$121.234 - j46.219$	$18.832 + j1.491$	$52.771 - j\ 4.177$
$10.113 + j1.428$	$96.946 - j13.689$	$14.720 - j0.850$	$67.711 + j\ 3.909$
$N = 20$		$N = 20$	
$11.322 + j2.266$	$84.924 - j16.998$	$14.668 - j0.855$	$67.946 + j\ 3.961$
$6.558 + j3.070$	$125.075 - j58.542$	$18.882 + j1.487$	$52.634 - j\ 4.145$
$6.008 + j4.532$	$106.088 - j80.025$	$17.645 + j1.450$	$56.294 - j\ 4.625$
$7.939 + j6.197$	$78.268 - j61.095$	$18.234 + j1.427$	$54.509 - j\ 4.267$
$8.713 + j4.980$	$86.512 - j49.440$	$17.883 + j1.452$	$55.553 - j\ 4.509$
$6.889 + j3.994$	$108.645 - j62.989$	$18.112 + j1.431$	$54.870 - j\ 4.336$
$6.640 + j4.415$	$104.430 - j69.432$	$17.957 + j1.447$	$55.328 - j\ 4.459$
$8.040 + j5.486$	$84.866 - j57.908$	$18.060 + j1.436$	$55.023 - j\ 4.374$
$8.225 + j5.215$	$86.717 - j54.984$	$17.997 + j1.443$	$55.210 - j\ 4.427$
$6.835 + j4.270$	$105.235 - j65.740$	$18.027 + j1.440$	$55.120 - j\ 4.402$
$6.835 + j4.270$	$105.235 - j65.740$	$18.027 + j1.440$	$55.120 - j\ 4.402$
$8.225 + j5.215$	$86.717 - j54.984$	$17.997 + j1.443$	$55.210 - j\ 4.427$
$8.040 + j5.486$	$84.866 - j57.908$	$18.060 + j1.436$	$55.023 - j\ 4.374$
$6.640 + j4.415$	$104.430 - j69.432$	$17.957 + j1.447$	$55.328 - j\ 4.459$
$6.889 + j3.994$	$108.645 - j62.989$	$18.112 + j1.431$	$54.870 - j\ 4.336$
$8.713 + j4.980$	$86.512 - j49.440$	$17.883 + j1.452$	$55.553 - j\ 4.509$
$7.939 + j6.197$	$78.268 - j61.095$	$18.234 + j1.427$	$54.509 - j\ 4.267$
$6.008 + j4.532$	$106.088 - j80.025$	$17.645 + j1.450$	$56.294 - j\ 4.625$
$6.558 + j3.070$	$125.075 - j58.542$	$18.882 + j1.487$	$52.634 - j\ 4.145$
$11.322 + j2.266$	$84.924 - j16.998$	$14.668 - j0.855$	$67.946 + j\ 3.961$

Endfire array (driving-point current specified)

$$\beta_0 h = 1.57079$$

Admittance $\beta_0 b = 1.57080$	Impedance	Admittance $\beta_0 b = 3.14159$	Impedance
$N = 1$		$N = 1$	
$10.449 - j3.889$	$84.059 + j\ 31.286$	$10.449 - j3.889$	$84.059 + j\ 31.286$
$N = 4$		$N = 4$	
$21.021 - j0.091$	$47.571 + j\ 0.205$	$5.122 - j3.827$	$125.295 + j\ 93.607$
$7.128 - j4.119$	$105.169 + j\ 60.777$	$4.399 - j3.396$	$142.437 + j109.966$
$7.400 - j4.432$	$99.455 + j\ 59.563$	$4.399 - j3.396$	$142.437 + j109.966$
$3.999 - j2.972$	$161.068 + j119.715$	$5.122 - j3.827$	$125.295 + j\ 93.607$
$N = 10$		$N = 10$	
$18.995 - j1.976$	$52.082 + j\ 5.419$	$3.533 - j3.361$	$148.595 + j141.338$
$7.821 - j4.265$	$98.548 + j\ 53.747$	$2.986 - j2.896$	$172.544 + j167.362$
$6.462 - j4.040$	$111.269 + j\ 69.563$	$2.754 - j2.736$	$182.750 + j181.551$
$4.552 - j3.546$	$136.734 + j106.504$	$2.640 - j2.661$	$187.898 + j189.374$
$4.458 - j3.576$	$136.500 + j109.499$	$2.591 - j2.629$	$190.138 + j192.923$
$3.461 - j3.112$	$159.762 + j143.670$	$2.591 - j2.629$	$190.138 + j192.923$
$3.638 - j3.296$	$150.965 + j136.780$	$2.640 - j2.661$	$187.898 + j189.374$
$2.863 - j2.783$	$179.606 + j174.555$	$2.754 - j2.736$	$182.750 + j181.551$
$3.232 - j3.222$	$155.171 + j154.712$	$2.986 - j2.896$	$172.544 + j167.362$
$2.450 - j2.360$	$211.705 + j203.950$	$3.533 - j3.361$	$148.595 + j141.338$
$N = 20$		$N = 20$	
$18.250 - j2.610$	$53.695 + j\ 7.678$	$2.755 - j2.976$	$167.509 + j180.951$
$8.076 - j4.268$	$96.793 + j\ 51.152$	$2.366 - j2.566$	$194.207 + j210.621$
$6.252 - j4.000$	$113.486 + j\ 72.618$	$2.175 - j2.404$	$206.910 + j228.769$
$4.693 - j3.602$	$134.084 + j102.908$	$2.060 - j2.313$	$214.737 + j241.090$
$4.300 - j3.500$	$139.895 + j113.871$	$1.984 - j2.254$	$220.090 + j249.939$
$3.591 - j3.196$	$155.387 + j138.288$	$1.932 - j2.213$	$223.909 + j256.432$
$3.467 - j3.172$	$156.992 + j143.647$	$1.895 - j2.184$	$226.647 + j261.167$
$3.019 - j2.926$	$170.781 + j165.519$	$1.870 - j2.165$	$228.549 + j264.491$
$2.993 - j2.945$	$169.756 + j167.026$	$1.855 - j2.152$	$229.753 + j266.612$
$2.659 - j2.729$	$183.170 + j187.995$	$1.847 - j2.146$	$230.338 + j267.646$
$2.683 - j2.778$	$179.903 + j186.225$	$1.847 - j2.146$	$230.338 + j267.646$
$2.405 - j2.573$	$193.846 + j207.450$	$1.855 - j2.152$	$229.753 + j266.612$
$2.465 - j2.651$	$188.140 + j202.324$	$1.870 - j2.165$	$228.549 + j264.491$
$2.211 - j2.443$	$203.633 + j225.004$	$1.895 - j2.184$	$226.647 + j261.167$
$2.305 - j2.556$	$194.603 + j215.766$	$1.932 - j2.213$	$223.909 + j256.432$
$2.053 - j2.325$	$213.437 + j241.665$	$1.984 - j2.254$	$220.090 + j249.939$
$2.190 - j2.497$	$198.569 + j226.323$	$2.060 - j2.313$	$214.737 + j241.090$
$1.912 - j2.196$	$225.537 + j258.968$	$2.175 - j2.404$	$206.910 + j228.769$
$2.119 - j2.521$	$195.412 + j232.494$	$2.366 - j2.566$	$194.207 + j210.621$
$1.787 - j1.965$	$253.380 + j278.536$	$2.755 - j2.976$	$167.509 + j180.951$

Broadside array (driving-point currents specified)

$$\beta_0 h = 2 \cdot 35620$$

Admittance $\beta_0 b = 1 \cdot 57080$	Impedance	Admittance $\beta_0 b = 3 \cdot 14159$	Impedance
\multicolumn{2}{c}{$N = 1$}			
$1 \cdot 416 - j1 \cdot 335$	$373 \cdot 800 + j352 \cdot 429$	$1 \cdot 416 - j1 \cdot 335$	$373 \cdot 800 + j352 \cdot 429$
$N = 4$		$N = 4$	
$1 \cdot 772 - j2 \cdot 166$	$226 \cdot 222 + j276 \cdot 570$	$1 \cdot 688 - j1 \cdot 860$	$267 \cdot 545 + j294 \cdot 791$
$2 \cdot 885 + j0 \cdot 558$	$334 \cdot 125 - j\ 64 \cdot 674$	$1 \cdot 761 - j2 \cdot 489$	$189 \cdot 461 + j267 \cdot 725$
$2 \cdot 885 + j0 \cdot 558$	$334 \cdot 125 - j\ 64 \cdot 674$	$1 \cdot 761 - j2 \cdot 489$	$189 \cdot 461 + j267 \cdot 725$
$1 \cdot 772 - j2 \cdot 166$	$226 \cdot 222 + j276 \cdot 570$	$1 \cdot 688 - j1 \cdot 860$	$267 \cdot 545 + j294 \cdot 791$
$N = 10$		$N = 10$	
$2 \cdot 546 - j1 \cdot 081$	$332 \cdot 848 + j141 \cdot 276$	$1 \cdot 668 - j1 \cdot 738$	$287 \cdot 394 + j299 \cdot 522$
$2 \cdot 273 - j0 \cdot 367$	$428 \cdot 840 + j\ 69 \cdot 202$	$1 \cdot 706 - j2 \cdot 798$	$158 \cdot 797 + j260 \cdot 557$
$4 \cdot 977 - j1 \cdot 391$	$186 \cdot 363 + j\ 52 \cdot 072$	$1 \cdot 767 - j2 \cdot 111$	$233 \cdot 130 + j278 \cdot 516$
$5 \cdot 319 - j2 \cdot 786$	$147 \cdot 544 + j\ 77 \cdot 279$	$1 \cdot 743 - j2 \cdot 477$	$189 \cdot 977 + j270 \cdot 003$
$2 \cdot 547 - j0 \cdot 541$	$375 \cdot 674 + j\ 79 \cdot 816$	$1 \cdot 766 - j2 \cdot 302$	$209 \cdot 859 + j273 \cdot 425$
$2 \cdot 547 - j0 \cdot 541$	$375 \cdot 674 + j\ 79 \cdot 816$	$1 \cdot 766 - j2 \cdot 302$	$209 \cdot 859 + j273 \cdot 425$
$5 \cdot 319 - j2 \cdot 786$	$147 \cdot 544 + j\ 77 \cdot 279$	$1 \cdot 743 - j2 \cdot 477$	$189 \cdot 977 + j270 \cdot 003$
$4 \cdot 977 - j1 \cdot 391$	$186 \cdot 363 + j\ 52 \cdot 072$	$1 \cdot 767 - j2 \cdot 111$	$233 \cdot 130 + j278 \cdot 516$
$2 \cdot 273 - j0 \cdot 367$	$428 \cdot 840 + j\ 69 \cdot 202$	$1 \cdot 706 - j2 \cdot 798$	$158 \cdot 797 + j260 \cdot 557$
$2 \cdot 546 - j1 \cdot 081$	$332 \cdot 848 + j141 \cdot 276$	$1 \cdot 668 - j1 \cdot 738$	$287 \cdot 394 + j299 \cdot 522$
$N = 20$		$N = 20$	
$1 \cdot 768 - j1 \cdot 569$	$316 \cdot 473 + j280 \cdot 864$	$1 \cdot 674 - j1 \cdot 677$	$298 \cdot 066 + j298 \cdot 738$
$3 \cdot 253 + j0 \cdot 018$	$307 \cdot 387 - j\ \ \ 1 \cdot 730$	$1 \cdot 622 - j2 \cdot 887$	$147 \cdot 942 + j263 \cdot 273$
$3 \cdot 757 + j1 \cdot 121$	$244 \cdot 412 - j\ 72 \cdot 928$	$1 \cdot 797 - j2 \cdot 006$	$247 \cdot 732 + j276 \cdot 631$
$2 \cdot 666 - j1 \cdot 466$	$288 \cdot 002 + j158 \cdot 418$	$1 \cdot 660 - j2 \cdot 600$	$174 \cdot 426 + j273 \cdot 219$
$2 \cdot 367 - j1 \cdot 387$	$314 \cdot 522 + j184 \cdot 290$	$1 \cdot 814 - j2 \cdot 142$	$230 \cdot 212 + j271 \cdot 886$
$3 \cdot 422 - j0 \cdot 000$	$292 \cdot 212 + j\ \ \ 0 \cdot 020$	$1 \cdot 690 - j2 \cdot 481$	$187 \cdot 563 + j275 \cdot 284$
$3 \cdot 590 + j0 \cdot 494$	$273 \cdot 339 - j\ 37 \cdot 627$	$1 \cdot 802 - j2 \cdot 227$	$219 \cdot 579 + j271 \cdot 342$
$2 \cdot 689 - j1 \cdot 268$	$304 \cdot 189 + j143 \cdot 435$	$1 \cdot 721 - j2 \cdot 406$	$196 \cdot 649 + j274 \cdot 916$
$2 \cdot 397 - j1 \cdot 327$	$319 \cdot 318 + j176 \cdot 796$	$1 \cdot 779 - j2 \cdot 291$	$211 \cdot 463 + j272 \cdot 272$
$3 \cdot 683 + j0 \cdot 277$	$269 \cdot 994 - j\ 20 \cdot 304$	$1 \cdot 751 - j2 \cdot 347$	$204 \cdot 192 + j273 \cdot 668$
$3 \cdot 683 + j0 \cdot 277$	$269 \cdot 994 - j\ 20 \cdot 304$	$1 \cdot 751 - j2 \cdot 347$	$204 \cdot 192 + j273 \cdot 668$
$2 \cdot 397 - j1 \cdot 327$	$319 \cdot 318 + j176 \cdot 796$	$1 \cdot 779 - j2 \cdot 291$	$211 \cdot 463 + j272 \cdot 272$
$2 \cdot 689 - j1 \cdot 268$	$304 \cdot 189 + j143 \cdot 435$	$1 \cdot 721 - j2 \cdot 406$	$196 \cdot 649 + j274 \cdot 916$
$3 \cdot 590 + j0 \cdot 494$	$273 \cdot 339 - j\ 37 \cdot 627$	$1 \cdot 802 - j2 \cdot 227$	$219 \cdot 579 + j271 \cdot 342$
$3 \cdot 422 - j0 \cdot 000$	$292 \cdot 212 + j\ \ \ 0 \cdot 020$	$1 \cdot 690 - j2 \cdot 481$	$187 \cdot 563 + j275 \cdot 284$
$2 \cdot 367 - j1 \cdot 387$	$314 \cdot 522 + j184 \cdot 290$	$1 \cdot 814 - j2 \cdot 142$	$230 \cdot 212 + j271 \cdot 886$
$2 \cdot 666 - j1 \cdot 466$	$288 \cdot 002 + j158 \cdot 418$	$1 \cdot 660 - j2 \cdot 600$	$174 \cdot 426 + j273 \cdot 219$
$3 \cdot 757 + j1 \cdot 121$	$244 \cdot 412 - j\ 72 \cdot 928$	$1 \cdot 797 - j2 \cdot 006$	$247 \cdot 732 + j276 \cdot 631$
$3 \cdot 253 + j0 \cdot 018$	$307 \cdot 387 - j\ \ \ 1 \cdot 730$	$1 \cdot 622 - j2 \cdot 887$	$147 \cdot 942 + j263 \cdot 273$
$1 \cdot 768 - j1 \cdot 569$	$316 \cdot 473 + j280 \cdot 864$	$1 \cdot 674 - j1 \cdot 677$	$298 \cdot 066 + j298 \cdot 738$

Broadside array (base voltages specified)

$$\beta_0 h = 2\cdot35620$$

Admittance $\beta_0 b = 1\cdot57080$	Impedance	Admittance $\beta_0 b = 3\cdot14159$	Impedance
N = 1		**N = 1**	
$1\cdot416 - j1\cdot335$	$373\cdot800 + j352\cdot429$	$1\cdot416 - j1\cdot335$	$373\cdot800 + j352\cdot429$
N = 4		**N = 4**	
$2\cdot378 - j1\cdot235$	$331\cdot264 + j172\cdot017$	$1\cdot667 - j1\cdot929$	$256\cdot499 + j296\cdot734$
$3\cdot440 - j0\cdot465$	$285\cdot452 + j\ 38\cdot575$	$1\cdot796 - j2\cdot409$	$198\cdot923 + j266\cdot764$
$3\cdot440 - j0\cdot465$	$285\cdot452 + j\ 38\cdot575$	$1\cdot796 - j2\cdot409$	$198\cdot923 + j266\cdot764$
$2\cdot378 - j1\cdot235$	$331\cdot264 + j172\cdot017$	$1\cdot667 - j1\cdot929$	$256\cdot499 + j296\cdot734$
N = 10		**N = 10**	
$2\cdot326 - j1\cdot136$	$347\cdot120 + j169\cdot481$	$1\cdot638 - j1\cdot901$	$260\cdot172 + j301\cdot832$
$3\cdot232 - j0\cdot566$	$300\cdot210 + j\ 52\cdot564$	$1\cdot845 - j2\cdot474$	$193\cdot668 + j259\cdot699$
$3\cdot597 - j0\cdot879$	$262\cdot334 + j\ 64\cdot099$	$1\cdot719 - j2\cdot288$	$209\cdot960 + j279\cdot353$
$3\cdot509 - j1\cdot021$	$262\cdot762 + j\ 76\cdot455$	$1\cdot782 - j2\cdot366$	$203\cdot171 + j269\cdot662$
$3\cdot341 - j0\cdot911$	$278\cdot568 + j\ 75\cdot945$	$1\cdot756 - j2\cdot334$	$205\cdot799 + j273\cdot575$
$3\cdot341 - j0\cdot911$	$278\cdot568 + j\ 75\cdot945$	$1\cdot756 - j2\cdot334$	$205\cdot799 + j273\cdot575$
$3\cdot509 - j1\cdot021$	$262\cdot762 + j\ 76\cdot455$	$1\cdot782 - j2\cdot366$	$203\cdot171 + j269\cdot662$
$3\cdot597 - j0\cdot879$	$262\cdot334 + j\ 64\cdot099$	$1\cdot719 - j2\cdot288$	$209\cdot960 + j279\cdot353$
$3\cdot232 - j0\cdot566$	$300\cdot210 + j\ 52\cdot564$	$1\cdot845 - j2\cdot474$	$193\cdot668 + j259\cdot699$
$2\cdot326 - j1\cdot136$	$347\cdot120 + j169\cdot481$	$1\cdot638 - j1\cdot901$	$260\cdot172 + j301\cdot832$
N = 20		**N = 20**	
$2\cdot266 - j1\cdot086$	$358\cdot924 + j172\cdot045$	$1\cdot630 - j1\cdot884$	$262\cdot611 + j303\cdot586$
$3\cdot186 - j0\cdot506$	$306\cdot136 + j\ 48\cdot569$	$1\cdot850 - j2\cdot475$	$193\cdot750 + j259\cdot242$
$3\cdot508 - j0\cdot761$	$272\cdot249 + j\ 59\cdot083$	$1\cdot707 - j2\cdot269$	$211\cdot723 + j281\cdot430$
$3\cdot359 - j0\cdot926$	$276\cdot676 + j\ 76\cdot317$	$1\cdot790 - j2\cdot370$	$202\cdot871 + j268\cdot669$
$3\cdot197 - j0\cdot897$	$289\cdot962 + j\ 81\cdot384$	$1\cdot738 - j2\cdot309$	$208\cdot073 + j276\cdot419$
$3\cdot189 - j0\cdot676$	$300\cdot102 + j\ 63\cdot641$	$1\cdot773 - j2\cdot350$	$204\cdot593 + j271\cdot231$
$3\cdot205 - j0\cdot709$	$297\cdot486 + j\ 65\cdot803$	$1\cdot749 - j2\cdot322$	$206\cdot964 + j274\cdot743$
$3\cdot184 - j0\cdot826$	$294\cdot272 + j\ 76\cdot330$	$1\cdot765 - j2\cdot341$	$205\cdot359 + j272\cdot381$
$3\cdot206 - j0\cdot798$	$293\cdot755 + j\ 73\cdot121$	$1\cdot755 - j2\cdot329$	$206\cdot355 + j273\cdot839$
$3\cdot267 - j0\cdot771$	$289\cdot930 + j\ 68\cdot405$	$1\cdot760 - j2\cdot335$	$205\cdot877 + j273\cdot139$
$3\cdot267 - j0\cdot771$	$289\cdot930 + j\ 68\cdot405$	$1\cdot760 - j2\cdot335$	$205\cdot877 + j273\cdot139$
$3\cdot206 - j0\cdot798$	$293\cdot755 + j\ 73\cdot121$	$1\cdot755 - j2\cdot329$	$206\cdot355 + j273\cdot839$
$3\cdot184 - j0\cdot826$	$294\cdot272 + j\ 76\cdot330$	$1\cdot765 - j2\cdot341$	$205\cdot359 + j272\cdot381$
$3\cdot205 - j0\cdot709$	$297\cdot486 + j\ 65\cdot803$	$1\cdot749 - j2\cdot322$	$206\cdot964 + j274\cdot743$
$3\cdot189 - j0\cdot676$	$300\cdot102 + j\ 63\cdot641$	$1\cdot773 - j2\cdot350$	$204\cdot593 + j271\cdot231$
$3\cdot197 - j0\cdot897$	$289\cdot962 + j\ 81\cdot384$	$1\cdot738 - j2\cdot309$	$208\cdot073 + j276\cdot419$
$3\cdot359 - j0\cdot926$	$276\cdot676 + j\ 76\cdot317$	$1\cdot790 - j2\cdot370$	$202\cdot871 + j268\cdot669$
$3\cdot508 - j0\cdot761$	$272\cdot249 + j\ 59\cdot083$	$1\cdot707 - j2\cdot269$	$211\cdot723 + j281\cdot430$
$3\cdot186 - j0\cdot506$	$306\cdot136 + j\ 48\cdot569$	$1\cdot850 - j2\cdot475$	$193\cdot750 + j259\cdot242$
$2\cdot266 - j1\cdot086$	$358\cdot924 + j172\cdot045$	$1\cdot630 - j1\cdot884$	$262\cdot611 + j303\cdot586$

Endfire array (driving-point current specified)
$$\beta_0 h = 2\cdot35620$$

Admittance $\beta_0 b = 1\cdot57080$	Impedance	Admittance $\beta_0 b = 3\cdot14159$	Impedance
$N = 1$		$N = 1$	
$1\cdot416 - j1\cdot335$	$373\cdot800 + j352\cdot429$	$1\cdot416 - j1\cdot335$	$373\cdot800 + j352\cdot429$
$N = 4$		$N = 4$	
$0\cdot768 - j2\cdot562$	$107\cdot348 + j358\cdot072$	$1\cdot009 - j0\cdot493$	$800\cdot083 + j391\cdot021$
$1\cdot181 - j0\cdot897$	$536\cdot798 + j407\cdot909$	$0\cdot932 - j0\cdot393$	$910\cdot638 + j384\cdot363$
$0\cdot995 - j1\cdot165$	$423\cdot857 + j496\cdot201$	$0\cdot932 - j0\cdot393$	$910\cdot638 + j384\cdot363$
$0\cdot811 - j0\cdot485$	$908\cdot772 + j543\cdot270$	$1\cdot009 - j0\cdot493$	$800\cdot083 + j391\cdot021$
$N = 10$		$N = 10$	
$0\cdot719 - j1\cdot989$	$160\cdot692 + j444\cdot648$	$0\cdot816 - j0\cdot100$	$1207\cdot569 + j147\cdot854$
$1\cdot399 - j0\cdot929$	$496\cdot162 + j329\cdot370$	$0\cdot718 - j0\cdot035$	$1390\cdot182 + j\ 68\cdot698$
$0\cdot823 - j0\cdot999$	$491\cdot292 + j596\cdot033$	$0\cdot671 - j0\cdot010$	$1489\cdot227 + j\ 22\cdot603$
$0\cdot942 - j0\cdot508$	$822\cdot512 + j443\cdot818$	$0\cdot646 - j0\cdot004$	$1547\cdot295 - j\ \ 8\cdot699$
$0\cdot751 - j0\cdot657$	$754\cdot637 + j659\cdot352$	$0\cdot635 - j0\cdot010$	$1573\cdot637 - j\ 23\cdot788$
$0\cdot750 - j0\cdot337$	$1108\cdot900 + j498\cdot327$	$0\cdot635 - j0\cdot010$	$1573\cdot637 - j\ 23\cdot788$
$0\cdot694 - j0\cdot473$	$984\cdot268 + j670\cdot369$	$0\cdot646 - j0\cdot004$	$1547\cdot295 - j\ \ 8\cdot699$
$0\cdot633 - j0\cdot227$	$1399\cdot044 + j502\cdot042$	$0\cdot671 - j0\cdot010$	$1489\cdot227 + j\ 22\cdot603$
$0\cdot656 - j0\cdot359$	$1173\cdot727 + j642\cdot474$	$0\cdot718 - j0\cdot035$	$1390\cdot182 + j\ 68\cdot698$
$0\cdot537 - j0\cdot150$	$1727\cdot200 + j481\cdot071$	$0\cdot816 - j0\cdot100$	$1207\cdot569 + j147\cdot854$
$N = 20$		$N = 20$	
$1\cdot317 - j1\cdot946$	$238\cdot516 + j352\cdot369$	$0\cdot757 + j0\cdot160$	$1263\cdot562 - j267\cdot384$
$2\cdot159 - j0\cdot457$	$443\cdot325 + j\ 93\cdot749$	$0\cdot652 + j0\cdot191$	$1412\cdot576 - j413\cdot497$
$1\cdot140 - j0\cdot949$	$518\cdot065 + j431\cdot501$	$0\cdot595 + j0\cdot206$	$1500\cdot532 - j519\cdot963$
$1\cdot353 - j0\cdot279$	$709\cdot164 + j146\cdot166$	$0\cdot558 + j0\cdot216$	$1558\cdot218 - j603\cdot151$
$1\cdot018 - j0\cdot603$	$727\cdot155 + j430\cdot732$	$0\cdot532 + j0\cdot223$	$1597\cdot740 - j669\cdot409$
$1\cdot079 - j0\cdot213$	$892\cdot289 + j176\cdot135$	$0\cdot514 + j0\cdot228$	$1625\cdot266 - j721\cdot834$
$0\cdot886 - j0\cdot440$	$905\cdot806 + j449\cdot935$	$0\cdot501 + j0\cdot232$	$1644\cdot340 - j762\cdot222$
$0\cdot899 - j0\cdot164$	$1076\cdot254 + j196\cdot561$	$0\cdot491 + j0\cdot235$	$1657\cdot144 - j791\cdot705$
$0\cdot781 - j0\cdot332$	$1084\cdot904 + j460\cdot938$	$0\cdot485 + j0\cdot236$	$1665\cdot037 - j811\cdot004$
$0\cdot754 - j0\cdot102$	$1302\cdot915 + j175\cdot574$	$0\cdot483 + j0\cdot237$	$1668\cdot794 - j820\cdot548$
$0\cdot719 - j0\cdot235$	$1255\cdot957 + j410\cdot737$	$0\cdot483 + j0\cdot237$	$1668\cdot794 - j820\cdot548$
$0\cdot652 - j0\cdot051$	$1523\cdot627 + j118\cdot167$	$0\cdot485 + j0\cdot236$	$1665\cdot037 - j811\cdot004$
$0\cdot684 - j0\cdot151$	$1394\cdot547 + j308\cdot085$	$0\cdot491 + j0\cdot235$	$1657\cdot144 - j791\cdot705$
$0\cdot588 - j0\cdot055$	$1686\cdot099 + j156\cdot772$	$0\cdot501 + j0\cdot232$	$1644\cdot340 - j762\cdot222$
$0\cdot603 - j0\cdot142$	$1571\cdot217 + j368\cdot572$	$0\cdot514 + j0\cdot228$	$1625\cdot266 - j721\cdot834$
$0\cdot518 - j0\cdot081$	$1884\cdot893 + j293\cdot189$	$0\cdot532 + j0\cdot223$	$1597\cdot740 - j669\cdot409$
$0\cdot494 - j0\cdot164$	$1824\cdot240 + j606\cdot325$	$0\cdot558 + j0\cdot216$	$1558\cdot218 - j603\cdot151$
$0\cdot445 - j0\cdot052$	$2218\cdot386 + j260\cdot922$	$0\cdot595 + j0\cdot206$	$1500\cdot532 - j519\cdot963$
$0\cdot462 - j0\cdot131$	$2003\cdot505 + j567\cdot693$	$0\cdot652 + j0\cdot191$	$1412\cdot576 - j413\cdot497$
$0\cdot383 - j0\cdot026$	$2597\cdot795 + j175\cdot311$	$0\cdot757 + j0\cdot160$	$1263\cdot562 - j267\cdot384$

Endfire array (base voltages specified)

$$\beta_0 h = 2.35620$$

Admittance $\beta_0 b = 1.57080$	Impedance	Admittance $\beta_0 b = 3.14159$	Impedance
N = 1		
$1.416 - j1.335$	$373.800 + j\ 352.429$	$1.416 - j1.335$	$373.800 + j352.429$
N = 4		
$1.587 - j1.820$	$272.125 + j\ 312.111$	$1.004 - j0.597$	$735.858 + j437.483$
$1.187 - j0.653$	$646.777 + j\ 355.615$	$0.931 - j0.300$	$972.956 + j313.400$
$1.038 - j0.426$	$824.490 + j\ 338.479$	$0.931 - j0.300$	$972.956 + j313.400$
$0.739 + j0.188$	$1271.479 - j\ 323.466$	$1.004 - j0.597$	$735.858 + j437.483$
N = 10		
$1.568 - j1.797$	$275.752 + j\ 316.025$	$0.833 - j0.377$	$996.584 + j451.392$
$1.220 - j0.689$	$621.279 + j\ 350.921$	$0.734 - j0.054$	$1355.209 + j\ 99.818$
$0.979 - j0.326$	$919.565 + j\ 305.944$	$0.670 + j0.055$	$1482.051 - j121.931$
$0.814 - j0.133$	$1195.813 + j\ 196.023$	$0.637 + j0.101$	$1529.885 - j243.575$
$0.736 - j0.056$	$1351.188 + j\ 103.261$	$0.624 + j0.120$	$1545.965 - j297.753$
$0.658 + j0.033$	$1516.022 - j\ \ 76.287$	$0.624 + j0.120$	$1545.965 - j297.753$
$0.641 + j0.053$	$1550.594 - j\ 128.146$	$0.637 + j0.101$	$1529.885 - j243.575$
$0.560 + j0.162$	$1646.217 - j\ 476.624$	$0.670 + j0.055$	$1482.051 - j121.931$
$0.605 + j0.072$	$1631.125 - j\ 193.964$	$0.734 - j0.054$	$1355.209 + j\ 99.819$
$0.487 + j0.468$	$1066.800 - j1025.906$	$0.833 - j0.377$	$996.584 + j451.392$
N = 20		
$1.881 - j1.656$	$299.529 + j\ 263.639$	$0.742 - j0.258$	$1201.877 + j418.102$
$1.445 - j0.580$	$595.950 + j\ 239.114$	$0.650 + j0.057$	$1526.390 - j132.914$
$1.149 - j0.242$	$833.526 + j\ 175.946$	$0.584 + j0.171$	$1576.159 - j462.011$
$1.004 - j0.058$	$993.159 + j\ \ \ 57.727$	$0.545 + j0.229$	$1560.743 - j655.797$
$0.837 + j0.001$	$1194.570 - j\ \ \ \ 1.535$	$0.520 + j0.264$	$1530.486 - j776.945$
$0.631 + j0.003$	$1583.581 - j\ \ \ \ 6.286$	$0.503 + j0.287$	$1500.443 - j856.095$
$0.607 + j0.054$	$1634.380 - j\ 145.843$	$0.491 + j0.303$	$1475.286 - j908.707$
$0.507 + j0.076$	$1929.067 - j\ 290.680$	$0.484 + j0.313$	$1456.283 - j943.151$
$0.516 + j0.121$	$1837.440 - j\ 429.232$	$0.479 + j0.320$	$1443.647 - j964.120$
$0.509 + j0.179$	$1748.511 - j\ 615.388$	$0.477 + j0.323$	$1437.399 - j974.022$
$0.491 + j0.195$	$1757.837 - j\ 698.636$	$0.477 + j0.323$	$1437.399 - j974.022$
$0.489 + j0.250$	$1621.102 - j\ 827.013$	$0.479 + j0.320$	$1443.647 - j964.120$
$0.442 + j0.240$	$1748.191 - j\ 947.672$	$0.484 + j0.313$	$1456.283 - j943.151$
$0.361 + j0.248$	$1880.501 - j1295.187$	$0.491 + j0.303$	$1475.286 - j908.707$
$0.335 + j0.238$	$1986.679 - j1409.322$	$0.503 + j0.287$	$1500.443 - j856.095$
$0.174 + j0.210$	$2335.692 - j2818.250$	$0.520 + j0.264$	$1530.486 - j776.945$
$0.249 + j0.214$	$2307.799 - j1986.459$	$0.545 + j0.229$	$1560.743 - j655.797$
$0.234 + j0.310$	$1552.071 - j2055.252$	$0.584 + j0.171$	$1576.159 - j462.011$
$0.282 + j0.211$	$2274.708 - j1702.714$	$0.650 + j0.057$	$1526.390 - j132.914$
$0.253 + j0.559$	$671.650 - j1484.457$	$0.742 - j0.258$	$1201.877 + j418.102$

Broadside array (driving-point currents specified)

$$\beta_0 h = 3 \cdot 14159$$

Admittance $\beta_0 b = 1 \cdot 57080$	Impedance	Admittance $\beta_0 b = 3 \cdot 14159$	Impedance
N = 1		*N* = 1	
$0 \cdot 985 + j1 \cdot 000$	$499 \cdot 710 - j507 \cdot 494$	$0 \cdot 985 + j1 \cdot 000$	$499 \cdot 710 - j507 \cdot 494$
N = 4		*N* = 4	
$1 \cdot 300 + j1 \cdot 158$	$428 \cdot 935 - j382 \cdot 052$	$1 \cdot 122 + j0 \cdot 538$	$724 \cdot 851 - j347 \cdot 681$
$2 \cdot 605 + j0 \cdot 524$	$368 \cdot 900 - j \ 74 \cdot 220$	$1 \cdot 051 + j0 \cdot 284$	$886 \cdot 543 - j239 \cdot 267$
$2 \cdot 605 + j0 \cdot 524$	$368 \cdot 900 - j \ 74 \cdot 220$	$1 \cdot 051 + j0 \cdot 284$	$886 \cdot 543 - j239 \cdot 267$
$1 \cdot 300 + j1 \cdot 158$	$428 \cdot 935 - j382 \cdot 052$	$1 \cdot 122 + j0 \cdot 538$	$724 \cdot 851 - j347 \cdot 681$
N = 10		*N* = 10	
$1 \cdot 417 + j1 \cdot 159$	$422 \cdot 724 - j345 \cdot 951$	$1 \cdot 133 + j0 \cdot 556$	$711 \cdot 388 - j349 \cdot 311$
$2 \cdot 785 + j0 \cdot 718$	$336 \cdot 688 - j \ 86 \cdot 845$	$1 \cdot 031 + j0 \cdot 271$	$907 \cdot 602 - j238 \cdot 628$
$2 \cdot 318 + j0 \cdot 788$	$386 \cdot 673 - j131 \cdot 498$	$1 \cdot 101 + j0 \cdot 324$	$835 \cdot 998 - j246 \cdot 013$
$2 \cdot 325 + j0 \cdot 865$	$377 \cdot 747 - j140 \cdot 553$	$1 \cdot 071 + j0 \cdot 304$	$864 \cdot 154 - j245 \cdot 716$
$2 \cdot 438 + j0 \cdot 876$	$363 \cdot 270 - j130 \cdot 614$	$1 \cdot 082 + j0 \cdot 311$	$853 \cdot 393 - j245 \cdot 529$
$2 \cdot 438 + j0 \cdot 876$	$363 \cdot 270 - j130 \cdot 614$	$1 \cdot 082 + j0 \cdot 311$	$853 \cdot 393 - j245 \cdot 529$
$2 \cdot 325 + j0 \cdot 865$	$377 \cdot 747 - j140 \cdot 553$	$1 \cdot 071 + j0 \cdot 304$	$864 \cdot 154 - j245 \cdot 716$
$2 \cdot 318 + j0 \cdot 788$	$386 \cdot 673 - j131 \cdot 498$	$1 \cdot 101 + j0 \cdot 324$	$835 \cdot 998 - j246 \cdot 013$
$2 \cdot 785 + j0 \cdot 718$	$336 \cdot 688 - j \ 86 \cdot 845$	$1 \cdot 031 + j0 \cdot 271$	$907 \cdot 602 - j238 \cdot 628$
$1 \cdot 417 + j1 \cdot 159$	$422 \cdot 724 - j345 \cdot 951$	$1 \cdot 133 + j0 \cdot 556$	$711 \cdot 388 - j349 \cdot 311$
N = 20		*N* = 20	
$1 \cdot 410 + j1 \cdot 161$	$422 \cdot 735 - j348 \cdot 012$	$1 \cdot 135 + j0 \cdot 561$	$708 \cdot 268 - j349 \cdot 951$
$2 \cdot 761 + j0 \cdot 700$	$340 \cdot 320 - j \ 86 \cdot 216$	$1 \cdot 028 + j0 \cdot 270$	$910 \cdot 264 - j239 \cdot 139$
$2 \cdot 337 + j0 \cdot 773$	$385 \cdot 711 - j127 \cdot 522$	$1 \cdot 105 + j0 \cdot 326$	$832 \cdot 672 - j245 \cdot 894$
$2 \cdot 343 + j0 \cdot 890$	$372 \cdot 969 - j141 \cdot 735$	$1 \cdot 066 + j0 \cdot 302$	$868 \cdot 043 - j246 \cdot 271$
$2 \cdot 405 + j0 \cdot 899$	$364 \cdot 852 - j136 \cdot 406$	$1 \cdot 088 + j0 \cdot 315$	$848 \cdot 027 - j245 \cdot 190$
$2 \cdot 409 + j0 \cdot 834$	$370 \cdot 623 - j128 \cdot 334$	$1 \cdot 074 + j0 \cdot 307$	$860 \cdot 805 - j246 \cdot 146$
$2 \cdot 381 + j0 \cdot 833$	$374 \cdot 226 - j130 \cdot 931$	$1 \cdot 084 + j0 \cdot 312$	$852 \cdot 290 - j245 \cdot 320$
$2 \cdot 378 + j0 \cdot 878$	$370 \cdot 031 - j136 \cdot 654$	$1 \cdot 077 + j0 \cdot 309$	$857 \cdot 942 - j245 \cdot 875$
$2 \cdot 390 + j0 \cdot 879$	$368 \cdot 540 - j135 \cdot 585$	$1 \cdot 081 + j0 \cdot 311$	$854 \cdot 451 - j245 \cdot 464$
$2 \cdot 392 + j0 \cdot 840$	$372 \cdot 168 - j130 \cdot 632$	$1 \cdot 079 + j0 \cdot 310$	$856 \cdot 102 - j245 \cdot 671$
$2 \cdot 392 + j0 \cdot 840$	$372 \cdot 168 - j130 \cdot 632$	$1 \cdot 079 + j0 \cdot 310$	$856 \cdot 102 - j245 \cdot 671$
$2 \cdot 390 + j0 \cdot 879$	$368 \cdot 540 - j135 \cdot 585$	$1 \cdot 081 + j0 \cdot 311$	$854 \cdot 451 - j245 \cdot 464$
$2 \cdot 378 + j0 \cdot 878$	$370 \cdot 031 - j136 \cdot 654$	$1 \cdot 077 + j0 \cdot 309$	$857 \cdot 942 - j245 \cdot 875$
$2 \cdot 381 + j0 \cdot 833$	$374 \cdot 226 - j130 \cdot 931$	$1 \cdot 084 + j0 \cdot 312$	$852 \cdot 290 - j245 \cdot 320$
$2 \cdot 409 + j0 \cdot 834$	$370 \cdot 623 - j128 \cdot 334$	$1 \cdot 074 + j0 \cdot 307$	$860 \cdot 805 - j246 \cdot 146$
$2 \cdot 405 + j0 \cdot 899$	$364 \cdot 852 - j136 \cdot 406$	$1 \cdot 088 + j0 \cdot 315$	$848 \cdot 027 - j245 \cdot 190$
$2 \cdot 343 + j0 \cdot 890$	$372 \cdot 969 - j141 \cdot 735$	$1 \cdot 066 + j0 \cdot 302$	$868 \cdot 043 - j246 \cdot 271$
$2 \cdot 337 + j0 \cdot 773$	$385 \cdot 711 - j127 \cdot 522$	$1 \cdot 105 + j0 \cdot 326$	$832 \cdot 672 - j245 \cdot 894$
$2 \cdot 761 + j0 \cdot 700$	$340 \cdot 320 - j \ 86 \cdot 216$	$1 \cdot 028 + j0 \cdot 270$	$910 \cdot 264 - j239 \cdot 139$
$1 \cdot 410 + j1 \cdot 161$	$422 \cdot 735 - j348 \cdot 012$	$1 \cdot 135 + j0 \cdot 561$	$708 \cdot 268 - j349 \cdot 951$

Broadside array (base voltages specified)

$$\beta_0 h = 3.14159$$

Admittance $\beta_0 b = 1.57080$	Impedance	Admittance $\beta_0 b = 3.14159$	Impedance
\multicolumn{4}{c}{N = 1}			
$0.985 + j1.000$	$499.710 - j507.494$	$0.985 + j1.000$	$499.710 - j507.494$
\multicolumn{4}{c}{N = 4}			
$1.568 + j0.815$	$502.274 - j260.992$	$1.078 + j0.563$	$728.585 - j380.868$
$2.530 + j1.186$	$324.034 - j151.902$	$1.085 + j0.255$	$873.469 - j205.141$
$2.530 + j1.186$	$324.034 - j151.902$	$1.085 + j0.255$	$873.469 - j205.141$
$1.568 + j0.815$	$502.274 - j260.992$	$1.078 + j0.563$	$728.585 - j380.868$
\multicolumn{4}{c}{N = 10}			
$1.624 + j0.957$	$457.034 - j269.359$	$1.059 + j0.600$	$714.585 - j405.086$
$2.329 + j1.180$	$341.631 - j173.109$	$1.109 + j0.198$	$873.705 - j156.039$
$2.562 + j0.859$	$350.885 - j117.687$	$1.058 + j0.363$	$845.679 - j290.370$
$2.452 + j0.762$	$371.908 - j115.630$	$1.092 + j0.285$	$857.310 - j223.673$
$2.310 + j0.894$	$376.473 - j145.732$	$1.076 + j0.317$	$855.336 - j252.234$
$2.310 + j0.894$	$376.473 - j145.732$	$1.076 + j0.317$	$855.336 - j252.234$
$2.452 + j0.762$	$371.908 - j115.630$	$1.092 + j0.285$	$857.310 - j223.673$
$2.562 + j0.859$	$350.885 - j117.687$	$1.058 + j0.363$	$845.679 - j290.370$
$2.329 + j1.180$	$341.631 - j173.109$	$1.109 + j0.198$	$873.705 - j156.039$
$1.624 + j0.957$	$457.034 - j269.359$	$1.059 + j0.600$	$714.585 - j405.086$
\multicolumn{4}{c}{N = 20}			
$1.626 + j0.929$	$463.503 - j264.895$	$1.053 + j0.609$	$711.428 - j411.334$
$2.368 + j1.183$	$337.980 - j168.790$	$1.115 + j0.190$	$871.458 - j148.401$
$2.558 + j0.901$	$347.775 - j122.478$	$1.051 + j0.374$	$844.861 - j300.449$
$2.400 + j0.756$	$379.037 - j119.387$	$1.101 + j0.272$	$856.088 - j211.894$
$2.319 + j0.832$	$382.028 - j136.976$	$1.065 + j0.334$	$855.035 - j268.393$
$2.393 + j0.910$	$365.151 - j138.824$	$1.090 + j0.294$	$855.212 - j230.435$
$2.428 + j0.879$	$364.115 - j131.830$	$1.072 + j0.321$	$855.872 - j256.175$
$2.376 + j0.826$	$375.550 - j130.485$	$1.084 + j0.303$	$855.462 - j238.823$
$2.362 + j0.838$	$376.066 - j133.411$	$1.077 + j0.314$	$855.814 - j249.498$
$2.408 + j0.883$	$366.082 - j134.306$	$1.081 + j0.309$	$855.664 - j244.371$
$2.408 + j0.883$	$366.082 - j134.306$	$1.081 + j0.309$	$855.664 - j244.371$
$2.362 + j0.838$	$376.066 - j133.411$	$1.077 + j0.314$	$855.814 - j249.498$
$2.376 + j0.826$	$375.550 - j130.485$	$1.084 + j0.303$	$855.462 - j238.823$
$2.428 + j0.879$	$364.115 - j131.830$	$1.072 + j0.321$	$855.872 - j256.175$
$2.393 + j0.910$	$365.151 - j138.824$	$1.090 + j0.294$	$855.212 - j230.435$
$2.319 + j0.832$	$382.028 - j136.976$	$1.065 + j0.334$	$855.035 - j268.393$
$2.400 + j0.756$	$379.037 - j119.387$	$1.101 + j0.272$	$856.088 - j211.894$
$2.558 + j0.901$	$347.775 - j122.478$	$1.051 + j0.374$	$844.861 - j300.449$
$2.368 + j1.183$	$337.980 - j168.790$	$1.115 + j0.190$	$871.458 - j148.401$
$1.626 + j0.929$	$463.503 - j264.895$	$1.053 + j0.609$	$711.428 - j411.334$

Endfire array (driving-point current specified)

$$\beta_0 h = 3.14159$$

Admittance $\beta_0 b = 1.57080$	Impedance	Admittance $\beta_0 b = 3.14159$	Impedance
\multicolumn{2}{}{$N = 1$}		\multicolumn{2}{}{$N = 1$}	
$0.985 + j1.000$	$499.710 - j507.494$	$0.985 + j1.000$	$499.710 - j507.494$
$N = 4$		$N = 4$	
$0.823 + j0.798$	$626.331 - j607.101$	$0.866 + j1.595$	$262.842 - j484.190$
$0.438 + j1.615$	$156.481 - j576.885$	$0.665 + j1.803$	$180.000 - j488.263$
$0.336 + j1.896$	$90.012 - j511.616$	$0.665 + j1.803$	$180.000 - j488.263$
$0.058 + j2.375$	$10.328 - j420.763$	$0.866 + j1.595$	$262.842 - j484.190$
$N = 10$		$N = 10$	
$0.842 + j0.819$	$610.374 - j593.897$	$0.822 + j1.788$	$212.309 - j461.649$
$0.453 + j1.600$	$163.769 - j578.767$	$0.585 + j2.028$	$131.295 - j455.109$
$0.470 + j1.926$	$119.479 - j489.972$	$0.528 + j2.138$	$108.895 - j440.821$
$0.464 + j2.050$	$104.931 - j464.035$	$0.504 + j2.187$	$100.117 - j434.080$
$0.439 + j2.119$	$93.669 - j452.559$	$0.495 + j2.207$	$96.704 - j431.406$
$0.424 + j2.191$	$85.172 - j439.884$	$0.495 + j2.207$	$96.704 - j431.406$
$0.403 + j2.220$	$79.130 - j436.040$	$0.504 + j2.187$	$100.117 - j434.080$
$0.402 + j2.279$	$75.038 - j425.585$	$0.528 + j2.138$	$108.895 - j440.821$
$0.363 + j2.280$	$68.208 - j427.764$	$0.585 + j2.028$	$131.295 - j455.109$
$0.241 + j2.621$	$34.754 - j378.347$	$0.822 + j1.788$	$212.309 - j461.649$
$N = 20$		$N = 20$	
$0.842 + j0.819$	$610.364 - j593.637$	$0.787 + j1.896$	$186.738 - j449.810$
$0.452 + j1.599$	$163.874 - j579.215$	$0.544 + j2.129$	$112.526 - j440.877$
$0.470 + j1.928$	$119.354 - j489.435$	$0.481 + j2.246$	$91.227 - j425.781$
$0.463 + j2.047$	$105.074 - j464.774$	$0.450 + j2.307$	$81.435 - j417.597$
$0.440 + j2.123$	$93.566 - j451.597$	$0.431 + j2.344$	$75.825 - j412.756$
$0.422 + j2.185$	$85.189 - j441.258$	$0.417 + j2.368$	$72.217 - j409.642$
$0.407 + j2.229$	$79.317 - j434.217$	$0.409 + j2.384$	$69.844 - j407.524$
$0.394 + j2.270$	$74.339 - j427.698$	$0.402 + j2.395$	$68.253 - j406.119$
$0.384 + j2.298$	$70.687 - j423.376$	$0.399 + j2.402$	$67.288 - j405.231$
$0.374 + j2.329$	$67.160 - j418.519$	$0.397 + j2.405$	$66.835 - j404.819$
$0.365 + j2.349$	$64.664 - j415.738$	$0.397 + j2.405$	$66.835 - j404.819$
$0.357 + j2.376$	$61.925 - j411.645$	$0.399 + j2.402$	$67.288 - j405.231$
$0.350 + j2.387$	$60.187 - j410.030$	$0.402 + j2.395$	$68.253 - j406.119$
$0.344 + j2.413$	$57.913 - j406.109$	$0.409 + j2.384$	$69.844 - j407.524$
$0.337 + j2.417$	$56.628 - j405.766$	$0.417 + j2.368$	$72.217 - j409.642$
$0.334 + j2.446$	$54.847 - j401.269$	$0.431 + j2.344$	$75.825 - j412.756$
$0.324 + j2.441$	$53.464 - j402.653$	$0.450 + j2.307$	$81.435 - j417.597$
$0.329 + j2.474$	$52.816 - j397.207$	$0.481 + j2.246$	$91.227 - j425.781$
$0.304 + j2.461$	$49.407 - j400.177$	$0.544 + j2.129$	$112.526 - j440.877$
$0.205 + j2.754$	$26.829 - j361.165$	$0.787 + j1.896$	$186.738 - j449.810$

Endfire array (base voltages specified)

$$\beta_0 h = 3\cdot14159$$

Admittance $\beta_0 b = 1\cdot57080$	Impedance	Admittance $\beta_0 b = 3\cdot14159$	Impedance
N = 1		*N = 1*	
$0\cdot985 + j1\cdot000$	$499\cdot710 - j507\cdot494$	$0\cdot985 + j1\cdot000$	$499\cdot710 - j507\cdot494$
N = 4		*N = 4*	
$1\cdot062 + j0\cdot567$	$732\cdot748 - j391\cdot436$	$0\cdot776 + j1\cdot612$	$242\cdot440 - j503\cdot755$
$0\cdot912 + j1\cdot492$	$298\cdot230 - j487\cdot943$	$0\cdot761 + j1\cdot800$	$199\cdot335 - j471\cdot228$
$0\cdot860 + j1\cdot706$	$235\cdot711 - j467\cdot372$	$0\cdot761 + j1\cdot800$	$199\cdot335 - j471\cdot228$
$0\cdot680 + j2\cdot257$	$122\cdot374 - j406\cdot108$	$0\cdot776 + j1\cdot612$	$242\cdot440 - j503\cdot755$
N = 10		*N = 10*	
$1\cdot015 + j0\cdot656$	$694\cdot918 - j449\cdot322$	$0\cdot636 + j1\cdot827$	$169\cdot953 - j488\cdot127$
$0\cdot940 + j1\cdot444$	$316\cdot569 - j486\cdot420$	$0\cdot608 + j2\cdot045$	$133\cdot598 - j449\cdot294$
$0\cdot826 + j1\cdot754$	$219\cdot728 - j466\cdot639$	$0\cdot585 + j2\cdot135$	$119\cdot452 - j435\cdot688$
$0\cdot719 + j1\cdot948$	$166\cdot737 - j451\cdot867$	$0\cdot570 + j2\cdot180$	$112\cdot239 - j429\cdot418$
$0\cdot663 + j2\cdot032$	$145\cdot060 - j444\cdot764$	$0\cdot562 + j2\cdot199$	$109\cdot080 - j426\cdot883$
$0\cdot595 + j2\cdot136$	$120\cdot953 - j434\cdot437$	$0\cdot562 + j2\cdot199$	$109\cdot080 - j426\cdot883$
$0\cdot577 + j2\cdot155$	$116\cdot025 - j433\cdot058$	$0\cdot570 + j2\cdot180$	$112\cdot239 - j429\cdot418$
$0\cdot512 + j2\cdot251$	$96\cdot063 - j422\cdot314$	$0\cdot585 + j2\cdot135$	$119\cdot452 - j435\cdot688$
$0\cdot547 + j2\cdot186$	$107\cdot662 - j430\cdot552$	$0\cdot608 + j2\cdot045$	$133\cdot598 - j449\cdot294$
$0\cdot506 + j2\cdot472$	$79\cdot476 - j388\cdot190$	$0\cdot636 + j1\cdot827$	$169\cdot953 - j488\cdot127$
N = 20		*N = 20*	
$1\cdot012 + j0\cdot660$	$693\cdot386 - j451\cdot865$	$0\cdot556 + j1\cdot946$	$135\cdot869 - j475\cdot068$
$0\cdot943 + j1\cdot439$	$318\cdot711 - j486\cdot074$	$0\cdot530 + j2\cdot155$	$107\cdot519 - j437\cdot577$
$0\cdot822 + j1\cdot760$	$217\cdot821 - j466\cdot334$	$0\cdot505 + j2\cdot249$	$95\cdot051 - j423\cdot302$
$0\cdot723 + j1\cdot940$	$168\cdot775 - j452\cdot604$	$0\cdot484 + j2\cdot305$	$87\cdot325 - j415\cdot556$
$0\cdot657 + j2\cdot042$	$142\cdot835 - j443\cdot665$	$0\cdot468 + j2\cdot340$	$82\cdot177 - j410\cdot927$
$0\cdot602 + j2\cdot123$	$123\cdot550 - j436\cdot002$	$0\cdot456 + j2\cdot363$	$78\cdot648 - j407\cdot944$
$0\cdot567 + j2\cdot172$	$112\cdot536 - j431\cdot044$	$0\cdot447 + j2\cdot379$	$76\cdot234 - j405\cdot963$
$0\cdot531 + j2\cdot223$	$101\cdot632 - j425\cdot476$	$0\cdot441 + j2\cdot390$	$74\cdot611 - j404\cdot642$
$0\cdot512 + j2\cdot251$	$96\cdot033 - j422\cdot381$	$0\cdot437 + j2\cdot397$	$73\cdot614 - j403\cdot834$
$0\cdot485 + j2\cdot291$	$88\cdot376 - j417\cdot712$	$0\cdot435 + j2\cdot400$	$73\cdot133 - j403\cdot451$
$0\cdot474 + j2\cdot307$	$85\cdot487 - j415\cdot897$	$0\cdot435 + j2\cdot400$	$73\cdot133 - j403\cdot451$
$0\cdot451 + j2\cdot343$	$79\cdot227 - j411\cdot531$	$0\cdot437 + j2\cdot397$	$73\cdot614 - j403\cdot834$
$0\cdot446 + j2\cdot349$	$78\cdot112 - j410\cdot911$	$0\cdot441 + j2\cdot390$	$74\cdot611 - j404\cdot642$
$0\cdot425 + j2\cdot386$	$72\cdot358 - j406\cdot303$	$0\cdot447 + j2\cdot379$	$76\cdot234 - j405\cdot963$
$0\cdot425 + j2\cdot380$	$72\cdot675 - j407\cdot110$	$0\cdot456 + j2\cdot363$	$78\cdot648 - j407\cdot944$
$0\cdot403 + j2\cdot423$	$66\cdot816 - j401\cdot672$	$0\cdot468 + j2\cdot340$	$82\cdot177 - j410\cdot927$
$0\cdot410 + j2\cdot404$	$68\cdot970 - j404\cdot249$	$0\cdot484 + j2\cdot305$	$87\cdot325 - j415\cdot556$
$0\cdot376 + j2\cdot462$	$60\cdot644 - j396\cdot977$	$0\cdot505 + j2\cdot249$	$95\cdot051 - j423\cdot302$
$0\cdot417 + j2\cdot388$	$70\cdot880 - j406\cdot343$	$0\cdot530 + j2\cdot155$	$107\cdot519 - j437\cdot577$
$0\cdot408 + j2\cdot615$	$58\cdot233 - j373\cdot316$	$0\cdot556 + j1\cdot946$	$135\cdot869 - j475\cdot068$

APPENDIX V

Programme for the Yagi-Uda array

Equations used in the programme:

$$W_{kiV}(z_k) = \int_{-h_i}^{h_i} M_{0zi}(z_i')K_{kid}(z_k, z_i')\, dz_i'$$

$$= \sin \beta_0 h_i [C_{b_{ki}}(h_i, z_k) - C_{b_{ki}}(h_i, h_k)]$$
$$- \cos \beta_0 h_i [S_{b_{ki}}(h_i, z_k) - S_{b_{ki}}(h_i, h_k)] \tag{A}$$
$$\text{(cf. 6.48a and 6.48b)}$$

$$W_{kiU}(z_k) = \int_{-h_i}^{h_i} F_{0zi}(z_i')K_{kid}(z_k, z_i')\, dz_i'$$

$$= C_{b_{ki}}(h_i, z_k) - C_{b_{ki}}(h_i, h_k) - \cos \beta_0 h_i [E_{b_{ki}}(h_i, z_k) - E_{b_{ki}}(h_i, h_k)] \tag{B}$$
$$\text{(cf. 6.49a and 6.49b)}$$

$$W_{kiD}(z_k) = \int_{-h_i}^{h_i} H_{0zi}(z_i')K_{kid}(z_k, z_i')\, dz_i'$$

$$= H_{b_{ki}}(h_i, z_k) - H_{b_{ki}}(h_i, h_k) - \cos \tfrac{1}{2}\beta_0 h_i [E_{b_{ki}}(h_i, z_k) - E_{b_{ki}}(h_i, h_k)] \tag{C}$$
$$\text{(cf. 6.50a and 6.50b)}$$

Numbered equations refer to equations in chapter 6.

A brief description of the Yagi-Array programme:
1. The programme was written in Fortran II language for the IBM 7094.
2. List of symbols used in the programme:

N   number of elements in the Yagi array.

HL   column matrix of N elements, each corresponding to the half-length of the antenna, beginning from element 1 to element N.

BKI   square matrix of order N. The ki^{th} element corresponds to the separation between the kth and the ith element.

A1 A_2 or A_2' in eq. (6.23) or eq. (6.64).

A   column matrix of $2N$ elements. The matrix element is arranged as $B_1, B_2, \dots B_N, D_1, D_2, \dots D_N$ with B_2 either primed or unprimed [eq. (6.33) or eq. (6.64)] at output.

NF   incremental angle of the field pattern desired = $10°/NF$.

NYESFD   if $NYESFD = 0$, omit field pattern evaluation; otherwise, the field pattern is evaluated.

NYESRS   if $NYESRS = 0$, omit evaluation of residuals; otherwise, residuals are evaluated.

3. The programme assumes that

$$V_2 = 1{\cdot}0 + j0{\cdot}0 \quad \text{and} \quad V_k = 0 \text{ for } k \neq 2.$$

4. The programme assumes that the radii of the antennas are all the same throughout the array. It also assumes that the separation between the two neighbouring director antennas is the same as that between the elements 2 and 3.

5. Input:
The programme requires 4 input cards for each case.

    Card 1: Read N

    Card 2: Read HD, HP, HR

        where HD is the half-length of the driven antenna (element No. 2)

        HP is the length of the parasitic director antennas (element No. 3 through No. N)

        HR is the half-length of the reflector antenna (element No. 1)

    Card 3: Read S1, S2, S3

        where S1 is the radius of the antenna

        S2 is the spacing between the reflector and the driven antenna

        S3 is the spacing between adjacent parasitic director antennas.

    HD, HP, HR, S1, S2 and S3 are in terms of wavelength

    Card 4: Read NYESFD, NF, NYESRS

6. Brief description of subroutines:
a. Main programme:
Purpose: To control the flow of the programme
b. FI Subroutine:
Purpose: To evaluate eqs. (A) through (C) and (6.76a)–(6.76d), (6.32)–(6.34), (6.15)–(6.17), (6.20)–(6.22), (6.53)–(6.62). (6.38), (6.39). $E_b(h, z)$, $C_b(h, z)$, $S_b(h, z)$, $H_b(h, z)$ are evaluated by Simpson's rule.
c. Yagi subroutine:
Purpose: To establish eqs. (6.40), (6.41) or (6.71), (6.72) and solve these simultaneous equations by Modified Gauss' method. The A1 and A obtained are the coefficients of the current distributions.
d. Curdis subroutine:
Purpose: To evaluate actual current distribution along the antennas. These are evaluated in 10 equal-distance points. The results are printed at output.
e. Field subroutine:
Purpose: To evaluate (6.84b), (6.89b), (6.90a), (6.90b), (6.97) and input admittance and impedance of the driving antenna of the Yagi array. The results are printed at the output. These are input admittance, input impedance of the driving antenna and field pattern, forward gain, backward gain and front to back ratio.
f. Resid subroutine:
Purpose: To check the result of the programme, evaluate the

differences between the left-hand side and the right-hand side of eq. (6.8) when the I_z obtained is substituted back. These differences are called residuals. When $h_k = \lambda/4$, eq. (6.8) is changed to the following form in this subroutine:

$$\sum_{i=1}^{N} \int_{-h_i}^{h_i} I_{zi}(z_i')K_{kid}(z_k, z_i')\,dz_i' = \frac{-j4\pi}{\zeta_0}[\tfrac{1}{2}V_{0k}^e S_{0zk} + C_k F_{0zk}],$$

$$k = 1, 2, \ldots N \qquad\qquad (D)$$

where

$$C_k = \frac{j\zeta_0}{4\pi}\sum_{i=1}^{N}\int_{-h_i}^{h_i} I_{zi}(z_i')K_{ki}(0, z_i')\,dz_i'.$$

The values of the right-hand side of eq. (6.8) or eq. (D) [which is $4\pi\mu_0^{-1}A_{zk}(z_k)$] together with the resultant difference and absolute value of ratio of the difference $4\pi\mu_0^{-1}A_{zk}$ are printed.

g. ECSH subroutine:

Purpose: To evaluate necessary integrals for Resid subroutine.

7. Maximum number of elements of the Yagi array is arbitrarily set to be 20. However, the programme does not use up all the storage positions. If one desires to increase the number of elements beyond 20, one has to redefine all the dimension statements in the programme.

REFERENCES

Preface

[1] M. A. Bontsch-Bruewitsch (1926), Die Strahlung der Komplizierten Rechtwinkeligen Antennen mit Gleichbeschaffenen Vibratoren, *Ann. Physik.* **81**, 425.

[2] G. C. Southworth (1930), Certain factors affecting the gain of directive antennas, *Proc. I.R.E.* **18**, 1502.

[3] E. J. Sterba (1931), Theoretical and practical aspects of directional transmitting systems, *Proc. I.R.E.* **19**, 1184.

[4] P. S. Carter, C. W. Hansell and N. E. Lindenblad (1931), Development of directive transmitting antennas by R.C.A. Communications, Inc., *Proc. I.R.E.* **19**, 1773.

[5] S. A. Schelkunoff (1943), A mathematical theory of linear arrays, *Bell System Tech. J.* **22**, 80.

[6] C. L. Dolph (1946), A current distribution for broadside arrays which optimizes the relationship between beam width and side-lobe level, *Proc. I.R.E.* **34**, 335.

[7] T. T. Taylor and J. R. Whinnery (1951), Application of potential theory to the design of linear arrays, *J. Appl. Phys.* **22**, 19.

[8] P. S. Carter (1932), Circuit relations in radiating systems and applications to antenna problems, *Proc. I.R.E.* **20**, 1004.

[9] G. H. Brown (1937), Directional antennas, *Proc. I.R.E.* **25**, 78.

[10] W. Walkinshaw (1946), Theoretical treatment of short Yagi aerials, *J. Inst. Elec. Engrs. (London)*, **93**, Part III(A), 598.

[11] C. R. Cox (1947), Mutual impedance between vertical antennas of unequal heights, *Proc. I.R.E.* **35**, 1367.

[12] G. Barzilai (1948), Mutual impedance of parallel aerials, *Wireless Engr.* **25**, 343 and (1949) **26**, 73.

[13] B. Starnecki and E. Fitch (1948), Mutual impedance of two center-driven parallel aerials, *Wireless Engr.* **25**, 385.

[14] H. Brückmann (1939), *Antennen*, S. Hirzel, Leipzig.

[15] C. T. Tai (1948), Coupled antennas, *Proc. I.R.E.* **36**, 487.

[16] R. King (1950), Theory of N-coupled parallel antennas, *J. Appl. Phys.* **21**, 94.

[17] R. W. P. King (1956), *Theory of Linear Antennas*, chapter 3, Harvard University Press, Cambridge, Mass.

[18] J. Aharoni (1946), *Antennae*, Clarendon Press, Oxford.

[19] J. Stratton (1941), *Electromagnetic Theory*, McGraw-Hill, New York.

[20] R. C. Hansen (1966), *Microwave Scanning Antennas*, vol. 2, Academic Press, New York.

[21] R. King (1959). Linear arrays: Currents, impedances and fields, *Trans. I.E.E.E.* **AP-7**, S440.

[22] R. B. Mack (1963), *A Study of Circular Arrays*, Technical Reports, Nos. 381–386, Cruft Laboratory, Harvard University.

[23] R. W. P. King and S. S. Sandler (1963), Linear arrays: Currents, impedances and fields, II, *Electromagnetic Theory and Antennas*, E. C. Jordan, Ed., 1307, Macmillan Co., New York.

[24] R. W. P. King and S. S. Sandler (1964), Theory of broadside arrays; Theory of endfire arrays, *Trans. I.E.E.E.* **AP-12**, 269, 276

[25] R. J. Mailloux (1966), The long Yagi-Uda array, *Trans. I.E.E.E.* **AP-14**, 128.

[26] I. L. Morris (1965), *Optimization of the Yagi Array*, Scientific Reports, Nos. 6 and 10 (Series 3), Cruft Laboratory, Harvard University.

[27] W.-M. Cheong (1967), *Arrays of Unequal and Unequally-Spaced Elements*, Ph.D. Thesis, Harvard University.

[28] R. W. P. King and T. T. Wu (1965), Currents, charges and near fields of cylindrical antennas, *Radio Science, Jour. Res. NBS*, **69D**, 429.

Chapter 1

[1] R. W. P. King (1956), *Theory of Linear Antennas*, Harvard University Press, Cambridge, Mass.

[2] R. W. P. King, H. R. Mimno and A. H. Wing (1965), *Transmission Lines, Antennas and Wave Guides*, Dover Publications, New York.

Chapter 2

[1] R. W. P. King (1956), *Theory of Linear Antennas*, Harvard University Press, Cambridge, Mass.

Chapter 4

[1] R. B. Mack (1963), *A Study of Circular Arrays: Part 4, Tables of Quasi-Zeroth-Order* $\Psi_{dR}^{(m)}(h)$, $T^{(m)}(h)$, $T'^{(m)}(\lambda/4)$, *Admittances, and Quasi-First-Order Susceptances*, Cruft Laboratory Technical Report 384, Harvard University.

[2] R. B. Mack (1963), *A Study of Circular Arrays: Part 2, Self and Mutual Admittances*, Cruft Laboratory Technical Report 382, Harvard University.

Chapter 5

[1] R. W. P. King, B. H. Sandler and S. S. Sandler (1964), *Tables for Curtain Arrays*, Cruft Laboratory Scientific Report 4 (Series 3), Harvard University.

[2] R. W. P. King (1956), *Theory of Linear Antennas*, Harvard University Press, Cambridge, Mass.

[3] R. B. Mack and E. W. Mack (1960), *Tables of E(h, z), C(h, z), S(h, z)*, Cruft Laboratory Technical Report 331, Harvard University.

Chapter 6

[1] R. J. Mailloux (1966), The long Yagi-Uda array, *Trans. I.E.E.E.* **AP-14**, No. 2, pp. 128–137.

[2] S. S. Sandler (1962), *A General Theory of Curtain Arrays*, Ph.D. Thesis, Harvard University.

[3] I. L. Morris (1965), *Optimization of the Yagi Array*, Ph.D. Thesis, Harvard University.

[4] R. W. P. King (1956), *Theory of Linear Antennas*, Harvard University Press, Cambridge, Mass.

[5] R. W. P. King, A. H. Wing and H. R. Mimno (1965), *Transmission Lines, Antennas and Wave Guides*, Dover Publications, New York.

[6] W.-M. Cheong (1967), *Arrays of Unequal and Unequally-Spaced Dipoles*, Ph.D. Thesis, Harvard University.

[7] D. E. Isbell (1960), Log-periodic dipole arrays, *Trans. I.R.E.* **AP-8**, 260–267.

[8] E. C. Jordan, G. A. Deschamps, J. D. Dyson and R. E. Mayes (1964), Developments in broadband antennas, *I.E.E.E. Spectrum*, **1**, 58–71.

[9] R. L. Carrell (1961), *Analysis and Design of the Log-Period Dipole Antenna*, Technical Report 52, Electrical Engineering Research Laboratory, University of Illinois, Urbana.

[10] V. H. Rumsey (1966), *Frequency-Independent Antennas*, Academic Press, New York.

[11] R. W. P. King (1965), *Transmission-Line Theory*, Dover Publications, New York.

[12] W.-M. Cheong and R. W. P. King (1967), Log-periodic dipole antenna, *Radio Science*, Vol. 2 (New Series), pp. 1315–1326.

[13] W.-M. Cheong and R. W. P. King (1967), Arrays of unequal and unequally spaced elements, *Radio Science*, Vol. 2 (New Series), pp. 1303–1314.

Chapter 7

[1] R. W. P. King and T. T. Wu (1965), The cylindrical antenna with arbitrary driving point, *Trans. I.E.E.E.* **AP-13**, pp. 710–718.

[2] R. W. P. King (1956), *Theory of Linear Antennas*, Harvard University Press, Cambridge, Mass.

[3] V. W. H. Chang and R. W. P. King (1968), Theoretical study of dipole arrays of N parallel elements, *Radio Science*, Vol. 3 (New Series), No. 5.

Chapter 8

[1] G. C. Montgomery (1947), *Technique of Microwave Measurements*, Vol. 11, Radiation Laboratory Series, McGraw-Hill, New York.

[2] E. L. Ginzton (1957), *Microwave Measurements*, McGraw-Hill, New York.

[3] H. Jasik (1961), *Antenna Engineering Handbook*, McGraw-Hill, New York.

[4] R. W. P. King, H. R. Mimno and A. H. Wing (1965), *Transmission Lines, Antennas and Wave Guides*, Dover Publications, New York.

[5] R. W. P. King (1965), *Transmission-Line Theory*, Dover Publications, New York.

[6] R. W. P. King (1956), *Theory of Linear Antennas*, Harvard University Press, Cambridge, Mass.

[7] D. D. King (1952), *Measurements at Centimeter Wavelengths*, Van Nostrand, New York.

[8] H. Wind and H. Rapaport (1954), *Handbook of Microwave Measurements*, Microwave Research Institute, Polytechnic Institute of Brooklyn, New York.

[9] T. T. Wu (1963), Input admittance of linear antennas driven from a coaxial line, *J. Res. NBS*, **67D**, 83.

[10] J. H. Richmond (1965), Digital computer solutions of the rigorous equations for scattering, *Proc. I.E.E.E.* **53**, 796.

[11] N. Marcuvitz (1964), *Wave-Guide Handbook*, Vol. 10, Radiation Laboratory Series, Boston Technical Publishers, Lexington, Mass.

[12] K. Iizuka and R. W. P. King (1962), *Terminal-Zone Corrections for a Dipole Driven by a 2-Wire Line*, Cruft Laboratory Technical Report 352, Harvard University.

[13] R. B. Mack and E. W. Mack (1960), *Tables of $E(h, z)$, $C(h, z)$ and $S(h, z)$*, Cruft Laboratory Technical Report 331, Harvard University.

[14] T. Moreno (1958), *Microwave Transmission Design Data*, Dover Publications, New York.

[15] R. B. Mack (1963), *A Study of Circular Arrays: Part 2, Self and Mutual Admittances*, Cruft Laboratory Technical Report 382, Harvard University.

[16] K. Iizuka and R. W. P. King (1962), The effect of an unbalance on the current along a dipole antenna, *Trans. I.E.E.E.* **AP-10**, 702.

[17] W. H. Huggins (1947), Broadband noncontacting short circuit for coaxial lines, *Proc. I.R.E.* **35**, 966.

[18] H. W. Andrews (1953), *Image-Plane and Coaxial Line Measuring Equipment at 600 MHz*, Cruft Laboratory Technical Report 177, Harvard University.

[19] R. W. P. King (1958), Quasi-stationary and nonstationary currents in electrical circuits, *Handbuch der Physik*, **16**, Springer-Verlag, Berlin.

[20] H. Whiteside (1962), *Electromagnetic Field Probes*, Cruft Laboratory Technical Report 377, Harvard University.

[21] H. Whiteside and R. W. P. King (1964), The loop antenna as a probe, *Trans. I.E.E.E.* **AP-12**, 291.

[22] R. W. P. King (1957), *The Loop Antenna as a Probe in Arbitrary Electromagnetic Fields*, Cruft Laboratory Technical Report 262, Harvard University.

[23] R. W. P. King (1959), The rectangular loop antenna as a dipole, *Trans. I.E.E.E.* **AP-7**, 53.

[24] R. W. P. King (1963), *Fundamental Electromagnetic Theory*, Dover Publications, New York.

[25] R. W. P. King and C. W. Harrison, Jr. *Electromagnetic radiation and antennas*, to be published by U.S. Atomic Energy Commission, Division of Technical Information.

[26] J. E. Storer (1956), Impedance of thin wire loop antennas, *Trans. I.E.E.E.* **75**, Part 1 (Communications and Electronics), p. 606.

[27] E. F. Knott (1965), A surface field measurement facility, *Proc. I.E.E.E.* **53**, 1105.

[28] L. Wetzel and D. B. Brick (1955), *An Experimental Investigation of High-Frequency Current Distributions on Conducting Cylinders*, Cruft Laboratory Scientific Report 4, Harvard University.

[29] R. W. Burton (1964), A coaxial line amplitude-insensitive phase detection system, *Microwave Journal*, **7**, 51.

[30] E. R. Wingrove, Jr. (1964), *Investigation of AFCRL Tri-Coordinate Radar Technique*, AFCRL-TR No. 65-8, General Electric Co., Syracuse, New York.

[31] A. Cohen and A. W. Maltese (1961), The Lincoln Laboratory Antenna Test Range, *Microwave Journal*, **4**, 57.

[32] S. Silver (1949), *Microwave Antenna Theory and Design*, Vol. 12, Radiation Laboratory Series, McGraw-Hill, New York.

[33] R. B. Mack (1963), *A Study of Circular Arrays: Part 1, Experimental Equipment*, Cruft Laboratory Technical Report 381, Harvard University.

[34] A. S. Meier and W. P. Summers (1949), Measured impedances of vertical antennas over finite ground planes, *Proc. I.R.E.* **37**, 609.

Appendix II

[1] R. B. Mack (1963), *A Study of Circular Arrays: Part 4, Tables of Quasi-zeroth-order $\Psi_{dR}^{(m)}(h)$, $T^{(m)}(h)$, $T'^{(m)}(\lambda/4)$, Admittances, and Quasi-first-order Susceptances*, Cruft Laboratory Technical Report 384, Harvard University, Cambridge, Mass.

[2] R. B. Mack and E. W. Mack (1963), *A Study of Circular Arrays: Part 6, Plan for Practical Application to Arrays of Twenty or Fewer Elements*, Cruft Laboratory Technical Report 386, Harvard University, Cambridge, Mass.

[3] J. B. Scarborough (1930), *Numerical Mathematical Analysis*, The Johns Hopkins Press, Baltimore, Md.

LIST OF SYMBOLS

D directivity, 14

$D_r(0)$ relative directivity, 206

$D_m(\Theta, \beta_0 h)$ field factor, 60

d_{12} distance between centres of collinear antennas, 288

$\mathbf{E}$ electric vector, 3

$\mathbf{E}^r$ electric vector in radiation zone, 6

E_r, E_Θ spherical components of electric field, 6

E_ρ, E_z cylindrical components of electric field, 6

E_z^{inc} incident electric field, 66

E_{0z} component in five-term current, 278

$E_a(h, z), E_b(h, z)$ integral functions, 55, 141

E_Θ^r Θ component of electric field in radiation zone, 6

F_{0z} component in two-term, three-term and five-term current, 127

$F_0(\Theta, \beta_0 h), F_m(\Theta, \beta_0 h)$ field functions of antenna with sinusoidal current, 7, 14

f_i discrete frequency applied to array, 259

$f(\Theta, \beta_0 h), f'(\Theta, \beta_0 h)$ field characteristics, 59, 60, 84, 85

$f_L(\Theta, \beta_0 h), f'_L(\Theta, \beta_0 h)$ field characteristics, 61, 62

G_c characteristic conductance of line, 248

$G_r(0)$ relative gain, 206

$G_m(\Theta, \beta_0 h)$ field factor, 60

$G_N(\pi/2, 0)$ absolute gain, 206

$G_{ki}(d_{ki}, z_k, z_i')$ kernel, 275

$G_{ki}^{even}(d_{ki}, z_k, z_i')$ even part of kernel, 275

$G_{ki}^{odd}(d_{ki}, z_k, z_i')$ odd part of kernel, 275

$g(\Theta, \beta_0 h), g'(\Theta, \beta_0 h)$ field functions of array, 84, 85

H_{0z} component in three-term and five-term current, 127

$H_m(\Theta, \beta_0 h)$ field factor, 60

h half-length of antenna, 3

h_{eD} effective length of each half of loop for dipole mode, 362

$h_e(\pi/2)$ effective length, 68

$h_{eN}(\Theta, \Phi)$ effective length of N-element array, 237

$I^{(m)}(z)$ phase sequence current, 78

$I_{1L}(w)$ current along transmission line, 325

$I_z(z)$ total axial current, 5

$I_{zi}^{\text{even}}(z_i),\ I_{zi}^{\text{odd}}(z_i)$ even and odd parts of current, 275

$I_D(\Theta, \Phi)$ current in receiving antenna, 236

$\mathbf{J}$ volume of density of current, 3

$J_{zi}^j(z_i)$ current functions, 282

$J_I(h, z),\ J_R(h, z)$ shorthand notation for imaginary and real parts of integral, 49

$\mathbf{K}$ surface density of current, 4

$K(z, z')$ kernel in integral equation, 49

$K^{(1)}(z, z'),\ K^{(2)}(z, z')$ kernels for phase sequences, 72

$K_d(z, z')$ difference kernel, 51

$K_{ki}(z, z')$ kernels involving antennas i and k, 99

$K_{kid}(z, z')$ difference kernels involving antennas i and k, 99

$K_R(z, z'),\ K_I(z, z')$ real and imaginary parts of $K(z, z')$, 49

$K_d^{(1)}(z, z'),\ K_d^{(2)}(z, z')$ difference kernels for phase sequences, 72

$K_{11}(z, z'),\ K_{12}(z, z')$ kernels for coupled antennas, 71

$K_{11d}(z, z'),\ K_{12d}(z, z')$ difference kernels for coupled antennas, 71

$K_{ki}(z_k, z_i')$ kernel, 276

$K_{kiR}(z_k, z_i'),\ K_{kiI}(z_k, z_i')$ real and imaginary parts of $K_{ki}(z_k, z_i')$

L_T lumped inductance for terminal zone, 328

$l^e(w)$ external inductance per unit length of line, 326

$l_L^e(w)$ part of $l^e(w)$ due to currents in line, 326

l_0^e external inductance per unit length of uniform line, 328

M_{0z} component in two-term, three-term and five-term current, 127

N number of elements in array, 17

n distance between elements in fractions of wavelength, 21

$\hat{n}$ unit normal, 4

P time-average power, 10

P_k time-average power transferred across surface of kth antenna, 29

P_{0z} component in five-term current, 279

Q denominator of T functions, 54

$Q^{(m)}$ denominator of $T^{(m)}$ functions, 78

Q_k complex coefficient, 279

$Q_m(\Theta, \beta_0 h)$ field factor, 285

$q(z)$ charge per unit length, 5

R distance from arbitrary point to field point, 3, 5, 6

R_c characteristic resistance of line, 247

R_k complex coefficient, 279

R_{ki} distance between elements on antennas i and k, 99

R_{kih} distance to end of antenna k from point on antenna i, 99

R_m^e radiation resistance, 16

$R_m(\Theta, \beta_0 h)$ field factor, 285

R_0 distance to origin, 3

R_0 distance to centre of antenna, 16

R_0 driving-point resistance, 25

R_{1h}, R_{2h} distances to ends of antenna, 16

R_{11}, R_{12} distances between elements of coupled antennas, 71

R_{11h}, R_{12h} distances to ends of coupled antennas, 71

R_0, Θ, Φ spherical coordinates, 286

r_{FB} front-to-back ratio, 207

r, Θ, Φ spherical coordinates, 6

$\hat{r}, \hat{\Theta}, \hat{\Phi}$ unit vectors in directions of spherical coordinate axes, 6

$\mathbf{S}$ Poynting vector, 9

S_B sensitivity constant of unloaded loop, 362

S_c sensitivity constant of short antenna, 358

S_{ki} Poynting vector on element k due to current in element i, 29

S_{0z} component in two-term, three-term and five-term current, 128

$S_B^{(1)}, S_E^{(1)}$ sensitivity constants for singly loaded loop, 363, 364

$S_B^{(2)}, S_E^{(2)}$ sensitivity constants for doubly loaded loop, 365

$S_a(h, z), S_b(h, z)$ integral functions, 24, 141

SWR   standing wave ratio, 340

$s^{(m)}, s'^{(m)}$ parameters, 102

s_k complex amplitude function, 103

T, T_D, T_U, T'_D, T'_U complex coefficients, 53, 56

$T_D^{(m)}, T_U^{(m)}, T_D'^{(m)}, T_U'^{(m)}$ complex coefficients for phase sequences, 75, 79

U function proportional to vector potential at end of antenna, 51

U_k function proportional to vector potential at end of antenna k, 98

$u(z)$ normalized current in unloaded receiving antenna, 67, 68

$V^{(0)}, V^{(1)}$ phase-sequence voltages, 72

$V(w)$ scalar potential difference along transmission line, 325

$V_L(w), V_T(w)$ parts of $V(w)$ due to charges in line and termination, 326

V_0 voltage at driving point, 23

$v(z)$ normalized current distribution, 80

$W_{kiD}(z_k)$ normalized vector potential difference of $\cos \frac{1}{2}\beta_0 z - \cos \frac{1}{2}\beta_0 h$, 238

$W_{kiU}(z_k)$ normalized vector potential difference of $\cos \beta_0 z - \cos \beta_0 h$, 238

$W_{kiV}(z_k)$ normalized vector potential difference of $\sin \beta_0(h - |z|)$, 238

$W_{pL}(w), W_{pT}(w)$ component of vector potential difference due to currents in line and load, 325

w distance from load along transmission line, 325

$w(z)$ normalized current distribution, 80

X_i, Y_i, Z_i Cartesian coordinates for centre of element i, 286

$Y^{(0)}, Y^{(1)}$ zero- and first-phase-sequence admittance of loop, 365

Y_a apparent admittance of load terminating line, 329

Y_{kin} driving-point (input) admittance of element k, 104

Y_{1in} driving-point (input) admittance, 82

Y_L admittance loading loop, 363

Y_{s1}, Y_{s2} self-admittance, 80

Y_T terminating admittance, 249

$Y(0), Y\left(\dfrac{l}{2}\right)$ driving-point admittance of loop, 365

$Y_0^{(m)}$ phase sequence admittance, 78

Y_0 admittance of circular loop and square loop with constant current, 362, 366

Y_1 driving-point admittance of array, 251

Y_{12}, Y_{21} mutual admittance, 80

y_T normalized terminating admittance, 249

$y(w)$ admittance per unit length of line, 325

y_0 admittance per unit length of line, 327

Z_a apparent impedance of load, 329

Z_c characteristic impedance of line, 246

Z_{ik} mutual impedance, 30

Z_{1in} driving-point (input) impedance, 82

Z_{kk} self-impedance, 32

Z_L load impedance, 68; for loop, 363

Z_L impedance of loop with constant current, 362

Z_{s1}, Z_{s2} self-impedances, 81

Z_T terminating impedance, 246

Z_0 impedance of antenna, 23

Z_1, Z_2 series impedances for antennas, 81

Z_{11}, Z_{22} impedances of primary and secondary circuits, 81

Z_{12}, Z_{21} mutual impedances, 81

z^i internal impedance per unit length, 361

484 LIST OF SYMBOLS

ρ volume density of charge, 3

ρ, ρ_a, ρ_g apparent terminal attenuation function in general, for load, and for generator, 339

ρ, Φ, z cylindrical coordinates, 6

$\hat{\rho}, \hat{\Phi}, \hat{z}$ unit vectors in directions of cylindrical coordinate axes, 6

Σ surface, 9

Σ_k surface of kth antenna, 29

σ spacing ratio for log-periodic antenna, 245

τ length ratio for log-periodic antenna, 245

Φ cylindrical and spherical coordinate, 6

Φ_{ki} matrix element, 138

$\Phi_{Tki}^{(m)}$ matrix element, 127

ϕ phase shift in section of line, 247

ϕ scalar potential, 3

ϕ, ϕ_a, ϕ_g apparent terminal phase function in general, for load and for generator, 339, 348

$\Psi_{dDI}, \Psi_{dI}, \Psi_{dUI}$ coefficients, 53

$\Psi_{dDR}, \Psi_{dR}, \Psi_{dUR}$ coefficients, 53

$\Psi_{kidI}, \Psi_{kidR}, \Psi_{kidu}, \Psi_{kidv}$ coefficients, 141

$\Psi_{kkdV}^f, \Psi_{kkdV}^h, \Psi_{kkdV}^m$ coefficients, 197

$\Psi_{kidV}^f, \Psi_{kidU}^f, \Psi_{kidD}^f$ coefficients, 197

$\Psi_{kidV}^h, \Psi_{kidU}^h, \Psi_{kidD}^h$ coefficients, 197, 198

$\Psi_{dDI}^{(m)}, \Psi_{dDR}^{(m)}, \Psi_{dR}^{(m)}$ coefficients for mth phase sequence, 76

$\Psi_{dD}^{(m)}, \Psi_{dI}^{(m)}, \Psi_{dUI}^{(m)}, \Psi_{dUR}^{(m)}, \Psi_{d\Sigma R}^{(m)}$ coefficients for mth phase sequence, 77

$\Psi_D(h), \Psi_U(h), \Psi_V(h)$ coefficients, 53

$\Psi_{kiu}(h), \Psi_{kiv}(h)$ coefficients, 141

$\Psi_D^{(m)}(h), \Psi_U^{(m)}(h), \Psi_V^{(m)}(h)$ coefficients for mth phase sequence, 77

Ω thickness parameter, 56

INDEX